ESSENTIALS OF COMPUTER NETWORKS

Dr. SANJAY SHARMA
B.E., M.Tech., Ph.D., MIETE

Edited by
DEEKSHA SHARMA
&
V. SHARMA

S.K. KATARIA & SONS®
Publisher of Engineering Books
4885/109, Prakash Mahal, Dr. Subhash Bhargav Lane,
Opposite Delhi Medical Association, Daryaganj, New Delhi–110002
Phone: +91-11-43551243, Mobile: +91-9871775858. 7678425209
e-mail: katariabook@gmail.com, katariabooks@yahoo.com
Website: www.skkatariaandsons.com
(e-books, subscriptions and bulk buying also available on our website)

Published by:

S.K. Kataria & Sons®

4885/109, Prakash Mahal, Dr. Subhash Bhargav Lane,
Opposite Delhi Medical Association, Daryaganj, New Delhi–110002
Phone: +91-11-43551243, Mobile: +91-9871775858. 7678425209
e-mail: katariabook@gmail.com, katariabooks@yahoo.com

First Edition: January, 2015, Reprint: 2025

Price: ₹ 595/-

Laser Type Setting:
Plus Computers, Shahdara, Delhi

Printed at:
New A.S. Printing Press, Delhi

Preface to the First Edition

Computer Networks is a core subject for undergraduate students in Electronics and Communication Engineering (ECE), Computer Science Engineering (CSE), and Information Technology streams etc. This book is primarily intended to serve as a textbook in accordance with the syllabus of Computer Networks offered by various Universities in India as well as abroad.

In this book, a significant effort has been made to find simple ways to develop theoretical aspects of Computer Networks. A strong emphasis has been given on numerical examples. Neat and Clear diagrams have been used for explanation.

Every effort has been made to make the treatments simple and comprehensive.

The author hopes that the book will fulfil the need of readers and would welcome any suggestions towards the improvement of the book.

Dr. Sanjay Sharma
January, 2015

Contents

UNIT – I

SYLLABUS:

Computer Networks: Introduction, applications, types of networks, network software, reference models- OSI model, TCP/IP model, comparison of OSI and TCP/IP models, example networks.

The Physical layer: Design Issues, review of data communication concepts (configuration, topology, transmission mode, media-guided and unguided, types of switching etc).

The above syllabus has been covered in following chapters:

Chapter 1 : Computer Networks

Chapter 2 : Example Networks

Chapter 3 : The Physical Layer

Chapter 1

Computer Networks

Syllabus

Computer Networks: Introduction, applications, types of networks, network software, reference models-OSI model, TCP/IP model, comparison of OSI and TCP/IP models.

Chapter Outline

- Introduction
- Basic Communication Model
- Introduction to Data Communication
- Components of Data Communication System
- A Simplified Data Communication Model
- Types of Communication: Simplex, Half Duplex, Full Duplex: (Data Flow)
- Computer Networks
- Computer Network Criteria
- Type of Connections (Topology)
- Network Topology Types
- Bus Topology
- Ring Topology
- Star Topology
- Mesh Topology
- Tree Topology
- Logical Topology
- Comparison of Ring and Star Topologies
- Comparison of Bus and Star Topologies
- Hybrid Topology
- Comparison of Star Bus and Star Ring Topologies
- Network Topology Design
- Types of Networks

1.1. INTRODUCTION

As a matter of fact, 'computer communications' refers to the electronic transmission of data from one system to another; it describes the manner in which computers exchange information with each other. Also, 'networking' refers to the concept of connecting a group of systems for the express purpose of sharing information. Data communications and networking are changing the world scenario. Today, business and other daily affairs rely on computer networks and internetworks.

A network is a set of devices often referred to as nodes connected by media links. A node can be a computer, printer or any other device capable of sending or receiving data generated by other nodes on the network. Also, the links connecting the devices are called communication channels.

There has been a tremendous development in computer industry. Infact, the development of the personal computer brought about tremendous changes for business, industry, science and education. A similar revolution is occurring in data communications and networking. Further, technological advances are making it possible for communication links to carry more and faster signals. Because of this, services are evolving to allow use of the expanded capacity, including the extension to established telephone services such as call waiting, voice mail and caller ID etc.

Thus, the basic aim of data communications and networking is to be able to exchange data such as text, audio and video from one point to another point

- LAN Topologies
- Metropolitan Area Network (MAN)
- Wide Area Network (WAN)
- Wireless Networks
- Internetworks
- Performance Comparison of LAN and WAN
- Comparison of LAN, WAN and MAN
- Goals and Applications of Computer Networks
- Network Hardware
- Network Scale
- Network Software
- Reference Models (Network Architectures)
- OSI Model
- Functions of Different Layers
- The TCP/IP Reference Model
- Comparison of OSI and TCP/IP Models
- Demerits of TCP/IP Model
- Hybrid (Internet) Reference Model

DO YOU KNOW?

The purpose of communication between the application entities is not served just by exchanging bits. The communication needs to be meaningful. Meaningful communication is always done with a purpose and aims at enlarging common understanding between communicating entities. Physical transfer of bits is transmission not communication.

in the world. This chapter gives an overview of the various topics surrounding computer communications and networking. Many of the concepts presented here are done so from a general perspective and are elaborated in subsequent chapters.

1.2. BASIC COMMUNICATION MODEL

Figure 1.1 shows a simple and most basic communication model.

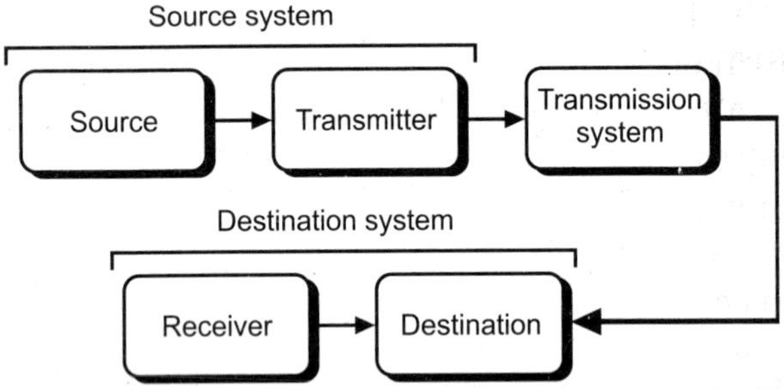

Fig. 1.1. A general communication model

The fundamental purpose of a communication system is the exchange of data between two points or parties. Depending upon the basic communication model, figure 1.2 shows one particular example, which establishes communication between a workstation and a server over a public telephone network.

Fig. 1.2. Establishment of communication

One more example can be the exchange of voice signals between two telephones over the same network.

The key elements of basic communication model in figure 1.1 can be described as under:

(i) Source

This device generates the data to be transmitted. Two examples can be telephones and personal computers.

(ii) Transmitter

As a matter of fact, the data generated by a source system are not transmitted directly in the form in which they are generated. In fact, a transmitter transforms and encodes the information in such a manner so as to produce electromagnetic waves or signals. These electromagnetic signals can be transmitted across some sort of transmission

Pioneer in Communications : Hernirich Rudolph Hertz (1857–1894) Born in Hamburg, Germany, and educated at the University of Berlin, Heinrich Rudolph Hertz was a physicist who verified experimentally British physicist James Clark Maxwell's Electromagnetic theory by demonstrating the nature of electromagnetic (EM) wave propagation.

systems. As an example, a modem takes a digital bit stream from an attached device such as a personal computer and transforms that bit stream into an analog signal which can be handled by the telephone network.

(iii) Transmission System

A transmission system can be a single transmission line or a complex network connecting source and destination.

(iv) Receiver

The receiver accepts the signal from the transmission system and converts it into a form which can be handled by the destination device.

As an example, a modem accepts an analog signal coming from a network or transmission line and converts it into a digital bit stream.

(v) Destination

The receiver passes information to the destination for processing. A destination takes the incoming data from the receiver.

1.3. INTRODUCTION TO DATA COMMUNICATION

In this chapter, we shall discuss data communication and networking. The aim of data communication and networking is to facilitate the exchange of data such as audio, text and video between any points in world. The transfer of data takes place over a computer network. A network is like a highway over which the data travels smoothly.

1.3.1. Definition of Data Communication

The word data refers to the information which is presented in a form that is agreed upon by the users and creators of data.

Data communication is the exchange of data between two devices via some form of transmission medium such as a co-axial cable.

1.3.2. Characteristics of Data Communication System

The characteristics of a data communication system are :

(i) Delivery
(ii) Accuracy
(iii) Timeliness

(i) Delivery

The data should be delivered to the correct destination. It should reach only to the intended user and not to any other.

(ii) Accuracy

There is a possibility of data alteration or corruption when it is travelling over a communication medium. This usually affects the accuracy of the received data. The data communication system should be such that it should deliver data accurately (without introducing any error).

(iii) Timeliness

For the audio and video data, the system should deliver the data in a timely manner i.e., as it is produced without any time delay. Such a data deliver is called as *real-time transmission* of data.

1.4. COMPONENTS OF DATA COMMUNICATION SYSTEM

If we specifically consider the communication between two computers, then, the data communication system will be as shown in figure 1.3.

It shall have following five components :

(i) Message (ii) Sender
(iii) Medium (iv) Receiver
(v) Protocol

> **DO YOU KNOW?**
>
> Communication networks, along with transportation networks, have become essential infrastructure in every society. Both types of networks provide flexible interconnectivity that allows the flow of people and goods in the case of transportation and information in the case of communications.

Description

1. Message

Message is nothing but information or data which is to be sent from one point to the other.

A message can be in the form of sound, text, number, pictures, video or combination of them.

2. Sender

Sender is a device which sends the message. Examples of sender are : computer, workstation video camera, telephone handset etc.

3. Medium

It is the physical path over which the message travels from the sender to the receiver.

The examples of transmission medium are coaxial cable, twisted pair wire, fiber optic cable, radio waves (used in terrestrial or satellite communication).

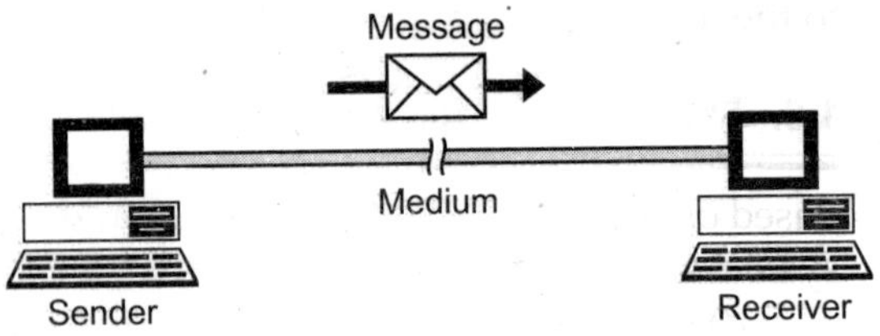

Fig. 1.3. Five components of a data communication system

4. Receiver

It is the device which receives the message. Examples of receiver are : computer, TV receiver, workstation, telephone handset etc.

5. Protocol

Protocol is defined as the set of rules which governs data communication. The connection of two devices takes place via the communication medium, but the actual communication between them shall take place with the help of a protocol.

1.5. A SIMPLIFIED DATA COMMUNICATION MODEL

Figure 1.4 shows the simplified data communication model.

Let us consider that the input device and transmitter are the components of a personal computer. The user of this PC wants to send a message m to another PC. Then, the message m will be in the form of a digital bit stream and labelled as **input data** as shown in figure 1.4. The sender's PC is connected to some **transmission medium** such as a local network or a telephone line, by an input/output device (i.e., transmitter) such as a local network transreceiver or a modem. The input data are then transferred to the transmitter as a sequence of voltage shifts that represent bits on a transmission cable or communication bus. The transmitter is connected directly to the medium and converts the incoming digital bit stream into an analog signal suitable for transmission over the communication

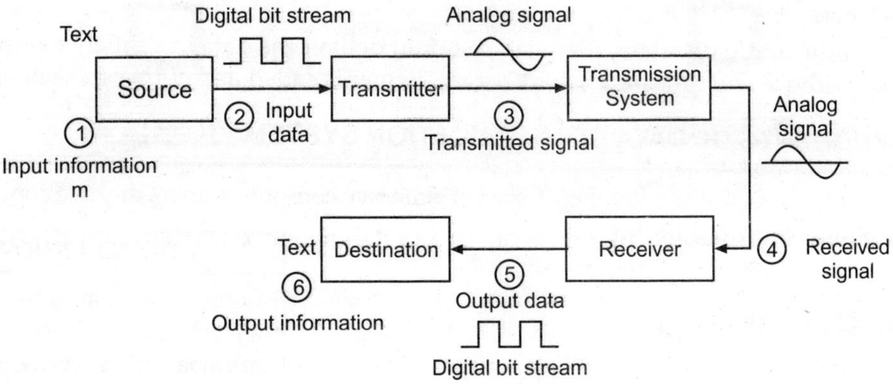

Fig. 1.4. Data communication model

cable. The analog transmitted signal travels on the transmission medium and is subjected to a number of impairments (distortions) before reaching the receiver. Due to these impairments (distortions), the received signal may be different from the transmitted signal. The receiver tries to estimate the original signal based upon the distorted received signal, and the knowledge of transmission medium. Thus, sequence of digital bits is produced at the output of the receiver. These bits are then sent to the destination which is another PC. The signal m represents the output information which is presented to the user and can be seen on the computer screen or printed using a printer.

1.6. TYPES OF COMMUNICATION: SIMPLEX, HALF DUPLEX, FULL DUPLEX: (DATA FLOW)

Based on whether the system communicates only in one direction or otherwise, the communication systems may be classified as under:

(i) Simplex systems
(ii) Half duplex systems
(iii) Full duplex systems.

1.6.1. Simplex Systems

In these systems, the information is communicated in only one direction. For example, the radio or TV broadcasting systems can only transmit. They cannot receive. In data communication system, the simplex communication takes place as shown in figure 1.5. The communication from CPU to monitor or keyboard to CPU is unidirectional. Keyboard and traditional monitors are examples of simplex devices.

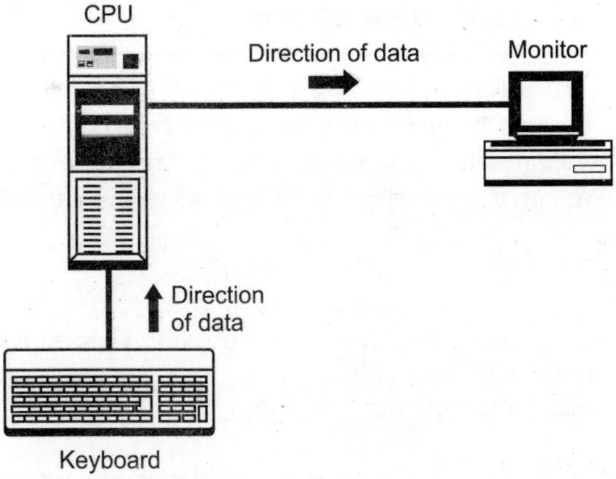

Fig. 1.5. Simplex mode of data transmission

1.6.2. Half Duplex Systems

These systems are bi-directional, i.e., they can transmit as well as receive but not simultaneously. At a time these systems can either transmit or receive, for example a transreceiver or walky talky set. A data communication system working in the half duplex mode is shown in figure 1.6. Each station can transmit and receive, but not at the same time. When one device is ending the other one is receiving and vice versa.

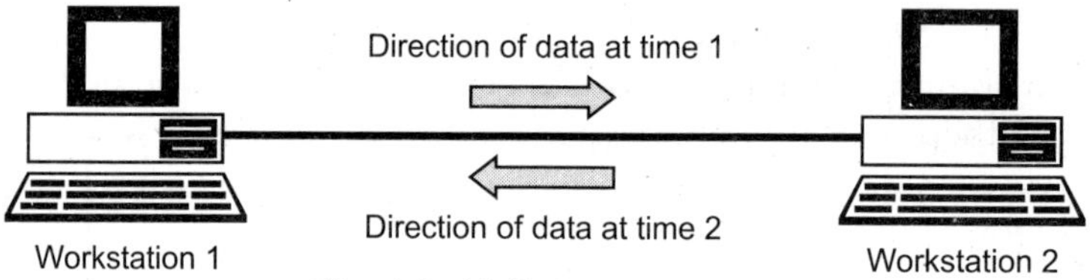

Fig. 1.6. Half duplex system

In half duplex transmission, the entire capacity of the channel is taken over by the transmitting (sending).

1.6.3. Full Duplex Systems

These are truly bi-directional systems as they allow the communication to take place in both the directions simultaneously. These systems can transmit as well as receive simultaneously. For example the telephone systems. A full duplex data communication system is shown in figure 1.7. Each station can transmit and receive simultaneously.

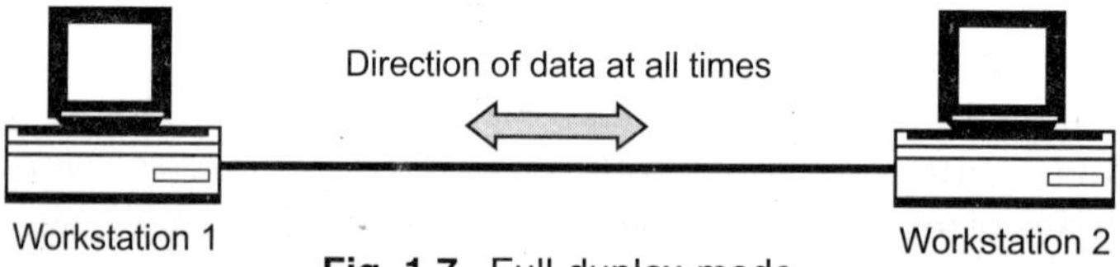

Fig. 1.7. Full duplex mode

In full duplex mode, signals going in either direction share the full capacity of link. The link may contain two physically separate transmission paths one for sending and another for receiving. Otherwise, the capacity of channel is divided between signals travelling in both directions.

1.7. COMPUTER NETWORKS

Computer networking is a business tool for companies for example, a bank can transfer funds between branches by using a network. People can access their bank accounts using automatic teller machines via a network. Travel agencies are using networks to make airline reservations. We can even make banking transactions and shop online using a network.

Network

Network is a broad term similar to *system*. Network is a communication system which supports many users. In context with the computers, we can say that a *computer network* is a system which allows communication among the computers connected in the network. There are various ways of interconnecting the computers.

Protocol

For successful communication to occur, it is not enough for the sender to simply transmit the message and assume that the receiver will receive it properly. There are certain rules that must be followed to ensure proper communication. A set of such rules is known as a protocol of the data communication system. Many different protocols are used in the modern data communication system. The interconnection of one station to many stations is called as networking. A network is any interconnection of two or more stations that wish to communicate.

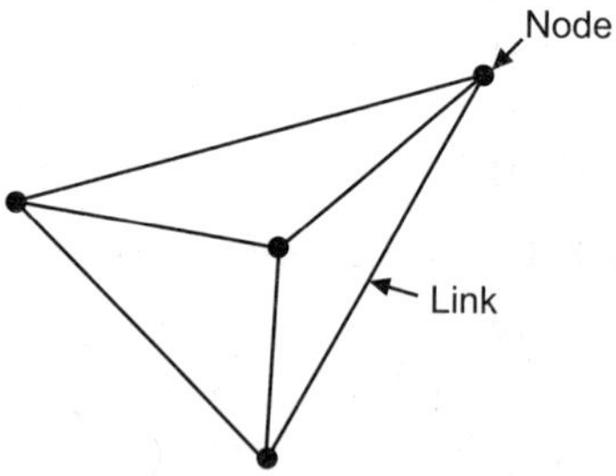

Fig. 1.8. A simple communication network

Node

Each station in a communication network is called as a node. The nodes are connected in different way to each other to form a network. One of such networks is shown in figure 1.8. Many other forms of interconnections are possible. The most familiar network is the telephone system. It is the largest and most sophisticated network of all.

1.7.1. Introduction to Computer Networks*

During 20th century the most important technology has been the information gathering, its processing and distribution. The computers and communications have been merged together and their merger has had a profound effect on the manner in which computer systems are organized. The old model in which a single computer used to serve all the computational needs of an organization has been replaced by a new one in which a large number of separate but interconnected computers do the job. Such systems are called as **computer networks.** Two computers are said to be interconnected if they interchange information. The connection between the separate computers can be done via a copper wire, fiber optics, microwaves or communication satellite.

Distributed System

If one computer can forcibly start, stop or control another computers are not autonomous. A system with one control unit and many slaves, or a large computer with remote printers and terminals is not called a **computer network, it is called a Distributed System.** In distributed system the existence of multiple autonomous computers is not visible to the user. The user can type a command to run a program and it runs. It is up to the operating system to select the best processor, find and transport all the input files to that processor and give the results to the user. With a computer network, the user must explicitly (definitely) log onto one machine, explicitly submit jobs remotely, explicitly move files around and generally handle all the network management personally. With a distributed system nothing has to be done explicitly, it is all automatically done by the system without the users knowledge. Basically a distributed system nothing has to be done explicitly, it is all automatically done by the system without the users knowledge. Basically a distributed system is a software system built on top of a network. The software gives it a high degree of cohesiveness (sticking together) and transparency.

1.7.2. Need of Computer Networks

The computer networks can prove to be useful in the following areas.

(i) Sharing the resources such as printers among all the users.
(ii) Sharing of expensive software and database.
(iii) Communication from one computer to the other.
(iv) Exchange of data and information amongst the users via the network.
(v) Sharing of information over the geographically wide areas.
(vi) For connecting the computers between various buildings of an organization.
(vii) For educational purposes.

1.7.3. Components of Computer Network

Following are some of the important components of a computer network.

(i) Two or more computers.
(ii) Cables (coaxial, twisted pair or fiber optic) as links between the computers.

* A network is a system of computers and related equipments connected by communication links in order to share data. The related equipment may be printers, fax machines, modems, copiers and so forth.

(iii) A network interfacing card (NIC) on each computer.
(iv) Switches
(v) A software called network operating system.

1.7.4. Advantages of Computer Networks*

Following are some of the advantages of computer networks :

1. Increased speed

(a) Network provide a very fast means for sharing and transfer of files.
(b) If the computer networks would not have been there, then we would have to copy the files on floppy discs and send them to the other computers.

2. Reduced cost

(a) Many popular versions of networkable softwares are now available at a considerably reduced costs as compared to individual licenced copies.
(b) In addition to this, it is also possible to share a program on a network. It is also possible to upgrade the program.

3. Improved security

(a) It is possible to protect the programs and files from illegal copying.
(b) By allotting password the access can be restricted to authorized users only.

4. Centralized software managements

(a) Due to the use of computer networks, all the softwares can be loaded on one computer.
(b) All the other computers can make use of this centralized software. It is not necessary to waste time and energy in installing updates and tracking files on independent computers.

5. Electronic-mail

(a) The computer network makes the hardware available which is necessary to install an e-mail system.
(b) The person to person communication is improved due to a presence of e-mail system.

6. Flexible access

It is possible for the authorized users to access their files from any computer connected on the network. This provides tremendous flexibilities in accessing.

1.7.5. Disadvantages of Networks

Following are some of the disadvantages of computer networks.

1. High cost of installation

(a) The initial cost of installation of a computer network is high. This is due to the cost of cables, network cards, computers, printers and various softwares that are required to be installed.
(b) The cost of services of technicians may also get added.

2. Requires time for administration

Computer networks need proper and careful administration and maintenance. This is a time consuming job.

---

* In networking technology, both the receiver and transmitter come in one unit and are usually installed inside the computer, such as network cards or modems.

3. Failure of server

(a) If the file servers *goes down* then the entire network comes to a standstill.

(b) If this happens then the entire organization can lose access to the necessary programs and files.

4. Cable faults

The computers in a network are interconnected with the help of connecting cables.

1.8. COMPUTER NETWORK CRITERIA*

Network is a broad term which is very much similar to system. Network is a communication system which supports many users. In context with the computer, we can say that a computer network is a system which allows communication among the computers connected in the network. A network must be able to meet certain criteria. The most important of them are :

(i) Performance
(ii) Reliability
(iii) Security

1. Performance

Performance can be measured in many ways. We can measure it in terms of transit time and response time.

(a) Transit time is defined as the amount of time required for a message to travel from one device to the other.

(b) Response time : It is the time elapsed between enquiry and response.

The other factors deciding the performance are as follows :

(i) Number of users
(ii) Type of transmission medium
(iii) Capability of connected hardware
(iv) Efficiency of software.

2. Reliability

The network reliability is important because it decides the frequency at which network failure takes place. It also decides the time taken by the network to recover and its robustness in the catastrophe.

3. Security

The network security refers to protection of data from the unauthorized user or access.

1.9. TYPE OF CONNECTIONS (TOPOLOGY)

The topology of a network describes the way computers are connected together. Topology is a major design consideration for cost and reliability. A network is two or more devices connected to each other through connecting links. There are two possible ways to connect the devices. They are as under :

(i) Point to point connection
(ii) Multipoint connection

* In networking technology, both the receiver and transmitter come in one unit and are usually installed inside the computer, such as network cards or modems.

1.9.1. Point-to-Point Connection

A point to point connection provides a dedicated link between two devices as shown in figure 1.9. Entire capacity of the link is reserved for transmission between these two devices only. It is possible to connect the two devices by means of a pair of wires as shown in figure 1.9(a) or using a microwave or satellite link as shown in figure 1.9(b).

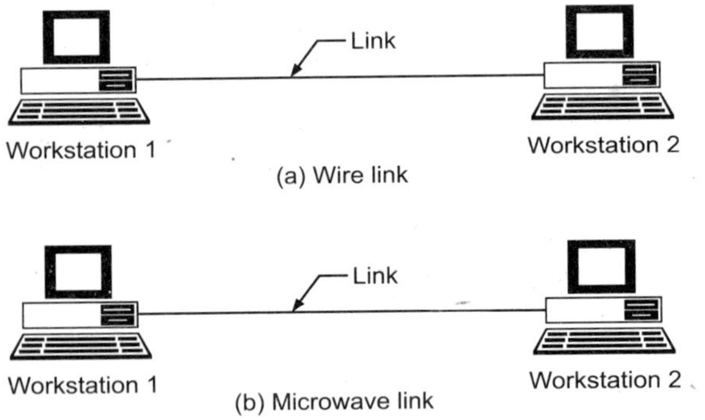

Fig. 1.9. Point to point connection

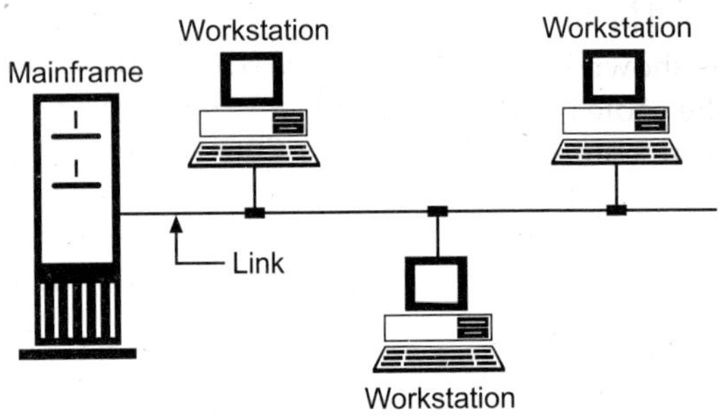

Fig. 1.10. Multipoint configuration

1.9.2. Multipoint Connection

A multipoint connection is also called a multidrop connection. In such a connection, more than two devices share a single link as shown in figure 1.10. In the multipoint connection, the channel capacity is shared. If many devices share the link simultaneously, it is called spatially shared connection. But, if users share it turn by turn then it is time sharing connection.

1.10. NETWORK TOPOLOGY TYPES*

(RGTU, Bhopal, Sem. Exam., June, 2014)

As a matter of fact, network topology describes the layout or appearance of a network that is, how the computers, cables, and other components within a data communication network are interconnected, both physically and logically. The physical topology describes the way in which a network is physically laid out, and the logical topology describes how data actually flow through the network. In a data communication network, two or more devices are connected to form a link whereas two or more links form a topology. The topology of a network is the geometric representation of the relationship of all the links connecting the devices (or nodes). The five basic network topologies are as shown in figure 1.11. While selecting one of the above five topologies, we have to consider the relative status of the device to be linked.

All the five topologies shown in figure 1.11, are examples of multipoint topologies. Two relationships on this basis are possible as under:

(i) Peer to peer network (work group)

(ii) Primary-secondary.

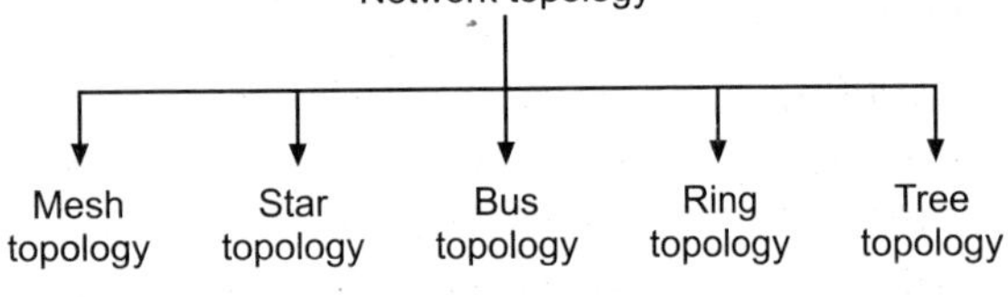

Fig. 1.11. Classification of network topology

Peer to peer is the relationship where the devices share the link equally. The examples are ring and mesh topologies. In Primary-secondary relationship, one device controls and the other devices have to transmit through it. For example, star and tree topology.

* What is network topology? Explain mesh, ring and star topologies bringing out their merits and demerits. *(RGTU, Bhopal, Sem. Exam., June, 2014)*

1.11. BUS TOPOLOGY

Basically, a bus network is a multi-point connection in which stations are connected to a single cable called a **bus**. A bus topology is a multipoint data communication circuit that makes it relatively simple to control data flow between and among the computers because this configuration allows all stations to receive every transmission over the network. In bus topology, all stations share one medium.

The bus topology is usually used when a network installation is small, simple or temporary as shown in figure 1.12. Bus topology is widely used in LAN networking*. On a typical bus network, the cable is just one or more wires, with no active electronics to amplify the signal or pass it along from computer to computer. This makes the bus a passive topology.

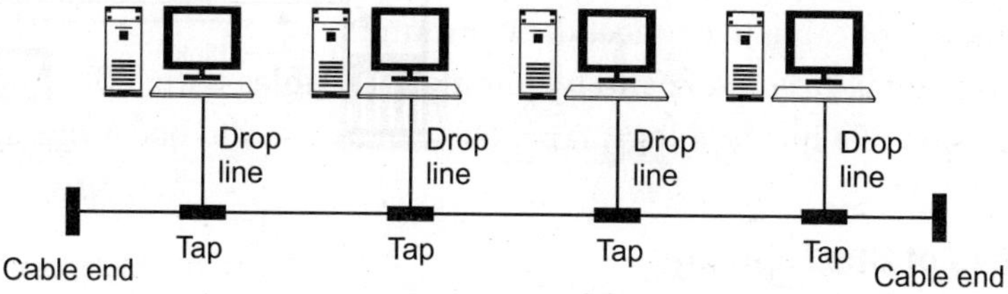

Fig. 1.12. Bus topology

When one computer sends a signal up to the cable, all the computers on the network receive the information, but the one with the address that matches the one encoded in the message accepts the information while all the others reject the message.

The speed of the bus topology is slow because only one computer can send a message at a time. A computer must wait until the bus is free before it can transmit. The bus topology requires a proper termination at both the ends of the cable. Since, the bus is a passive topology, the electrical signal from a transmitting computer is free to travel the entire length of the cable. Without termination, when the signal reaches the end of the cable, it returns back and travels break up the cable. Without termination when the signal reaches the end of the cable, it returns back and travels back up the cable. The transmitted waves and reflected waves, if they are in phase add and if they are out of phase cancel. This, adding and cancellation of the wave leads to what is called a standing wave. The standing waves can distort the normal signals which are travelling along the cable. The standing waves are overcome by attaching terminators (load impedance) equal to the characteristics impedance of the cable e.g. a 50 Ω cable is terminated by a 50 Ω load impedance at both the ends of the cable. The terminators absorb the electrical energy and stop reflections.

1.11.1. Salient Features of the Bus Topology

Following are some of the important characteristics of the bus topology :

(i) This is multipoint configuration.

(ii) Adequate signal to noise ratio should be maintained.

(iii) The signal should not be too strong. This is necessary to avoid possibility of signal distortion.

(iv) This is called as signal balancing which is not an easy task at all. Specially, the signal balancing becomes increasingly difficult with increase in the number of stations.

(v) The solution to this problem is to divide the transmission medium into small segments and within each such segment, a pairwise balancing is done using amplifiers or repeaters between the segments.

* Ethernet is one of the most popular LANs that use bus topology.

1.11.2. Transmission Media for Bus LANs

We can use the following transmission media for the bus LANs :

(i) Twisted pair
(ii) Baseband co-axial cable
(iii) Broadband co-axial cable
(iv) Optical fiber.

> **DO YOU KNOW?**
>
> There is need to understand distinction between data transmission and data communication. While transmission refers to transport of information, communication implies meaningful exchange of information.

1.11.3. Advantages of Bus Topology

(i) The bus topology is easy to understand, install, and use for small networks.
(ii) The cabling cost is less as the bus topology requires the least amount of cable to connect the computers.
(iii) The bus topology is easy to expand by joining two cables with a BNC barrel connector.
(iv) In the expansion of a bus topology, repeaters can be used to boost the signal and increase the distance.

1.11.4. Drawbacks of Bus Topology

(i) Heavy network traffic slows down the bus speed. In bus topology, only one computer can transmit and others have to wait till their turn comes and there is no co-ordination between computers for reservation of transmitting time slot.
(ii) The BNC connectors used for expansion of the bus attenuates the signal considerably.
(iii) A cable break or loose BNC connector causes reflection and brings down the whole network causing all network activity to stop.

Important Point:
Ethernet 10 base 2 also known as thinnet is an inexpensive network based on the bus topology. A bus network behaves erratically if it is not terminated or improperly terminated. Token bus networks are defined by the IEEE 802.4 standard.

1.12. RING TOPOLOGY

(RGTU, Bhopal, Sem. Exam., June, 2014)

IBM invented ring topology, which is well known as IBM token ring. In a ring topology, all the stations are connected in cascading order to make a ring. This means that in a ring topology, each computer is connected to the next computer, with the last one connected to the first as shown in figure 1.13. Rings are used in high-performance networks where large bandwidth is essential, e.g. time attractive features such as video and audio.

In other words, a ring topology is a multipoint data communication network where all stations are interconnected in tandem (series) to form a closed loop or circle. A ring topology is sometimes called a loop. Each station in the loop is joined by point-to-point links to two other stations, (the transmitter of one and the receiver of the other).

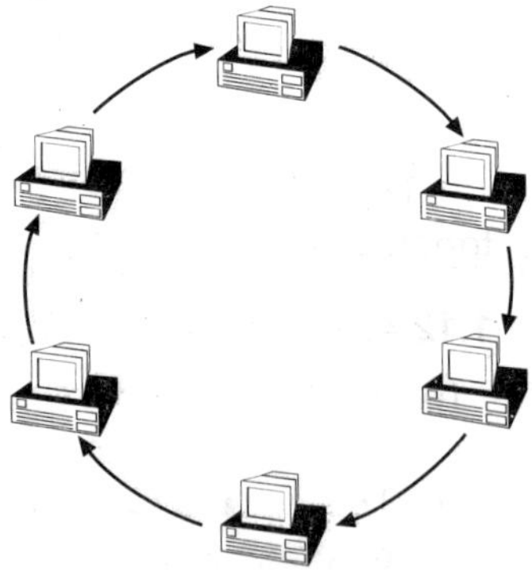

Fig. 1.13. Ring topology

The messages flow around the ring in one direction. There is no termination because there is no end to the ring. Some ring networks do token passing. A short message called a token, is passed around the ring until a computer wishes to send information to another computer. That computer modifies the token, adds an electronic address and data and sends it around the ring. Each computer in sequence receives the token and the information passes them to the next computer until either the electronic address matches the address of a computer or the token returns to its origin. The receiving computer returns a message to the originator indicating that the message has been received. The sending computer then creates another token and places it on the network, allowing another station to capture the token and begin transmitting. The token circulates until a station is ready to send and capture the token. Faster networks circulate several tokens at once.

1.12.1. Salient Features of Ring LANs

(i) The basic ring topology is shown in figure 1.14 which shows that a number of repeaters are used and that the transmission is unidirectional.

(ii) The data are transferred in a sequential manner bit by bit around the ring. It may be noted that each repeater will regenerate and retransmit each bit.

1.12.2. Functions of a Ring

A ring can operate as a communication network if it performs the following three functions:

(i) Data insertion

(ii) Data reception

(iii) Data removal

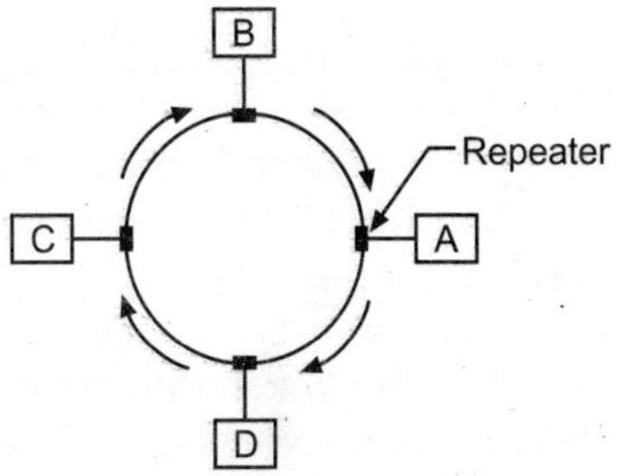

Fig. 1.14. Illustration of ring Topology

The above mentioned functions are actually provided by the repeaters. Each repeater also acts as the device attachment point. Hence, the function of data insertion is accomplished by the repeaters. Data are transmitted in the form of packets. Each packet consists of a destination address field. As this packet by a repeater, the destination address field is copied by the repeater. If the destination address field corresponds to the address of a device then that repeater copies the remaining contents of the packet as well. Data insertion and reception can be done easily by the repeaters but data removal is more difficult on a ring. As the ring is a closed loop, a packet will circulate on it indefinitely if it is not removed. A packet can be removed by the addressed repeater or each packet can be removed by the transmitting repeater itself after the packet has made one trip around the ring. The second approach is more desirable.

1.12.3. Problems Faced in the Ring Topology

(i) If any link breaks or if any repeater fails then the entire network will be disabled.

(ii) To install a new repeater for supporting a new device, it is necessary to have the identification of two nearby, topologically adjacent repeaters.

(iii) It is necessary to take preventive measures to deal with the time jitter.

(iv) Due to the closed nature of the ring topology, it is necessary to remove the circulating packets.

These problems except for the last one can be rectified by simply refinements of the ring topology.

1.12.4. Advantages of Ring Topology

(i) No one computer can monopolise the network because every computer is given equal access to the token.

(ii) The fair sharing of the network allows the network to continue function in a useful, if slower, manner rather than fail once capacity is exceeded as more users are added.

1.12.5. Drawbacks of Ring Topology

(i) Failure of one computer on the ring can affect the whole network.

(ii) It is difficult to troubleshoot the ring.

(iii) Adding or removing the computers disturbs the network activity.

Important Point:
It may be noted that token ring networks are defined by the IEEE 802.5 standard. Fiber Distributed Data interface (FDDI) is a fast fiber-optic network based on the ring topology.

DO YOU KNOW?
Communication networks provide a service much like other ubiquitous utilities such as the water supply or electricity power systems. On the other hand, communication networks exhibit tremendous flexibility in their use and in this respect they are closest to transportation networks.

1.13. STAR TOPOLOGY

(RGTU, Bhopal, Sem. Exam., June, 2014)

In a star topology, all the cables run from the computers to a central location where they are all connected by a device called a hub as shown in figure 1.15.

Stars are used in concentrated networks, where the end points are directly reachable from a central location when network expansion is expressed and when the greater reliability of a star topology is required.

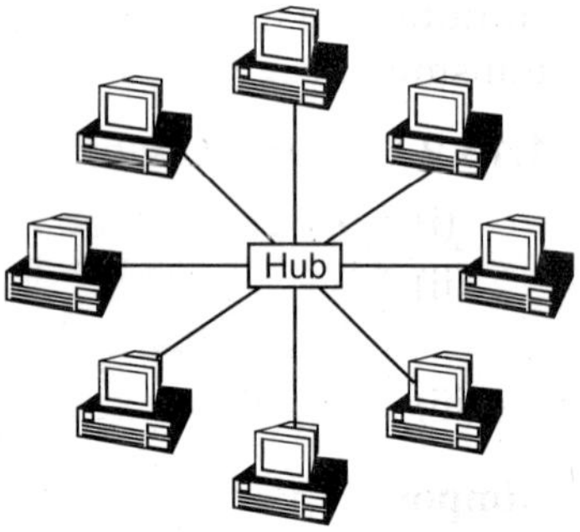

Fig. 1.15. Illustration of Star topology

Each computer on a star network communicates with a central hub that re-sends the message either to all the computers in a broadcast star network or only to the destination computer in a switched star network. The hub in a broadcast star network can be active or passive. An active hub generates the electrical signal and sends it to all the computers connected to it. This type of hub is usually called a multiport repeater. Active hubs require external power supply.

A passive hub is a wiring panel or punch down block which acts as a connection point. It does not amplify or regenerate the signal. Passive hubs do not require electrical power supply. Several types of cables can be used to implement a star network. A hybrid hub can use different types of cable in the same star network. A star network can be expanded by placing another star hub as shown in figure 1.16. This arrangement allows several more computers or hubs to be connected to that hub. This creates a hybrid star network.

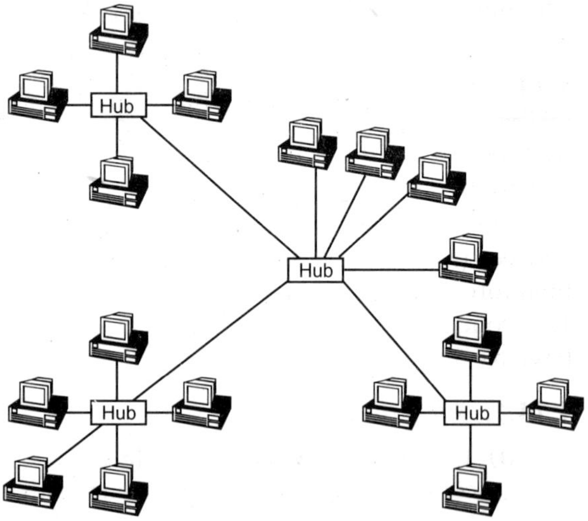

Fig. 1.16. Illustration of expansion of star topology

1.13.1. Star LANs

In the star type LANs, the unshielded twisted pair is used as the transmission medium*. This is because unshielded twisted pair is a telephone

* Transmission medium shall be discussed in detail in chapter 3.

wire which is available in each and every office building. The other advantages of using twisted wires are as follows :

(i) No additional installation cost is required for the installation of LAN.

(ii) The telephone wires cover the entire building it is possible to spread the network in every part of each building.

The basic star topology is as shown in figure 1.17. This is called single level start topology. As shown in figure 1.17, the central element of the star topology is an active element called hub or repeater. Each station (A, B, C, ...) is connected to the hub with the help of two links one for transmitting and the other for reception of the data. When a single station transmits, the hub repeats the signal and sends it to each station. Typically the length of each link is 100 m. If the twisted pair is used and the length may increase upto 500 m if the optical fiber is used as transmission medium.

It is important to note that if two stations transmit simultaneously, then there will be a collision between their transmitted signals.

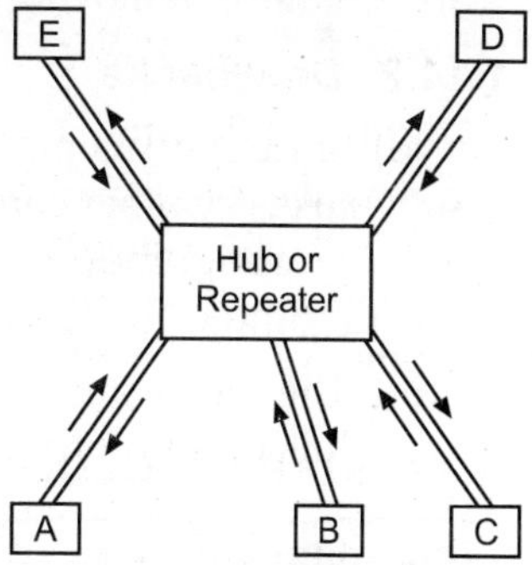

Fig. 1.17. Illustration of single level star topology

1.13.2. Drawbacks of Star Topology

(i) If the central hub fails, the whole network fails to operate.

(ii) Many star networks require a device at the central point to rebroadcast or switch the network traffic.

(iii) The cabling cost is more since cables must be pulled from all computers to the central hub.

Important Point:

Ethernet 10 base T is a popular network based on the star topology. Intelligent hubs with microprocessor that implement features in addition to repeating network. Signals provide for centralized monitoring and management of the network. It is the most flexible and the easiest to diagnose when there is a network fault.

1.14. MESH TOPOLOGY

(RGTU, Bhopal, Sem. Exam., June, 2014)

In a mesh topology, every device has a dedicated point-to-point link to every other device as shown in figure 1.18.

The term dedicated means that the link carries traffic only between two devices it connects. A fully connected mesh network therefore has $n(n-1)/2$ physical channels to link h_n devices. To accommodate those links, every device on the network must have n – 1 input/output ports.

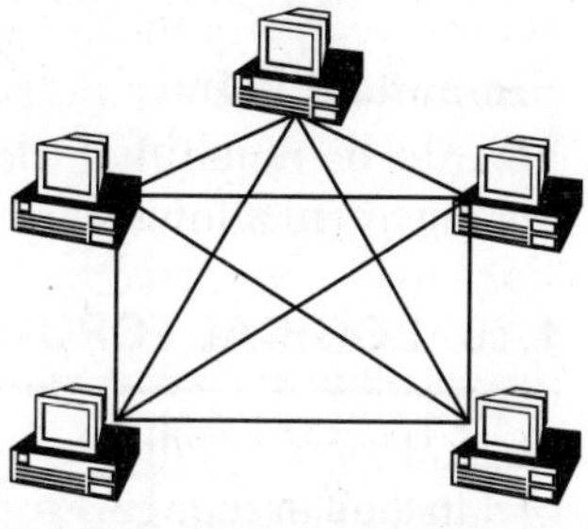

Fig. 1.18. Illustration of Mesh topology

1.14.1. Advantages

(i) The use of dedicated links guarantees that each connection can carry its own data load, thus eliminating traffic problems.

(ii) A mesh topology is robust because the failure of single computer does not bring down the entire network.

(iii) It provides security and privacy because every message sent travels along a dedicated line.
(iv) Point to point links make fault diagnose easy.

1.14.2. Drawbacks

(i) Since every computer must be connected to every other computer installation and reconfiguration is difficult.
(ii) Cabling cost is more.
(iii) The hardware required to connect each link input/output and cable is expensive.

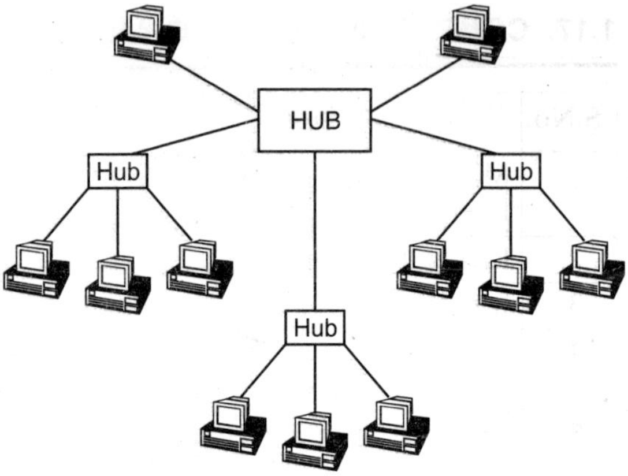

Fig. 1.19. Illustration of tree topology

Important Point:
It may be noted that mesh topology is usually implemented as a backbone connecting the main computers of a hybrid network that can include several other topologies.

1.15. TREE TOPOLOGY

A tree topology is the variation of a star. As in a star, nodes in a tree are linked to a central hub that controls the traffic to the network.

However, not every computer plugs into the central hub, majority of them are connected to a secondary hub which, in turn, is connected to the central hub as shown in figure 1.19. The central hub in the tree is an active hub which contains repeater. The repeater amplifies the signal and increases the distance a signal can travel. The secondary hubs may be active or passive. A passive hub provides a simple physical connection between the attached devices.

1.15.1. Advantages

(i) It allows more devices to be attached to a single hub and can therefore increase the distance of a signal can travel between devices.
(ii) It allows the network to isolate and priorities communications from different computers.

1.15.2. Drawbacks

(i) If the central hub fails, the system breaks down.
(ii) The cabling cost is more.

Important Point:
It may be noted that the advantages and disadvantages of a tree topology are generally the same as those of a star.

1.16. LOGICAL TOPOLOGY

Logical topology means the way the stations are logically connected to each other for the purpose of data unit exchange. Physical topology discussed earlier can be different from the logical topology of the network. For example, for a bus topology, the bus acts as the central controller. It receives data and forwards it to the various nodes. Thus, the stations have a logical connection to the bus which acts as a controller. Therefore, the logical topology is star topology, even though the physical topology is bus.

1.17. COMPARISON OF RING AND STAR TOPOLOGIES

S.No.	Ring	Star
1.	Media failure on unidirectional or single loop causes complete network failure.	Media faults are automatically isolated to the failed segment.
2.	Relatively difficult to reconfigure.	Relatively easy to configure.
3.	It is difficult to troubleshoot.	Easy to troubleshoot.
4.	The failure of one computer can affect the whole network.	The failure of single computer or cable doesn't bring the network down.
5.	One computer cannot monopolize the network.	Failure of the central hub causes the whole network failure.
6.	Adding and removing computers disrupts the network.	Adding and removing the computers is relatively easier.
7.	A single node failure leads to the collapse of the full network.	Failure of a single node involves disconnecting only that node and the remaining nodes remain fully functional.
8.	Fault diagnosis is difficult.	Fault diagnosis is relatively easy.

1.18. COMPARISON OF BUS AND STAR TOPOLOGIES

S.No.	Bus	Star
1.	Uses a cable as bus or backbone to connect all nodes.	Uses a central hub to connect the nodes to each other.
2.	Baseband or broadband coaxial cable is used.	Twisted pair, coaxial cables or optical fiber cables are used.
3.	If a part of bus fails, the whole network fails.	Failure of the central hub will make the entire network collapse.
4.	Adding an new node is difficult.	Adding and removing a node is relatively easy.
5.	Fault diagnosis is relatively difficult.	Fault diagnosis is easy.

1.19. HYBRID TOPOLOGY

We have discussed various basic topologies such as bus, ring, mesh, star etc. Basically, hybrid topology is a combination of different topologies connected together by a backbone cable. Each network is connected to the backbone cable by a device called a bridge. Hybrid topology is the one which makes use of two or more basic topologies mentioned above together.

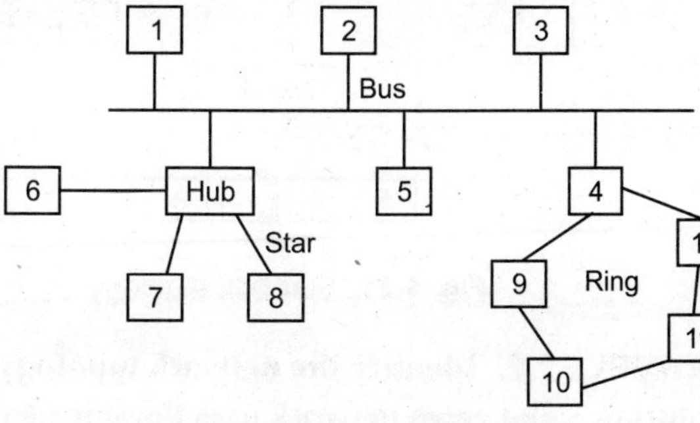

Fig. 1.20. Hybrid topology

There are different ways in which a hybrid network is created. Figure 1.20 shows the hybrid topology in which bus, star and ring topologies are used simultaneously. In figure 1.20, the nodes 1, 2, 3, 4 and 5 are connected in the bus topology, node 6, 7 and 8 form a star and nodes 4, 9, 10, 11, 12 are arranged in a ring topology.

The practical networks generally possess a hybrid topology. Many complex networks can be reduced to some form of hybrid topology.

The hybrid topology which is to be used for a particular application depends on the requirements of that application.

1.20. COMPARISON OF STAR BUS AND STAR RING TOPOLOGIES

S.No.	Parameter	Star Bus	Star Ring
1.	Topology	Hybrid. It is a combination of star and bus topologies.	Hybrid. It is combination of star and ring topologies.
2.	Configuration	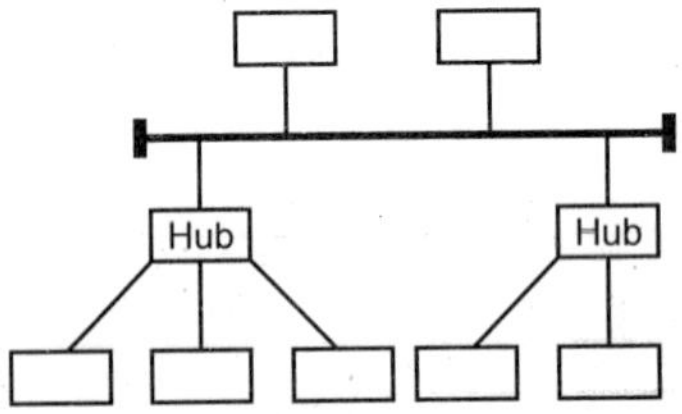	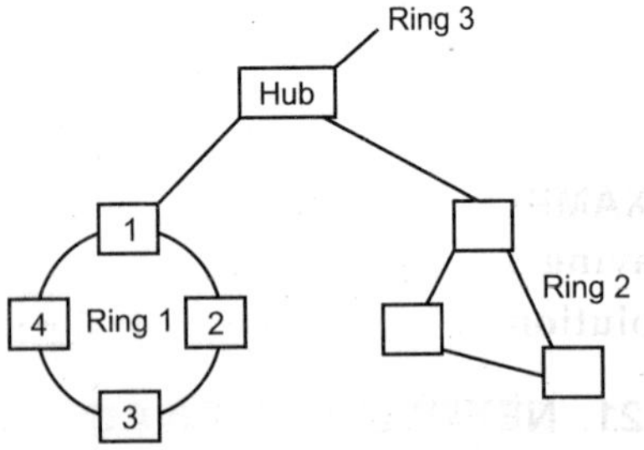
3.	Peculiarity	Many star are connected to a bus.	Many rings are connected in star.
4.	Applications	Suitable for large networks.	Suitable for interconnection of many small network.

EXAMPLE 1.1. Draw the star bus topology connecting three star networks consisting of four computers.

Solution: Figure 1.21 shows the required star bus topology.

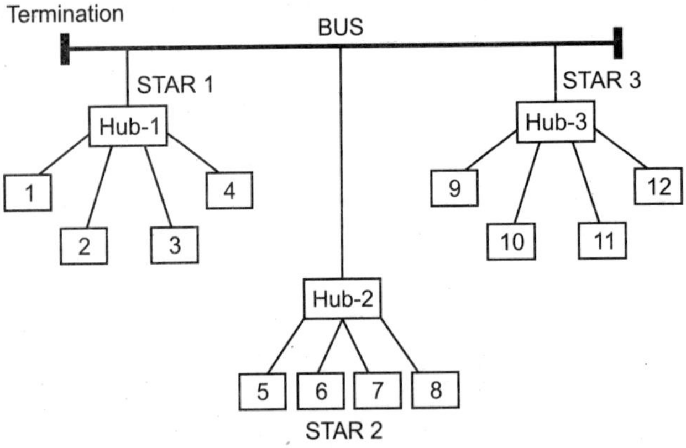

Fig. 1.21. Star bus topology

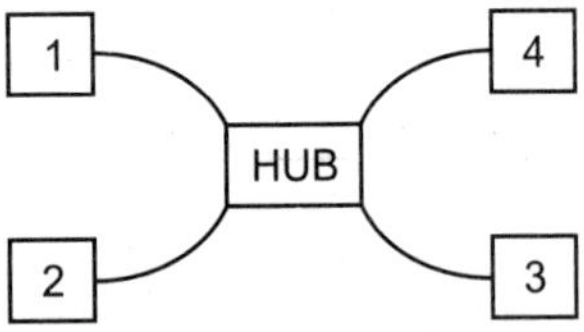

Fig. 1.22.

EXAMPLE 1.2. Identify the network topology shown in figure 1.22.

Solution : The given network uses the star topology.

EXAMPLE 1.3. Identify the topology shown in figure 1.23. Comment on whether this topology is active or passive.

Solution: This is the ring topology. The ring topology is an active network.

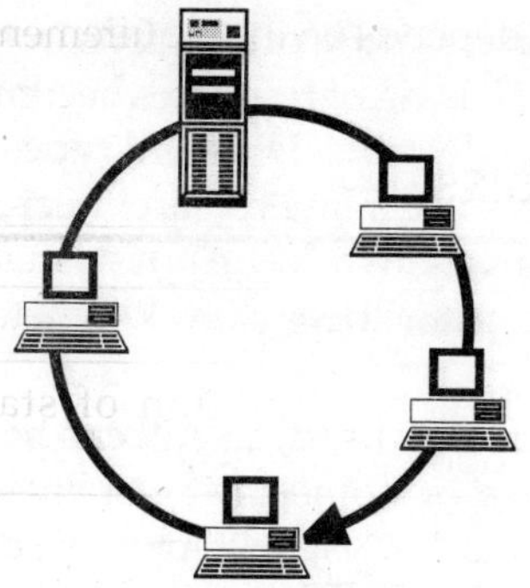

Fig. 1.23.

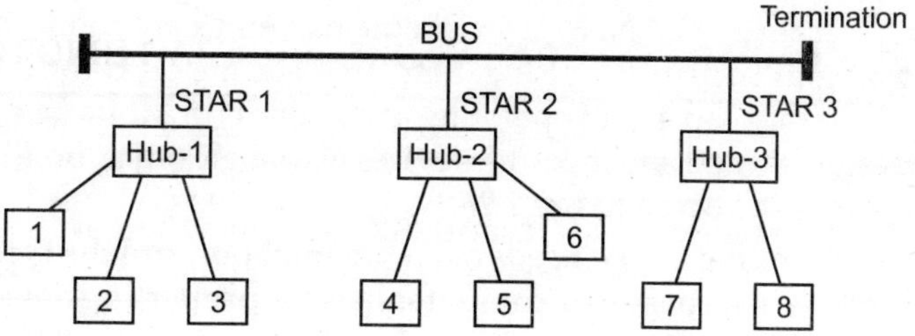

Fig. 1.24. Star-bus network

EXAMPLE 1.4. Draw with neat-labelled sketch of star-bus topology connecting 3 star networks having 3 computers in 2 stars and 2 computers in one star.

Solution : Figure 1.24 shows the required star-bus topology.

1.21. NETWORK TOPOLOGY DESIGN

The topology design problem involves the following aspects:

(i) Placement of network nodes.
(ii) Placement of routers.
(iii) To decide how to connect these nodes.
(iv) Network reliability and link capacity constraints.

A single node with a very large capacity directly serving all the potential clients is the cheapest design, as far as costs are connected. But this model is not suitable for the latest applications which operate at higher data rates and require small response time. The single node design even though cost effects, has a disadvantage of long response times.

1.22. TYPES OF NETWORKS

In this section, let us discuss the following categories of networks. The distance between computers that are connected as a network determines the type of network, such as a Local Area Network (LAN), Metropolitan Area Network (MAN), and Wide Area Network (WAN).

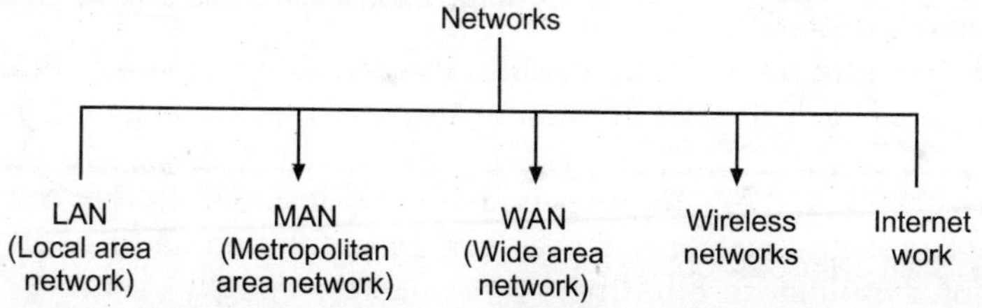

Fig. 1.25. Illustration of network categories

1.22.1. Local Area Networks (LAN)*

The local area network (LAN) is a network which is designed to operate over a small physical area such as an office, factory or a group of buildings. LANs are very widely used in a variety of applications.

LANs are easy to design and troubleshoot. The personal computers and workstations in the offices are interconnected via LAN. The exchange of information and sharing of resources becomes easy because of LAN. In LAN, all the machines are connected to a single cable. Different types of topologies such as Bus, Ring, Star, Tree etc. are used for LANs. LANs use a layered architecture and they are capable of operating at hundreds of Mbits/sec. A local area network (LAN) is usually privately owned and links the devices in a single office, building or campus of upto a few kilometers in size as shown in figure 1.26.

Depending on the needs of an organization and the type of technology used, a LAN can be as simple as two personal computers and a printer in someone's office or home or it can extend throughout a company and include voice, sound and video peripherals. LANs are widely used to allow resources to be shared between personal computers or workstations. The resources to be shared can be hardware like a printer or softwares or data. A common example of a LAN found in many business organizations, links a work group of task related computers, e.g. accounting and finance PCs, administrative PCs or engineering workstations. One of the computer in a network can become a server serving all the remaining computers called clients. Software can be stored on the server and it can be used by the remaining clients. In a LAN, its size can be determined by licensing restrictions on the number of users per copy of software or by restricting the number of users licensed to access the operating system. LANs are also distinguished from MANs and WANs based on the transmission media they use as topology. In general, a given LAN will use only one type of transmission medium. The most common topologies used are bus, ring and star.

The data rates for LAN range from 4 to 16 Mbps with the maximum of 100 Mbps**.

1.23. LAN TOPOLOGIES

Various topologies are possible for the broadcast LANs such as bus topology or ring topology as shown in figure 1.26.

1. Bus Topology

Bus topology is shown in figure 1.26(a). In this topology, at any instant, only one computer acts as master and it is allowed to transmit (broadcast). The others are supposed to listen. If two or more machines want to transmit simultaneously then an arbitration mechanism has to be used for resolving the conflict. It is possible to have a centralized or distributed type arbitration mechanism. The most popular example of bus topology is Ethernet (IEEE 802.3).

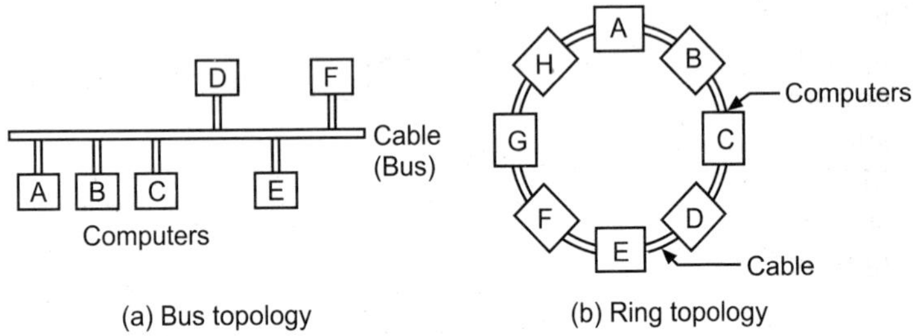

Fig. 1.26. Illustrations of LAN topologies

* A local area network (LAN) is a high-speed network designed to link computers and other data communication systems together within a small geographic area such as an office department, or a single floor of a multistory building.

** Several LANs can be connected together in a building or campus to extend the connectivity. A LAN is considered a private network. The most popular LANs in use today are Ethernet, Token Ring, and Gigabit Ethernet.

It has a decentralized control and it operates at 10 or 100 Mbps. Computers on Ethernet can transmit whenever they want. If collision of their packets takes place, then they wait for a random time and retransmit their packets.

2. Ring Topology

This is another broadcast topology and it is shown in figure 1.26(b). In a ring, each bit propagates around on its own without waiting for the rest of the packet to which it belongs. Since it is a broadcast system, some rules are essential for arbitrating the simultaneous access to the ring. An example of ring based LAN is IEEE 802.5 (IBM token ring) operating at 4 and 16 Mbps.

3. Static and Dynamic Broadcast Networks

The broadcast networks are further classified into two types namely,

(i) Static networks and

(ii) Dynamic networks.

This classification is based on how the channel is allocated. In static allocation, each machine is allowed to broadcast only in its allotted time slot. But static allocation wastes the channel capacity when a machine does not want to transmit in its allotted time slot. Hence, most of the systems try to allocates the channel dynamically i.e. on demand.

1.24. METROPOLITAN AREA NETWORK (MAN)

A MAN is basically a bigger version of a LAN and normally uses similar technology. It is designed to extend over an entire city. It can be single network such as a cable TV network, or it may be a means of connecting a number of LANs into a larger network so that resources can be shared LAN to LAN as well as device to device as shown in figure 1.27. A MAN can cover approximately 30 to 100 miles, connecting multiple networks that are located in different locations of a city or town. A MAN may be wholly owned and operated by a private company or it may be a service provided by a public company, such as local telephone company (telco). A MAN is distinguished by the IEEE 802.6 standard or it is also known as **Distributed Queue Dual Bus (DQDB).**

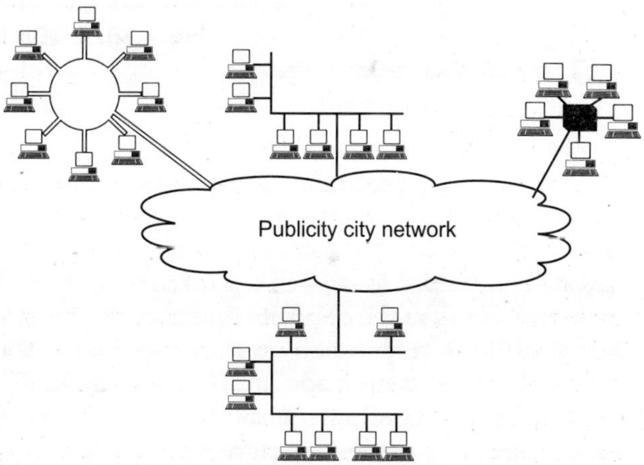

Fig. 1.27. Metropolitan area network (MAN)

The DQDB consists of two unidirectional cables (buses) to which all the computers are connected as shown in figure 1.28.

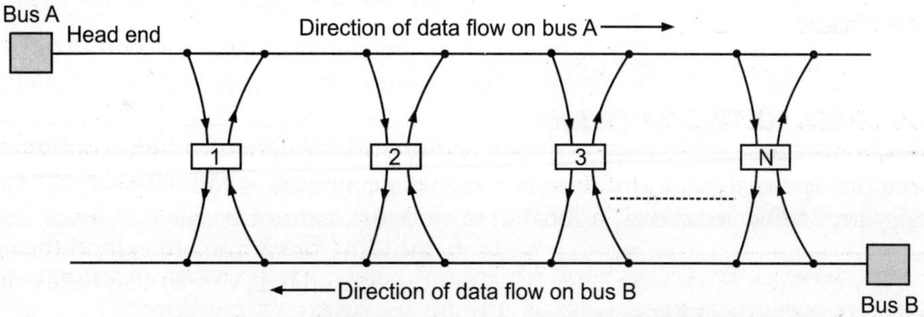

Fig. 1.28. Distributed queue dual bus architecture

Each bus has a device which initiates the transmission activity called as the *head-end.* Traffic that is destined for a computer to the right of the sender uses the upper bus and to the left uses the lower bus as shown in figure 1.28.

1.24.1. Distributed Queue Dual Bus (DQDB) Standard

IEEE 802.6 is a metropolitan area network (MAN) standard. It has an ability to integrate individual LANs by providing the interconnecting backbone structure. This makes formation of larger networks possible. After trying various techniques, finally a standard has been agreed using the Distributed Queue Dual Bus (DQDB). The architecture consists of a slotted system using a dual unidirectional user system operating at 155 Mbit/s. The frame synchronizing signals are generated by the network controller (or head end) which is at one end of system. The frame sync signal is transmitted on one bus, out of the two buses available as shown in figure 1.29. The controlling station at the other end generates a frame pattern with the same synchronization on the other bus. Figure 1.30 shows the slot format of DQDB.

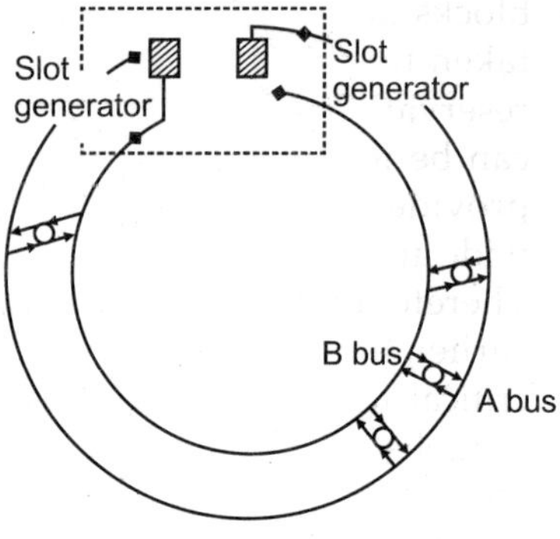

Fig. 1.29. MAN dual bus

1 byte	52 byte
Access control field	Protocol Data Unit (PDU)

Fig. 1.30. DQDB slot format

There are two fields namely :

(i) Access control field

(ii) Protocol data unit (PDU)

The access control field contains three bits used for the purpose of request. A station can queue a request to talk on a bus by putting a reservation on the other bus. Every station has two count registers, one for each bus. A count register is incremented by 1 everytime an upstream stations reservation is seen in a passing slot. This register is decremented by 1 for every empty slot which flows in the opposite direction on the other bus. If a station is waiting to talk on a bus, then it makes a reservation on the other bus in the first free reservation field. Then a copy of request count register is made, and it is decremented for each empty slot passing by on the alternative bus. This process continues until this copied register contents reduce to zero. The next empty slot is then used. The request count register continues in a normal way and takes the requests made further.

1.24.2. Advantage

The performance is better than that of token ring at low data levels.

1.24.3. Drawback

The performance degrades at high data levels.

1.25. WIDE AREA NETWORK (WAN)*

When a network spans a large distance or when the computers to be connected to each other are at widely separated locations, a local area network cannot be used. A wide area network

* A wide area network (WAN) is used for long-distance transmission of information. WANs cover a large geographical area, such as an entire country or continent.

(WAN) must be installed. The communication between different users of WAN is established using leased telephone lines or satellite links and similar channels. It is cheaper and more efficient to use the phone network for the links. Most wide area networks are used for transferring large blocks of data between its users. As the data is from existing records or files, the exact time taken for this data transfer is not a critical parameter. Another example of WAN is an airline reservation system. Terminals are located all over the country through which the reservation can be made. It is important to note here that all the terminals use the same common data provided by the central reservation computer. Because of the large distances involved in the wide area networks, the propagation delays and variable signal travel times are major problems. Therefore, most wide area networks are not used for time critical applications. As explained earlier, they are more suitable for transfer of data from one user to the other which is not a time critical application. Wide area networks are basically packet switching networks. A WAN provides long distance transmission of data, voice image and video information over large geographical areas that may comprise a country, a continent or even the whole world as shown in figure 1.31.

WAN contains a collection of machines used for running user (i.e. application) programs. All the machines called hosts are connected by a communication subnet as shown in figure 1.32. The function of the subnet is to carry messages from host to host. The subnet consist of two important components, transmission lines and switching elements.

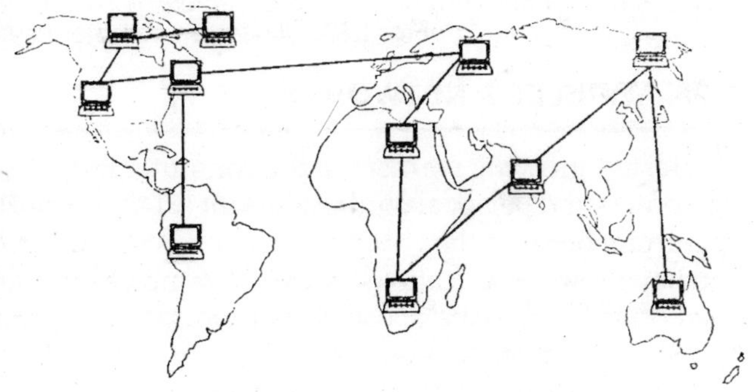

Fig. 1.31. Wide area network (WAN)

Transmission lines move bits from the machine to another. The switching elements are specialized computers used to connect two or more transmission lines. When data arrive on an incoming line, the switching element must choose an outgoing line to forward them. The switching elements are either called as packet switching nodes, intermediate systems, data switching exchanges or routers. When a packet is sent from one router to another via one or more intermediate routers, the packet is received at intermediate router. It is stored in the routers until the required output line is free and then forwarded. A subnet using this principle is called a point to point, store-forward or packet switched subnet.

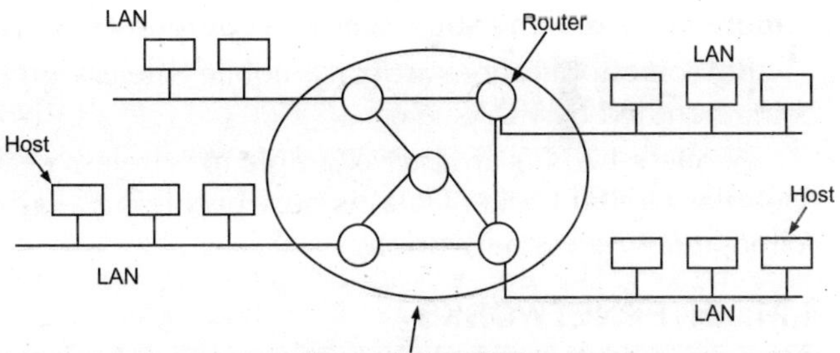

Fig. 1.32. Illustration of communication subnet and hosts

WANs may use public, leased or private communication devices, and can spread over a wide geographical area. A WAN that is wholly owned and used by a single company is often called as an enterprise network. In most WANs, the network contains a large number of cables or telephone lines each one connecting a pair of routers. If two routers which are not connected to each other via a cable want to communicate, then they have to do it indirectly via other routers.

1.25.1. Packet Switched Subnet

When a packet is sent from one router to the other, via some intermediate routers, the packet received at each intermediate router is stored until the required output line is free. Once the line becomes free, the packet is forwarded. A subnet working on this principle is called as point to point or store and forward or packet switched subnet. If the packets are small and of same size they are called as cells.

1.25.2. Router Interconnection Topologies*

Figure 1.33 shows some of the possible router interconnection topologies in a point to subnet. The LANs have a symmetric topology while WANs have irregular topologies. The WANs can also be formed using satellite or ground radio system. Satellite networks are inherently broadcast type so they are useful when the broadcast property is important.

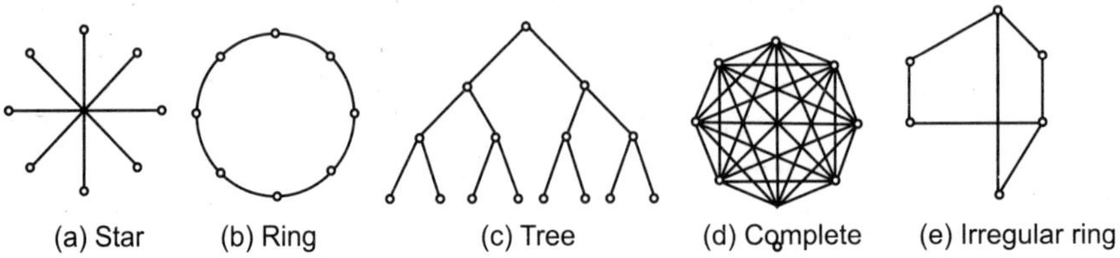

Fig. 1.33. Illustration of router interconnection topologies

1.26. WIRELESS NETWORKS

The fastest growing segment of the computer industry is the mobile computers such as notebook computers and personal digital assistant (PDAs). The wireless networks are becoming increasingly important because the wired connection is not possible in cars or aeroplanes. Wireless networks can have many applications. A very common example is the portable office. People travelling on road often want to make use of their portable electronic equipment for telephone calls, e-mails, faxes, read remote files etc. Wireless networks can exist on trucks, buses, taxies, aeroplanes etc. They are used where the telephone systems are destroyed in the event of disasters such as fires, floods and earthquakes etc. The wireless networks are important for military. Wireless networks and mobile computing are related but they are not identical because portable computers are sometimes wired and some wireless computers are not portable.

But some applications are truly mobile wireless applications such as a portable office, inventories being handled by PDAs, etc.

Wireless LAN is another example of wireless network. Direct digital cellular service CDPD (Cellular Digital Packet Data) is now becoming available. It is possible to have combinations of wired and wireless networking.

1.27. INTERNETWORKS

The Internet is a collection of networks located all around the world that are connected by gateways. Each gateway has a routing table containing information about the networks that are connected to the gateway. Several networks may be connected to one gateway. The gateway accepts information from the source network and checks its routing table to see if the destination is in a network connected to the gateway. When two or more networks are connected they are called internetwork or internet as shown in figure 1.34. Individual networks are joined into internetworks by the used internetworking devices like bridges, routers and gateways. The term internet (lower case i) is a generic term used to mean an interconnection of networks and Internet (uppercase I) is the name of a specific worldwide network. A common form of internet is a collection of LANs connected by a WAN.

* WANs may use leased lines from telephone companies, Public Switch Data Networks (PSDN), or satellites for communication links.

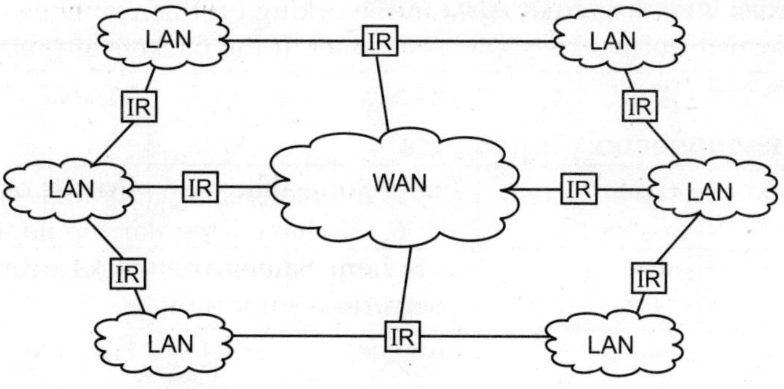

Fig. 1.34. Illustration of Internetwork

1.27.1. Difference between a subnet and WAN

If the system within a closed curve contains only routers then it is called a subnet. But if it contains routers as well as hosts then it is a WAN.

1.28. PERFORMANCE COMPARISON OF LAN AND WAN

S. No.	LAN	WAN
1.	The LAN is owned by a person, college factory etc. It is a privately owned network.	WAN can be private or it can be public leased type network.
2.	LAN is designed to operate over a small physical area such as office, factory or group of buildings.	WAN is used for the network that spans over a large distance such as system, spanning states, countries etc.
3.	LANs are easy to design and easy to maintain.	WAN is not so easy to design and maintain.
4.	The communication medium used for interconnection is a simple co-axial cable.	The communication medium used in WAN can be PSTN or satellite links due to longer distances involved.
5.	Due to shorter distances, problems such as propagation delay do not exist. So, LAN can be used for time applications.	Due to long distances involved, the problems such as propagation delay, variable signal travel time do exist. So, WAN cannot be used for the time critical applications.
6.	LAN can operate on very high data rates.	WAN operates on low data rates.
7.	In a LAN, each station can transmit and receive over the communication medium.	In WAN, each station cannot transmit.
8.	Local area network operates on the principle of broadcasting.	WAN operates on the principle of switching.

1.29. COMPARISON OF LAN, WAN AND MAN

S. No.	Parameter	LAN	WAN	MAN
1.	Ownership of network	Private	Private or public	Private or public
2.	Area covered	Small	Very large (states or countries)	Moderate (city)
3.	Design and maintenance	Easy	Not easy	Not easy
4.	Communication medium	Coaxial cable	PSTN or satellite links	Coaxial cables, PSTN, optical fiber cables, wireless.
5.	Data rates	High	Low	Moderate
6.	Mode of communication	Each station can transmit and receive	Each station cannot transmit	Each station can transmit or receive.
7.	Principle	Operates on the principle of broadcasting	Switching	Both
8.	Propagation delay	Short	Long	Moderate

1.30. GOALS AND APPLICATIONS OF COMPUTER NETWORKS*

As a matter of fact, the computer networks are playing an important role in providing services to large organizations as well as to the individual common man.

1.30.1. Service Provided by the Network for Companies

Many organizations have a large number of computers in operation. These computers may be within the same building, campus, city or different cities. Even though the computers are located in different locations, the organizations want to keep track of inventories, monitor productivity, do the ordering and billing etc. The computer networks are useful to the organizations in the following ways :

(i) Resource sharing
(ii) For providing high reliability
(iii) To save money
(iv) It can provide a powerful communication medium.

1. Resource Sharing

It allows all programs, equipment and data available to anyone on the network irrespective of the physical location of the resource and the user. Figures 1.35(a) and (b) show a printer being shared and different information being shared.

* Computer communications and networking technologies involve the study of (a) Computer-based communication networks, (b) key factors necessary to effect such communications, and (c) various technologies that define specific networks such as Ethernet.

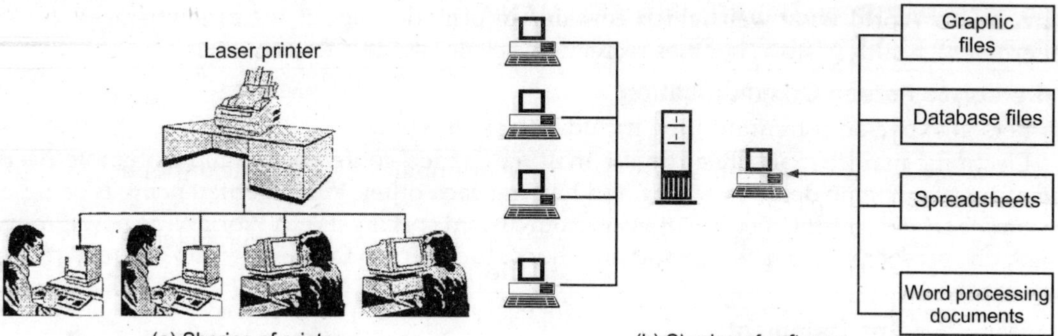

(a) Sharing of printer (b) Sharing of software

Fig. 1.35.

2. High Reliability due to the Alternative Sources of Data

It provides high reliabilty by having alternative sources of data. For e.g. all files could be replicated on more than one machines, so if one of them is unavailable due to hardware failure or any other reason, the other copies can be used. The aspect of high reliability is very important for military, banking, air traffic control, nuclear reactor safety and many other applications where continuous operations is a must even if there are hardware or software failures..

> **DO YOU KNOW?**
>
> Networks are increasingly being connected together to form very large communication systems. The largest of these, the Internet, is worldwide in scope.

3. Money Saving

Computer networking is an important financial aspect for organizations because it saves money. Organizations can use separate personal computer one per user instead of using mainframe computer which are expensive. The organizations can use the workgroup model (peer to peer) in which all the PCs are networked together and each one can have the access to the other communicating or sharing purpose. The organization, if it wants security for its operation it can go in for the domain model in which there is a server and clients. All the clients can communicate and access data through the server. The whole arrangement is called as client-server model, and it is shown in figure 1.36.

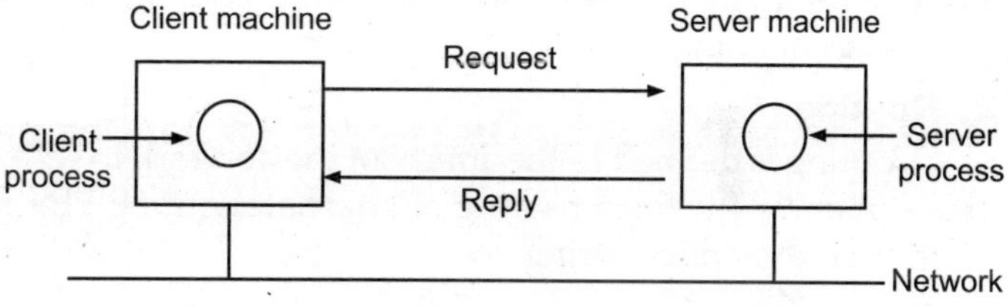

Fig. 1.36. Illustration of client server model

1.30.2. Networks for People

Starting in 1990s, the computer networks began to start delivering services to the private individual at home. The computer networks offer the following services to an individual person:

1. Access to remote information
2. Person to person communication
3. Interactive entertainment..

1. Access to Remote Information

Access to remote information involves interaction between a person and remote database. Access to remote information comes in many forms like :

Home shopping, paying telephone, electricity bills, e-banking, on the share market etc. Newspaper is on-line and is personalized, digital library consisting of books, magazines, scientific journals etc. World wide web which contains information about the arts, business, cooking, government, health, history, hobbies recreation, science sports etc..

2. Person to Person Communication

Person to person communication includes the following:

Electronic-mail (e-mail), Real time e-mail, i.e., video conferencing allows remote users to communicate with no delay by seeing and hearing each other. Video-conferencing is being used for remote school, getting medical opinion from distant specialists etc. Worldwide newsgroups in which one person posts a message and all other subscribers to the news group can read it or give their feedback.

3. Interactive Entertainment

Interactive entertainment includes Multiperson real-time simulation games; Video on demand; Participation in live TV programmes like quiz, contest, discussions, etc.

In short, the ability to merge information, communication and entertainment will surely give rise to a massive new industry based on computer networking.

1.30.3. Network Functions

Following are some of the important functions that a network needs to perform:

1. Switching
2. Routing
3. Flow control
4. Speed
5. Security
6. Backup
7. Failure monitoring
8. Traffic monitoring
9. Accountability
10. Internetworking
11. Network management.

1. Switching

Switching is defined as the ability of a network to connect different channels attached to each node to each other. This is essential for moving the traffic from incoming channel to the desired outgoing channels.

2. Routing

Routing is defined as the ability of the network to select a path. The routing can be of different types, namely the fixed routing or alternate routing.The routing can also be classified as static routing or dynamic routing.

3. Flow Control

Flow control is the control over the rate of traffic. It is necessary in order to reduce the network congestion.

4. Speed and Code Conversion

The speed of different devices is different. Also the codes used by different devices are different from each other. In the digital networks, we have to allow the communication between all such devices.

5. Security

Network security is defined as the ability of a network to disallow any unauthorized access to the network and the data travelling over it.We have to take the measures such as using passwords or use of data encryption and physical security.

6. Back up

It is defined as the ability of a network to react to the component failures. Back up also includes sending to indicate failures or to route the traffic via some other path to avoid a failed component.

7. Failure Monitoring

It is the ability of the network to keep track of faulty and working components.

8. Traffic Monitoring

It is defined as the ability of the network to keep track of traffic levels. It is useful for the network design.

9. Accountability

It is the ability of the network to keep a track of who is actually using the network. This is possible through the billing and charge back process. It gives us an understanding of the various users of a network. This is different from the traffic monitoring.

10. Internetworking

When two or more networks are connected they are called internetwork or internet. Individual networks are joined into internetworks by the internetworking devices like bridges, routers and gateways.

The term internet (lower case i) is a generic term used to mean an interconnection of networks and Internet (uppercase I) is the name of a specific worldwide network. A common form of internet is a collection of LANs connected by a WAN. Internetworking is defined as performing the function needed for communicating with other networks in an internetwork. The internetworking includes the following functions :

(i) To provide routes for traffic.

(ii) To allocate resources such as buffers and links.

11. Network Management

The network management is not a one single function. It is instead a combination of many functions. It includes the following functions :

(i) To maintain the user's list.

(ii) To maintain the addresses of the devices.

(iii) To keep an eye on changes in schedules of network.

(iv) Fault isolation.

Some of the network functions are carried out at each node and the others are needed to be carried out only by the first and the last node.

1.31. NETWORK HARDWARE

Now, let us discuss the technical issues involved in the network design. Two important dimensions of a computer network are :

(i) Transmission technology and

(ii) Scale.

1.31.1. Types of Transmission Technology

The transmission technology can be categorized broadly into two types :

(1) Broadcast networks and

(2) Point-to-point networks.

1. Broadcast Networks

Broadcast networks have a single communication channel that is shared or used by all the machines on the network. Short messages called *packets* sent by any machine are received by all the others.

Broadcast systems generally use a special code in the address field for addressing a packet to all the concerned computers. This mode of operation is called **broadcasting**. Some broadcast systems also support transmission to a subset of the machines known as multicasting. Upon receiving a packet, a machine checks the address field. If the packet is addressed to it then the packet is processed, otherwise the packet is ignored.

2. Point-to-point Networks

Point to point networks consist of many connections between individual pairs of machines. To go from the source to the destination, a packet on this types of network may have to go through intermediate computers before they reach the desired computer. Often the packets have to follow multiple routes of different lengths. Hence, routing algorithms are very important in the point-to-point networks. An important general rule is as follows :

Small, localized networks (e.g. LAN) tend to use the broadcasting, whereas networks located over wide geographical areas (such as WAN) use point-to-point transmission.

1.32. NETWORK SCALE

This is an alternative criterion for classification of networks. Figure 1.37 gives the network classification based on their physical size. All these systems are multiprocessor systems.

Interprocessor distance	Processors are located in same circuit board	Example of networks
0.1 m	Same system	Data flow machine
1 m	Same system	Multicomputer
10 m	Same room	LAN
100 m	Same building	LAN
1 km	Same campus	LAN
10 km	Same city	MAN
100 km	Same country	WAN
1,000 km	Same continent	WAN
10,000 km	Same planet	Internet

Fig. 1.37. Network classification according to scale

Beyond the multicomputers are the true networks, in which the computers communicate by exchanging messages over long cables. Such netwoks are divided into following categories :

(i) Local area networks

(ii) Metropolitan networks and

(iii) Wide area networks.

DO YOU KNOW?

The network architecture does not encompass the application software which are, in fact, users of the services provided by the network.

Internetwork

The connection of two or more networks is called as an internetwork. The best example of internetwork is the Internet.

1.33. NETWORK SOFTWARE

As a matter of fact, the first computer networks were designed with the hardware as the primary concern and the software as secondary concern. However, this strategy no longer works in modern times. Network software is now highly structured. In the present article, we shall examine the software structuring techniques in details.

1.33.1. Protocol Hierarchies (Layered Architecture)

The language used by members of a network is called a network communication protocol. Protocols facilitate understanding of a communication by providing members with a common language. From a general perspective, a network communication protocol is an accepted or established set of procedures rules, or formal specifications governing specific behaviour or language. As a matter of fact, most networks are organized as a series of layers or levels. To reduce the design complexity, networks are organized as a series of layer or levels, one above the other as shown in figure 1.38. The number of layers, the name of each layer, the contents of each layer and the function of each layer differ from network to network. The purpose of each layer is to offer certain services to the higher layers. Layer n on one machine (source) carries on a conversation with layer n on another machine (destination).

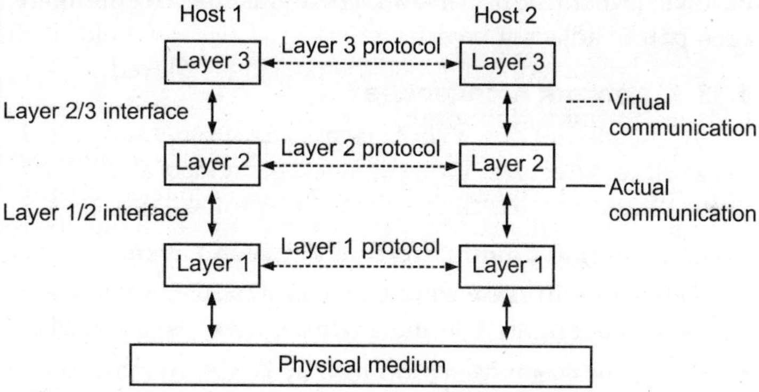

Fig. 1.38. Layers, protocols and interfaces

> **DO YOU KNOW?**
>
> Systems achieve openness by following certain architecture and obeying standard protocols. These standards are open for anybody to use and implement unlike proprietary architectures whose implementation details are always either trade secrets or covered by patent rights.

The rules and conventions used in this conversation are collectively known as the layer *n* protocol.

Basically, a protocol is an agreement between the two machines as how communication link should be established, maintained and released. Violation of the protocol will make the communication difficult. When a bunch of related and interoperating protocols are put together in a package on a system, we call it a **suite**.

Peer

A three layers network is shown in figure 1.38. The entities comprising the corresponding layers on different machines are called **peers**. The communication actually takes place between the peers using the protocol. The dotted lines in figure 1.38 show the virtual communication, and physical communication is shown by solid lines.

1.33.2. How does Data Transfer take Place?

As a matter of fact, data do not get transferred directly form layer n of one machine to layer n of the other machine. Instead, the data transfer takes place as explained below.

The data and control information are passed on to the lower layers until the lowest layer (layer 1) is reached. Below layer 1 lies the physical medium such as coaxial cable, through which the actual communicatiion takes place. This is shown in figure 1.39.

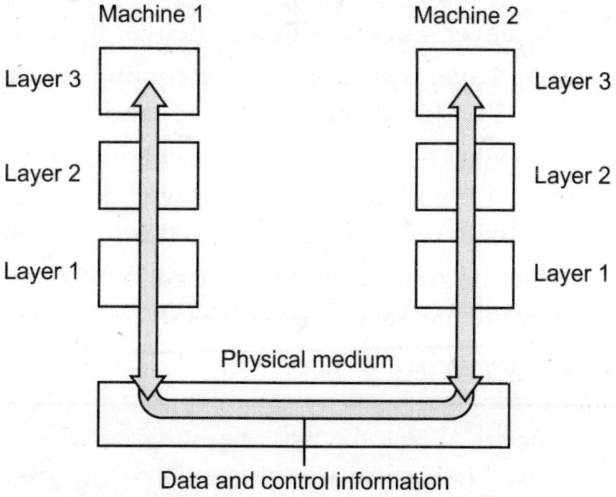

Fig. 1.39. Illustration of data transfer

Interface

An interface defines the operations and services offered by lower layer to the upper layer. There is an interface between each pair of adjacent layers.

> **DO YOU KNOW?**
>
> The architecture of a computer network, or simply network architecture, specifies a complete set of rules for the connections and interactions of its physical and logical components for providing and utilizing communication services.

1.33.3. Network Architecture

As a matter of fact, a set of layers and protocols is called **network architecture**. Protocol stack is defined as a list of protocols used for a certain system, one protocol per layer.

1.33.3.1. Virtual Communication Between Layers

Let us now go into technical details of the communication between (say) layer 3 of two machines. Let us consider figure 1.40 and go through the steps given below to understand the communication.

Step 1 : A messages M is produced by layer 5 of machine 1 and given to layer 4 for transmission.

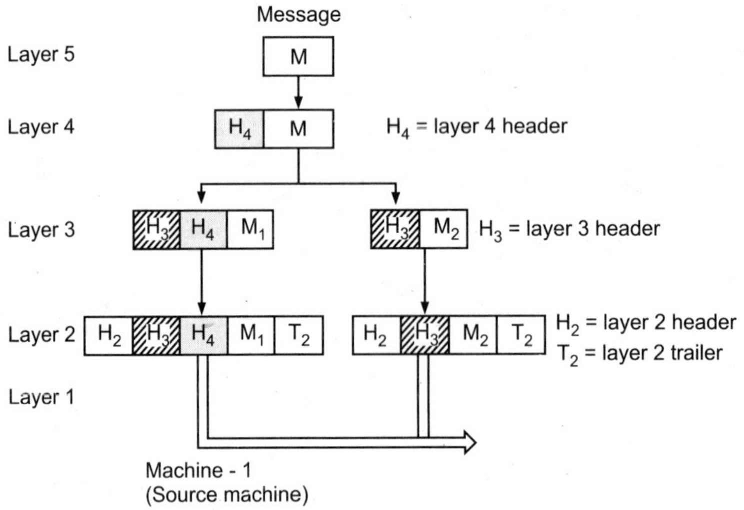

Fig. 1.40. Information flow for virtual communication between layers 5

Step 2 : Layer 4 adds a header in front of the header so as to identify the message and passes the (header + message) to layer 3. A header includes the control information and it allows a layer 4 in machine 2 to deliver the messages in right order.

Step 3 : Layer 3 breaks up the incoming messages into small units, packets and appends a layer 3 adder to each packet M_1 and M_2 as shown in figure 1.40 and passes these packets to layer 2.

Step 4 : Layer 2 adds header as well as trailer to each packet obtained from layer 3 and hands over the resultant unit to layer 1 for physical transmission.

The control information placed in headers is used at the destination machine (machine 2) to convey the message to layer 5 as shown in figure 1.41.

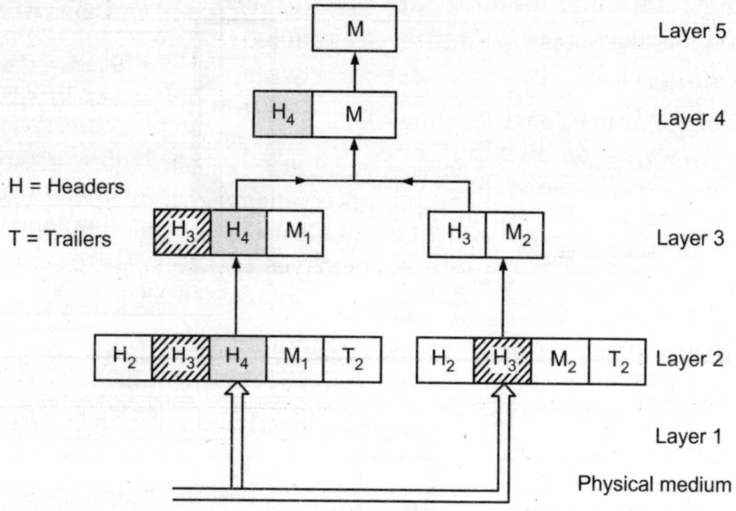

Fig. 1.41.

1.33.4. Design Issues for the Layers

In this section, let us discuss some of the key design issues that occur in computer networking.

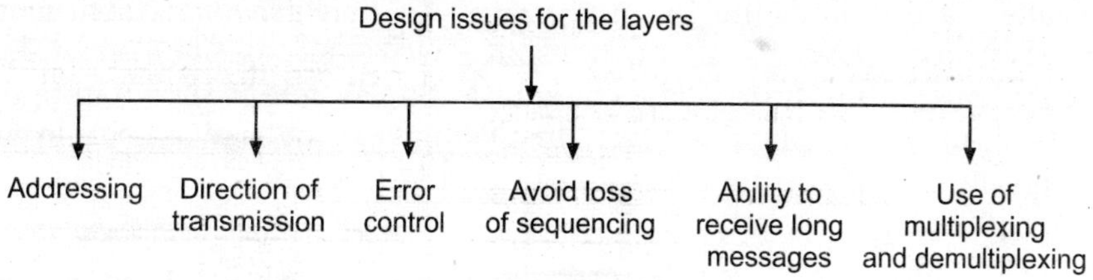

Fig. 1.42. Design issues

1. Addressing

For every layer, it is necessary to have a mechanism to identify senders and receivers. Since there are multiple possible destinations, some form of addressing is required in order to specify a specific destination.

2. Direction of Transmission

Another point in design issues is the direction of data transfer. Based on whether the system communicates only in one direction or otherwise, the communication systems are classified as under:

(i) Simplex systems

(ii) Half duplex systems

(iii) Full duplex systems.

> **DO YOU KNOW?**
>
> The architecture of a system, whether it is a building, organization or a computing system, describes how the system has been assembled using various components of the system. It defines the specifications of the components and their inter-relationships.

(i) Simplex Systems

In these systems, the information is communicated in only one direction. For example, the radio or TV broadcasting systems can only transmit. They cannot receive. In data communication systems the simplex communication takes place as shown in figure 1.43. The communication from CPU to monitor or keyboard to CPU is unidirectional.

Keyboard and traditional monitors are examples of simplex devices.

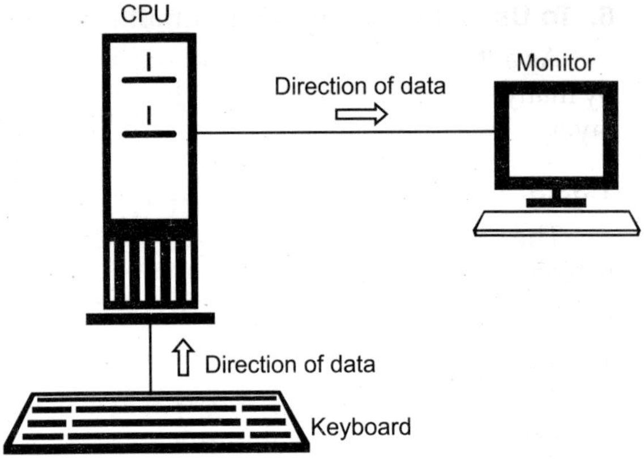

Fig. 1.43. Simplex mode of data transmission

(ii) Half Duplex Systems

These systems are bidirectional, i.e., they can transmit as well as receive but not simultaneously. At a time, these systems can either transmit or receive, for example, a transreceiver or walky talky set.

A data communication system working in the half duplex mode is shown in figure 1.44. Each station can transmit and receive, but not at the same time. When one device is sending the other one is receiving and vice versa.

In half duplex transmission, the entire capacity of the channel is taken over by the transmitting (sending).

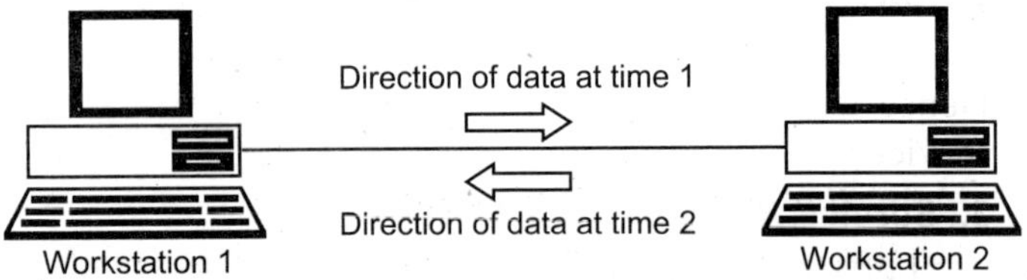

Fig. 1.44. Half duplex system

(iii) Full Duplex Systems

These are truly bidirectional systems as they allow the communication to take place in both the directions simultaneously. These systems can transmit as well as receive simultaneously, for example, the telephone systems. A full duplex data communication system is shown in figure 1.45. Each station can transmit and receive simultaneously. In full duplex mode, signals going in either direction share the full capacity of link. The link may contain two physically separate transmission paths one for sending and another for receiving. Otherwise the capacity of channel is divided between signals travelling in both directions. Many networks provide atleast two logical channels per connection, one for the normal data and the other for urgent data.

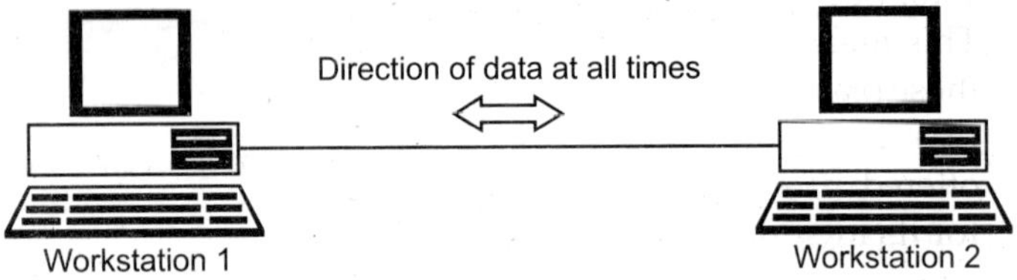

Fig. 1.45. Full duplex mode

3. Error Control

Another important issue is the error control because physical communication circuits are not perfect. Error detection and connection both are essential. Many error detecting and correcting codes are known out of which those agreed by sender and receiver should be used. The receiver should be able to tell the sender by some means, that it has received a correct message or a wrong message.

4. Avoid Loss of Sequencing

All the communication channels cannot preserve the order in which messages are sent on it. Therefore, there is a possibility of loss of sequencing. To avoid this, all the pieces should be numbered so that they can be put back together at the receiver in the appropriate sequence.

5. Ability of Receiving Long Messages

At several levels, another problem should be solved, which is inability of all processes to accept arbitrarily long messages. Hence, a mechanism needs to be developed to disassemble, transmit and then reassemble message.

6. To Use Multiplexing and Demultiplexing

As a matter of fact, multiplexing and demultiplexing is to be used to share the same channel by many sources simultaneously. It can be used for any layer. Multiplexing is needed in the physical layer.

1.33.5. Interfaces and Services

The basic function of each layer in the layered structure is to provide service to the layer above it. Now, let us discuss exactly what service does it provide. But before that let us define some important terms.

1. Entities and Peer Entities

An entity is defined as the active element in each layer. An entity can be either a software entity or a hardware entity. The example of software entity is a process and that of a hardware entity is an intelligent I/O chip. Entities in the same layer but on different machines are called **peer entities**.

2. Service Provider and Service User

The entities at layer n implement services for the layer (n + 1) which is above the nth layer. Hence, layer n which provides service is called **service provider** and layer (n + 1) which takes this service is called **service user**.

3. Service Access Points (SAPs)*

Let us consider figure 1.46 to understand the definition of SAPs. The long form of SAP is service access point. They are available at the interface of n and n + 1 layer as shown in figure 1.46. Services are available at SAPs. This means that the layer n SAPs are those places at the interface where layer (n + 1) can access the services being offered. Each SAP has a unique address for its identification.

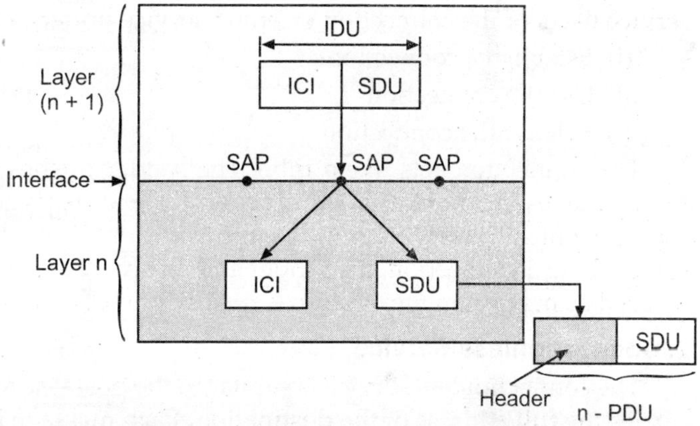

Fig. 1.46. Relation between layers at an interface

4. Interface Data Unit (IDU)

For successful exchange of information between two layers, a set of rules about the interface should be present. As shown in figure 1.47, the layer (n + 1) entity passes an IDU (interface data unit) to the layer n entity through the SAP.

An IDU consists of two parts namely SDU (service data unit) and ICI (interface control information).

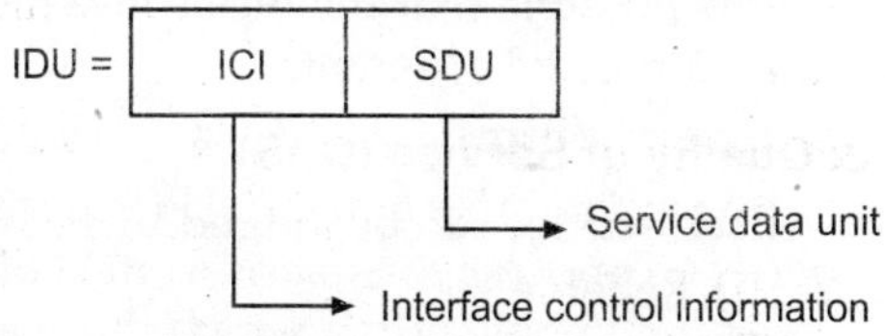

Fig. 1.47. Illustration of IDU

5. Service Data Unit (SDU)

SDU is a part of IDU. The SDU is the information passed across the network to the peer entity and then upto layer (n + 1). ICI contains the control information which is necessary to help the lower layer (n) to do the necessary job.

6. Protocol Data Unit (PDU)

In order to transfer the SDU, the layer n entity has to divide it into many smaller pieces. Each piece is given a header and sent as a separate PDU (Protocol Data Unit) such as a packet. The PDU headers are used by the peer entities to carry out their peer protocol.

* In UNIX, an SAP is called a socket, and an SAP address is a socket number.

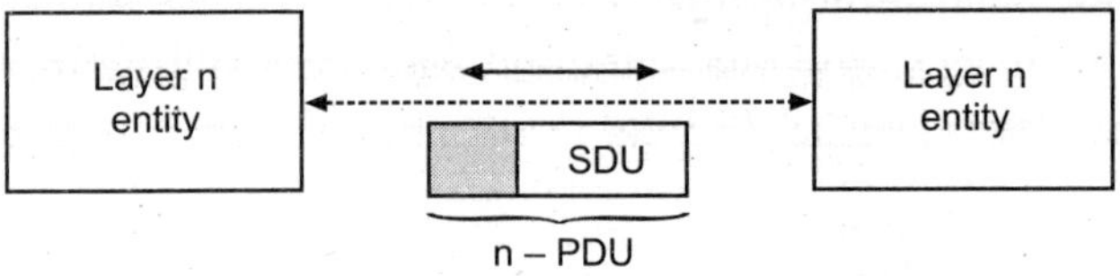

Fig. 1.48. Layer n entities exchange n-PDUs in their layer n protocol

Some PDUs contain data while other PDUs contain the control information. The PDU headers will identify or differentiate between different types of PDUs. They also provide sequence numbers and counts.

1.33.6. Connection Oriented and Connectionless Services

As a matter of fact, layers can offer two types of services to the layers above them,

(i) Connection oriented service

(ii) Connectionless service

1. Connection Oriented Service*

The connection oriented service is similar to the one provided in the telephone system. The service users of the connection oriented service undergo the following sequence of operation:

(i) Establish a connection.

(ii) Use the connection

(iii) Release the connection

The connection acts like a tube. The sender pushes bits from one end of the tube and the receiver takes them out from the other end. The order is generally preserved. This means that the order in which the bits are sent is same as the order in which they are received. Sometimes after establishing a connection, the sender and receiver can discuss and negotiate about parameters to be used as maximum message size, quality of service and some other issues.

2. Connectionless Service

The connectionless service is similar to the postal service. Each message (analogus to a letter) carries the full address of the destination. Each message is routed independently from source to destination through the system.

It is possible that the order in which the messages are sent and the order in which they are received may be different.

3. Quality of Service (QoS)

Each service can be judged by its quality of service. Services can be of two types:

(i) Reliable

(ii) Unreliable

Reliable services are those which never lose data. Generally, reliable services are implemented with receiver sending acknowledgements of the received messages. But the acknowledgements introduce overheads and delays which are sometimes undesirable. For applications like digitized voice traffic, the delays due to acknowledgements are unacceptable. Applications such as electronic mail do not require any connections. The cost associated , complexity and overheads of reliable services is not required here. Such applications require high reliability of message arrival but no guarantee i.e., unreliable service will be ok. The unacknowledged services are unreliable

* Addressing on an information packet is not necessary for connection-oriented service because a physical, dedicated link is established between sending and receiving nodes before transmitting the information packets.

connectionless services. Such services are called datagram service which is similar to telegram service. However, not that acknowledged datagram service can also be provided. One more type of service is the request-reply service. In this type, the sender transmits a single datagram which contains a request and a reply obtained from the receiver contains the reply. Table 1.1 lists various types of services and their examples.

Table 1.1. Six different types of services

S.No.	Service	Type	Example
1.	Reliable message stream	Connection oriented	Sequence of pages.
2.	Reliable byte stream	Connection oriented	Remote login.
3.	Unreliable connection	Connection oriented	Digitized voice
4.	Unreliable datagram	Connectionless	Electronic mail
5.	Acknowledged datagram	Connectionless	Registered e-mail
6.	Request-Reply	Connectionless	Database query.

The unreliable service is used only if the reliable service is not available or is too expensive.

1.33.7. Service Primitives

Primitives means operation. A service is specified by a set of primitives. A user process can access these primitives to access the service.

Primitives of connection oriented service are different from those of connectionless service. The service primitives required to implement a reliable byte stream in a client server environment are given in Table 1.2.

Table 1.2. Service Primitives

S.No.	Name	Meaning
1.	LISTEN	Block waiting for an incoming connection
2.	CONNECT	Establish a connection
3.	RECEIVE	Block waits for a message
4.	SEND	Send the message
5.	DISCONNECT	Terminate the connection

1. LISTEN

When a server is prepared to accept an incoming connection it executes the LISTEN primitive. After executing the LISTEN, the server process is blocked until request for connection arises.

2. CONNECT

The next step is that the client process executes CONNECT for establishing a connection with the server. The client process is suspended until there is a response.

3. RECEIVE

The server executes RECEIVE as the next step. The RECEIVE call will blocks the server.

4. SEND

The client then executes the SEND primitive in order to transmit its request followed by the execution of RECEIVE to get the reply. The request packet reaches the server machine and unblocks the server process so that it can process the request. After doing the work, it uses SEND for returning the answer to the client.

5. DISCONNECT

This primitive is for terminating and releasing the connection. An initial DISCONNECT is a blocking call which suspends the client and sends a packet to the server to tell it that this connection is no longer required. After receiving this packet, the server also issues a DISCONNECT to acknowledge the client and terminate the connection. When the servers packet reaches the client's machine, the client process is released and the connection is broken.

1.33.8. Relationship of Services to Protocols

Services and protocols are two completely different concepts and should not be mixed up.

1. Service

It is defined as a set of operations that a layer can provide to the layer above it. A service defines, states the operations a layer is ready to perform but does not specify anything about the implementation of these operations.

2. Protocol

It is a set of rules which governs the format and meaning of frames, packets or messages that are being exchanged between the peer entities. The entities use protocols so as to implement their services. Ensuring their predecided services, they are free to change the protocol.

> **DO YOU KNOW?**
>
> In the case of computers, protocols must be precise and unambiguous as they involve communications between machines. For this reason, protocols involve defining the messages that are to be exchanged between machines as well as the actions that are to be taken when certain events take place and when certain messages are transmitted or received.

1.34. REFERENCE MODELS (NETWORK ARCHITECTURES)

Network architecture is a formal, logical structure that defines how network devices and software interact and function. It defines communication protocols, message formats, and standards required for interoperability. New hardware or software products created within a specific architecture are generally compatible with other products created within the same architecture. Network architectures are designed by standards organizations and manufacturers.

For example, IBM designed the systems Network Architecture (SNA), Digital Equipment Corporation designd the Digital Network Architecture (DNA), and the International Organization For standardization (ISO) designed the open systems Interconnect (OSI) architecture.

After discussing about the layered networks, now, let us discuss two network architectures or reference models. The two most important reference models are as under :

(i) The OSI reference model and

(ii) The TCP/IP reference model

The International Standards Organization (ISO) covers all aspects of network communication in the Open Systems Interconnection (OSI) model. An OSI model is a layered framework for the design of network systems that allows for communication across all types of computer systems. The purpose of each layer is to offer certain services to the higher layers. Layer n on one machine (source) carries on a conversation with layer n on another machine (destination). The rules and conventions used in this conversation are collectively known as the **layer n protocol**. Basically, a protocol is an agreement between the two machines as how communication link should be established, maintained and released.

1.35. OSI MODEL

The users of a computer network are located over a wide physical range i.e. all over the world. Therefore, to ensure that nationwide and worldwide data communication systems can be developed and are compatible to each other, an international group of standards has been developed. These standards will fit into a framework which has been developed by the International organization of standardization (ISO). This framework is called as Model for open system interconnection (OSI) and it is normally referred to as OSI reference model.

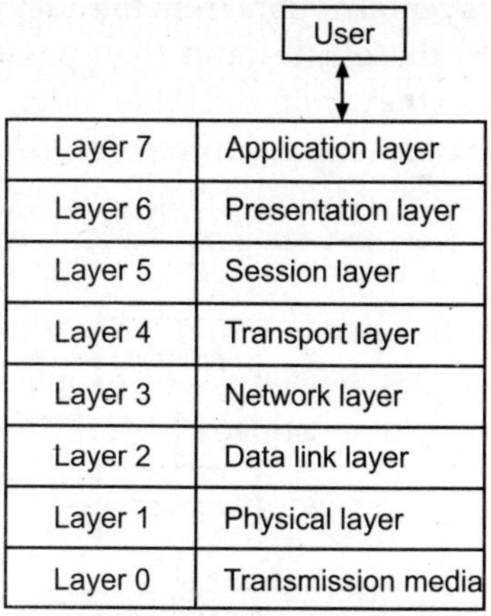

Fig. 1.49. A seven layer ISO-OSI reference model

One of OSI's most redeeming features is that it formally defines and codifies the concept of layered network architecture. It uses well-defined operationally descriptive layers that describe what happens at each stage in the processing of data for transmission. This layering concept is extremely important because networks are nontrivial systems.

Figure 1.49 shows the seven layer architecture of ISO-OSI reference model. It defines seven levels or layers in a complete communication system. The lowest layer is physical layer and highest one is called the application layer.

A more detailed OSI reference model is shown in figure 1.50. The OSI model shown in figure 1.49 does not contain the physical medium. This model is based on a proposal developed by the International Standards Organization (ISO). It is called as ISO-OSI (Open System Interconnection) reference model because it is designed to deal with open systems i.e., the systems which are open for communication with other systems.

Table 1.3. Functions of the layers of ISO-OSI model

Level	Name of the Layer	Functions
1.	Physical Layer	Make and break connections, define voltages and data rates, convert data bits into electrical signal. Decide whether transmission is simplex, half duplex or full duplex.
2.	Data link Layer	Synchronization, error detection and correction. To assemble outgoing messages into frames.
3.	Network Layer	Routing of the signals; to divide the outgoing message into packets; to act as network controller for routing data.
4.	Transport Layer	Decides whether transmission should be parallel or single path, multiplexing, splitting or segmenting the data, to break data into smaller units for efficient handling.
5.	Session Layer	To manage and synchronize conversation between two systems. It controls logging on and off, user identification, billing and session management.
6.	Presentation Layer	It works as a translating layer.
7.	Application Layer	Retransferring files of information. LOGIN, password checking etc.

All the applications need not use all the seven layers shown in figure 1.50. The lower three layers are enough for most of the applications. Each layer is built from electronic circuits and/

or software and has a separate existence from the remaining layers. Each layer is supposed to handle message or data from the layers which are immediately above or below it. This is done by following the protocol rules. Thus, each layer takes data from the adjacent layer, handles it according to these rules and then passes the processed data to the next layer on the other side.

> **DO YOU KNOW?**
>
> The open systems International (OSI) model is a way of specifying communications protocols to make it easier to design them for compatibility among different systems. There are seven layers to the OSI model.

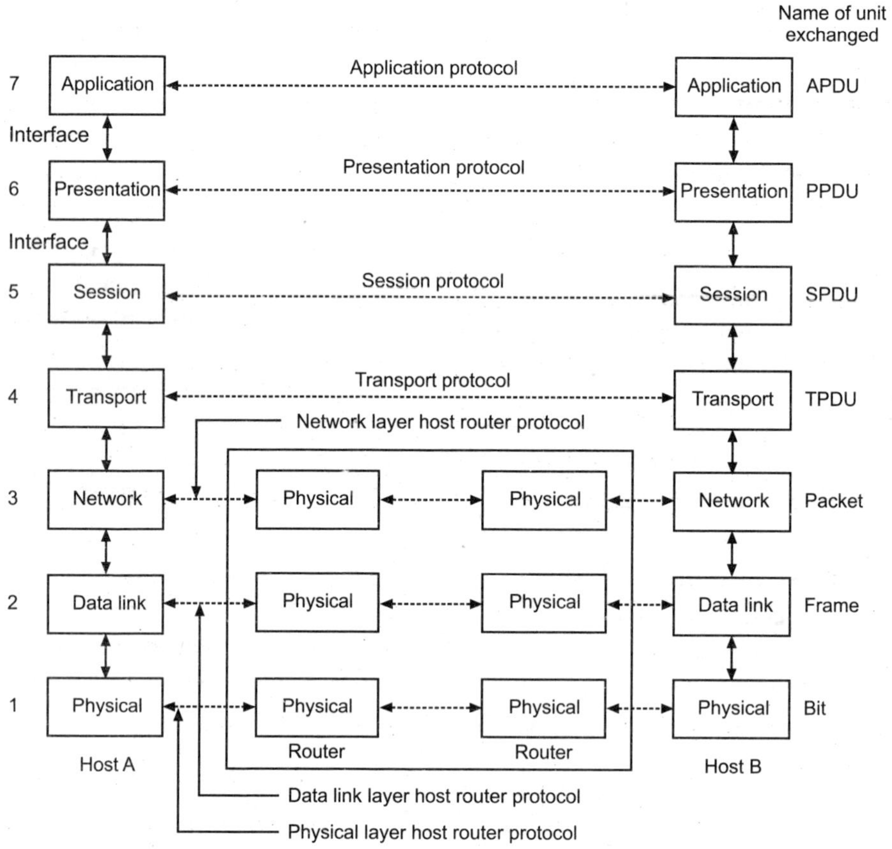

Fig. 1.50. The OSI reference model

1.36. FUNCTIONS OF DIFFERENT LAYERS

Layer 1 : The Physical Layer

Functions of the physical layer are as follows:

(i) To activate, maintain and deactivate the physical connection.

(ii) To define voltages and data rates needed for transmission.

(iii) To convert the digital bits into electrical signal.

(iv) To decide whether the transmission is simplex, half duplex or full duplex.

> **DO YOU KNOW?**
>
> Open system Interconnection (OSI) represents a generalization of concepts of inter-process communication so that any open system may be technically able to communicate with another open system.

A physical layer does not perform the following operations:

(i) It does not detect or correct errors.

(ii) It does not decide the medium or modulation.

The examples of the physical layer protocols are RS-232 or RS-449 standards.

Layer 2 : Data Link Layer

(i) Functions of the data link layer are synchronization and error control for the information which is to be transmitted over the physical link.

(ii) To enable the error detection, it adds error detection bits to the data which are to be transmitted.

(iii) The encoded data are then passed to the physical layer.

(iv) These error detection bits are used by the data link layer on the other side to detect and correct the errors.

(v) At this level, the outgoing messages are assembled into frames, and the system waits for the acknowledgements to be received after every frame transmitted.

(vi) Correct operation of the data link layer ensures reliable transmission of each message. Examples of data link layer protocols are HDLC, SDLC and X.25 protocols.

Layer 3 : The Network Layer

The functions of network layer are as follows :

(i) To route the signals through various channels to the other end.

(ii) To act as the network controller by deciding which route data should take.

(iii) To divide the outgoing messages into packets and to assemble incoming packets into messages for the higher levels.

In short, the network layer acts as a network controller for routing data.

Layer 4 : Transport Layer

As the name suggests, this layer provides the transport services. The functions of the transport layer are as listed below :

(i) It decides if the data transmission should take place on parallel paths or single path.

(ii) It does the functions such as multiplexing, splitting or segmenting on the data.

(iii) Transport layer guarantees transmission of data from one end to the other.

(iv) It breaks the data groups into smaller units so that they are handled more efficiently by the network layer.

Layer 5 : The Session Layer

(i) This layer manages and synchronizes conversations between two different applications. This is the level at which the user will establish system to system connection.

(ii) It controls logging on and off, user identification, billing and session management.

(iii) In the transmission of data from one system to the other, at session layer streams of data are marked and resynchronized properly so that the ends of messages are not cut prematurely and data loss is avoided.

Layer 6 : The Presentation Layer

(i) The presentation layer makes it sure that the information is delivered in such a form that the receiving system will understand and use it.

(ii) The form and syntax (language) of the two communicating systems can be different e.g. one system is using the ASCII code for file transfer and the other one uses IBM's EBCDIC.

(iii) Under such conditions, the presentation layer provides the *translation* from ASCII to EBCDIC and vice versa.

Layer 7 : Application Layer

(i) Application layer is at the top of all as shown in figure 1.50. It provides different services such as manipulation of information in various ways, retransferring the files of information, distributing the results etc. to the user who is sitting above this layer.

(ii) The functions such as LOGIN, or password checking are also performed by the application layer.

Now, let us go into the details of each and every layer.

1.36.1. Exchange of Information Using the OSI Model

At the physical layer, communication is direct i.e. machine X sends a stream of bits to machine Y. At higher layers, each layer in the sending machine adds its own information to the message it receives from the layer just above it and passes the whole package to the layer just below it as shown in figure 1.51. The information added by each layer is in the form of headers or trailers. Headers are added to the messages at the layers 6, 5, 4, 3, and 2. A trailer is added at layer 2. At layer 1, the entire package is converted to a form that can be transferred to the receiving machine. At the receiving machine, the message is unwrapped layer by layer with each process receiving and removing the data meant for it.

The upper OSI layers are always implemented in software (4, 5, 6 and 7) and lower layers are a combination of hardware and software (2, 3) except for the physical layer which is mostly hardware. Layers 1, 2 and 3 (i.e. physical, data link and network) are the network support layers. They deal with the physical aspects of moving data from one device to another such as electrical specifications, physical connections, physical addressing and transport timing and reliability. Layer 4, the transport layer ensures end to end reliable data transmission. Layers 5, 6 and 7 (i.e. session, presentation and application) they allow interoperatibility among unrelated software systems.

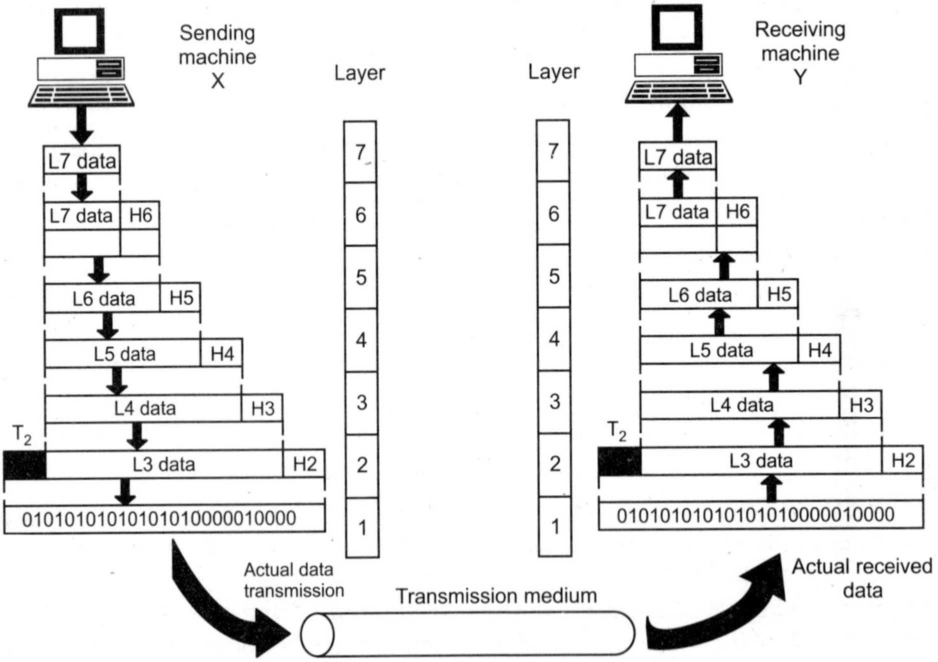

Fig. 1.51. An exchange using the OSI model

1.36.2. Physical Layer

The physical layer is responsible for sending bits from one computer to another. The physical layer is not concerned with the meaning of the bits, but it deals with physical connection to the network and with transmission and reception of signals. The physical level is used to define physical and electrical details such as what will represent a 1 or a 0, how many pins a network will have, how data will be synchronized and when the network adapter may or may not transmit the data. The position of the physical layer with respect to the transmission medium and the data link layer is shown in figure 1.52.

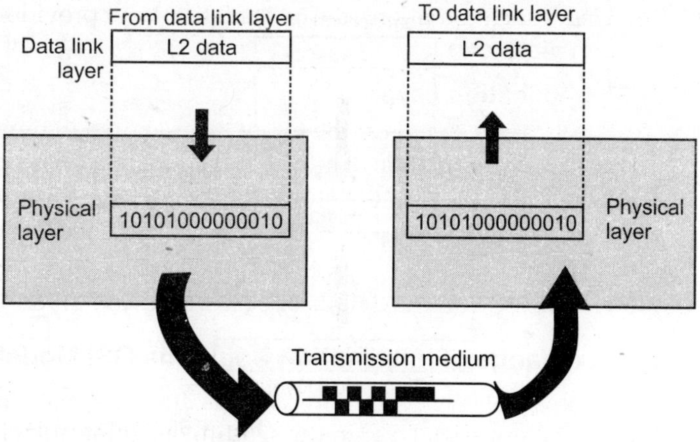

Fig. 1.52. Physical layer

Following are the functions of the physical layer :

(i) To define the type of encoding i.e. how 0's and 1's are changed to signals.
(ii) To define the transmission rate i.e. the number of bits transmitted per second.
(iii) To deal with the synchronization of the transmitter and receiver.
(iv) To deal with network connection types, including multipoint and point to point connections.
(v) To deal with physical topologies i.e. bus, star, ring, or mesh.
(vi) To deal with the media bandwidth i.e. baseband and broadband transmission.
(vii) Multiplexing which deals with combining several data channels into one.
(viii) To define the characteristics between the device and the transmission medium.
(ix) To define the transmission mode between two devices i.e. whether it should be simplex, half duplex or full duplex.

Connecting devices associated with physical layer

Passive hubs, simple active hubs, terminators, couplers, cable and cabling, connectors, repeaters, multiplexers, transmitters, receivers, transceivers are associated with the physical layers.

1.36.3. Data Link Layer

It is responsible for reliable node to node delivery of the data. It accepts packets from the network layer and forms frames and gives it to the physical layer as shown in figure 1.53.

Following are the functions of data link layer :

1. Framing

It divides the streams of bits received from the network layer into manageable data units called frames.

2. Physical Addressing

It adds a header to the frame to define the physical address of the sender and /or receiver of the frame.

3. Flow Control

It provides a flow control mechanism to avoid a fast transmitter from over-running a slow receiver by buffering the extra bits.

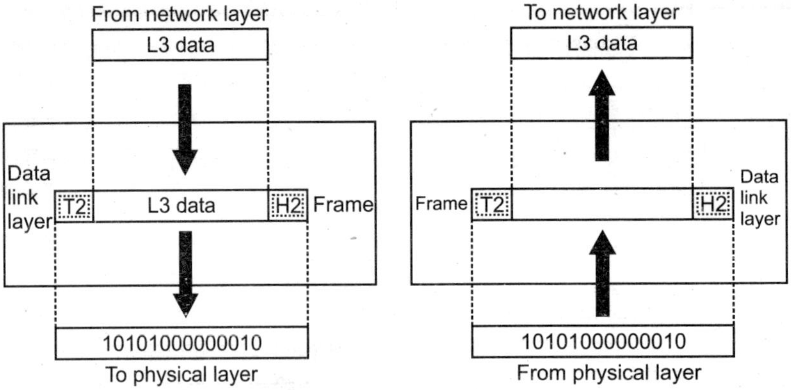

Fig. 1.53. Data link layer

4. Error Control

Error control is achieved by adding a trailer at the end of the frame. It also uses a mechanism to prevent duplication of frames.

5. Access Control

The data link layer protocol determines which of the devices has control over the link at any given time, when two or more devices are connected to the same link. The Institution of Electrical and Electronics Engineers (IEEE) felt the need to define the data link layer in more details, so they split it into two sub-layers.

(i) Logical Link Control (LLC)

It establishes and maintains links between the communicating device.

(ii) Media Access Control (MAC)

It controls the way multiple devices share the same media channel. The logical link control sub-layer provides Service Access Points (SAPs) that the other computers can refer to and use to transfer information from LLC to the network layer. The MAC sub-layer provides for shared access to the network adapter and communicates directly with the network interface cards. Network interface cards (NIC) have a unique 12-digit hexadecimal MAC address assigned before they leave the factory where they are manufactured.

DO YOU KNOW?

The data link layer encapsulates the data bits into frames having a header and a trailer. Data link protocols specify the format of the frames and procedures for exchange of the frames. Data link protocols can be bit oriented or byte oriented.

The MAC addresses are used to establish logical link between two computers on the same LAN. Bridges, intelligent hubs and network interface cards are devices associated with the data link layer.

1.36.4. Network Layer

The main function of this layer is to deliver packets from source to destination across multiple networks (links). If two systems are connected on the same link, there is no need for a network layer. The relationship of the network layer to the data link and transport is shown in figure 1.54.

Functions of the Network Layer

(i) It translates logical network address into physical machine addresses i.e. the numbers used as destination IDs in the physical network cards.

(ii) It determines the quality of service by deciding the priority of message and the route a message will take if there are several ways a message can get to its destination.

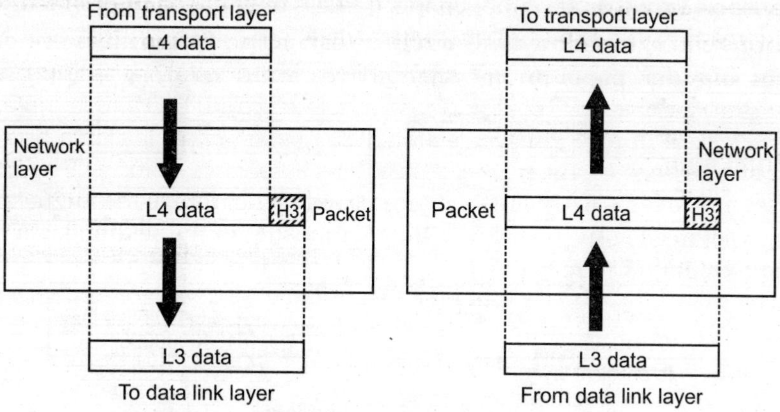

Fig. 1.54. Network layer

(iii) It breaks the larger packets into smaller packets if the packet is larger than the largest data frame the data link will accept.

(iv) It is concerned with the circuit, message or packet switching.

(v) It provides connection services, including network layer flow control, network layer error control and packet sequence control.

(vi) Routers and gateways operate in the network layer.

1.36.5. Transport Layer

It is responsible for source to destination delivery of the entire message. It ensures that the whole message arrives intact and in order, ensuring both error control and flow control at the source to destination level. Figure 1.55 shows the relationship of the transport layer to the network layer and session layer.

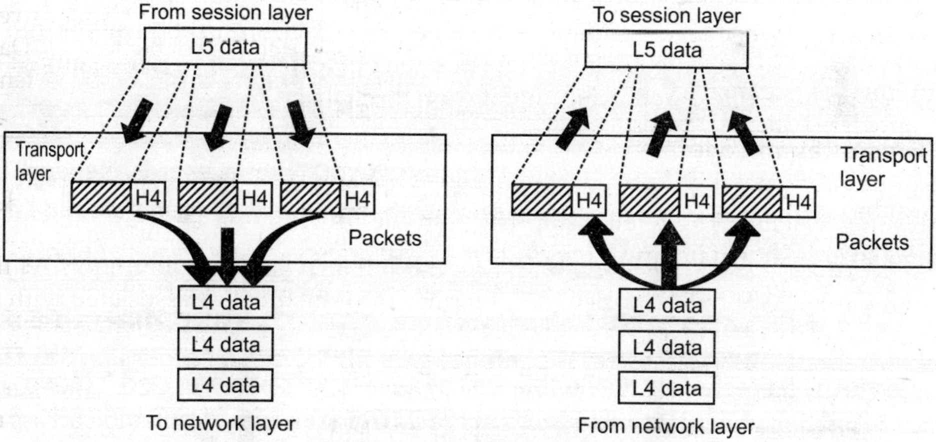

Fig. 1.55. Transport layer

Functions of Transport Layer

The transport layer performs the following functions :

(i) It divides each message into packets at the source and re-assembles then at the destination.

(ii) The transport layer header H4 includes a service point address to deliver a specific process from source to a specific process at the destination.

(iii) The transport layer can be either connectionless or connection-oriented.

(iv) It performs end to end flow control rather than across a single link.

(v) It makes sure that the entire message arrives at the receiving transport layer without error.

1.36.6. Session Layer

The job of session layer is to establish, maintain and synchronize the interaction between the communicating systems. Figure 1.56 shows the relationship of the session layer to the transport layer and the presentation layer.

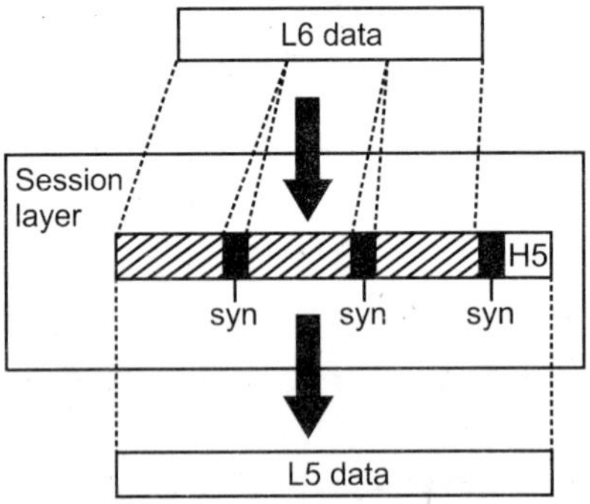

Fig. 1.56. Session layer

Functions of Session Layer

The session layer performs the following functions :

(i) It allows two systems to start a dialogue with each other. The communication initiated between two processes can be either in half duplex or full duplex.

(ii) The session layer allows addition of check points i.e. synchronization points into a stream of data. In case of crash during the transmission of the data can be retransmitted from the check point inspite of retransmitting it from the start.

1.36.7. Presentation Layer

This layer takes care of the syntax and semantics of the information exchanged between two communicating systems. The relationship between the presentation layer and the application and session layer is shown in figure 1.57.

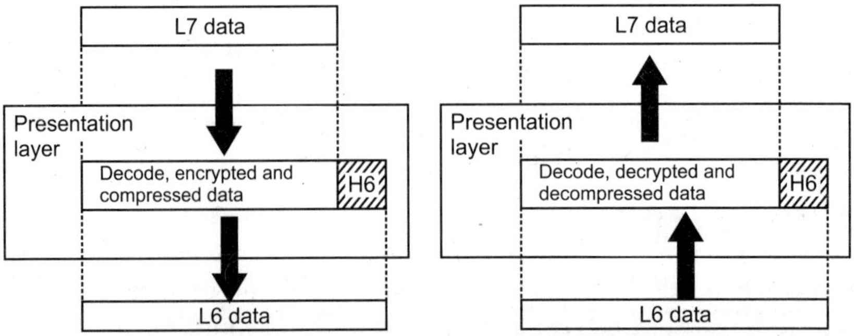

Fig. 1.57. Presentation layer

Functions of Presentation Layer

The presentation layer performs the following function :

(i) It translates data between the formats, the network requires and the format the computer expects (e.g. ASCII or EBCDIC).

(ii) It does the protocol conversion.

(iii) For security and privacy purpose, it carries out encryption at the transmitter and decryption at the receiver.

(iv) It carries out data compression to reduce the bandwidth of the data to be transmitted.

1.36.8. Application Layer

It is the topmost layer of OSI model. It provides services that directly support user application such as database access, e-mail and file transfer. It allows applications to communicate with applications on other computers as though they were on the same computer. The relationship of the application layer to the user and the presentation layer is shown in figure 1.58.

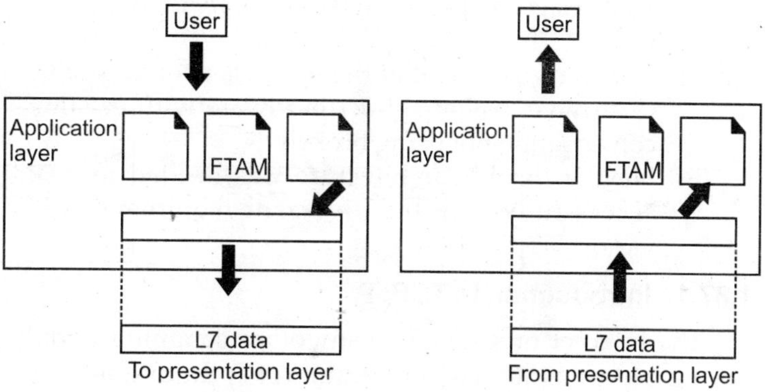

Fig. 1.58. Application layer

Functions of Application Layer

The application layer performs the following functions :

(i) The application layer creates a software emulation of a terminal at the remote host (Network Virtual Terminal). The user's computer talks to the software terminal, then the software terminal talks to the host and vice-versa. The remote host feels that it is communicating with one of its own terminals and allows you to log on.

(ii) The application layer provides file transfer access and management (FTAM) which allows user to access files in a remote computer to retrieve files from a remote computer and to manage or control files in a remote computer.

(iii) It provides a basis for e-mail forwarding and string.

(iv) It provides distributed database sources and access to the worldwide information about various objects and services.

1.36.9. Merits of OSI Reference Model

It distinguishes very clearly between the services, interfaces and protocols. The protocols in OSI model are better hidden. So, they can be easily replaced by new protocols as the technology changes. OSI model is truly a general model. This model supports connection oriented as well as connectionless services.

1.36.10. Demerits of OSI Model

Sessions and presentation layers are not of much use. This model was devised before the protocols were invented. So, in real life, there is a problem of fitting protocol into a model.

1.37 THE TCP/IP REFERENCE MODEL* *(RGTU, Bhopal, Sem. Exam., June, 2014)*

This is the other reference model which was used earlier by ARPANET and then it is being used in the Internet. TCP/IP is a short form of transmission control protocol and internet protocol. ARPANET was a research network sponsored by the U.S. department of Defence. It included many universities and government installations using the leased telephone lines. Later on, the satellites and radio networks were added to it. This inclusion could not be handled by the existing protocols at that time. So, a new reference architecture was needed. This new architecture is known as TCP/IP reference model due to the use of the two protocols TCP and IP. While designing the new model, certain goals were to be achieved. Some of them were as follows:

(i) First design goal was to have an ability to connect multiple networks together in a seamless way.

(ii) Another goal was that the network should be able to survive loss of subnet hardware with existing conversation not being broken.

(iii) Next, a flexible architecture was needed to deal successfully with the divergent requirements of various applications.

DO YOU KNOW?

The TCP/IP protocol suite is a very robust and flexible set of standards for large diverse networks.

1.37.1. Introduction to TCP/IP

The Internet protocols like any other communication protocol, is a set of rules which will govern every possible communication over the internet. Since the development of the ARPANET, TCP/IP together has emerged as the controlling body, it is being used in computers of not only in the U.S. but all over the world for all the types and sizes of computers. It has become the language of the Internet. TCP/IP are two protocols: Transmission control protocol and Internet protocol. These two protocols describe the movement of data between the host computers or Internet. The protocol, however is a suite of many other protocols which provide for reliable communications across the Internet and the web. In the TCP/IP protocol suite, there are various layers, with each layer being responsible for different facets of communication.

The Internet Protocols (IP) and Transmission Control Protocol (TCP) are together known as TCP/IP protocol. TCP/IP offers a simple naming and addressing scheme whereby different resources on Internet can be easily located. Information on Internet is carried in *Packets*. The IP protocol is used to put a message into a *packet*. Each packet has the address of the sender and the recipient's address. These addresses are known as the IP addresses. Using the TCP protocol, a single large message is divided into a sequence of packets and each is put into an IP packet. The packets are passed from one network to another until they reach their destination. At the destination, the TCP software reassembles the packets into a complete message. It is not necessary for all the packets in a single message to take the same route each time it is sent.

1.37.2. Overview of TCP/IP Architecture

Transmission Control Protocol and the Internet Protocol (TCP/IP) was developed by the Department of Defence's Project Research Agency (ARPA, later DARPA) under its project on network interconnection. It is a set of protocols that allow communication across multiple diverse network. ARPA originally created TCP/IP to connect military networks together, but later on this protocol was also given to government agencies and universities free of cost.

Since the TCP/IP was developed for military use, it became robust to failures and flexible to

* TCP/IP is inextricably linked to the Internet.

diverse networks. TCP/IP is the most widely used protocol for interconnecting computers and it is the protocol of the Internet. TCP/IP became the standard for interoperating Unix Computers, especially in military and university environments. With the development of the Hypertext Transfer Protocol (HTTP) for sharing Hypertext Markup Language (HTML) documents freely on the Internet, the World Wide Web (WWW) was born and soon TCP/IP came into much use. Figure 1.59 shows the TCP/IP reference model along with the OSI model used for comparison.

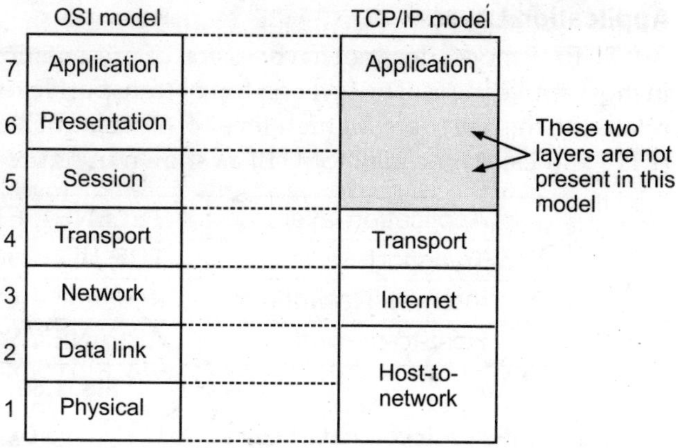

Fig. 1.59. TCP/IP reference model

1.37.3. Description of TCP/IP Model*

As shown in figure 1.59, the TCP/IP model has only four layers.

Internet Layer

> **DO YOU KNOW?**
>
> Intranets are private networks that use the same TCP/IP protocol suite as the Internet and are often connected to it.

The goals requirements discussed earlier led to the selection of a packet switching network which is based on a connectionless internetwork layer. This layer is called as the internet layer and it holds the whole architecture together. The task of this layer is to allow the host to insert packets into any network and then make them travel independently to the destination. The order in which the packets are received can be different from the sequence in which they were sent. Then the higher layers are supposed to arrange them in the proper order. Note that *internet* is being used as a generic term. The internet layer defines (specifies) a packet format and a protocol called internet protocol (IP). The internet layer is supposed to deliver IP packets to their destinations. So routing of packets and congestion control are important issues related to this layer. Hence TCP/IP internet layer is very similar to the network layer in OSI model as shown in figure 1.59.

Transport Layer

This is the layer above the internet layer. Its functions are same as those of a transport layer in OSI layer. This layer allows the per entities of the source and destination machines to converse with each other. The end to and protocols used here are TCP and UDP (User datagram protocol). TCP is a reliable connection oriented protocol. It allows a byte stream transmitted from one machine to be delivered to the other machine without introducing any errors. TCP also handles the flow control. UDP (user datagram protocol) is the protocol used in the transport layer. It is an unreliable, connectionless protocol and used for the applications which do not want the TCP's sequencing or flow control. UDP is also preferred over TCP in those applications in which prompt delivery is more important than accurate delivery. It is used in transmitting a video.

* Compared to OSI, TCP/IP is simple and dependable, it has a proven track history (more than 25 years), it is nonproprietary, its developers have a pragmatic approach to its enhancement, and it meets the networking needs of a diverse population, including researchers, educators, and business personnel.

Application Layer

TCP/IP model does not have session or presentation layers, because they are of little importance in most applications. The layer on top of transport layer is called as application layer. The protocols related to this layer are all high level protocols such as virtual terminal (TELENET), file transfer (FTP) and electronic mail (SMTP) as shown in figure 1.60.

Application layer	TELENET, FTP, SMTP, DN, HTTP, NNTP
Transport	TCP UDP
Internet (Network)	IP
Host-to-network	ARPANET, SATNET LAN, packet radio

Fig. 1.60.

Many other protocols have been added to these, over the years such as Domain Name Service (DNS), NNTP and HTTP etc.

Host-to-network Layer

This is the lowest layer in TCP/IP reference model. The host has to connect to the network using some protocol, so that it can send the IP packets over it. This protocol varies from host to host and network to network.

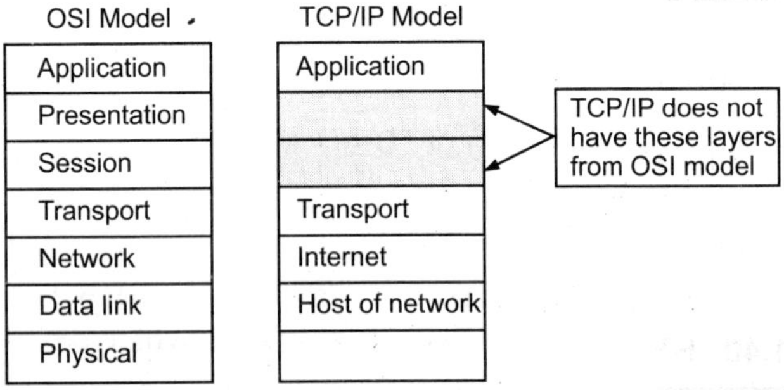

Fig. 1.61. Relationship between OSI and TCP/IP models

1.38. COMPARISON OF OSI AND TCP/IP MODELS*

We consider figure 1.61 and table 1.4, for the comparison of the two reference models.

Table 1.4. Difference between OSI and TCP/IP model

S.No.	OSI	TCP/IP
1.	Has 7 layers	Has 4 layers
2.	Transport layer guarantees delivery of packets.	Transport layer does not guarantee delivery of packets.
3.	Horizontal approach.	Vertical approach.
4.	Separate session layer.	No session layer, characteristics are provided by transport layer.
5.	Separate presentation layer.	No presentation layer, characteristics are provided by application layer.
6.	Network layer provides both connectionless and connection oriented service.	Network layer provides only connection less services.

* Nonetheless, the OSI model has had a lasting impact on networks, including TCP/IP. Although its acceptance has diminished considerably during the last half of the 1990s, the OSI model continues to provide a detailed standard for describing a network. It is from this perspective that the network design community continues to regard the OSI model as a theoretical framework for the development of networks and their architecture.

7.	It defines the service, interfaces and protocols very clearly and makes a clear distinction between them.	It does not clearly distinguish between service, interfaces and protocols.
8.	The protocols are better hidden and can be easily replaced as the technology changes.	It is not easy to replace the protocols.
9.	OSI is truly a general model.	TCP/IP cannot be used for any other application.
10.	It has a problem of protocol fitting into a model.	The model does not fit any other protocol stack.

1.39. DEMERITS OF TCP/IP MODEL

(i) TCP/IP model does not clearly distinguish the concepts of service, interface and protocol.

(ii) This model is not at all general and it cannot describe any protocol stack other than TCP/IP.

(iii) The host-to-network layer is not a layer at all in the normal sense. It is simply an interface.

(iv) The TCP/IP model does not even mention the physical and data link layers. A proper model should include both as separate layers.

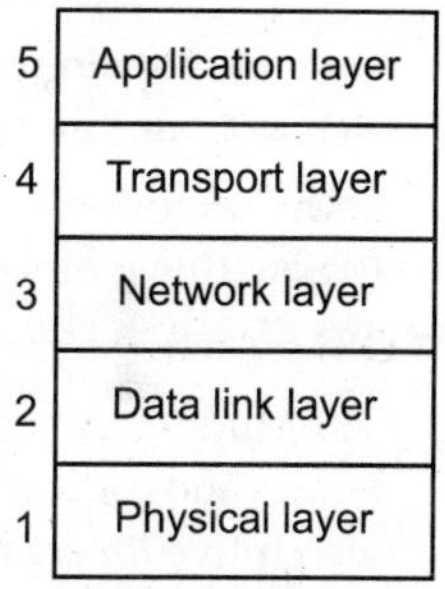

Fig. 1.62. Hybrid model

1.40. HYBRID (INTERNET) REFERENCE MODEL

In spite of many problems associated with the OSI model, it has proven to be very useful one practically. But the OSI protocols have not become popular. On the other hand, the TCP/IP model is practically non existing but the TCP/IP protocols are used widely. So sometimes a modified OSI model with primary concentration TCP/IP is used which is called as the hybrid model. The hybrid model is shown in figure 1.62. It is also called as the **Internet model**.

GLOSSARY

- **Commercial Internet Exchange (CIX)** A subscription organization that consists of a consortium of commercial and non-profit regional network providers that started offering Internet service independent of the NSFNET backbone and without NSF's restriction on traffic type Today, CIX serves as an Internet site very similar to a NAP.
- **Computer Network** A collection of computers and other devices that makes use of a common network protocol to share resources with each other over a network medium.
- **Decentralized System** Computer systems which are independent of each other and maintain separate databases germane to specific activities.
- **Desktop** Another name for a networked device.
- **Device** Any entity which is connected to a network. Few examples include terminals, printers, computers, or special network-related hardware units such as communication servers, repeaters, bridges, switches and routers. Local or sending devices originate communication; remote or receiving devices are the recipients of such communication.
- **Internet work** A collection of interconnected networks that functions as a single network. The individual networks comprising an Internet work are known as *subnetworks.*
- **Local Area Network (LAN)** A network which interconnects computing resources within a moderately sized geographical area. This may include a room, several rooms within a building, or several buildings on a campus. A LAN's range usually is no more than 10 km in radius.

- **Loop** A network configuration in which nodes are connected via dedicated wiring instead of through a centralized hub (as is the case of *star* design). Loops can be either *simple* (only one connection between any two nodes), *partial* (some nodes are interconnected by more than one link), or *complete* (every node has a connection to every other node). A loop is also referred to as a *meshed* design.
- **Metropolitan Area Network (MAN)** A network which interconnects computing resources that span a metropolitan area such as buildings located throughout a local county or city. MANs generally refer to networks which span a larger geographical area than LANs but a smaller geographical area than WANs.
- **Network Architecture** A formal, logical structure which defines how network devices and software interact and function; defines communication protocols, message formats, and standards required for interoperability.
- **Network Protocol** A formal specification which defines the vocabulary and rules of data communication.
- **Network Protocol Suite** A set of related and interoperating network protocols. An example is the TCP/IP protocol suite, that consists of protocol's for a e-mail, web service, file transfers, and routing.
- **Network Topology** The basic design of a computer network which details how key network components such as nodes and link are interconnected.
- **Node** Another name for a *device*. It is usually used to identify computers which are network hosts. workstations, or servers.
- **OSI** An acronym for open system interconnection.
- **OSI Model** A network architecture to develop network protocol standards. The OSI model basically defines and codifies the concept of *layered* network architecture. It makes use of well-defined operationally descriptive layers that describe what happens at each stage in the processing of data for transmission. The OSI model consists of the following seven layers, which are numbered in descending order : application (7), presentation (6), session (5), transport (4), network (3), data link (2), and physical (1).
- **Point-to-Point Network** A network design in which only adjacent nodes (nodes that are next to each other and only one hop away) can communicate with one another.
- **Protocol** An accepted or established set of procedures, rules, or formal specifications governing specifics behaviour or language. When applied to networks, a *network protocol* is a formal specification which defines the vocabulary and rules of data communication.
- **Server** A networked device which provides resources to *client* machines. Few examples include print servers, mail servers, file servers, and Web servers, Servers are shared by more than one user; clients have only a single user.
- **Star** A network configuration characterized by the presence of a central processing hub, which serves as a wire center for connecting nodes. All data should pass through the hub in order for nodes to communicate with each other.
- **Wide Area Network (WAN)** A network which interconnects computing resources that are widely separated geographically (usually over 100 km). This includes towns, cities, states, and countries. A WAN generally spans an area greater than 5 miles (8 km) and can be thought of as consisting of a collection of *LANs*.

SUMMARY

- Data communications and networking are changing the world scenario. Today, business and other daily affairs rely on computer networks and internetworks.
- A network is a set of devices often referred to as nodes connected by media links. A node can be a

computer, printer or any other device capable of sending or receiving data generated by other nodes on the network. Also, the links connecting the devices are called communication channels.

- There has been a tremendous development in computer industry. Infact, the development of the personal computer brought about tremendous changes for business, industry, science and education.
- A transmission system can be a single transmission line or a complex network connecting source and destination.
- The receiver accepts the signal from the transmission system and converts it into a form which can be handled by the destination device. As an example, a modem accepts an analog signal coming from a network or transmission line and converts it into a digital bit stream.
- A destination takes the incoming data from the receiver.
- The wide area networks cover a large geographical area. Typically, a WAN consists of a number of interconnected switching nodes. A transmission from any one device is routed through these internal nodes to the specified destination device. These nodes are not concerned with the content of the data, rather, their purpose is to provide a switching facility that can move the data from node to node until they reach their destination.
- A protocol is a set of rules that governs data communications. A protocol defines what is communicated, how it is communicated and when it is communicated.
- In general sense, communications can be assumed to involve three agents or entities. These are applications, computers and networks.
- Applications execute on computers that typically support multiple simultaneous applications.
- Computers are connected to networks and the data to be exchanged are transferred by the network from one computer to another computer.
- The network access layer is primarily concerned with the exchange of data between a computer and the network to which it is attached.
- When communication is desired among computers from different vendors, the software development effort can not be appropriate. Infact, different vendors use different data formats and data exchange protocols. Even within one vendor's product line, different model computers may communicate in unique ways.
- An ISO standard that cover all aspects of networks communications is the **Open System Interconnection (OSI)** model. An **open system** is a model that allows any two different systems to communicate regardless of their underlying architecture.
- The Open System Interconnection (OSI) model is a layered framework for network systems designing. This model allows for communication across all types of computer systems.
- The passing of the data network information down through the layers of sending machine and back up through the layers of receiving machine is made possible by an **interface** between each pair of adjacent layers. Each interface defines what information and services a layer must provide for the layer above it.
- The **physical layer** co-ordinates the functions required to transmit a bit stream over a physical medium. It deals with the mechanical and electrical specifications of the interface and transmission medium. It also defines the procedures and functions to be performed by physical devices and interfaces for transmission to occur.
- The **data link layer** transforms the physical layer to a reliable link and is responsible for **node-to-node** delivery. It makes the physical layer appear error free to the upper layer (*i.e.*, network layer).
- The **transport layer** is responsible for source-to-destination (end-to-end) **delivery** of the entire message. Whereas the network layer oversees end-to-end delivery of individual packet; it does not recognize any relationship between those packets.
- The services provided by the first three layers (physical, data link, and network) are not sufficient for some processes. The **session layer** is the network *dialogue controller*.
- The **presentation layer** is concerned with the syntax and semantics of the information exchanged between systems.

SHORT QUESTIONS WITH ANSWERS

Q.1. What do you mean by a network?

Ans. A network is a set of devices often referred to as nodes connected by media links. A node can be a computer, printer or any other device capable of sending or receiving data generated by other nodes on the network. Also, the links connecting the devices are called communication channels.

Q.2. What do you mean by Wide Area Networks? Explain.

Ans. The wide area networks cover a large geographical area. Typically, a WAN consists of a number of interconnected switching nodes. A transmission from any one device is routed through these internal nodes to the specified destination device. These nodes are not concerned with the content of the data, rather, their purpose is to provide a switching facility that will move the data from node to node until they reach their destination. Traditionaliy, WANs have been implemented using one of following two technologies:
(*i*)Circuit switching, and (*ii*) Packet switching.

Q.3. Write the Disadvantages of a Network.

Ans. **Crashes:** The biggest disadvantages is on a server-based network. When the server cashes, work gets disrupted as all network resources and its benefits are lost. Unless proper precautions are taken to ensure regular backups, the crash may result in the loss of days and even months of critical data and time.

Data Security: As all the data resources are pooled together, it is possible for unauthorised personnel to access classified information if network security is weak or poorly implemented.

Privacy: A network may also mean loss of privacy as anyone, especially your boss, with the right network privileges may be in a position to read your private e-mail.

Q.4. What do you mean by network protocol?

Ans. Network protocols are standards that allow computers to communicate. A protocol defines how computers identify one another on a network, the form that the data should take in transit, and how this information is processed once it reaches its final destination. Protocols also define procedures for handling lost or damaged transmissions or "packets." TCP/IP (for UNIX, Windows NT, Windows 95, and other platforms), IPX (for Novell Netware), DECnet (for networking Digital Equipment Crop. computers), Apple Talk (for Macintosh computers), and NetBIOS/NetBEUI (for LAN Manager and Windows NT networks) are the main types of network protocols in use today.

Q.5. What do you mean by OSI Reference Model? Explain.

Ans. An ISO standard that cover all aspects of networks communications is the **Open System Interconnection (OSI)** model. An **open system** is a model that allows any two different systems to communicate regardless of their underlying architecture. Vendor-specific protocols close off communication between unrelated systems. The purpose of the OSI model is to open communication between different systems without requiring changes to the logic of the underlying hardware and software. The OSI model is not a protocol rather it is a model for understanding and designing a network architecture which is flexible, robust, and interoperable.

Q.6. Explain why it is necessary to have standards for networking products.

Ans. Developing standards for computers enables hardware and software products to be compatible.

Q.7. Define an open system.

Ans. An open system is a set of protocols that allows two computers to communicate with each other, regardless of the manufacturer, CPU, or operating system.

Q.8. Explain the function of the application layer.

Ans. The application layer enables users to access networks through tools such as Telnet, FTP, and Network Neighbourhood.

Q.9. Explain the function of the Session layer.

Ans. To set up a session between two applications, such as login in Telnet.

Q.10. List the functions of the Network layer.

Ans. Setting up a physical connection, disconnecting a connection, and routing information.

Q.11. Define the functions of the Physical layer.

Ans. Electrical and mechanical interface with communications mdeia, converting electrical and optical signals to bits and vice versa.

Q.12. What are the frame transmission methods?

Ans. Asynchronous and synchronous.

Q.13. What layers of the OSI model does frame relay operate on?

Ans. Both the Data link layer and Physical layer of the OSI model.

REVIEW QUESTIONS

1. Draw the block diagram of a general communication model and explain the function of each block.
2. Explain a simplified data communication model.
3. Explain the concept of Networks and types of Computer Networks.
4. Write the advantages of Networking.
5. Explain OSI model in detail.
6. What is TCP/IP protocol architecture? Explain.
7. State various services provided by the network for companies and people.
8. What is the difference between broadcast and point to pint networks ?
9. Write a short technical note on MAN.
10. Write a short technical note on WAN.
11. How can you compare LAN and WAN?
12. How does the actual data transfer take place between two machines ?
13. Write a short technical note on Virtual communication between layers.
14. Discuss the important design issues for various layers.
15. Write a short technical note on connection oriented and connectionless services.
16. What is the relationship between services and protocols ? Explain
17. Draw the OSI reference model and explain the functions of different layers.
18. Write a short note on TCP/IP.
19. State advantages and disadvantages of TCP/IP.
20. Write a short technical note on Wireless LAN
21. What are the different types of transmission technology ? Explain different types of networks on the basis of transmission technology.
22. Distinguish between TCP/IP and OSI reference models. Which model is more popular and why?
23. Define topology and explain the advantages and disadvantages of Bus, Star and Ring topologies.
24. Discuss DQDB standard in context to MAN.
25. Write a short technical note on : Topology.
26. Write short technical note on : Connection oriented and connection less.
27. Explain the working of Transport, Session, Presentation and Physical layer of OSI model given by IEEE.
28. Discuss the TCP/IP protocol suite on the basis of protocol layering principle.

Chapter 2

Example Networks

Syllabus

Example Networks

Chapter Outline

- Networks Classification
- Novell Netware
- The Arpanet
- Internet
- Example Data Communication Services
- Switched Multimegabit Data Service (SMDS)
- The X.25 Networks
- Frame Relay
- Performance Comparison of X.25 and Frame Relay
- Asynchronous Transfer Mode (ATM)
- Ethernet
- Network Standardization

2.1. NETWORKS CLASSIFICATION

As a matter of fact, a large number of networks are currently operating around the world. These networks may be classified into different types as depicted in figure 2.1.

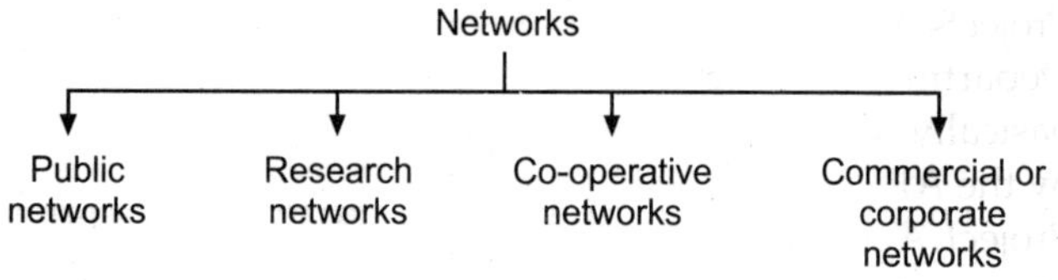

Fig. 2.1. Networks classification

2.1.1. Salient Features of a Network

Few important salient features of a network may be listed as under :

(i) Administration

(ii) Facilities offered.

(iii) Technical design.

(iv) User communities.

Various networks differ in these features. In the next few articles, we shall study few important Examples of Networks.

2.2. NOVELL NETWARE

This is the most popular netware in the PC world. It was specially designed to be used by a network of PCs. In such a network, each user has a desktop PC which functions as a client and a number of powerful PCs operate the servers. The server PCs provide file services, database services and other services to a group of clients.

Hence, Novell Netware is based upon the client-server model. Figure 2.2 illustrates the proprietary protocol stack used by the Netware. It looks more like TCP/IP model than OSI.

Layer			
Application	SAP	File server	.....
Transport	NCP		SPX
Network		IPX	
Data link	Ethernet	Token ring	ARC net
Physical	Ethernet	Token ring	ARC net

Fig. 2.2. The Novell Netware reference model

The physical layer can be chosen from among various standards such as Ethernet, Token ring and ARC net and so does the data link layer. The protocol run by the network layer is IPX (Internet Packet eXchange) which is an unreliable connectionless protocol. Above IPX, comes the transport layer which uses a connection oriented transport protocol called NCP (Network Core Protocol). At this layer, a second protocol called SPX (Sequenced Packet exchange) is also available. The session and presentation layers are not present. Various application protocols are present in the application layer. About once a minute, each server broadcasts a packet giving its address and the information about what services it offers. The SAP (Service Advertising Protocol) is used for these broadcasts.

2.3. THE ARPANET

The Internet's roots can be traced back to 1957 when the United States formed the Advanced Research Projects Agency (ARPA) within the Department of Defense (DoD). It is basically a WAN. It was developed by the ARPA (Advanced Research Project Agency) in 1968 which is the research arm of DoD. ARPANET was designed to service even a nuclear attack. Before ARPANET, the networks were basically the telephone networks which operated on the circuit switching principle. But, this network was too vulnerable, since the loss of even one line or switch would terminate all the conversations. ARPANET used the concept of packet switching network consisting of subnet and host computers. The subnet was a datagram subnet and each subnet consists of minicomputers called IMPs (Interface Message Processors). Each node of the network used to have an IMP and a host connected by a short wire. The host could send messages of upto 8063 bits to its IMP which would break them into packets and forward them independently toward the destination. Originally, the Internet meant ARPANET, and access to ARPANET was restricted to the military, defense contractors, and University personnel involved in defense research.

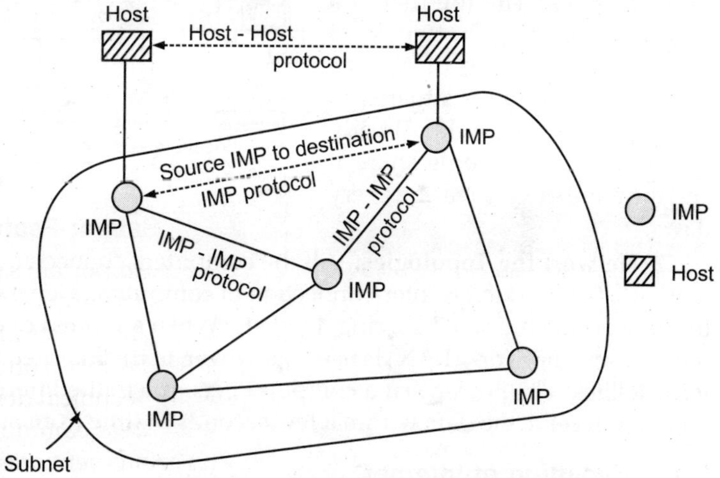

Fig. 2.3. Illustration of ARPANET

The subnet was the first electronic store and forward type packet switched network. Hence, each packet was stored before it was forwarded. The original ARPANET design has been shown in figure 2.3.

The software for ARPANET was split into two parts namely subnet and host. In 1974, the TCP/IP model and protocol were invented specifically to handle communication over Internetworks since more and more networks were getting connected to ARPANET. The TCP/IP made the connection of LANs to ARPANET easy. During 1980s, so many LANs were connected to ARPANET that finding hosts became increasingly difficult and expensive. Hence, DNS (Domain Naming System) was created for organizing machines into domains and map host names onto IP address. In 1983, the management of ARPANET was handed over to the Defence Communications Agency (DCA) which separated the military portion into a separate MILNET. By 1990, the ARPANET was shut down and dismantled, however, MILNET continues to operate. ARPANET technology was based on packet-switching, and in 1969, with the connection of its first four nodes-SRI, UCSB, UCLA and University of Utah-ARPANET heralded the era of packet switching networking.

2.4. INTERNET

1. Basic Concepts: The Internet is a collection of computer networks based on a specific set of network standards-TCP/IP. Internet was the next step of ARPANET and NSFNET. Sometimes, in mid 1980s, people started imagining the interconnection of many networks and the Internet came into existence. It started growing exponentially with all the types of existing networks getting connected to Internet. This was possible because of the use of TCP/IP reference model and TCP/IP protocol stack. The internet is a globally existing network of networks consisting of a huge number of computers situated in all the parts of the world. Already about 20 million people share it and the number is increasing every day.

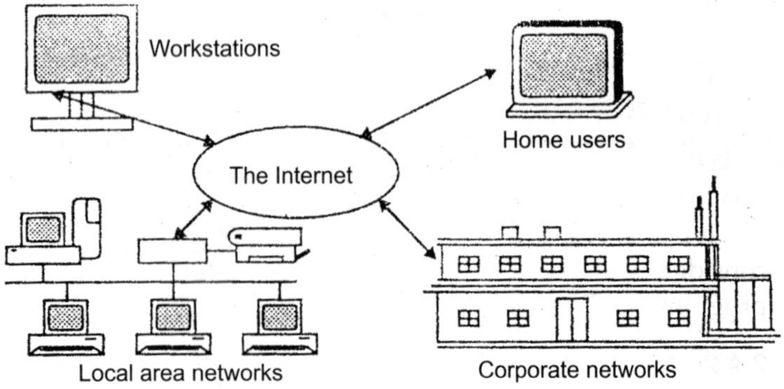

Fig. 2.4.

2. Networking Topologies: All the computers connected to the Internet are part of this huge network. Networking is inter-connection of computers. Generally, the networking topologies used for networking are star, bus, ring, loop etc. When a limited of computers are to be interconnected, the local area network (LAN) is used. However, in the Internet, the interconnection is achieved even via satellites. The power that a computer gets due to the Internet is amazing. It is now possible to send and receive the data within a few seconds of time from any part of the world.

2.4.1. Evolution of Internet

1. Origin of Internet: The origin of Internet can be traced to a U.S. Department of Defence (DoD) organization called Advanced Research Projects Agency (ARPA). ARPA developed a four node packet switching network called ARPANET in 1969. The network was intended to support the military research on fault tolerant computer networks. DoD wanted to ensure a reliable data transfer in the event of a nuclear war, even if parts of the network has been destroyed. As the years pass by, the network grew in size and by January 1983, the ARPANET no longer remained an experimental network but its control was passed over to Defence Communications Agency (DCA). The network then became available for the academic research, government employee and contractors. It was not until 1986, however, that the dramatic growth of the Internet began. At this time, the National Science Foundation (NSF) which formed the National Science Foundation Network (NSFNET) linked five

Pioneer in Communications: Claude Elwood Shannon (1916–)

Born in Gaylord, Michigan, Claude Shannon graduated from Michigan University with degrees in mathematics and engineering and, in 1940, from the Massachusetts Institute of Technology (MIT) with master's and doctoral degrees in mathematics. In 1938, as a student at MIT, Shannon published the paper, "A Symbolic Analysis of Relay and Switching Circuits,".

of the regional super-computer centres together to provide a national high speed backbone network across the United Sates. The world's fastest and most powerful computers were made available to an academic and scientific community. In 1990, the ARPANET was officially decommissioned.

2. Few Major Networks: Some of the major networks contributing to the growth of the Internet may be listed as under :

(i) ARPANET
(ii) USENET (User's network)
(iii) CSNET (Computer Science Network)
(iv) BITNET (Because it's Time Network)
(v) NSFNET (National Science Foundation Network)
(vi) WWW (World Wide Web)
(vii) NREN (National Research and Education Network)
(viii) Intranet

It is said that the Internet is growing at a rate of 20% per month. The data speeds have gone up considerably, which makes the access even more fast. The event which started as a military assistance program is now largely a private enterprise.

2.4.2. Net Structure*

1. Basic Concepts: To understand how the Internet works, let us imagine complex network of roads. It consists of superhighways, highways, to the small roads on the countryside, all connected to each other in one way or the other. The nature of the Internet is much similar to this. The high speed and powerful super computers connected to the Internet is the backbone of the system. The data that is being moved by these supercomputers is a few million bits per second. The regional networks are then connected to these supercomputers. The small computer network and individual users are then connected to this regional network. The data and information available on the Internet is increasing every day since every user can contribute to it. Because of this, the entire net will never crash down. Even if a part of it closes down the remaining network keeps working. There is no central computer to control the Internet. There is no central body which governs the Internet structure, though volunteer groups of individuals have set some standards for Internet technologies. No particular person, group or organization owns the Internet.

2. Services on the Internet: The services that are available on the Internet may be classified into following two categories :

(i) Communication services
(ii) Information retrieval services.

They may be further classified as depicted in figure 2.5.

* Regardless of the definition or perspective, the Internet interconnects individual, autonomous, heterogeneous computer networks and enables them to function and appear as a single global network.

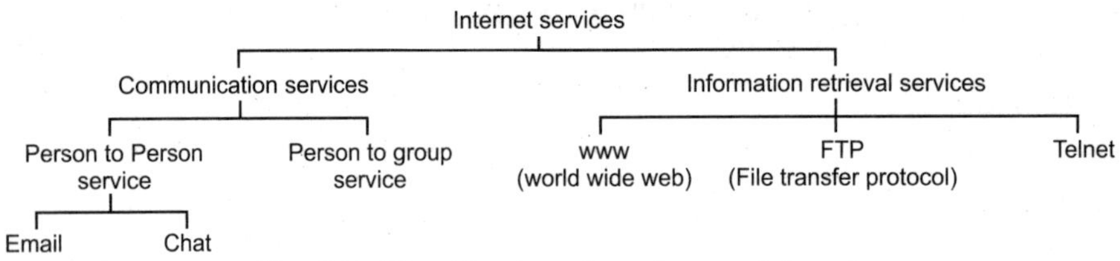

Fig. 2.5. Classification of services on Internet

The person to person include E-mail. This is the most popular service and most of the Internet traffic corresponds to the E-mail. The person to group service includes on line discussions between a user and many other participants from around the globe. A user can choose his topic of discussion from a list of more than 25,000 topics. The fast spreading popularity of the Internet is basically because of the information retrieval services. These services as shown in figure 2.5, include,

(i) The world wide web (www), which is world's largest and ever-growing data base.

(ii) FTP or file transfer protocol, which allows remote accessing to the files which contain programs, technical handouts, reports etc.

(iii) Telnet or remote login to commerical services.

2.4.3. Internet Protocols

The Internet protocol is like any other communication protocol. It is a set of rules which will govern every possible communication over the Internet. Since the development of the ARPANET, TCP/IP together has emerged as the controlling body, it is being used in computers of not only in the U.S., but all over the world for all types and sizes of computers. It has become the language of the Internet. Actually, TCP/IP are two protocols : Transmission control protocol and Internet protocol. These two protocols describe the movement of data between the host computers on Internet. The protocol, however, is a suite of many other protocols that provide for reliable communications across the Internet and the web. In the TCP/IP protocol suite, there are various layers, with each layer being responsible for different facets of communication. The Internet Protocol (IP) and Transmission Control Protocol (TCP) are together known as TCP/IP protocol. TCP/IP offers a simple naming and addressing scheme-whereby different resources on Internet can be easily located. Information on Internet is carried in *packets*. The IP protocol is used to put a message into a *packet*. Each packet has the address of the sender and the recipient's address. These addresses are known as the IP addresses. Using the TCP protocol, a single large message is divided into a sequence of packets and each is put into an IP packet. The packets are passed from one network to another until they reach their destination. At the destination, the TCP software reassembles the packet into a complete message. It is essential for all the packets in a single message to take the same route each time it is sent. The Internet is thus a *packet switched*

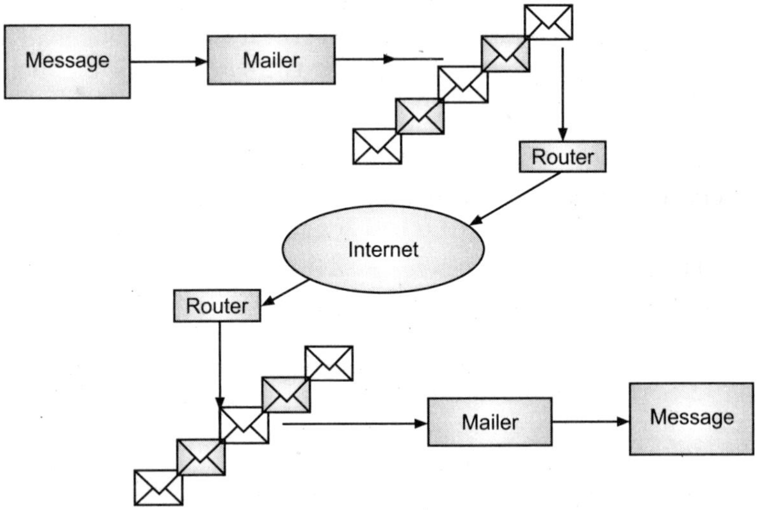

Fig. 2.6. The route followed by the packets

network. The route followed by the packets of information are as shown in figure 2.6. The packets as shown in figure 2.6, are read by routers and then are sent down to the destination address.

2.4.4. Internet Address

1. Basic Concepts: Internet addresses (called IP addresses for short) are node addresses that identify their location within the Internet. With so many computers and users using the Internet, it would be impossible to differentiate them without a proper addressing system. This addressing system assigns names and numbers to identify the computers on the Internet. The names are known as *domain names* and the numbers are known as *IP addresses*. Every computer on the Internet shall have both a domain name and an IP address.

2. An IP Address*: Each computer on the Internet is identified by its unique IP address. An IP address is made of four numbers and each one must be less than 256. Thus, each set of numbers can have values from 1 to 225. An example of an IP address is

202.64.254.10

The numbers between the dots are known as *octets*. The leftmost octet *i.e.*, 202 represents the largest network and the rightmost octet 10 describes the specific machine which is being addressed. Although the IP addressing scheme can identify the computers on Internet, the numbers are impossible to remember. Therefore, a method of recognizing the computers by names was devised.

3. Domain Name Addressing: A smaller network making up the Internet and having many computers within it is known as a *domain*. The domain may represent a type of organization or a geographical location. For example, all the computers on the Internet which belong to the educational institutions form a part of the *educational* or *edu* domain. Commonly used domain types are as under :

Domain name	Description
com	A company or commercial organization
edu	Educational Institutes
gov	Government organization
mil	Military sites
net	Network resources or Internet service provider.
org	Non profit or non commercial organizations.

In order to accommodate different countries all over the world, a two letter abbreviation for every country in the world is added at the end of the domain name.

e.g. katariapub@pn2.vsnl.net.in

Here, the last two letters *i.e.*, 'in' represent *India*. Commonly used abbreviations as country suffix are as under:

Abbreviation	Country Name
au	Australia
fr	France
ca	Canada
uk	United Kingdom
in	India

*IP addresses play an important role in the successful delivery of data across the Internet. If a node cannot be located, then data cannot be delivered to it.

Let us consider again the sample example :

katariapub@pn2.vsnl.net.in

Where reading from right to left,
in - India
net - Network Provider
vsnl - organization name, in this case, it is VSNL
pn2 - server or computer name
katariapub - users name

The actual Internet protocol, however, will recognize only the IP number. The domain naming convention is introduced only for the user's convenience. Therefore, there must be some mechanism to convert the domain names to the corresponding IP addresses, and vice versa. This is done by an electronic server called *domain name server* (DNS). It is an electronic directory which is maintained on almost all the host computers.

2.4.5. Getting Connected : What do You Need ?

1. Primary Requirements: To be a part of the Internet community, we need to have the following:

(i) A dial up telephone line.
(ii) A modem (an interface cable is required to connect it with the computer, in case of an external modem).
(iii) A 486 or pentium P.C. The P.C. must have an adequate RAM capacity, windows 95/98 as operating system.
(iv) Communication software (TCP/IP, Web browsers, E-mail software).
(v) An Internet access account with an Internet Service Provider (ISP).
(vi) The PC must have microphone sound card and loud speakers.

There are several ways of gaining access to the Internet. Most business, educational Institutes and government agencies provide access to the Internet and the web for E-mail and business or educational correspondence. Most of the home connections to the Internet use the standard voice grade telephone lines connected to a modem and a P.C. In most of the cases, a *SLIP* (Serial Line Internet Protocol) or *PPP* (Point to Point Protocol) connection is necessary for the home connection. The user's software must support the SLIP or PPP communication protocol. The PPP is a more recent protocol which corrects may of the deficiencies in SLIP.

2. Internet Service Providers (ISPs): Internet Service Providers (ISPs) all over the world offer various options and packages to the general public for Internet access. With the privatization of the Internet services, various ISPs, are offering their services in India now. The major players currently are as under :

(i) Videsh Sanchar Nigam Limited (VSNL).
(ii) Mahanager Telephone Nigam Limited (MTNL).
(iii) Satyam Infoway Limited (Satyam Online).
(iv) Bharti BT Internet Pvt. Limited (Mantra Online).

2.4.6. Who Owns the Internet

The Internet is a decentralized network and no one runs or owns it. There is an organization or agency that controls the activity on the Internet. However, there are a few organizations which co-ordinate and guide the technical parts of Internet. They are as under :

(i) Internet Engineering Task Force (IETF).
(ii) Internet Research Task Force (IRTF).
(iii) Internet Architecture Board (IAB).

The Internet communication protocols are developed and maintained by the IETF. The protocols are the methods by which computers on the Internet are connected. The long term research problem which may be critical to the Internet are looked into by the IRTF. Any major changes coming from the IETF are ratified by the IAB.

2.4.7. Applications of Internet

Traditionally, the Internet has the following applications :

1. E-mail
2. News
3. Remote login
4. File transfer
5. World wide web (www)

> **DO YOU KNOW?**
>
> Although frame relay began as a data only service, it is capable of supporting voice, fax, and video transmissions.

1. Electronic Mail (E-mail): This is possibly the most popular use of the Internet. It is possible to send and receive messages to anybody, who is connected to the net and whose E-mail address is known to us. The E-mail address is composed of the user's account name followed by an '@' symbol. Technically speaking, the common form of E-mail address is :

username@location.organization.type.country

Although there are more than 50 million users, and the number is growing exponentially, this naming scheme ensures that exchange of messages between users is extremely simple. We can exchange E-mail without incurring the long distance charges of a telephone call, therefore, it can be an economical alternative to telephone usage. It has added advantage that the user can send the same message to a number of people.

2. News: News groups are specialized forums in which users with a common interest can exchange messages. a large number of news groups on technical or non-technical topics.

3. Remote Login: Users anywhere on the Internet can log into any other machine on which they have an account. This is possible using the programs like Telnet or R login. Telnet is another application which allows the user to login or connect to a remote computer system, from anywhere in the world. Telnet is a network application that is used to log on to one computer on the Internet from another. It is similar to establish a connection between two computers using modems and telephone lines. However, via. Telnet and not the telephone network. This aves the long distant telephone charges. Telnet provides a convenient way of accessing your E-mail server from anywhere in the world. For example, an Internet user having an account on a VSNL server at Delhi may want to check his E-mail when he is out of the city. In such a case, we can use somebody elses account to get connected to the Internet and then Telnet to his server. Once the Telnet session is established, he will be asked for his *username* and *password*.

4. File Transfer: The application on Internet that allows user to transfer files and form hosts or servers is called as file transfer protocol (FTP). FTP offers a convenient method of publishing and maintaining web pages. In order to transfer files using FTP, the user requires a client software. The FTP client establishes a connection with a FTP server. Depending on the type of server, the user may have a special access or login account to the server. There are also a large number of anonymous FTP servers which do not impose login restrictions and the informations available on then can be downloaded by all the users. A large number of information articles, database etc. becomes available this way.

5. The World Wide Web (www): Uptill early 1990s, the Internet was populated by information related to academic, government and research. But www, a new application brought millions of new, non-academic users to the net.

This is the most user friendly application of the Internet which is responsible for its rapid growth. The www is analogous to a bulletin board or a notice board. The notices, articles, pictures etc. displayed on the notice noard can be seen, read by anybody. However, everyone is not authorized to put up the

Pioneer in Communications: Vinton G. Cerf

Vinton G. Cerf is widely known as a Father of the Internet. Cerf initially became involved with the development of the Internet when, as a graduate student at UCLA from 1967-72, he was part of a team that worked on one component of the U.S. Department of Defence Advanced Research Projects Agency's (DARPA) packet-switching network project, ARPANET. After completing his doctorate in 1972, Cerf became an assistant professor in computer science and electrical engineering at Stanford University.

information on the notice board. Only authorized persons can do it. Extending this to the Internet a 'web site' is such a publically accessible notice board on the *server* or *host computer* connected to the Internet. Any user can view or read the contents of this area. The available information may be in the form of text, pictures, photographs, images or graphics. It can be on any subject or topic. Each document on the 'website' is called as *web page* or simply a *page*. Further, www has become the most popular application on the Internet and the large amount of data that it contains is growing continuously.

One can find information on just about any subject from news to sports, to music or films etc. The information provided is not static but changes continuously. The www is fast emerging as a new and important advertising medium. Different companies are putting up information pages about their products and services on www. They also use the interactive features of www to get the feedback forms and questionnaires for customer support and product development. A new development on www is electronic commerce or e.commerce. It is possible to shop for books, music, software, even vegetable, from home on the Internet. The www is thus an example of a 'client server' technology. The protocol used for the exchage of information between the host and the client is called as the Hyper Text Transfer Protocol (HTTP). The technique used to address the documents on the web is called as the *Uniform Resource Locator* (URL). The URL is there to provide an addressing scheme.

2.5. EXAMPLE DATA COMMUNICATION SERVICES

The telephone companies started offering networking services to the organizations which were interested in subscribing the services. Such systems are called as *public networks*. They are similar to and sometimes a part of the public telephone system. DQDB is one of such services which we have already discussed. Now, we shall discuss the following other example services:

(i) SMDS
(ii) Frame relay
(iii) X.25
(iv) ATM and broadband ISDN.

2.6. SWITCHED MULTIMEGABIT DATA SERVICE (SMDS)

SMDS stands for switched multimegabit data service. It is a wide area network service designed for LAN-to-LAN connectivity. The SMDS as shown in figure 2.7, is designed to connect the multiple LANs together. This is the first high speed broadband service offered to the public. Thus, SMDS is a cell-based, connection less, high-speed, public, packet-switched, broadband, metropolitan area data network. SMDS is also a data service, which means that it can only transmit data although it is capable of transmitting voice and video.

Figure 2.7 (*a*) shows the interconnection of four LANs using six high speed leased lines. However, this interconnection is expensive. The same LANs can be interconnected using SMDS as shown in figure 2.7 (*b*). It allows the packets from one LAN to pass to any other LAN. The SMDS network is in the telephone company's office. SMDS is designed to handle bursty service. The type of traffic in

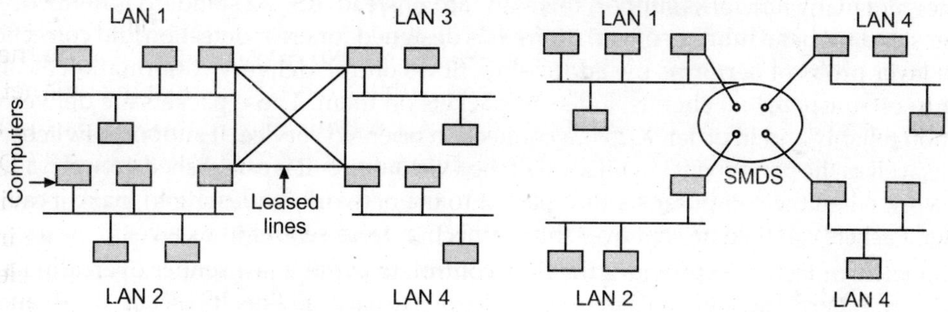

(a) Four LANs interconnected using leased lines

(b) Interconnection using SMDS

Fig. 2.7.

interconnected LANs is not continuous but bursty type *i.e.*, once in a while a packet will be transferred from one LAN to other but otherwise there is no LAN to LAN traffic. Hence, the option of leased lines will be expensive. It will result in high monthly bills. The SMDS is much cheaper solution to this problem. SMDS are supposed to be fast enough. Standard speed is 45 Mbps. With SMDS, each LAN gets conencted to *n* telephone company switch which will route the packets through the SMDS network toward the destination. The SMDS is a connectionless packet delivery service with a packet format as shown in figure 2.8.

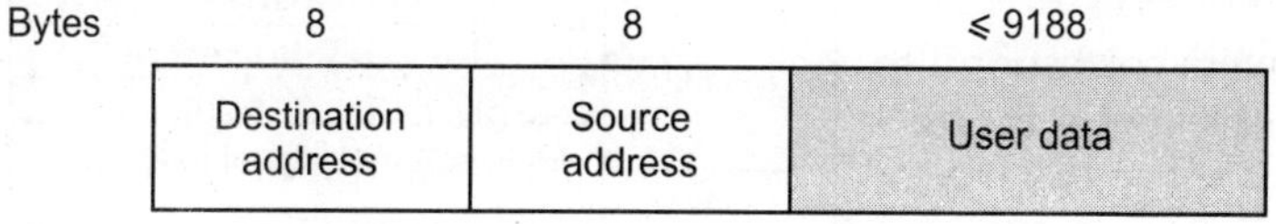

Fig. 2.8. Illustration of SMDS packet format

Basically, the three fields in the packet are destination, source and user data. The destination address and source field are of 8 bytes whereas the user data field has a variable length (upto 9188 bytes). In the source and destination, addresses contain a 4-bit code followed by a telephone number upto 15 decimal digits. As soon as a packet arrives at the SMDS network, the first router will check the source address. If it is not correct, then the packet is discarded. If it is correct, then the packet is routed to the destination. It is possible to broadcast a packet to many members, by using SMDS in the broadcasting feature. One more important feature is address screening on both outgoing and incoming packets. If the outgoing screening is used, then the customer can list a few telephone numbers and specify that no packets should be sent to them. On the contrary, if the incoming screening is used, then packets from only certain predecided numbers will be accepted.

2.7. THE X.25 NETWORKS

1. Basic Concepts: X.25 is a standard used by many older public networks specially outside the U.S. This was developed in 1970s by CCITT for providing an interface between public packet switched network and their customers. Prior to frame relay's development, LANs were primarily interconnected by dedicated (i.e., private) leased lines using point-to-point protocols or X.25. The packet switching networks use X.25 protocol. The X.25 recommendations were first prepared in 1976 and then revised in 1978, 1980 and 1984. This protocol is based on the protocols used in early packet switching networks such as ARPANET, DATAPAC, TRANSPAC etc. A protocol X.21 which is a physical layer protocol is used to specify the physical electrical and procedural interface between the host and network. The problem with this standard is that it needs digital signal rather than analog signals on telephone lines.

Hence, not many networks support this standard. Instead, RS-232 standard is defined. The data link layer standard has a number of variations. It is designed for error detection and corrections. The network layer protocol performs the addressing, flow control, delivery confirmation etc. It allows the user to establish virtual circuits and send packets on them. These packets are delivered to the destination reliably and in order. X.25 is a connection oriented service. It supports switched virtual circuits as well as the permanent circuits. a switched virtual circuit is established between a computer and network when the computer sends a packet to the network requesting to make a call to other computer. Packets can then be sent over this connection from sender to receiver.

As a matter of fact, X.25 provides the flow control, to avoid a fast sender overriding a slow or busy receiver. A permanent virtual circuit is analogous to a leased line. It is set up in advance with a mutual agreement between the users. Since it is always present, no call set up is required for its use. In order to allow the computers which do not use the X.25 to communicate with the X.25 network, a Packet Assembler Disassembler (PAD) is used. PAD is required to be installed alongwith each computer which does not use X.25. Further, X.25 defines the interface for exchange of packets between a DTE and switch data subnetwork node.

2. Three Layers of X.25: The X.25 interface is defined at three levels.

The three levels may be listed as under:

(i) Physical layer (level 1)

(ii) Data link layer (level 2)

(iii) Packet layer (level 3).

These three layers correspond to the three lower most layers of the ISO-OSI reference model. the physical layer takes care of the interface between a computer terminal and the link which attaches it to the packet switching node. The X.25 defines the interface for exchange of packets between the user's machine (DTE) and the packet switching node to which this DTE is attached which is called as DCE. The three layers of X.25 interface have been shown in figure 2.9. At the physical level, X.21 physical interface is being used which is defined for circuit switched data network. At the data link level, X.25 specifies the link access procedure-B (LAP-B) protocol which is a subset of HDLC protocol.

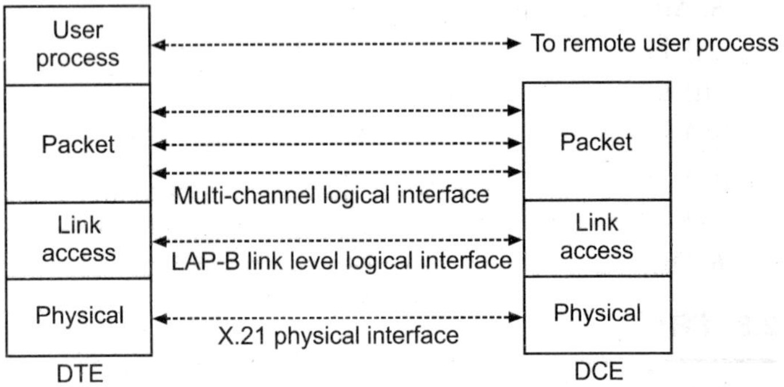

Fig. 2.9 Different layers of X.25 and interface between DTE and DCE

At the network level, X.25 defines a protocol for an access to packet data subnetwork. This protocol defines the format, content and procedures for exchange of control and data transfer packets. The packet layer provides an external virtual circuit service.

3. Relationship between the levels of X.25: Figure 2.10 shows the relationship between the levels of X.25. User data is passed down to X.25 level 3.

This data then appends the control information as a header to form a packet. This control information is then used in the operation of the protocol. The entire X.25 packet formed at the packet level is then passed down to the second layer i.e., the data link layer. The control information is appended at the front and back of the packet forming a LAP-B frame. The control information in LAP-B frame is required for the operation of the LAP-B protocol. This frame is then passed to the physical layer for transmission.

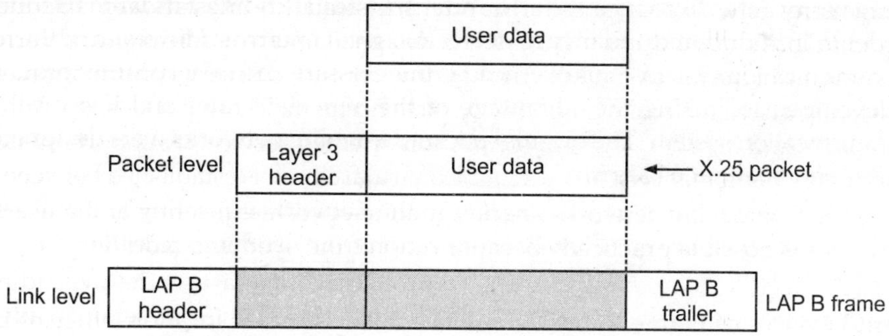

Fig. 2.10 Relationship between the levels of X.25

4. Virtual Circuit Service: With the X.25 packet layer, data are transmitted in packets over external virtual circuits. The virtual circuit service of X.25 provides for two types of virtual circuits. The virtual circuit service of X.25 provides for two types of virtual circuits *i.e., virtual call* and *permanent virtual circuit*. A virtual call is a dynamically established virtual circuit using a call set up and call clearning procedure. A permanent virtual circuit is a fixed, network assigned virtual circuit. Data transfer takes place as with virtual calls, but no call set up or clearing is required.

5. Advantages of X.25: Advantages of X.25 are as

(i) Frame delivery is more reliable.

(ii) Frames are delivered in order.

(iii) Retransmission of frames is possible.

(iv) Flow control is provided.

(v) X.25 supports the switched virtual circuits and permanent circuits.

6. Drawback: X.25 is much slower than Frame relay.

2.8. FRAME RELAY

1. Connection Oriented Service: Packet switching was developed when the long distance digital communication showed a large error rate.

> **DO YOU KNOW?**
>
> Frame relay uses virtual links instead of permanent, dedicated connections.

To reduce the error rate, additional coding bits were introduced in each packet in order to introduce redundancy to detect and recover errors. But in the modern high speed telecommunication systems, this overhead is unnecessary and infact counter productive. Frame relay was developed for taking the advantage of the high data rates and low error rates in the modern communication system. The original packet switching networks were designed with a data rate at the user end of about 64 kbps. But the frame relay networks are designed to operate efficiently at the user's data rates upto 2 Mbps. This is possible practically because most of the overhead (additional bits) are striped off. Frame relay is a virtual circuit wide area network which was designed in early 1990s.

Frame relay is a public WAN packet-switching protocol that provides LAN-to-LAN connectivity. Frame relay is a connection oriented servicve. It can be imagined to be equivalent to a virtual leased line. On a virtual line, data bursts can be sent at full speed. Frame relay provides a service to determine the start and end of each frame. In fact, a frame relay relays frames across a network between two sites. Frame relay was originally part of the ISDN standard. It also detects the transmission errors. But, the frame delay does not have error control or flow control. If a frame contains errors then the frame relay service discards it. Unlike X.25, frame relay does not provide acknowledgements or normal flow control. Packet switching was developed when the long distance digital communication showed a large error rate. To reduce the error rate, additional coding bits were introduced in each packet in order to introduce redundancy to detect and recover errors. However, in the modern high speed telecommunication systems, this overhead is unnecessary and infact counter productive. Frame relay was developed for taking the advantage of the high data rates and low error rates in the modern communication system. The original packet switching networks were designed with a data rate at the user end of about 64 kbps.

However, the frame relay networks are designed to operate efficiently at the user's data rates upto 2 Mbps. This is possible practically because most of the overhead (additional bits) are striped off.

2. Salient Features of Frame Relay: Frame relay also is meant for more efficient transmission scheme than the X.25 protocol. Various key features of this protocol are as under :

(i) Frame relay operates at a high speed (1.544 Mbps to 44.376 Mbps).

(ii) Frame relay operates only in the physical and data link layers. Hence, it can be easily used in Internet.

(iii) It allows the bursty data.

(iv) It has a large frame size of 9000 bytes. Hence, it can accommodate all local area network frame sizes.

(v) Frame relay can only detect errors (at the data link layer). But, there is no flow control or error control.

(vi) The damaged frame is simply dropped. There is no retransmission. This is to increase the speed. Hence, frame relay needs a reliable medium and protocols having flow and error control.

(vii) The cell control packets are used for setting up and clearing virtual circuits.

(viii) In band signaling is used.

(ix) Multiplexing of virtual circuits takes place at layer 3.

(x) Layers 2 and 3 include flow control and error control.

FRAD (Frame Relay Assembler/Disassembler) is a device used by frame relay to handle frames arriving from other protocols. A FRAD assembles and disassembles frames coming from other protocols.

2.8.1. Advantages of Frame Relay

Some of the advantages of frame relay may be listed as under :

(i) Streamlined communication process.

(ii) The number of functions of a protocol at the user-network interface is reduced.

(iii) Lower delay.

(iv) Higher throughput.

(v) Frame delay can be used at access speeds upto 2 Mbps.

2.8.2. Drawbacks of Frame Relay

(i) Frames are delivered unreliably.

(ii) Packets may not be delivered in the same sequence as that at the sending end.

(iii) Packets having errors are simply discarded.

(iv) Frame relay does not provide flow control.

(v) It does not provide the acknowledgement of received packets.

2.9. PERFORMANCE COMPARISON OF X.25 AND FRAME RELAY

S.No.	X.25	Frame Relay
1.	X.25 networks work at speed upto 64 kbps.	Frame relay operates at higher speed of 1.5 Mbps
2.	Here frames are delivered more reliably than frame relay.	Here, frames are delivered unreliably than X.25.
3.	Frames are delivered in order here.	Frames delivered here are not in proper order.
4.	It allows the user to establish virtual circuits and then send packets of upto 128 bytes on.	It allows user to establish virtual circuit and then send packets of upto 1600 bytes between them.
5.	Bad frames can be received back by sending acknowledgement signal.	Bad frame is discarded by frame relay.
6.	X.25 provides flow control.	Frame relay does not provide flow control.
7.	X.25 provides more services, a way to determine the start and end of each frame, detection of transmission error and recovering from those errors by sending acknowledgement signal and maintaining flow control.	Frame relay provides minimal services, primarily a way to determine the start and end of each frame and detection of transmission errors.
8.	X.25 supports both switched virtual circuits and permanent ones.	Frame relay suppots virtual leased line.
9.	X.25 provides acknowledgement signal.	Frame relay does not provide acknowledgement signal.

2.10. ASYNCHRONOUS TRANSFER MODE (ATM)

1. Definition and Basic Concept: The long form of ATM is asynchronous transfer mode. It is a streamlined packet transfer interface. ATM also is a connection oriented network. ATM uses packets of fixed size for the communication of data. These packets are called as ATM cells. ATM is used for efficient data transfer over high speed data networks. ATM provides real time and non real time services.

ATM operates as a 'network within a network' concept. It has its own internal handshaking and management protocols, quality of service (QoS) facilities, performance and flow-control facilities, and many other components that are usually separated on network technologies such as frame relay.

2. Services Provided by ATM: The services provided are as under :

(i) Synchronous TDM streams such as T-1.

(ii) Services using the constant bit rates.

(iii) Compressed voice and video.

(iv) Traffic with specific quality requirement using the non-real time variable bit rate.

(v) IP based services using available bit rate (ABR) and unspecified bit rate (UBR) services.

The ATM requires an adaption layer for supporting the information transfer protocols that are not based on ATM.

3. Cell Relay: ATM is also known as cell relay. It has advantages of better reliability and fidelity which allows faster packet switching than X.25.

4. Applications:

(i) ATM was developed as a part of work on ISDN.

(ii) However, now, it is used in non-ISDN systems where the data rates are very high.

2.10.1. ATM versus Frame Relay

An ATM network fundamentally resembles a frame relay network. The cell relay is a culmination of all the developments in the circuit switching and packet switching techniques in the past 30 years. ATM is the next step of frame relay. The basic difference between the ATM and frame relay is that in frame relay, variable length packets called frames are transmitted where ATM uses packets of fixed lengths, called cells. Cells are transmitted from source to destination via an ATM switched subnet.

Similar to the frame relay, the cell relay also provides a small number of bits for the purpose of error control. The number of bits depends on the inherent reliability of the transmission system. Due to a fixed packet length, the processing overhead is reduced even further for ATM as compared to frame relay. Therefore, ATM is designed to work in the range of 10s and 100s of Mbps. ATM also allows the definition of multiple virtual channels with the data rates which are dynamically defined at the time the virtual channel is created. By using small, fixed length cells, the ATM becomes so efficient that it can offer a constant data rate even though it is using a packet switching technique. Thus, using ATM, it is possible to allow multiple channels with the data rate of each channel dynamically set on demand.

2.10.2. ATM Protocol Architecture (ATM Reference Model)

Figure 2.11 shows the ATM protocol architecture. ATM is a streamlined protocol. It has minimal error and flow control capabilities. Hence, the number of overhead bits required with each cell is reduced which enables ATM to operate at high data rates. Also, due to the ATM cells of fixed size, the processing required at each node is simplified. This also supports the sue of ATM at high data rates. Figure 2.11 shows the ATM protocol architecture for an interface between user and network. The standards issued for ATM by ITU-T are based on this protocol architecture.

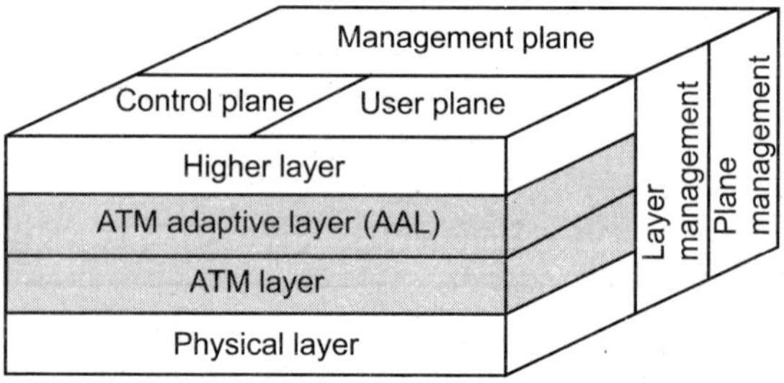

Fig. 2.11 ATM protocol architecture

1. Physical layer: The physical layer of the protocol involves the specifications of a transmission medium and signal encoding scheme. The data rates specified at this layer are between 25.6 Mbps and 622.08 Mbps, but, data rates higher and lower than these are possible.

2. Layers Related to ATM Functions: The two shaded layers in figure 2.11 correspond to the ATM functions. The two layers are as under:

(i) ATM layer

(ii) ATM Adaption Layer (AAL)

(i) ATM layer: This layer is common to all the services that provide the packet transfer capabilities. This layer defines the transmission of data in fixed size cells and it also defines the use of logical connections.

(ii) ATM Adaption Layer (AAL): This layer is a service dependent layer, it is used for supporting the information transfer protocol not based on ATM. The AAL maps the higher layer information into the ATM cell and is transported over the ATM network.

3. Various planes in ATM protocol model: The ATM protocol architecture of figure 2.11 consists of following three separate planes :

(i) User plane (ii) Control plane

(iii) Management plane

(i) User Plane: It is used for transferring user information alongwith associated controls such as flow control, error control etc.

(ii) Control Plane: It is supposed to perform the call control and connection control functions.

(iii) Management Plane: It includes the plane management. The management plane performs management functions related to a system. They include,

(a) Provision of co-ordination between all planes

(b) Layer management

(c) Management functions relating to resources and parameters residing in its protocol entities.

DO YOU KNOW?

Asynchronous transfer mode is a high-speed switching network architecture created in the late 1980s/early 1990s to apply circuit switching concepts to data networks.

2.10.3. ATM Virtual Circuits*

1. Definition and Basic Concept: ATM networks are connection oriented. Hence, before sending data, first it is necessary to send a packet to set up a connection. The set up packet passes through the subnet. All the routers on its path make entry in their table about this new connection. Connections are called as virtual circuits. Most ATM networks also support the permanent virtual circuits. They are permanent conenctions between two distant hosts. They are similar to leased telephone lines. A virtual circuit is shown in figure 2.12.

2. ATM Cells: Once a connection is established, both the sides can start transmitting the data. In ATM, the information is transmitted in the form of small packets of fixed size. They are called *ATM cells*.

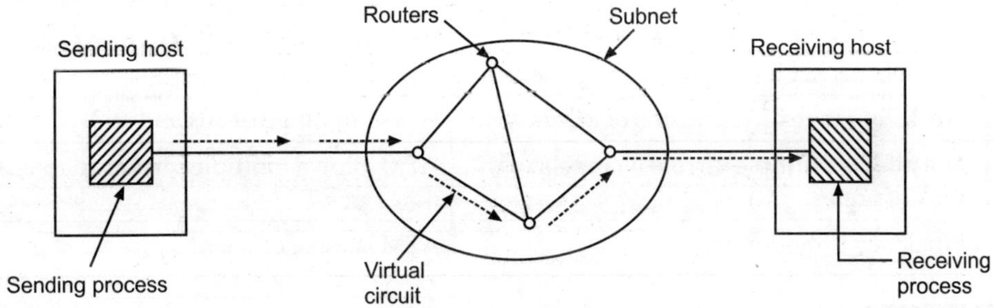

Fig. 2.12. A virtual circuit

3. Format of ATM Cells: Figure 2.13 shows the format of an ATM cells. Each ATM cell is 53 byte long with 5 bytes for header and 48 bytes reserved for data. Cell routing is done in hardware at a high speed. All the cells follow the same route to destination. Hence, their order (sequence) does not change in transit. Cell delivery is no guaranteed. Common speeds of ATM networks are 155 Mbps and 622 Mbps.

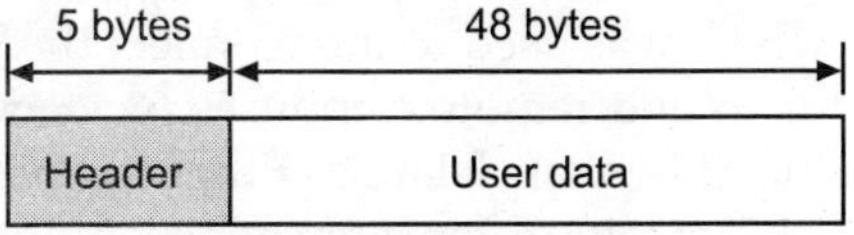

Fig. 2.13. An ATM cell

* If viewed from the perspective of other LAN technologies such as Ethernet/802.3, ATM can provide many services that are somewhat equivalent to Ethernet/802.3. It can also be used in a LAN for desktop connectivity instead of Ethernet/802.3.

2.10.4. Advantages of Virtual Path*

Some of the advantages of virtual path are as under :

(i) Simplified network structure.

(ii) Improved network performance and reliability.

(iii) Reduction in processing and short connection set-up time.

(iv) Enhancement in network services.

2.10.5. ATM Service Categories

The classification of ATM service categories is given in figure 2.14.

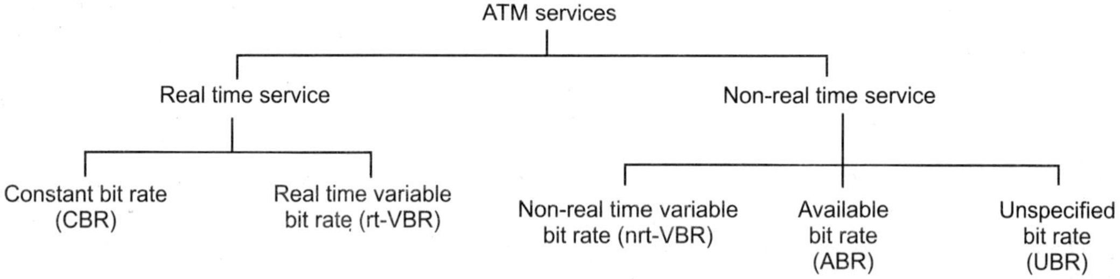

Fig. 2.14. ATM services

2.10.6. Performance Comparison of Frame Relay and ATM

S. No.	Frame relay	ATM
1.	Frame relay is next step of packet switching technique.	ATM is next step of frame relay.
2.	Variable length packets are called frames.	ATM uses fixed length packets called cells.
3.	Processing overhead is more due to variable packet length.	Processing overhead is reduced due to fixed packet length.
4.	Works in range of 1s and 10s of Mbps.	Works in range of 10s and 100s of Mbps.
5.	Not allows multiple virtual channels with the data rates.	ATM allows multiple virtual channels with the data rates.
6.	Efficiency is moderate.	ATM is most efficient.

2.11. ETHERNET

1. Definition and Basic Concepts: Ethernet's roots began in the late 1960s at the University of Hawaii with a networking project called ALOHA, which was used to interconnect the University's main and remote campuses located throughout the Hawaiian Islands. Ethernet evolved into a LAN protocol for interconnecting office devices and is now the dominant LAN technology, claiming more than 85% of all installed network connections. Both Internet and ATM were

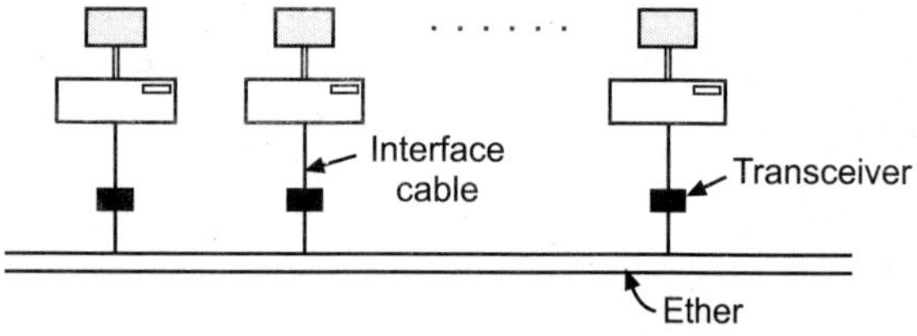

Fig. 2.15. Architecture of original Ethernet

* In terms of functionality, ATM can be like Ethernet/802.3 or, more specifically, can be used where Ethernet/802.3 might be used.

designed for wide area networking. But, in many applications, a large number of computers are to be connected to each other. For this, the local area network (LAN) was introduced. The most popular LAN is called Ethernet.The IEEE 802.3 standard is popularly called as Ethernet. It is a bus based broadcast network with decentralized control. It can operate at 10 Mbps or 100 Mbps or above. Computers on an Ethernet can transmit whenever they want to do so. If two or more machines transmit simultaneously, then their packets collide. Then the transmitting computers just wait for an arbitrary time and retransmit their signal. There are various technologies available in the LAN market but the most popular one of them is Ethernet. In this article, let us discuss three generations of Ethernet :

(i) Traditional Ethernet (10 Mbps)
(ii) Fast Ethernet (100 Mbps)
(iii) Gigabit Ethernet (1000 Mbps)

Traditional Ethernet was created in 1976 and has a data rate of Mbps. The fast Ethernet is its next version and has a data rate of 100 Mbps. The Gigabit Ethernet operates at the data rate of 1000 Mbps.

2. Why is it called Ethernet?: This system is called as Ethernet after the luminiferous ether through which the electromagnetic radiation was once thought to propagate.

3. Transmission Medium: The transmission medium is thick co-oxial cable (called ether) upto 2.5 km long. Repeaters are placed after every 500 meters. Upto 256 machines can be attached to the multidrop cable. The architecture of the original Ethernet is shown in figure 2.15. The original Ethernet was standardized as IEEE 802.3 standard. The committee also stadardized a tokan bus (802.4) and token ring (802.5).

DO YOU KNOW?
ATM is also known as cell relay to distinguish it from frame relay.

4. Computer Connected to Internet via LAN: When a computer is connected to Internet via LAN, it requires all the five layers of the Internet model. The three upper layers (networks, transport and application) are common to all the LANs. The data link layer is divided into two sublayers namely the logical link control (LLC) and the medium access control sublayer (MAC). The LLC sublayer was originally designed to be the same so that all the LANs become interoperable. This means that the LANs differ only in their MAC sublayer and their physical layer. As far as the different types of Ethernets are concerned, the MAC sublayer is slightly different but the physical sublayer is quite different.

2.11.1. Traditional Ethernet

1. Definition: The traditional Ethernet is the oldest version of Ethernet created in 1976 which is designed to operate at 10 Mbps. The access to the network by a device is through the CDMA/CD and the media are shared between all the stations.

2. Finally, Why Ethernet has been so Successful?: First, an Ethernet is extremely easy to administer and maintain. There are no switches, which can fall, no routing or both tables that have to be kept up-to-date. We can add new host easily to this network, second, it is inexpensive, cable is cheap, only network adapter is little costly.

2.11.2. Bridged Ethernet

We can divide a LAN using bridges. Bridges have two effects on an Ethernet LAN. They are as under :

(i) They increase the bandwidth
(ii) They separate the collision domains.

2.11.3. Switched Ethernet

Improving the realistic constraints of segmentation led to the development of Ethernet switches and the concept of switched Ethernet. The concept of bridged LAN can be extended to the switched

LAN. An N port switch is used to connect N number of stations on the LAN. The bandwidth is shared only between the stations and the switch. The collision domain is divided into N domains. Due to the use of a layer 2 switch, faster handling of packets is also possible.

2.11.4. Full Duplex Ethernet

The introduction of private Ethernet switches effectively eliminates contention since each node connected to the switch has its own dedicated segment. As a direct consequence, it is possible for nodes to transmit and receive data simultaneously. That is, private Ethernet switches promote full-duplex Ethernet. The 10 base 5 and 10 Base 2 Ethernets have a serious limitation. The communication on them is always half duplex. This means that a station can either transmit or receive at a time. It cannot send and receive simultaneously. Hence, the full duplex switched Ethernet evolved from the switched Ethernet. Figure 2.16 shows the full duplex switched Ethernet. Due to the full duplex mode, the capacity of each domain increases from 10 to 20 Mbps. We have to use two links between each station and the switch, one to send and other to receive. The full duplex switched Ethernet eliminates the need of CSMA/CD because there is no more need of carrier sensing.

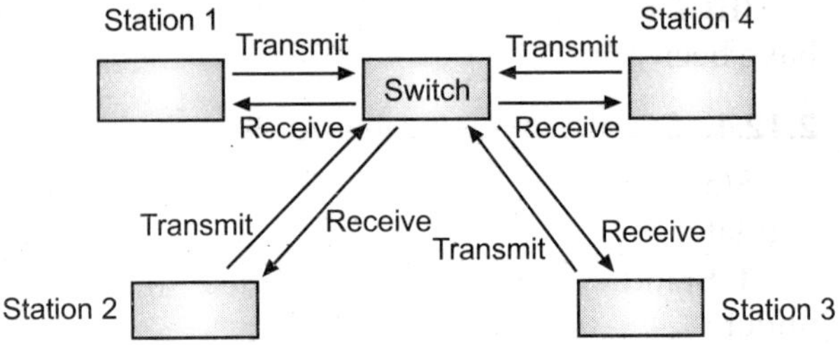

Fig. 2.16. Full duplex switched Ethernet

2.11.5. Fast Ethernet

Fast Ethernet is the protocol designed to work upto 100 Mbps. The traditional Ethernet can operate only upto 10 Mbps. Hence, for higher data rates, fast Ethernet has been developed.

Auto-negotiation: This is the new feature of the fast Ethernet. The auto-negotiation will make the negotiation on the mode or data rate of operation between devices possible.

2.11.6. Gigabit Ethernet

The gigabit Ethernet protocol has been designed in order to support the data rates upto 1000 Mbps. The MAC layer was supposed to remain unchanged throughout the evolution of the Ethernet, but, it does not remain so when the rate of 1 Gbps is to be supported. The Gigabit Ethernet is capable of operating in either half duplex or full duplex modes. If it operates in the half duplex mode, then the access method used is CSMA/CD. But, if the full duplex mode is used, then, CAMA/CD is not required. Almost all the implementations in Gigabit Ethernet use the full duplex modes.

2.11.7. Wireless LANs

The wireless LANs are becoming more and more popular because they can satisfy the requirements like mobility, relocation of user, ad-hoc networking and coverage of locations which are difficult to wire. Earlier, the wireless LANs were costly, could support only low data rates, a license was required. Hence, there were limitations on the practical utility of wireless LANs. But, all these problems are being addressed now which is increasing the popularity of wireless LANs day by day.

2.12. NETWORK STANDARDIZATION

Network standards are essential in order to coordinate the number of vendors and suppliers and their ideas. Standards avoid the possible chaos, they allow different computers to communicate, increases the sales of products adhering to standards (e.g. ISI standards). Data communication standards are classified into following two categories :

(i) De facto standards (ii) De jure standards

(i) De Facto: The meaning of this word is *by fact* or *by convention*. These are the standards that have not been approved by an organized body, but have been adopted as standards through widespread use. These standards are often established originally by the manufactures.

(ii) De jure: The meaning of this word is *by law* or *by regulation*. These are the standards that have been approved by an officially recognized body.

2.12.1. Standard Organizations

Standards are developed by the standards creation committees, forums and government regulatory agencies.

1. Standard Creation Committees: Some of the standard creation committees may be listed as under :

(i) International organization for standardization (ISO)
(ii) International Telecommunication Union – Telecommunication standards
(iii) American National Standards Institute (ANSI)
(iv) Institute of Electrical and Electronics Engineers (IEEE)
(v) Electronic Industries Association (EIA)

2. Regulatory Agencies: Federal communication commission (FCC) is the government regulator body in U.S. for all communication technology.

GLOSSARY

- **CCITT** Please go through *Consultative Committee for International Telephony and Telegraphy.*
- **Cell** A unit of data that is transmitted across a network. It is similar to a data *frame*. When used in the context of *ATM*, a cell contains exactly 53-bytes, with 48 bytes for user data and 5 bytes for overhead.
- **Institute of Electrical and Electronics Engineers (IEEE)** A professional society of engineers, scientists, and students. One of its several activities is to act as a coordinating body for computing and communication standards.
- **International Organization for Standardization (ISO)** An international organization which develops and promotes networking standards worldwide.
- **International Telecommunication Union (ITU)** A global standards organization. In fact, ITU is the former *CCITT.*
- **Internet** When used as a noun and spelled with a lower case *i*, internet is an abbreviation for *internet work*, that refers to a collection of interconnected networks which functions as a single network. When used as a proper noun and spelled with an uppercase I, Internet refers to the world's largest internet work, which consists of hundreds of thousands of interconnected networks worldwide and based on a particular set of network standards *(TCP/IP).*
- **Inverse Multiplexing** The reverse of *multiplexing*. Instead of partitioning a single communication medium into several channels, an inverse *multiplexer* combines several *smaller* channels (i.e., low-speed circuits) into a single high-speed circuit. This method is also sometimes generically called *line aggregation.*
- **Network Standards** A formal set of rules, developed by and agreed upon by different organizations, defining hardware interfaces, communication protocols and network architectures. Several standards exist, including *de jure, de facto, proprietary*, and *consortia.*
- **Network Termination Unit (NTU)** A device which terminates *E-1* circuits. An NTU provides broadly similar *CSU/DSU* functionality.
- **Wireless Communication** A type of communications in which signals travel through space inplace of through a physical cable. There are two general types of wireless communication; *radio transmission* and *infrared transmission.*
- **X.25** An early networking protocol which defines the interface between a public packet-switching data network and the device used to access this network.

SUMMARY

- A large number of networks are currently operating around the world. These networks may be classified into different types as depicted in figure 2.17.

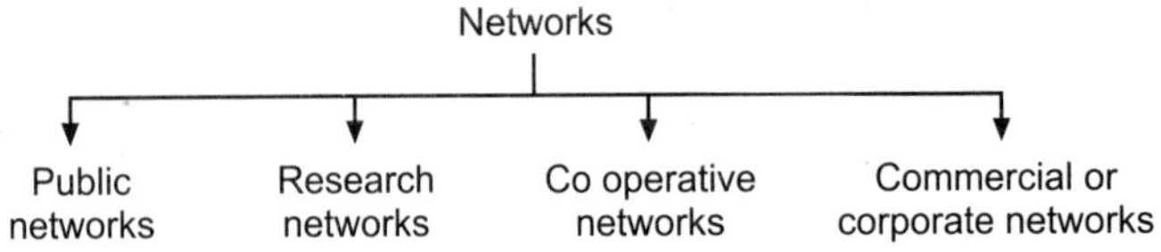

Fig. 2.17. Networks classification

- Few important salient features of a network may be listed as under :
  (i) Administration (ii) Facilities offered.
  (iii) Technical design. (iv) User communities.
- This is the most popular netware in the PC world. It was specially designed to be used by a network of PCs. In such a network, each user has a desktop PC which functions as a client and a number of powerful PCs operate the servers.
- The physical layer can be chosen from among various standards such as Ethernet, Token ring and ARC net and so does the data link layer. The protocol run by the network layer is IPX (Internet Packet exchange) which is an unreliable connectionless protocol. Above IPX, comes the transport layer which uses a connection oriented transport protocol called NCP (Network Core Protocol).
- The services that are available on the Internet may be classified into following two categories :
  (i) Communication services; (ii) Information retrieval services
- The Internet protocol is like any other communication protocol. It is a set of rules which will govern every possible communication over the Internet. Since the development of the ARPANET, TCP/IP together has emerged as the controlling body, it is being used in computers of not only in the U.S., but all over the world for all types and sizes of computers. It has become the language of the Internet. Actually, TCP/IP are two protocols : Transmission control protocol and Internet protocol.
- With so many computers and users using the Internet, it would be impossible to differentiate them without a proper addressing system. This addressing system assigns names and numbers to identify the computers on the Internet.
- Each computer on the Internet is identified by its unique IP address. An IP address is made of four numbers and each one must be less than 256. Thus, each set of numbers can have values from 1 to 225.
- A smaller network making up the Internet and having many computers within it is known as a *domain*. The domain may represent a type of organization or a geographical location. For example, all the computers on the Internet which belong to the educational institutions form a part of the *educational* or *edu* domain.
- The Internet is a decentralized network and no one runs or owns it. There is an organization or agency that controls the activity on the Internet. However, there are a few organizations which co-ordinate and guide the technical parts of Internet.
- The Internet communication protocols are developed and maintained by the IETF. The protocols are the methods by which computers on the Internet are connected. The long term research problem which may be critical to the Internet are looked into by the IRTF. Any major changes coming from the IETF are ratified by the IAB.
- Traditionally, the Internet has the following applications :
  1. E-mail, 2. News, 3. Remote login, 4. File transfer, 5. World wide web (www).
- Uptill early 1990s, the Internet was populated by information related to academic, government and research. But www, a new application brought millions of new, non-academic users to the net.
- The telephone companies started offering networking services to the organizations which were interested in subscribing the services. Such systems are called as *public networks*. They are similar to and sometimes a part of the public telephone system. DQDB is one of such services which we have already discused.

- The three fields in the packet are destination, source and user data. The destination address and source field are of 8 bytes whereas the user data field has a variable length (upto 9188 bytes). In the source and destination, addresses contain a 4-bit code followed by a telephone number upto 15 decimal digits.
- These three layers correspond to the three lower most layers of the ISO-OSI reference model. The physical layer takes care of the interface between a computer terminal and the link which attaches it to the packet switching node.
- At the network level, X.25 defines a protocol for an access to packet data subnetwork. This protocol defines the format, content and procedures for exchange of control and data transfer packets. The packet layer provides an external virtual circuit service.
- Frame relay is a connection oriented service. It can be imagined to be equivalent to a virtual leased line. On a virtual line, data bursts can be sent at full speed. Frame relay provides a service to determine the start and end of each frame. It also detects the transmission errors.
- Some of the advantages of frame relay may be listed as under :
  (i) Streamlined communication process.
  (ii) The number of functions of a protocol at the user-network interface is reduced
  (iii) Lower delay.
  (iv) Higher through put out
  (v) Frame delay can be used at access speeds upto 2 Mbps.

SHORT QUESTIONS WITH ANSWERS

Q.1. Write the salient features of a network.

Ans. Few important salient features of a network may be listed as under :

(i) Administration (ii) Facilities offered.
(iii) Technical design. (iv) User communities.

Q.2. Write the services on the Internet.

Ans. The services that are available on the Internet may be classified into following two categories :

(i) Communication services (ii) Information retrieval services

Q.3. Write the applications of Internet.

Ans. Traditionally, the Internet has the following applications :

1. E-mail
2. News
3. Remote login
4. File transfer
5. World wide web (www).

Q.4. What are example data communication services?

Ans. The telephone companies started offering networking services to the organizations which were interested in subscribing the services. Such systems are called as *public networks*. They are similar to and sometimes a part of the public telephone system.

Q.5. What are the three layers of X.25?

Ans. The three levels may be listed as under :

(i) Physical layer (level 1) (ii) Data link layer (level 2)
(iii) Packet layer (level 3).

Q.6. Write advantages of X.25.

Ans. (i) Frame delivery is more reliable. (ii) Frames are delivered in order.
(iii) Retransmission of frames is possible. (iv) Flow control is provided.
(v) X.25 supports the switched virtual circuits and permanent circuits.

Q.7. Write advantages of frame relay.

Ans. Some of the advantages of frame relay may be listed as under :

(i) Streamlined communication process.
(ii) The number of functions of a protocol at the user-network interface is reduced.
(iii) Lower delay.
(iv) Higher through put out.
(v) Frame delay can be used at access speeds upto 2 Mbps.

Q.8. What is traditional Ethernet?

Ans. The traditional Ethernet is the oldest version of Ethernet created in 1976 which is designed to operate at 10 Mbps. The access to the network by a device is through the CDMA/CD and the media are shared between all the stations.

Q.9. What is meant by Bridged Ethernet ?

Ans. We can divide a LAN using bridges. Bridges have two effects on an Ethernet LAN. They are as under :
(i) They increase the bandwidth (ii) They separate the collision domains.

Q.10. Name the organization that sets standards for cable and connectors.

Ans. Electronic Industry Association (EIA).

REVIEW QUESTIONS

1. Write the classification of networks.
2. State salient features of a network.
3. Write a short note on : Novell Netware.
4. Write a short note on : (a) ARPANET, (b) Ethernet, (c) X.25, (d) Internet address
5. Write about the net structure of the Internet.
6. Explain various services provided by the Internet.
7. What is meant by the Internet protocols.
8. What is meant by domain name addressing ?
9. Explain about various applications of Internet.
10. With the help of structure, explain SMDS.
11. Explain about the packet format of SMDS.
12. What is frame relay? Is it connection oriented or connectionless ?
13. State advantages of frame relay.
14. Compare ATM and frame relay.
15. Write a note on ATM.
16. Explain the ATM reference model.
17. Explain ATM virtual circuits.
18. What are the advantages of virtual path.
19. What do you understand by network standardization ?

□□□

Chapter 3

The Physical Layer

Syllabus

The Physical Layer: Design Issues, review of data communication concepts (configuration, topology, transmission mode, media-guided and unguided, types of switching etc).

Chapter Outline

3.1. INTRODUCTION

In this chapter, let us discuss about the lowest layer in the OSI model which is called the *physical layer*. The physical layer is the lowest layer (layer 1) of the OSI model. Before sending data on the network, the physical layer on the local node must process the raw data stream, translating frames received from the data link layer (layer 2) into electrical, optical, or electromagnetic signals representing 0 and 1 values, or bits. The physical layer defines the mechanical, electrical and timing interfaces to the network. From the most general perspective, the physical layer can be described by one word; wire. This chapter also deals with the fundamental limits put by the nature on data transmission over a channel. After that we shall discuss various types of transmission media such as guided, wireless and satellite. At the end of the chapter, we shall briefly discuss about cable TV and use of cable TV for internet.

3.2. PHYSICAL LAYER DESIGN ISSUES

1. Basic Aspect: The major task of physical layer is to provide services for the data link layer. Following are the services provided or duties of the physical layer or the design issues of the physical layer.

2. Transformation from Bit to Signal: The data link layer consists of 0's and 1's in the bit form. The bit stream cannot travel as it is on the transmission medium. Hence, the physical layer converts the bit stream into a signal which is suitable for the transmission medium.

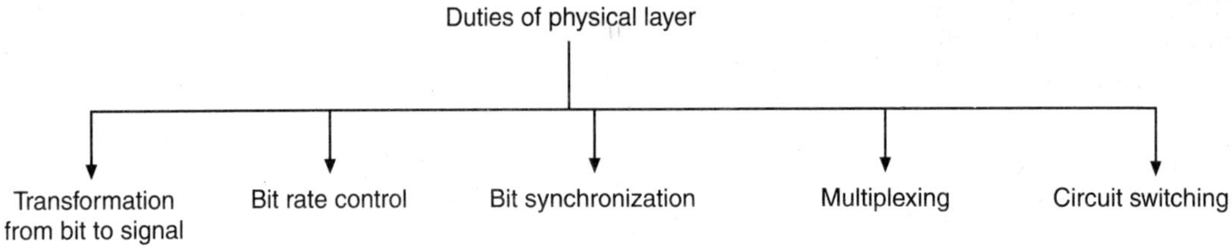

Fig. 3.1. Duties of physical layer

3. Bit Rate Control: The transmission medium decides the highest value of bit rate and physical layer acts as a bit rate controller. The design of the physical layer hardware and software will determine the data rate.

4. Bit Synchronization: The timing related to the data bit transfer is very important in computer communication. The physical layer governs the synchronization of the bit by providing a clock which controls the transmitter as well as receiver.

5. Multiplexing: Physical layer can use different techniques of multiplexing, in order to improve the channel efficiency.

6. Switching: There are three switching methods, namely, circuit switching, message switching and packet switching. Out of which, circuit switching is the function of physical layer.

3.2.1. Transmission Media and Physical Layer

In the OSI model (discussed in chapter 1), the transmission media are located below the physical layer. It is directly controlled by the physical layer. Hence, we can say that transmission media belong to the lowest layer (layer 0) of the OSI model, as shown in figure 3.2. Basically, a transmission medium is the cable or air over which data can travel from sender to receiver.

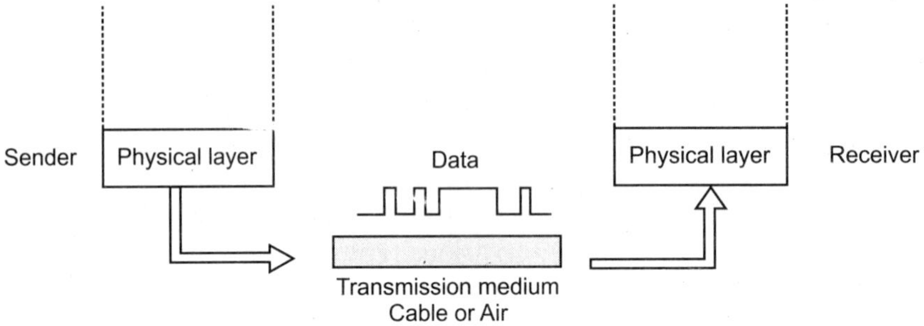

Fig. 3.2. Relation between transmission medium and physical layer

3.3. REVIEW OF DATA COMMUNICATION CONCEPTS

As discussed earlier, analog signals are used to represent analog data and digital signals are used to represent digital data. Usually, analog data are a function of time and occupy a limited frequency spectrum. Such data can be represented by an electromagnetic signal occupying the same spectrum. On the other hand, digital data can be represented by digital signals, with a different voltage level for each of the two binary digits. As shown in figure 3.3, these are not the only possibilities. Digital data can also be represented by analog signals by use of a modem (modulator/demodulator). The modem converts series of binary (two-valued) voltage pulses into an analog signal by encoding the digital data onto a carrier frequency. The resulting signal occupies a certain spectrum of frequency centered about the carrier and may be propagated across a medium suitable for that carrier. The most common modems represent digital data in the voice spectrum and hence allow those data to be propagated over ordinary voice grade telephone lines. At the other end of the line, another modem demodulates the signal to recover the original data.

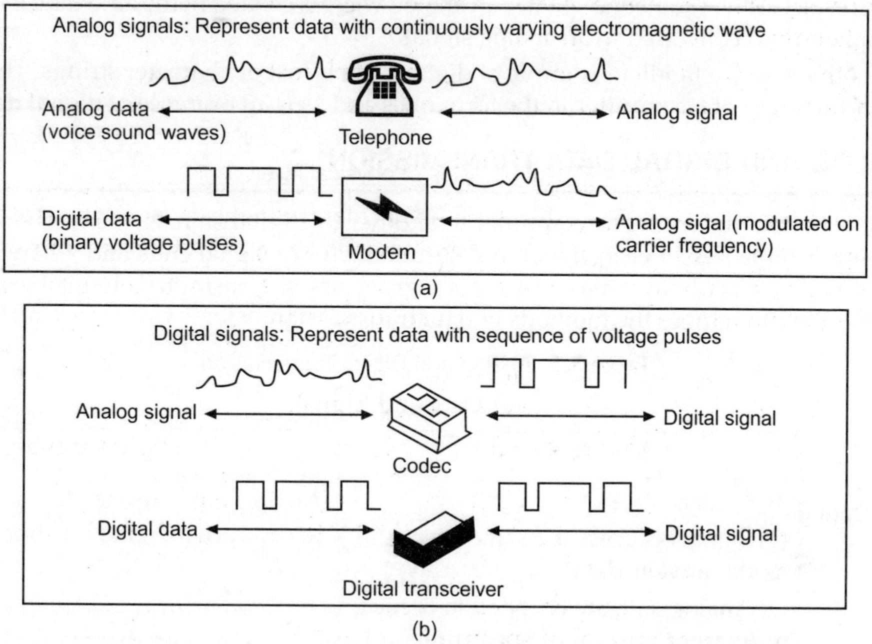

Fig. 3.3. Analog and digital signaling of analog and digital data

In an operation very similar to that performed by a modem, analog data can be represented by digital signals. The device that performs this function for voice data is a codec (coder-encoder). In essence, the codec takes an analog signal that directly represents the voice data and approximates that signal by a bit stream. At the receiving end, the bit stream is used to reconstruct the analog data.

Figure 3.3 illustrates that data may be encoded into signals in a variety of ways.

3.4. ANALOG AND DIGITAL DATA

In most simple manner, the terms analog and digital correspond to continuous and discrete respectively. In data communications, often these two terms are used for the following entities :

(i) Data
(ii) Signaling
(iii) Transmission

(i) Data : Data may be defined as entities that convey meaning or information.

(ii) Signaling: Signals are electric or electro-magnetic representations of data. Signaling is the physical propagation of the signal along a suitable medium.

(iii) Transmission: Transmission is the communication of data by the propagation and processing of signals.

3.4.1. Classification of Data

Data may be classified as analog and digital data. Analog data take on continuous values in some interval. As an example, voice and video are continuously varying patterns of intensity. Moreover, most data collected by sensors, such as temperature and pressure, are continuous valued. On the other hand, digital data take on discrete values. Examples of digital data are text and integers.

The most familiar example of analog data is audio, which is in the form of sound waves and can be perceived directly by human beings. In other words, we can say that an example of analog data is the human voice. When someone speaks, an analog wave is created in the air. This can be captured by a microphone and converted to an analog signal.

On the other hand, a familiar example of digital data is text or character strings. Thus, the data stored in the memory of a computer in the form of 0s and 1s is an example of digital data.

3.5. ANALOG AND DIGITAL DATA TRANSMISSION*

When data is transferred from one computer to another by digital signals, the receiving computer has to distinguish the size of each signal to determine when a signal ends and when the next one begins. As a matter of fact, both analog and digital signals may be transmitted on suitable transmission media. Table 3.1 summarizes the methods of data transmission.

Table 3.1. Analog and Digital Transmission

(a) Data and Signals

	Analog signal	Digital signal
Analog Data	Two alternatives (1) Signal occupies the same spectrum as the analog data; (2) Analog data are encoded to occupy a different portion of spectrum.	Analog data are encoded using a codec to produce a digital bit stream.
Digital Data	Digital data are encoded using a modem to produce analog signal.	Two alternatives: (1) Signal consists of a two voltage levels to represent the two binary values; (2) Digital data are encoded to produce a digital signal with desired properties.

(b) Treatment of Signals

	Analog Transmission	Digital Transmission
Analog Signal	It is propagated through amplifiers; same treatment whether signal is used to represent analog data or digital data	Assumes that the analog signal represents digital data. Signal is propagated through repeaters; at each repeater digital data are recovered from inbound signal and used to generate a new analog outbound signal.
Digital Signal	Not used	Digital signal represents a stream of 1s and 0s, which may represent digital data or may be an encoding of analog data. Signal is propagated through repeaters; at each repeater, stream of 1s and 0s is recovered from inbound signal and used to generate a new digital outbound signal.

1. Analog Transmission: Analog transmission is a means of transmitting analog signals without regard to their content. The signals may represent analog data (i.e., voice) or digital data (i.e., binary

* What is the difference between Digital and analog data communication system?

data that pass through a modem). In either case, the analog signal will become weaker (attenuate) after a certain distance. To achieve longer distances, the analog transmission system includes amplifiers that boost the energy in the signal. Further, the amplifier also boosts the noise components. With amplifiers cascaded to achieve long distances, the signal becomes more and more distorted. For analog data, such as voice, quite a bit of distortion can be tolerated and the data remain intelligible. However, for digital data, cascaded amplifiers will introduce errors.

2. Digital Transmission: In contrast to analog transmission, the digital transmission is concerned with the content of the signal. A digital signal can be transmitted only a limited distance before attenuation, noise and other impairments distorts the integrity of the data. To achieve greater distances, repeaters are used. A repeater receives the digital signal, recovers the pattern of 1s and 0s and retransmits a new signal. Hence, the attenuation is overcome. We can use same technique with an analog signal if it is assumed that the signal carries digital data. At appropriately spaced points, the transmission system has repeaters rather than amplifiers. The repeater recovers the digital data from the analog signal and generates a new, clean analog signal. In this way, the noise does not become cumulative.

3.6. WHY DIGITAL TRANSMISSION?

In this article, we shall compare the performance of analog transmission with that of digital transmission systems based on the following factors :

1. Effect of noise
2. Distance to be covered
3. Services provided

3.6.1. Comparison of Analog and Digital Transmission: Effect of Noise

In analog transmission, the objective is to transmit a waveform as shown in figure 3.4(a). But in doing so, the shape of transmitted signal gets distorted as shown in figure 3.4(b) due to non-adequacies of the communication channel.

Since all the information is contained in the shape of the waveform, it is necessary to preserve the shape of waveform. But the noise distorts the waveform as shown in figure 3.4(b). Therefore, the received information will also be distorted.

Now, let us consider the transmission of digital signal as shown in figure 3.4(c). To transmit such a binary signal, the objective is to transmit either 0 or 1.

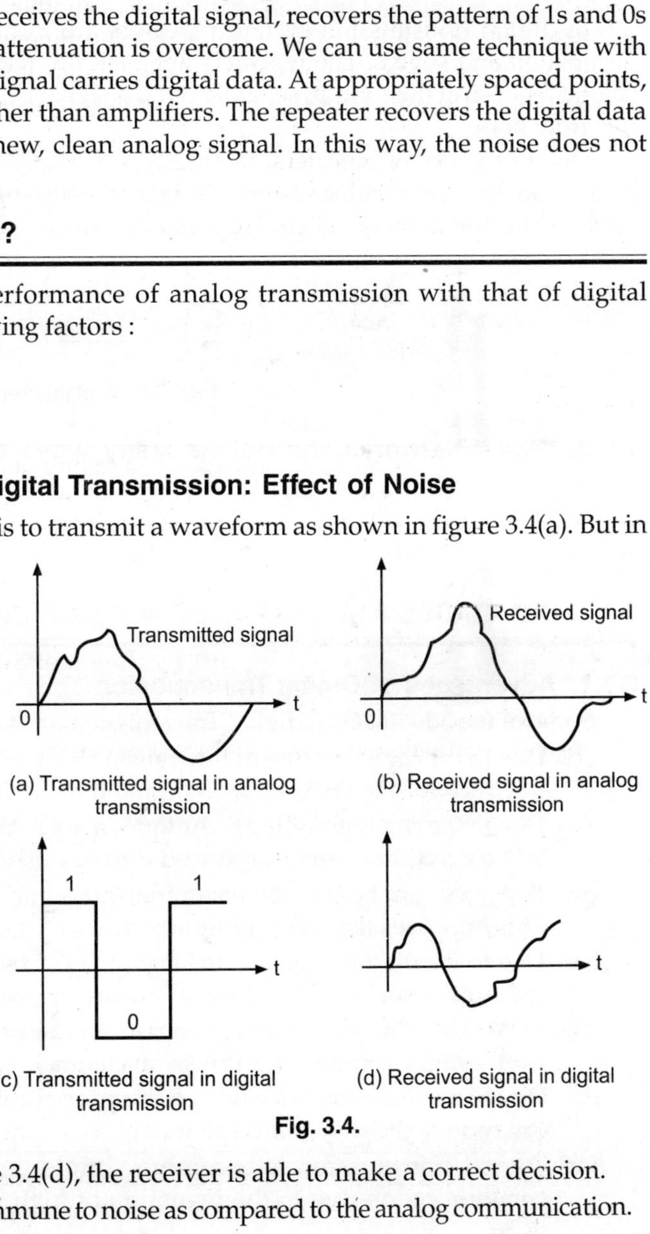

(a) Transmitted signal in analog transmission
(b) Received signal in analog transmission
(c) Transmitted signal in digital transmission
(d) Received signal in digital transmission

Fig. 3.4.

The task of the receiver is to identify whether the received symbol is 0 or 1. Note that even though the received waveform is distorted as shown in figure 3.4(d), the receiver is able to make a correct decision.

Therefore, digital transmission is more immune to noise as compared to the analog communication.

3.6.2. Digital Communication is Suitable for Long Distances

The digital transmission becoems cost effective over analog transmission if the distance involved is large. Let us consider a long distance transmission link shown in figure 3.5. As the signal from the source travels a distance from source, its amplitude reduces due to attenuation. In addition to that, the signal becomes increasingly distorted. The noise from various sources gets added to the signal.

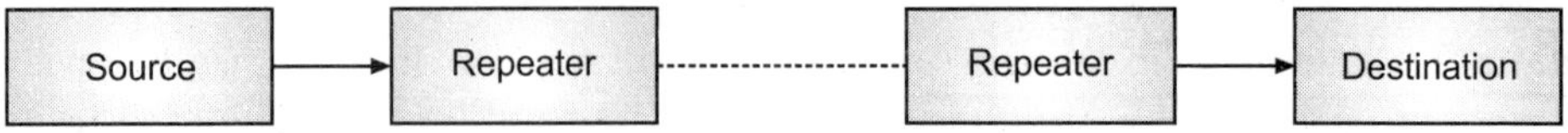

Fig. 3.5. A long distance transmission link

In digital transmission system repeaters are induced as shown in figure 3.5 between the destination and source. The repeaters are basically regenerators of the signals. They receive the (signal + noise) at their input, separate out the signal from noise and regenerate the signal which is free from noise.

Due to the use of repeaters, the noise performance of digital transmission systems is much better than that of the analog systems. Note that repeaters cannot be used for the analog transmission system. The operation of a digital repeater becomes clear from figure 3.6.

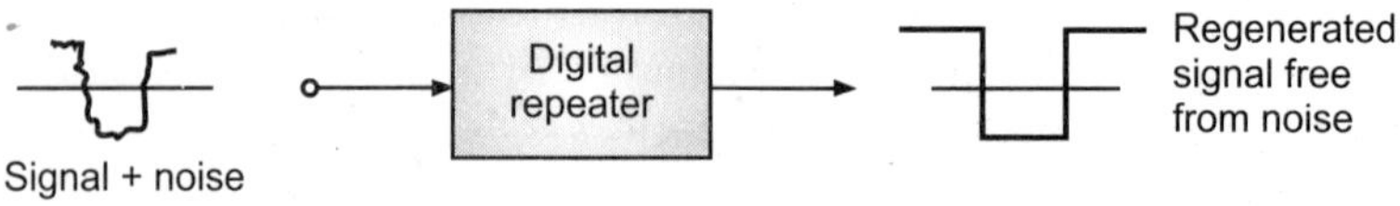

Fig. 3.6. A digital repeater

3.6.3. Digital Networks can Handle Many Types of Services

The networks based on digital transmission are capable of handling any type of information which can be represented in the digital form. Hence, such networks can handle many types of services.

3.7. ADVANTAGES AND DISADVANTAGES OF DIGITAL TRANSMISSION

3.7.1. Advantages of Digital Transmission

Some of the advatages of digital Transmission are as follows :

(i) Due to the digital nature of the transmitted signal, the interference of additive noise does not introduce many errors. So digital transmission has a better noise immunity.

(ii) Due to the channel coding techniques used in digital transmission , it is possible to detect and correct the errors introduced during the data transmission.

(iii) Repeaters can be used between transmitter and receiver to regenerate the digital signal. This improves the noise immunity further.

(iv) Due to the digital nature of the signal, it is possible to use the advanced data processing techniques such as digital signal processing, image processing, data compression etc.

(v) TDM (Time Division Multiplexing) technique can be used to transmit many voice channels over a single common transmission channel.

(vi) Digital transmission is useful in military applications where only a few permitted receivers can receive the transmitted channel.

(vii) Digital transmission is becoming simpler and cheaper as compared to the analog communication due to the invention of high speed computers and integrated circuits (ICs).

3.7.2. Disadvantages of Digital Transmission

Some of the important disadvantages of digital transmission are as follows:

(i) The bit rates of digital systems are high. Therefore, they require a larger channel bandwidth as compared to analog systems.

(ii) Digital modulation needs synchronization in case of synchronous modulation.

3.8. PERFORMANCE COMPARISON OF ANALOG AND DIGITAL TRANSMISSION*

S.No.	Analog transmission	Digital transmission
1.	Transmitted modulated signal is analog in nature.	Transmitted signal is digital, *i.e.*, train of digital pulses
2.	Amplitude, frequency or phase variations in the transmitted signal represent the information or message.	Amplitude, width or position of the transmitted pulses is constant. The message is transmitted in the form of code words.
3.	Noise immunity is poor for AM, but improved for FM and PM.	Noise immunity is excellent
4.	It is not possible to separate out noise and signal. Therefore, repeaters cannot be used.	It is possible to separate signal from noise. So repeaters can be used.
5.	Coding is not possible	Coding techniques can be used to detect and correct the errors.
6.	Bandwidth required is lower than that for the digital modulation methods.	Due to higher bit rates, higher channel bandwidth is needed.
7.	FDM is used for multiplexing	TDM is used for multiplexing
8.	Not suitable for transmission of secret information in military applications.	Due to coding techniques, it is suitable for military applications.
9.	Analog modulation systems are AM, FM, PM, PAM, PWM, etc.	Digital modulation systems are PCM, DM, ADM, DPCM, etc,.

3.9. COMPOSITE SIGNAL AND TRANSMISSION MEDIUM

The data is generally in the form of pulses and pulse is a composite signal which contains many frequencies. Here, note that the peculiar shape of a pulse is due to the sum of specific frequencies at specific amplitudes and phases. If there is any change in the amplitudes or phases of these frequency components, then the shape of the pulse will not remain the same.

3.9.1. Medium (Channel)

The signal always travels over some medium or channel. The medium can be a coaxial cable or optical fiber etc. A medium does not pass all frequencies equally. It may pass some frequencies and weaken or block the other frequencies. Hence, when a composite signal is passed over such a transmission medium, at the receiving end, we get a wave, having a different-shape as shown in figure 3.7.

To avoid the signal distortion, the medium must pass all the frequencies present at the input without any change. However, no medium is perfect and hence some signal distortion always takes place. Every medium has its own characteristic properties which in turn affect the shape of the signal in transit.

* Explain the difference between Analog and digital communication.

Apart from the inherent properties of the medium, there are some external factors which affect the shape and strength of the signal.

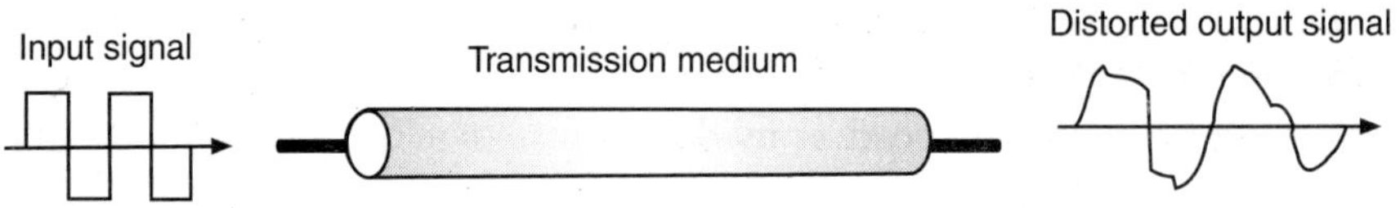

Fig. 3.7. Signal distortion on a transmission medium.

3.9.2. Channel Bandwidth*

The range of frequencies that contain the information is called as the bandwidth. However, the term channel bandwidth is used to describe the range of frequencies required to transmit the desired information. As an example, the amplitude modulation (AM) system requires a channel bandwidth of 10 kHz to transmit a signal of 5 kHz bandwidth. But the single sideband system (SSB) only 5 kHz channel bandwidth to transmit the same signal. Actually, the efforts must be made to reduce the required channel bandwidth so that we can accommodate in more number of channels in the same available EM spectrum.

3.9.3. Review of Digital Signals

The data can also be represented by a digital signal. A digital signal is a discrete time signal having finite number of amplitudes. For example, let us consider the digital signal shown in figure 3.8. Here, 0 is represented by zero volts and a 1 by some positive voltage.

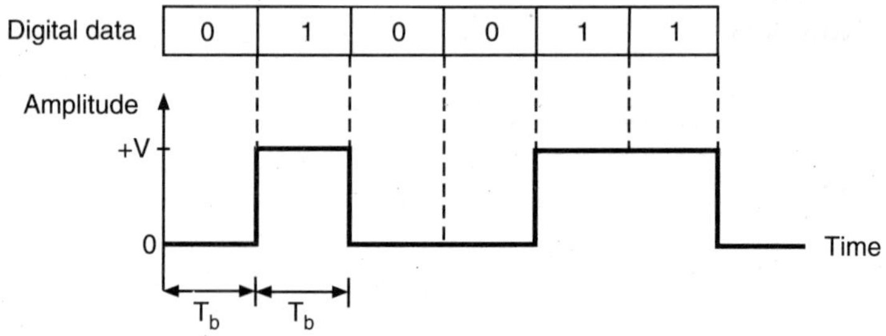

Fig. 3.8 Digital signal.

3.9.4. Bit Interval

The bit interval is the time required to send one single bit. As shown in figure 3.20, the time required to send a 0 or 1 is T_b which is the bit interval.

3.9.5. Bit Rate (Data Rate)

Bit rate is the number of bits transmitted or sent in one second. It is expressed in bits per second (bps). Relation between bit rate and bit interval may be expressed as under :

$$\text{Bit rate} = \frac{1}{\text{Bit interval}}$$

Bit rate or significant rate is defined as the number of bits which can be transmitted in a second. If the bit duration is T_b, then bit rate will be $1/T_b$. Let us consider figure 3.9. Here, we observe that the bit duration is necessarily equal to the pulse duration. In figure 3.9 the first pulse is of two bit duration.

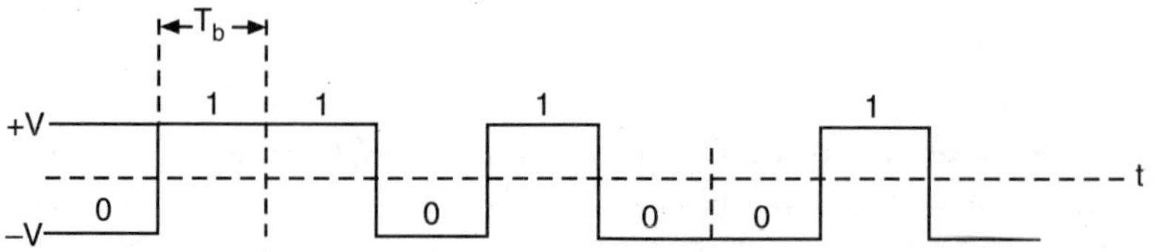

Fig. 3.9. A bit stream.

* Bandwidth represents the range of frequencies that can pass through a medium.

Bit rate is also called as signaling rate and it must be as high as possible. However, with increase in bit rate, the bandwidth of transmission medium must be increased, in order to transmit the signal without any distortion.

3.9.6. Bauds (or Baud Rate)*

Baud is the unit of signaling speed or modulation rate or the rate of symbol transmission. It indicates the rate at which a signal level changes over a given period of time. When binary bits are transmitted as an electrical signal with two levels 0 and 1, the bit rate and the modulation rate *i.e.*, baud rate are same. This has been shown in figure 3.10.

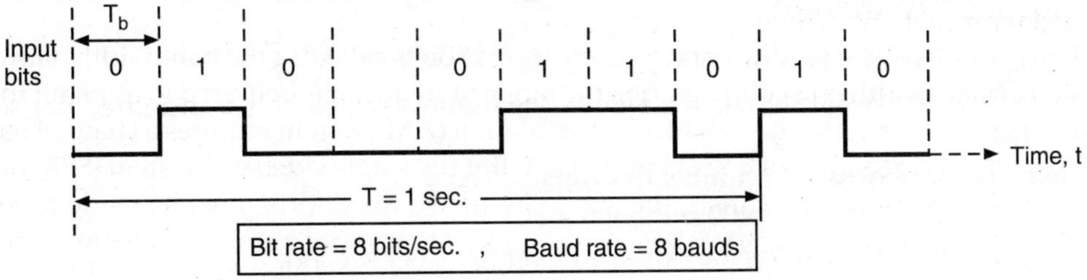

Fig. 3.10. Baud rate for two level modulation

Here, it may be noted that for a two level signal, (binary signal) the bit rate and bauds are equal. Now, let us consider figure 3.11 where four different levels are used to represent the data.

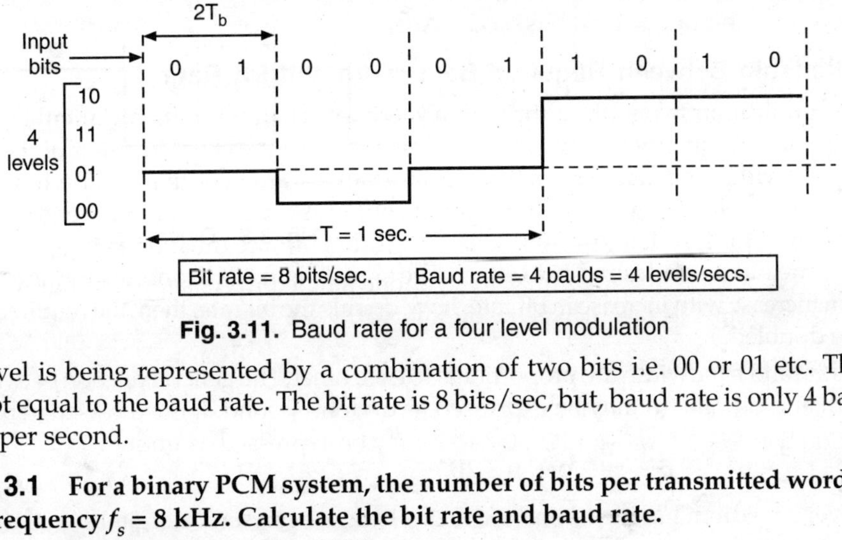

Fig. 3.11. Baud rate for a four level modulation

Each level is being represented by a combination of two bits i.e. 00 or 01 etc. The bit rate is therefore not equal to the baud rate. The bit rate is 8 bits/sec, but, baud rate is only 4 bauds as there are 4-levels per second.

EXAMPLE 3.1 For a binary PCM system, the number of bits per transmitted word is 8 and the sampling frequency f_s = 8 kHz. Calculate the bit rate and baud rate.

Solution: Given that $N = 8$, $f_s = 8$ kHz

We have Bit rate = $N \times f_s = 8 \times 8$ kHz = 64 k bits/sec.

Baud rate = Bit rate = 64 kHz (since transmission is binary) **Ans.**

EXAMPLE 3.2 For the same data as in previous example, find the bit rate and baud rate if a QPSK system is used.

Solution: In a QPSK system, two successive bits are clubbed together to form one message. Hence, one symbol corresponds to 2 bit duration.

* Differentiate between bit rate and baud rate. A modem constellation diagram has data point at co-ordinates : (1, 1) (1, – 1) (– 1, 1) and (– 1, – 1). How many bps can a modem with these parameters achieve at 1200 baud ?

Therefore, baud rate = $\frac{1}{2}$ × bit rate = 32 k bits/sec. **Ans.**

Bit rate does not change.

EXAMPLE 3.3 A system sends a signal that can assume 8 different voltage levels. It sends 400 of these signals per second. What are the baud and bit rates ?

Solution:

(i) As the signal assumes 8 different voltage levels, we required 3 bit digital signal to have 8 different combinations. Hence, the number of bits; per voltage level is 3. Let each voltage level represent one symbol.

Hence, number of bits/symbol = 3.

(ii) The system sends 400 signal/sec. Hence, the number of symbols transmitted per second is also 400.

Therefore, symbol rate = Number of symbols/sec.

Symbol rate = 400 symbols/sec.

(iii) The baud rate is defined as the number of symbols per second.

Hence, Baud rate = Symbol rate.

or Baud rate = 400 symbols /sec. **Ans.**

(iv) We are using 3 bit to represent each symbol.

Therefore, bit rate = 3 × symbol rate = 3 × 400

bit rate = 1200 bits/sec. **Ans.**

3.9.7. Relationship Between Required Bandwidth and Bit Rate

Data communication takes place through a medium (channel) which can either be wired or wireless. Therefore, the quality of transmission depends upon the characteristics of the medium. In this regard, bandwidth and data rate are the important parameters that decide how the data is, transmitted from one place to another. In computer communication, we have to send as many bits as possible per second for fast data transfer. This means that bit rate must be as high as possible. But increase in bit rate has an undesired side effect. The signal bandwidth and the required bandwidth of the medium increase with increase in bit rate. If we double the bit rate, then, the required bandwidth requires to be doubled.

Thus, bit rate and bandwidth are proportional to each other. The general relation between required bandwidth (B) and bit rate (n) may be expressed as under:

$$B \geq \frac{n}{2} \quad \text{or} \quad n \leq 2B.$$

Hence, over a medium having a bandwidth of 4 kHz, we can send a digital signal with a bit rate upto 8 kbps. In practice, the maximum bit rate can be more than 30 kbps using the traditional MODEMS.

3.10. DATA RATE LIMITS (MAXIMUM DATA RATE OF A CHANNEL)

In data communication, a large data is required to be transferred from one place to the other. It is essential to transfer it as quickly as possible. In other words, the data rate in bits per second over a channel must be very high.

The data rate depends on the following factors.

(i) Available bandwidth (ii) The levels of signal we can use.

(iii) The quality of the channel (the level of noise).

Basically, two theorems have been developed to evaluate the data rate. These are:

(i) Nyquist's theorem for a noiseless channel.

(ii) Shannon's theorem for a noisy channel.

3.10.1. Nyquist's Limit

As we know that a transmission channel is a medium over which the electrical signals from a transmitter travel to the receiver. Two important characteristics of a transmission channel are as under:

(i) Signal to Noise ratio (SNR), and

(ii) Bandwidth

These two characteristics will ultimately decide the maximum capacity of a channel to carry information. Nyquist and Shannon worked on finding the maximum channel capacity of a band limited channel.

Nyquist theorem states that if the bandwidth of a transmission channel is B which carries a signal having L number of levels, then the maximum data rate R on this channel is given by,

$$R = 2B \log_2 L \qquad \text{... (3.1)}$$

As maximum data rate for reliable transmission is defined as channel capacity C, the above expression gets modified as under :

$$C = 2\,B \log_2 L \qquad \text{... (3.2)}$$

The above expression indicates that the data rate can be increased by increasing the number of different signal elements (L).

3.10.2. Shannon's Channel Capacity

1. Statement: The statement of the Shannon's theorem is : Given that a source of M equally likely messages with M>> 1, which is generating information at a rate R. Given that a channel of capacity C exists. Then if,

$$R \leq C$$

then there exists a coding technique such that the output of the source may be transmitted over the channel with a probability of error in the received message which may be made arbitrarily small.

2. Meaning

(i) The theorem is concerned with the rate of transmission of information (R) over a communication channel.

(ii) The channel capacity C is a rate of transmission in bits/sec. According to the theorem for $R \leq C$, transmission can be accomplished with coding without error in the presence of noise.

There is a negative statement associated with the Shannon's theorem. It's statement is as under :

3. Negative Statement of Shannon's theorem: Given a source of equally likely messages with M >> 1, which is generating an information at a rate R, and if

$$R > C$$

then the probability of error is close to unity for every possible set of M transmitted signals.

4. Meaning

(i) If the information rate *R* exceeds a specific value C, then the error probability will increase towards unity as *M* increases.

(ii) When $R > C$, complexity of coding is increased which results in an increase in the probability of error.

5. Few Important Points

(i) We have seen the Nyquist bandwidth in the previous section.

(ii) Shannon extended Nyquist's work. He included the effect of noise present on the transmission channel.

(iii) According to Shannon's theorem, if (S/N) is the signal to noise ratio then the maximum data rate is given by

$$R = B \log_2 \left[1+\frac{S}{N}\right] \text{ bits/sec.} \qquad (3.3)$$

(iv) Shannon's theorem puts a limit on the maximum number of levels for a given (S/N) ratio and bandwidth. This expression shows that the maximum data rate for a communication channel is dependent on the channel bandwidth B and signal to noise ratio (S/N).

Importance of Channel Bandwidth

Bandwidth of the communication channel should be higher than the bandwidth of the signal that is to be transmitted over it. This is essential in order to preserve the shape of the signal being transmitted. If the channel bandwidth is less than the signal bandwidth, then, the signal shape will be distorted when it travels over this channel.

EXAMPLE 3.4. A channel has a bandwidth of 5 kHz and a signal to noise power ratio of 63. Evaluate the bandwidth required if the S/N power ratio is reduced to 31. What will be the signal power required if the channel bandwidth is reduced to 3 kHz ?

Solution: It is given that B = 5 kHz and $\frac{S}{N} = 63$. Hence, using the Shannon Hartley theorem the channel capacity is given by,

or $$C = B \log_2 \left(1+\frac{S}{N}\right) = 5 \times 10^3 \log_2 [1 + 63]$$

Therefore, $$C = 30 \times 10^3 \text{ bits/sec.} \qquad \text{... (i)}$$

The new value of $\frac{S}{N} = 31$. Assuming the channel capacity C to be constant, we can write.

$$30 \times 10^3 = B \log_2 (1 + 31)$$

or $$B = \frac{30 \times 10^3}{5} = 6 \text{ kHz} \qquad \text{... (ii)}$$

Given that the new bandwidth is 3 kHz. We know that noise power $N = N_0 B$.

Let the noise power corresponding to a bandwidth of 6 kHz be $N_1 = 6 N_0$ and the noise power corresponding to the new bandwidth of 3 kHz be $N_2 = 3 N_0$.

Therefore, $$\frac{N_1}{N_2} = \frac{6N_0}{3N_0} = 2 \qquad \text{... (iii)}$$

The old signal to noise ratio $= \frac{S_1}{N_1} = 31$

so, $$S_1 = 31 N_1 \qquad \text{... (iv)}$$

The new signal to noise ratio $= \frac{S_2}{N_2}$. We do not know its value, hence let us find it out.

We have $30 \times 10^3 = 3 \times 10^3 \log_2\left(1+\frac{S_2}{N_2}\right)$

or $\frac{S_2}{N_2} = 1023$ ... (v)

or $S_2 = 1023\, N_2$

But from equation (iii), $N_2 = \frac{N_1}{2}$, substituting, we obtain

$$S_2 = 1023\,\frac{N_1}{2} \quad \text{... (vi)}$$

Dividing equation (vi) by equation (iv), we obtain

$$\frac{S_2}{S_1} = \frac{1023\,N_1}{2 \times 31\,N_1} = 16.5$$

or $S_2 = 16.5\, S_1$ **Ans.**

Therefore, if the bandwidth is reduced by 50%, then the signal power must be increased 16.5 times i.e., 1650% to get the same capacity.

EXAMPLE 3.5. Evaluate the maximum bit rate for a channel having bandwidth 3100 Hz and S/N ratio of 20 dB.

Solution : Given that $B = 3100$ Hz, $\frac{S}{N} = 20$ dB

But, we have 20 dB = 10 log (S/N)

or S/N = 100

We know that maximum bit rate is given by,

$$R_{max} = B \log_2\left(1+\frac{S}{N}\right) = 3100 \log_2 (1 + 100)$$

or
$$R_{max} = \frac{3100 \log_{10} 101}{\log_{10} 2} = 20{,}640 \text{ bits/sec.}$$ **Ans.**

EXAMPLE 3.6. Find the maximum bit rate for a channel having bandwidth 3100 Hz and S/N ratio of 10 dB.

Solution: Given that B = 3100 Hz

We have $\left(\frac{S}{N}\right)_{dB} = 10$

or $10 = 10 \log_{10}\left(\frac{S}{N}\right)$

or $\frac{S}{N} = 10$

We know that maximum bit rate $= R_{max} = B \log_2\left(1+\frac{S}{N}\right)$

or
$$R_{max} = 3100 \log_2 (1 + 10) = \frac{3100 \log_{10} 11}{\log_{10} 2} = 10{,}724 \text{ bits/sec.}$$ **Ans.**

EXAMPLE 3.7. Determined the maximum bit rate for a channel having bandwidth equal to 1600 Hz if:

(i) S/N ratio is 0 dB

(ii) S/N ratio is 20 dB

Solution: Given that $B = 1600$ Hz

(i) R_{max} for S/N = 0 dB

$$\left(\frac{S}{N}\right)_{dB} = 10 \log_{10}\left(\frac{S}{N}\right)$$

or $$\frac{S}{N} = 1$$

Therefore, $$R_{max} = B \log_2\left(1+\frac{S}{N}\right)$$

or $$R_{max} = 1600 \log_2 (1 + 1)$$

or $$R_{max} = 1600 \text{ bits/sec.} \quad \textbf{Ans.}$$

(ii) R_{max} for S/N = 20 dB

$$\left(\frac{S}{N}\right)_{dB} = 10 \log_{10}\left(\frac{S}{N}\right)$$

or $$20 = 10 \log_{10}\left(\frac{S}{N}\right)$$

or $$\frac{S}{N} = 100$$

Therefore, $$R_{max} = B \log_2\left(1+\frac{S}{N}\right)$$

$$R_{max} = 1600 \log_2 (101) = \frac{1600 \log_{10}(101)}{\log_{10}(2)} = 10{,}654 \text{ bits/sec.} \quad \textbf{Ans.}$$

Pioneer in Communications : Stephanie Kwolek (1923—)

Stephanie Kwolek was born in New Kensingtion, Pennsylvania, in 1923. She had earned a BS in chemistry in 1946 from Carnegie Institute of Technology. She contributed to the development of DuPont's Kapton® polymide film and Nomex aramid fiber.

Using Both the Limits

In practice, we have to use both the methods to evaluate the required bandwidth and signal level. Let us consider the following example for this particular aspect.

EXAMPLE 3.8. The bandwidth of a channel is 2 MHz and its signal to noise ratio is 63. Determine the appropriate bit rate and signal level.

Solution: First we, calculate the upper limit using Shannon's theorem as under:

$$C = B \log_2\left(1+\frac{S}{N}\right) = 2 \times 10^6 \log_2 (1 + 63) = 12 \text{ M bits/sec.}$$

This is the upper limit. For ensuring a better performance we select a somewhat lower value 8 Mbps.

Next, we calculate number of signal levels using Nyquist's theorem as under :

$$C = 2 B \log_2 L$$

or $$8 \times 10^6 = 2 \times 2 \times 10^6 \log_2 L$$

or $$2 = \log_2 L$$

or $$L = 4$$

3.11. TRANSMISSION IMPAIRMENTS

In any communication system, the received signal is never identical to the transmitted one due to some transmission impairments. The quality of analog signals will deteriorate due to the transmission impairment whereas errors get introduced if the signal is digital.

The most important impairments may be listed as under :

(i) Attenuation and attenuation distortion
(ii) Delay distortion
(iii) Noise

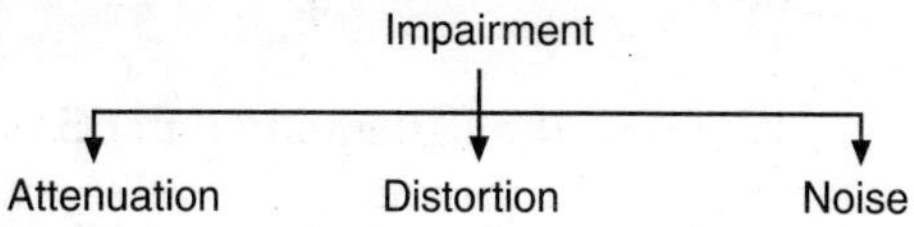

3.11.1. Attenuation

1. Definition: Signals carry energy which can be sensed at the receiver's end. The receiver detects a signal because of its energy. When the signal is transmitted, it carries the maximum energy. But, it continues to lose energy in transit. After a while, the reduction in energy reaches an extent where the receiver cannot detect the signal. The energy loss depends on the distance the wave has travelled and the media in which the signal travels and also the bit rate at which, it is travelling. The strength of a signal decreases with increase in distance travelled over a medium. Attenuation means loss of energy. When any signal travels over a medium or channel, it loses some of its energy in the form of heat in the resistance of the medium.

Attenuation may be expressed in decibels (dB) as under :

$$\text{Attenuation (dB)} = 10 \log_{10} \frac{P_{out}}{P_{in}} \qquad \text{... (3.4)}$$

where P_{in} = Power at the sending end
P_{out} = Power at the receiving end

It is important to note that attenuation decides the signal to noise ratio (SNR) and hence, the quality of received signal.

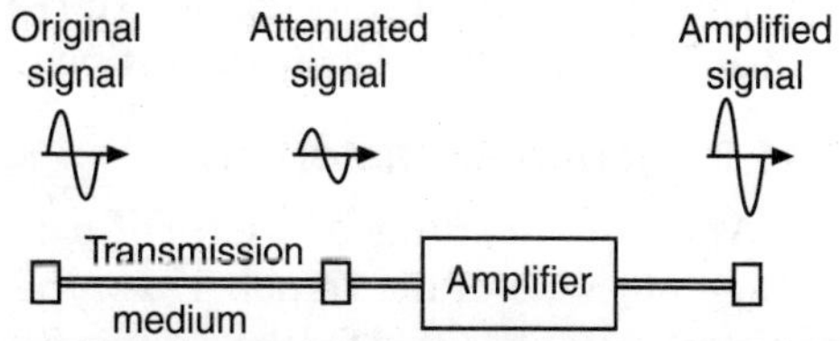

Fig. 3.12. Concept of attenuation

2. Remedy to Reduce Attenuation: We can introduce amplifiers to compensate for such loss as shown in figure 3.12. It may be noted that amplification and attenuation are exactly opposite to each other.

3. Effect of Frequency on Attenuation

(i) Attenuation increases with increase in frequency.
(ii) Attenuation increases with increase in signal frequency.
(iii) The SNR can be improved by using amplifiers or repeaters to boost the signal strength.

The effect of frequency on attenuation is of concern if the signal is analog. This effect can be reduced by using equalizers. Another way to do this is to use an amplifier that amplifies the higher frequencies more than lower frequencies. The effect of equalization is illustrated in figure 3.13.

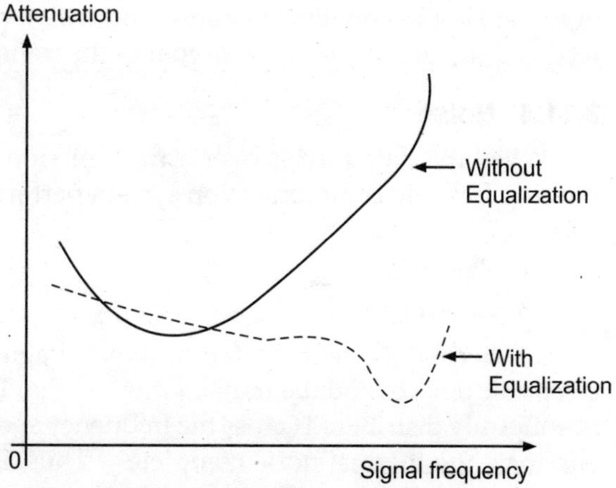

Fig. 3.13. Effect of equalizer on attenuation

EXAMPLE 3.9. Evaluate the attenuation in dB if input power is three times higher than the received power.

Solution: We have $P_{in} = 3\ P_{out}$

or $$\frac{P_{out}}{P_{in}} = \frac{1}{3}$$

We know that attenuation in dB $= 10 \log_{10} \frac{P_{out}}{P_{in}} = 10 \log_{10} (1/3)$

Attenuation = – 4.8 dB **Ans.**

3.11.2. Delay Distortion

This problem is particularly present in the wired media. This distortion is caused due to a property which states that the velocity of propagation of a signal through a medium varies with frequency. The velocity is maximum near the center frequency and reduces as the signal frequency deviates away from the center frequency. Hence, various frequency components of a signal arrive at the receiver at different instants of time resulting in phase shifts between different frequencies. This distorts the signal and the distortion is called as **delay distortion**. Delay distortion is particularly important for digital data. Due to delay distortion, a digital pulse transmitted over a medium tends to spread out as shown in figure 3.14 causing the intersymbol interference.

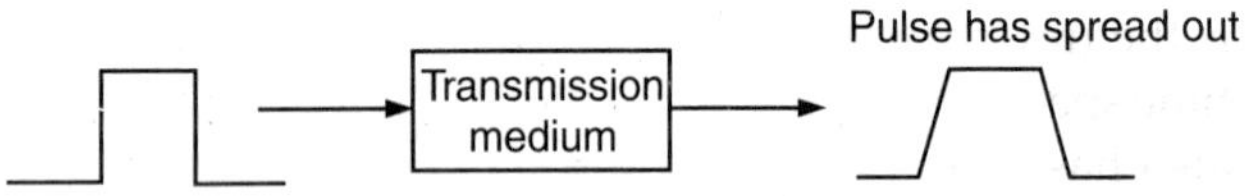

Fig. 3.14. Delay distortion

The Intersymbol interference (ISI) put a major limitation on the maximum bit rate over a transmission channel. Delay distortion can be reduced by using equalizing techniques.

3.11.3. Harmonic Distortion

Another meaning of distortion is change in shape of the signal. This type of distortion is observed for the composite signals made of different frequencies. If the medium is not prefect, then all the frequency components present at the input will not be equally attenuated and will not be proportionally delayed. Hence, the shape of the received signal is different from that of the transmitted signal as shown in figure 3.15.

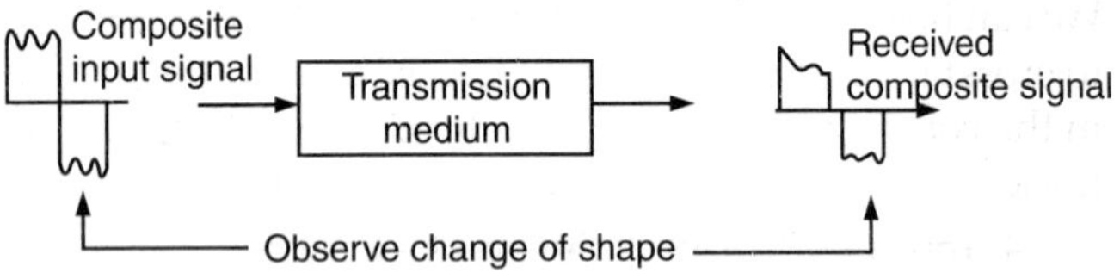

Fig. 3.15. Distortion

3.11.4. Noise

When the data travels over a transmission medium, noise gets added to it. Noise is a major limiting factor in communication system performance. Noise can be categorized into four types as follows:

1. Thermal noise
2. Intermodulation noise
3. Crosstalk
4. Impulse noise

1. Thermal noise: It is due to thermal agitation of electrons. Thermal noise is present in all electronic devices and the transmission media. Thermal noise is proportional to temperature and it is uniformly distributed across the frequency spectrum. Hence, it is called as **white noise**. We cannot eliminate the thermal noise completely. Thus, it places an upper limit on the performance of the communication system. The thermal noise power is given by,

$$N = kTB \text{ watts}$$

where, k = Boltzman's constant

T = Absolute temperature °K

B = Bandwidth

It can be expressed in dBW as

$$N = 10 \log k + 10 \log T + 10 \log B.$$

Thermal noise can be reduced by reducing temperature or bandwidth.

2. Intermodulation Noise: If signals at different frequencies are transmitted simultaneously on a common transmission medium (e.g. FDM), then it results in intermodulation noise. The intermodulation produces frequency components at sum and difference frequencies of the frequencies travelling on the medium. Suppose signals of frequencies f_1 and f_2 are travelling simultaneously over the same medium, then due to intermodulation produces frequency components $(f_1 + f_2)$ and $(f_1 - f_2)$, which are called **intermodulation noise**. This is similar to a mixer. Hence, whenever there is any non-linearity in the transmitter or channel, the intermodulation noise is produced.

3. Crosstalk: Crosstalk basically means interference between the adjacent telephone channels. It is the unwanted coupling of information from one channel to the other. The guard time (T_g) is the time spacing introduced between the adjacent telephone channels. A signal without baseband filtering also has crosstalk if the pulse tail or post-cursor overlaps into the next time slot of the frame as shown in figure 3.16.

Pulse overlap shown in figure 3.16 can be controlled by using guard time (T_g) between the pulses. The guard time (T_g) is analogous to the guard bands between the channels in the FDM systems. In practice, crosstalk can be experienced while using a telephone. We can hear another conversation due to crosstalk. It can occur due to electrical coupling between nearby twisted pairs or sometimes even in the coaxial cables. Typically, the magnitude of crosstalk is almost the same as that of the thermal noise.

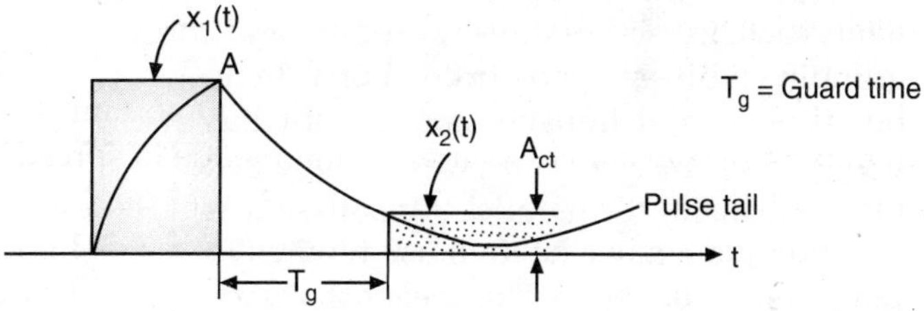

Fig. 3.16. Illustration of crosstalk

4. Impulse Noise: All the types of noise discussed so far are reasonably predictable and their magnitude relatively constant. Hence, it is a bit easier for the system engineers to deal with them. But, impulse noise is completely different from the other types. It is non-continuous, it consists of irregular short pulses or noise spikes of short duration of high amplitude. Impulse noise gets generated due to many reasons such as external electromagnetic disturbances, lightning etc.

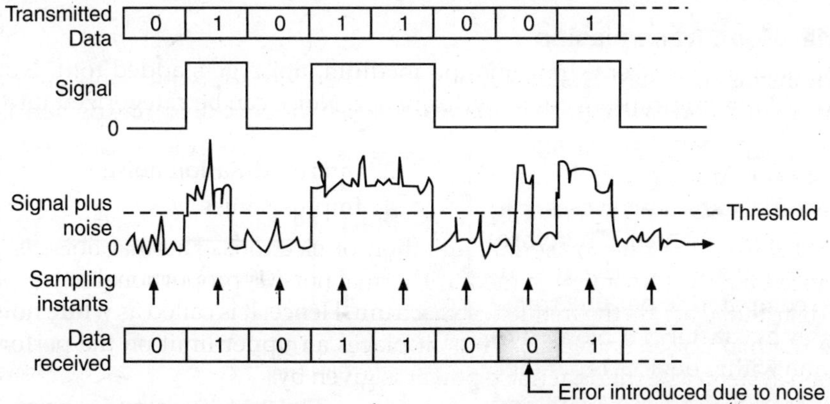

Fig. 3.17. Effect of impulse noise

Impulse noise does not affect the quality of analog signal to a great extent but it affects the digital data badly. Impulse noise introduces errors in the digital data. It is possible to recover the original data by means of sampling as shown in figure 3.17.

3.12. TRANSMISSION MODE

In this section, let us discuss how we can transmit binary data. While transmitting data from one device (i.e., transmitter or sender) to another device wiring is of almost importance. Further, dealing with wiring, the data stream is of utmost importance. Do we send 1 bit at a time; or do we group bits into larger groups and, if so, how? We can accomplish the transmission of binary data across a link in either parallel or serial mode. In parallel mode, multiple bits are sent with each clock pulse. In serial mode, 1 bit is sent with each clock pulse. There is only one way to send parallel data whereas there are two subclasses of serial transmission. These two subclasses are synchronous and asynchronous.

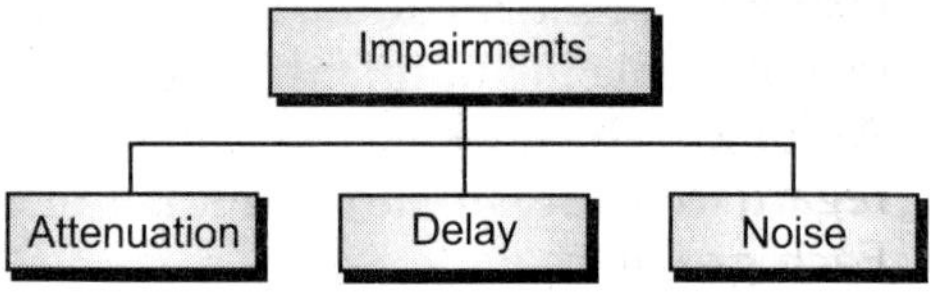

Fig. 3.18. Data transmission

3.12.1. Parallel Mode Transmission

We know that binary data consisting of 1s and 0s binary data can be organized into groups of n bits each. In fact computers produce and consume data in groups of bits. We can send data by grouping n bits at a time instead of a single bit. This type of transmission is knownas parallel transmission. Thus, we conclude that the mechanism for parallel transmission is quite simple i.e., we have to use n wires to send n bits at one time. This means that each bit has its own wire and all n bits of one group can be transmitted with each clock pulse from one device to other device. Figure 3.19 illustrates how parallel transmission works for n = 8. Typically, the eight wires are bundled in a cable with a connector at each end.

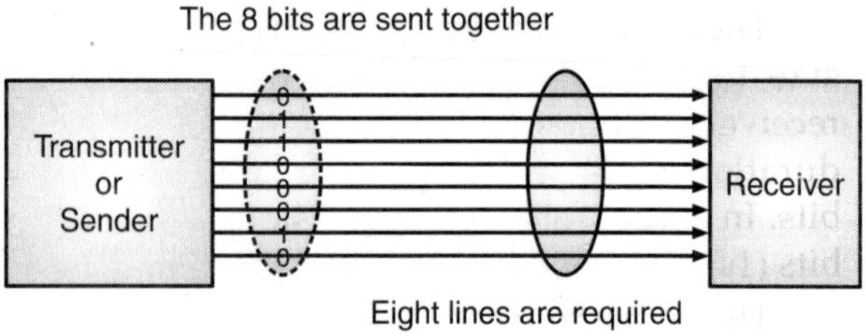

Fig. 3.19. Parallel transmission

The primary advantage of parallel transmission is speed. Thus, all else being equal, parallel transmission can increase the transfer speed by a factor of n compared to serial transmission. However, a parallel transmission has an obvious disadvantage also. This is manufacturing cost since a parallel transmission requires n communication lines to transmit the data stream. Because this is expensive, parallel transmission is usually limited to short distances.

3.12.2. Serial Mode Transmission

In serial transmission, one bit follows another bit. Therefore, we require only one communication channel rather than channels *n* to transmit data between two communicating devices i.e., transmitter and receiver. The primary advantage of serial transmission over parallel transmission is that serial transmission uses only one communication channel. Therefore, serial transmission reduces the cost of transmission compared to parallel transmission roughly by a factor of n. Now, since communication within devices is parallel, therefore, conversion devices are required

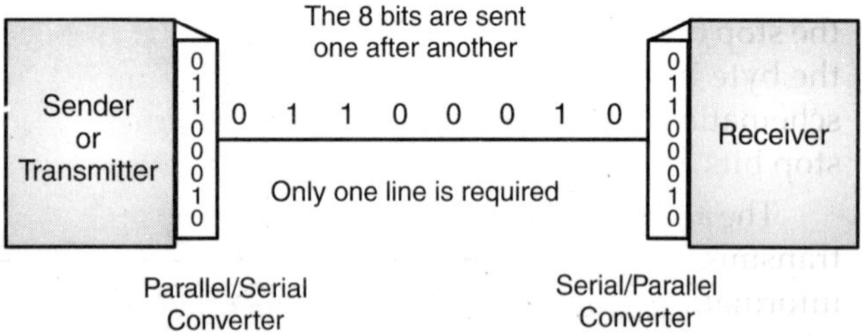

Fig. 3.20. Serial transmission

at the interface between the transmitter and the line (parallel-to-serial) and between the line and the receiver (serial-to-parallel).

Further serial transmission can occur in following two ways:

(i) Asynchronous or

(ii) Synchronous

3.12.2.1. Asynchronous Transmission

Asynchronous transmission occurs character by character and is used for serial communication, such as by a modem or serial printer. In asynchronous transmission, each data character has a start bit that identifies the start of the character, and one or two bits that identify the end of the character. Asynchronous transmission is so called because in this kind of transmission, the timing of a signal is not important. Instead information is received and translated by agreed upon patterns. As long as those patterns are followed, the receiving device can retrieve the information without regard to the rhythm in which it is sent. Patterns are based on grouping the bit stream into bytes. Each group, usually 8 bits, is sent along the link as a unit. The sending system handles each group independently, relaying it to the link whenever ready, without regard to a timer. Without synchronization, the receiver cannot use timing to predict when the next group will arrive. To alert the receiver to the arrival of a new group, therefore, an extra bit is added to the beginning of each byte. This bit usually a 0, is called the *start bit*. To let the receiver know that the byte is finished, 1 or more additional bits are appended to the end of the byte.

These bits, usually 1s, are called *stop bits*. By this method, each byte is increased in size to at least 10 bits, of which 8 are information and 2 or more are signals to the receiver. In addition, the transmission of each byte may then be followed by a gap of varying duration. This gap can be represented by either an idle channel or by a stream of additional stop bits. In asynchronous transmission, we send 1 start bit (0) at the beginning and 1 or more stop bits (1s) at the end of each byte. There may be a gap between each byte.

The start and stop bits and the gap alert the receiver to the beginning and end of each byte and allow it to synchronize with the data stream. This mechanism is called asynchronous because, at the byte level, sender and receiver do not have to be synchronised. But within each byte, the receiver must still be synchro-nized with the incoming bit stream. That is, some synchronization is required, but only for the duration of a single byte. The receiving device resynchronizes at the onset of each new byte, when the receiver detects a start bit, it sets a timer and begins counting bits as they come in. After n bits, the receiver looks for a stop bit. As soon as it detects the stop bit, it waits until it detects the next start bit. Asynchronous here means "asynchronous at the byte level," but the bits are still synchronized; their durations are the same. Figure 3.21 is a schematic illustration of asynchronous transmission. In this example, the start bits are 0s, the stop bits are 1s, and the gap is represented by an idle line rather than by additional stop bits.

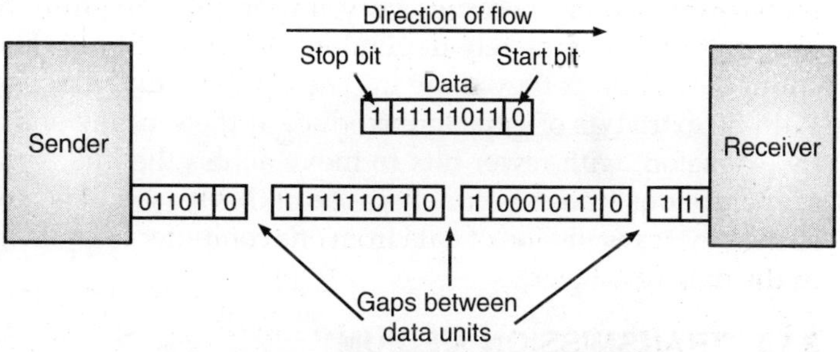

Fig. 3.21. Asynchronous transmission

The addition of stop and start bits and the insertion of gaps into the bit stream make asynchronous transmission slower than forms of transmission that can operate without the addition of control information. But it is cheap and effective, two advantages that make it an attractive choice for situations such as low-speed communication. For example, the connection of a keyboard to a computer is a natural application for asynchronous transmission. A user types only one character at a time, types extremely slowly in data processing terms, and leaves unpredictable gaps of time between each character.

3.12.2.2. Synchronous Transmission*

Some applications require transferring large block of data, such as a file from disk or a print job from a computer to a printer. Synchronous transmission is an efficient method of transferring large blocks of data by using time intervals for synchronization. In synchronous transmission, the bit stream is combined into longer frames, which may contain multiple bytes. Each byte, however, is introduced onto the transmission link without a gap between it and the next one. It is left to the receiver to separate the bit stream into bytes for decoding purposes. In other words, data are transmitted as an unbroken string of 1s and 0s, and the receiver separates that string into the bytes, or characters, it needs to reconstruct the information. In synchronous transmission, we send bits one after another without start/stop bits or gaps. It is the responsibility of the receiver to group the bits. Figure 3.22 gives a schematic illustration of synchronous transmission. We have drawn in the divisions between bytes. In reality, those divisions do not exist; the sender puts its data onto the line as one long string. If the sender wishes to send data in separate bursts, the gaps between bursts must be filled with a special sequence of 0s and 1s that means *idle*.

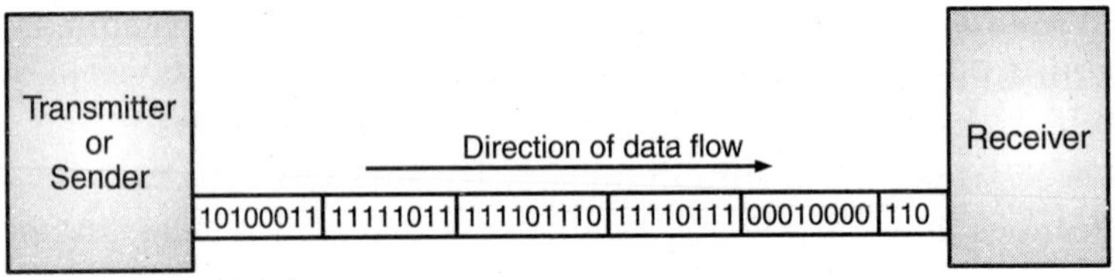

Fig. 3.22. Illustration of synchronous transmission

The receiver counts the bits as they arrive and groups them in 8-bit units. Without gaps and start/stop bits, there is no built-in mechanism to help the receiving device adjust its bit synchronization midstream. Timing becomes very important, therefore, because the accuracy of the received information is completely dependent on the ability of the receiving device to keep an accurate count of the bits as they come in. The primary advantage of synchronous transmission is speed. With no extra bits or gaps to introduce at the sending end and remove at the receiving end and, by extension, with fewer bits to move across the link, synchronous transmission is faster than asynchronous transmission. Because of this reason, it is more useful for high-speed applications such as the transmission of data from one computer to another. Byte synchronization is accomplished in the data link layer.

3.13. TRANSMISSION MEDIUM** *(RGTU, Bhopal, Sem. Exam., June, 2014)*

As a matter of fact, media are what the message is transmitted over. In other words, a communication channel is also called as a medium. Different media have different properties and used in different environments for different purpose. The purpose of the physical layer is to transport a raw bit stream from one computer to another.

3.13.1. Classification of Transmission Medium

As a matter of fact, we can classify the transmission media as shown in figure 3.23 into two categories.

* One method of synchronizing the transmitter and receiver is through the use of an external connection that carries a clock pulse.

** What are the different transmission media used in computer communication? *(RGTU, Bhopal, Sem. Exam., June, 2014)*

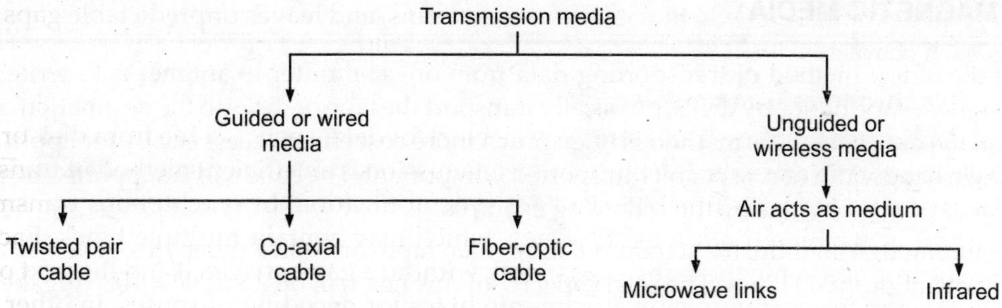

Fig. 3.23. Classification of transmission media

Media are roughly grouped into two classes:

(i) Guided media

(ii) Unguided media

> **DO YOU KNOW?**
>
> The physical layer supports the relaying function in the transmission medium.

(i) **Guided Media:** Guided media is a communication medium which allows the data to get guided along it. For this the media need to have a point to point physical connection.

(ii) **Unguided Media :** The wireless media is also called unguided media.

The Examples of Guided Media are Copper Wire and Fiber-Optics. Similarly radio and lasers through the air are examples of unguided media as shown in figure 3.24.

In computer networking, the medium affects nearly every aspect of communication. Most important, it determines how quickly and to whom a computer can talk and how expensive the process is.

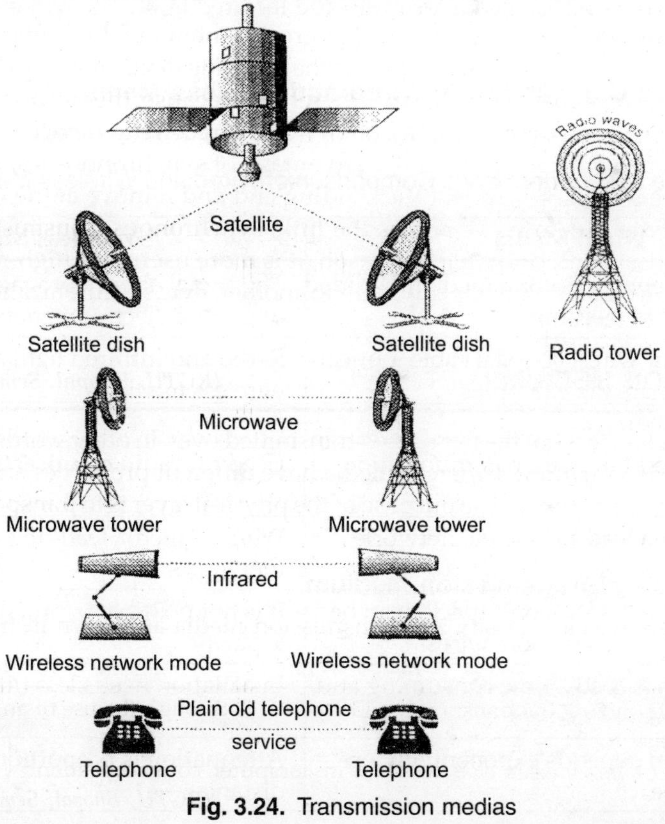

Fig. 3.24. Transmission medias

3.14. MAGNETIC MEDIA

One of the oldest method of transporting data from one computer to another is to write the data onto magnetic tape or floppy disks, physically transport the tape or disks to the destination computer and read the data back. This method is often much more cost effective, especially for applications in which high bandwidth or cost per bit transported is important. The bandwidth supported by magnetic tapes can be made obvious by the following example.

Assuming that an industry standard 8 mm video tape can hold 7 gigabytes. A box 50 × 50 × 50 cm can hold about 1000 tapes. The total capacity of this box will be 7 × 1000 = 7000 gigabytes or 56 × 10^{12} bits (1 byte = 8 bits). A box of tapes can be delivered anywhere in India in 24 hours by any courier company. The effective bandwidth of this transmission will be 56 × 10^{12} bits/86400 sec = 648 Mbps. This speed is slightly better than the high-speed version of asynchronous Transfer Mode (ATM) which is 622 Mbps.

3.15. TYPES OF TRANSMISSION MEDIA *(RGTU, Bhopal, Sem. Exam., June, 2014)*

3.15.1. Wired Media

In this type of media, the signal energy is contained and guided within a solid media. The examples of wired media are copper pair wires, co-axial cables and fiber optic cables. The wired media is used for point to point communication.

3.15.2. Wireless Media

In the wireless media, the signal energy propagates in the form of unguided electromagnetic waves. The examples of wireless media are radio and infrared light. The wireless media is used for radio broadcasting in all the directions.

3.15.3. Performance Comparison of Wired and Wireless Media

Table 3.2 gives the performance comparison of wired and wireless media.

Table 3.2. Performance Comparison of Wired and Wireless Media

S.No.	Wired Media	Wireless Media
1.	The signal energy is contained and guided within a solid medium.	The signal energy progagates in the form of unguided electromagnetic waves.
2.	Twisted pair wires, co-axial cable, optical fiber cables are the examples of wired media.	Radio and infrared light are the examples of wireless media.
3.	Used for point to point communication. directions.	Used for radio broadcasting in all.
4.	Wired media lead to discrete network topologies.	Wireless media leads to continuous network topologies.
5.	Additional transmission capacity can be procured by adding more wires.	It is not possible to procure additional capacity.
6.	Installation is costly, time consuming and complicated.	Installation needs less time and money.
7.	Attenuation depends exponentially on the distance.	Attenuation is proportional to square of the distance.

3.15.4. Types of Wired Media

The wired media can be classified as under:

(i) Co-axial cable
(ii) Twisted pair cable
(iii) Optical fiber cable

The selection of networking media depends on various factors such as cost, connectivity, width, performance in presence of noise, geographical coverage etc.

> **DO YOU KNOW?**
> As the signal travels along a transmission medium, it gets attenuated and distorted. A repeater regenerates the digital signal and makes it as good as the signal transmitted by the source.

3.16. TWISTED PAIR CABLES

(RGTU, Bhopal, Sem. Exam., June, 2014)

The construction of twisted pair cable has been shown in figure 3.25. This is also commonly used medium and it is cheaper than the co-axial cable.

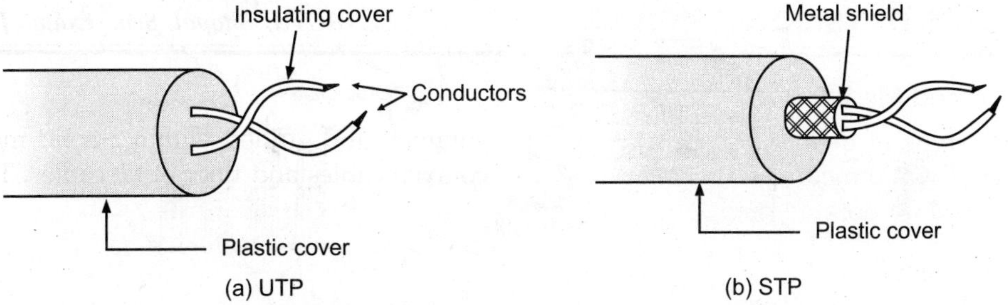

Fig. 3.25. Construction of twisted pair cables

3.16.1. Types of Twisted Pair Cables

The two commonly used types of twisted pair cables are as follows :

(i) Unshielded twisted pair (UTP)
(ii) Shielded twisted pair (STP)

(i) UTP : A twisted pair consists of two insulated conductor twisted together in the spiral form as shown in figure 3.25 (a). It can be shielded or unshielded. The unshielded twisted pair cables are very cheap and easy to install. But, they are badly affected by the noise interference.

(ii) STP: Figure 3.25 (b) shows STP cable. It has a metal foil or braided mesh to cover each pair of insulating conductors. This is known as the metal shield. It reduces the interference of the noise but makes the cable bulky and expensive. So, practically UTP is more used than STP. The STP was developed by IBM and is used primarily for the IBM only. Applications of the twisted pair cables are in point to point and point to multipoint communications, telephone systems. Twisted pairs can be used for either analog or digital transmission. The supported by the wire depends on the thickness of the wire and the distance travelled. Twisted pairs support several megabites/sec for a few kilometres and have less cost.

Why to Twist the Wires ?

Twisting of wires will reduce the effect of noise or external interference. Number of twists per unit length will determine the quality of cable. More twists means better quality.

> **DO YOU KNOW?**
> The physical medium interface is defined in terms of its mechanical, electrical, functional, and procedural specifications.

3.16.2. Categories of UTP

Table 3.3 shows various categories of the unshielded twisted pair cales. These categories are decided by EIA, i.e., electronic industries association.

Table 3.3. Categories of UTP Cables

Category	Bandwidth	Data rate	Digital/Analog	Application
1	Very low	< 100 kbps	Analog	Telephone
2	< 2 MHz	2 Mbps	Analog/Digital	T_1 lines LANS
3	16 MHz	10 Mbps	Digital	LANs
4	20 MHz	20 Mbps	Digital	LANs
5	100 MHz	100 Mbps	Digital	LANs
6	200 MHz	200 Mbps	Digital	LANs
7	600 MHz	600 Mbps	Digital	LANs

These cables have less crosstalk and a better quality signal over longer distances. These cables are mostly suitable for high speed computer communication. A connection diagram using the UTP is shown in figure 3.26.

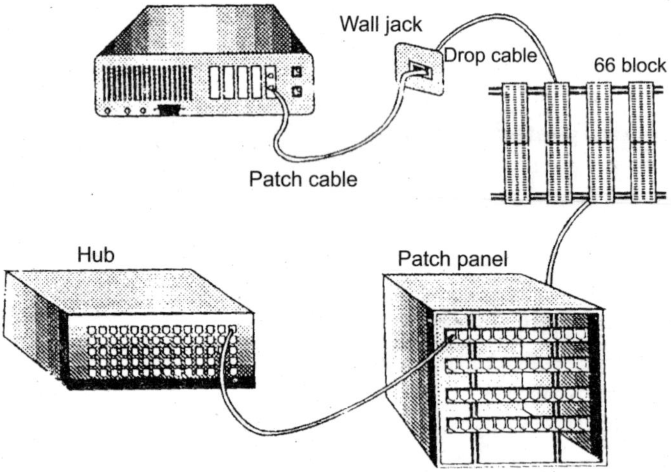

Fig. 3.26. A common UTP installation

3.16.3. Applications of Twisted Pair Cables

Some of the applications of twisted pair cables are as follows:

(i) In telephone lines to carry voice and data channels.
(ii) In the local loop.
(iii) In the DSL line (ADSL).
(iv) Local area networks such as 10 Base-T and 100 Base-T use the twisted pair cables.
(v) In the ISDN (Integrated Services Digital Network).

DO YOU KNOW?

The voice channels of telephone network are not clear channels. They incorporate devices like companders to improve signal to noise ratio and echo suppressors to counteract the effect of propagation delay that causes echo.

Important Point:
Category 1 and 2 cables were originally used for voice communication with low data rates. Category 4 offers data rates upto 20 mbps. A modular RJ-45 telephone connector is used to connect a four-pair cable. A modular RJ-11 telephone connector is used to connect a two pair cable. Shielded twisted pair (STP) cables were introduced by IBM corporation.

3.16.4. Performance Comparison of Cables

S. No.	Factors	UTP	STP
1.	Bandwidth capacity (typically 10 mbps)	1–155 mbps (typically 16 mbps)	1–155 mbps
2.	Node capacity per segment	2	2
3.	Attenuation	High	High
4.	EMI	Very high	High
5.	Installation	Easy	Fairly easy
6.	Cost	Lowest	Moderate

3.16.5. Connectors

For connecting one computer to the other, we need to use some transmission medium such as a cable. The cables are of different types such as twisted pair cables, co-axial cables or fiber optic cables. Further, for connecting these cables between two computers we have to use connectors on both ends of a cable. Generally, the connectors are male-female type to ensure reliable connection.

3.16.6. Connector for Twisted Pair Cable

The unshielded twisted pair (UTP) cable is the most commonly used cable in computer communication. The most common UTP connector is RJ45 where RJ is the short form of Registered Jack. It is a male-female type keyed connector as shown in figure 3.27. This connector can be inserted in only one way.

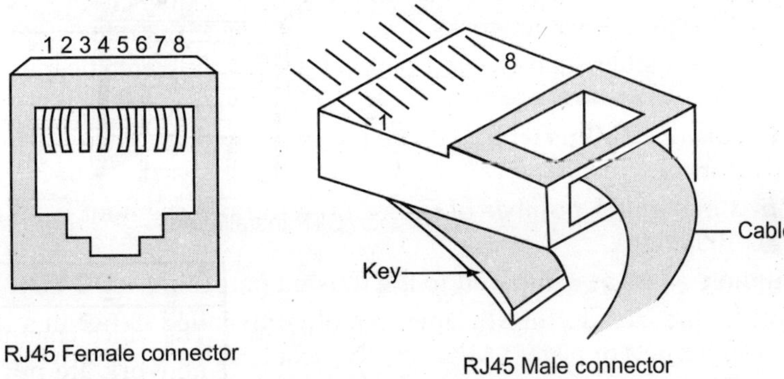

Fig. 3.27. UTP RJ 45 connector

3.17. CO-AXIAL CABLES

(RGTU, Bhopal, Sem. Exam., June, 2014)

Figure 3.28 shows the construction of co-axial cable. It consists of two concentric conductors separated by a dielectric material. The external conductor is metallic braid and used for the purpose of shielding. The co-axial cable may contain one or more co-axial pairs.

The construction of a co-axial cable with other accessories is shown in figure 3.29. The wire mesh protects the wire from Electromagnetic Interference (EMI). It is often called a shield. A tough plastic jacket forms the cover of the cable as shown in figure 3.29 providing insulation and protection.

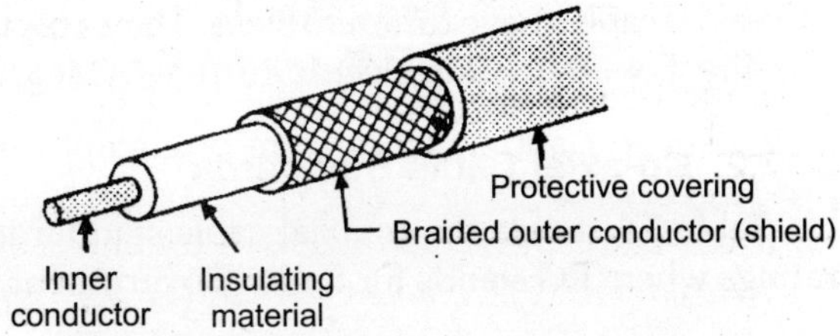

Fig. 3.28. Construction of a co-axial cable

The co-axial cable was initially developed as the backbone of analog telephone networks where a single telephone cable would be used to carry more than 10,000 voice channels at a time. The digital transmission systems using the co-axial cable were developed in 1970s. These systems operated in the range of 8.5 Mb/s to 565 Mb/s. The most popular application of a co-axial cable is in the cable TV system. The existing co-axial cable system has a range from 54 MHz to 500 MHz. Other important application is cable modem, with the cable modem termination system (CMTS). One more application is Ethernet LAN using the co-axial cable. The co-axial cable is used for its large bandwidth and high noise immunity.

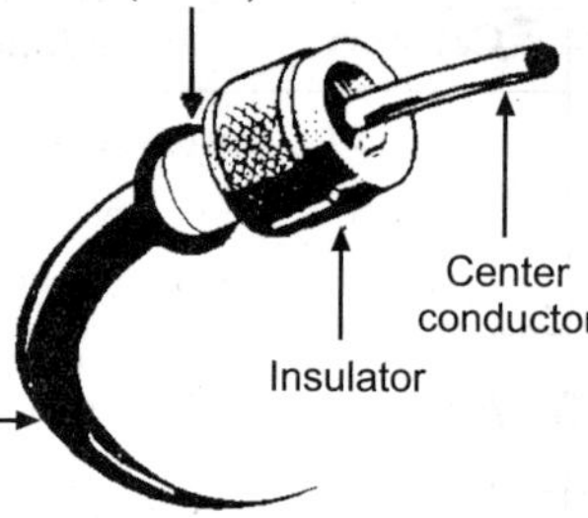

Fig. 3.29. Co-axial cable

> **DO YOU KNOW?**
>
> Coaxial cables are used as the long distance transmission medium in telephone networks. Coaxial cables also find application in cable television networks and local area networks for computer communications. They are not used at frequencies lower than 60 KHz due to their degraded phase and cross-talk properties at low frequencies.

3.17.1. Characteristics of a Co-axial Cable

The important characteristics of a co-axial cable are as under :

(i) Two types of cables having 75 W and 50 W impedance are available.

(ii) Due to the shield provided, this cable has excellent noise immunity.

(iii) It has a large bandwidth and low losses.

(iv) This cable is suitable for point to point or point to multipoint applications. In fact, this is the most widely used medium for local area networks.

(v) These cables are costlier than twisted pair cables but they are cheaper than the optical fiber cables.

(vi) It has a data rate of 10 Mbps which can be increased with the increase in diameter of the inner conductor.

(vii) The specified maximum number of nodes on a thinnet segment is 30 nodes and on a thicknet it is 100 nodes.

(viii) The attenuation is less as compared to the twisted pair cable.

(ix) Co-axial cables are easy to install. They are often installed either in a device to device daisy chain (Ethernet) or a star (ARC net). Co-axial cables are relatively inexpensive.

> **Important Point:**
>
> 50 Ω cable is commonly used for digital transmission and 75 Ω cable is used for analog transmission. Two types of co-axial cables based on the frequency they support - Baseband (0-4 KHz) and Broadband (above 4 kHz).
>
> Baseband co-axial is used for telephone networks and Broadband co-axial is used for cable TV. Co-axial cables have different sizes. They are classified by their size (RG) and by their resistance to direct or alternating electric currents. (Measured in ohms, also called impedance).

3.17.2. Co-axial Cable Standards

Table 3.4 shows the co-axial cable standards. The co-axial cables are categorised by their RG ratings where RG stands for Radio Government.

Table 3.4. Categories of Co-axial Cables

Category	Impedance	Application
RG-11	50 Ω	Thick Ethernet
RG-58	50 Ω	Thin Ethernet
RG-59	75 Ω	Cable T.V.

DO YOU KNOW?

Coaxial cable was initially deployed in the backbone of analog telephone networks where a single cable could be used to carry in excess of 10,000 simultaneous analog voice circuits. Digital transmission systems using coaxial cable were also deployed in the telephone network in the 1970s. These systems operate in the range of 8.448 Mbps to 564.922 Mbps. However, the deployment of coaxial cable transmission systems in the backbone of the telephone network was discontinued because of the much higher bandwidth and lower cost of optical fiber transmission systems.

3.17.3. Applications of Co-axial Cables

(i) Analog telephone networks.
(ii) Digital telephone network
(iii) Cable TV
(iv) Traditional Ethernet LANs
(v) Digital transmission
(vi) Thick and thin Ethernet

3.17.4. Baseband Co-axial Cable

1. Definition: The baseband co-axial cable is the one that makes use of digital signaling. The original Ethernet scheme makes use of baseband co-axial cable.

2. Caracteristics of Baseband Co-axial Cable: The baseband co-axial cables are used for digital signaling. The digital signal inserted on these cables is encoded using Manchester or Differential Manchester coding. The digital signal consumes the entire frequency spectrum of the cable. Hence, it is not possible to transmit multiple channel using FDM. The transmission of digital signal on the cable is bi-directional. The baseband co-axial cable was originally used for the Ethernet system that operates at 10 Mbps. These cables have a characteristic impedance of 50 W rather than 75 W of the cable TV co-axial cables. This is because for 50 W cables, the reflections taking place are less intense, they provide better noise immunity against low frequency electromagnetic noise as compared to the 75 W cable. The maximum length of baseband co-axial cable between two repeaters is dependent on the data rates. Lower the data rate longer is the cable. The length has to be reduced with increased data rates so as to reduce the probability of errors getting introduced. There are two systems that use the bus LANs namely 10 BASE 5 and 10 BASE 2 which are compared based on various factors in table 3.5.

Table 3.5. IEEE 802.3 Specifications for 10 Mbps Baseband Co-axial Cable Bus LAN

S. No.	Parameter	10 BASE 5	10 BASE 2
1.	Data rate	10 Mbps	10 Mbps
2.	Maximum segment length	500 m	185 m
3.	Network span	2500 m	1000 m
4.	Nodes per segment	100	30
5.	Node spacing	2.5 m	0.5 m
6.	Cable diameter	1 cm	0.5 cm

3.17.5. Broadband Co-axial Cable

This is the co-axial cable which is used in the cable TV system. The type of signaling is analog at radio frequencies. This system has certain drawbacks such as it is more expensive, more difficult to install and maintain as compared to the baseband co-axial cable. IEEE 802.3 standards have specified this as an option but practically, the broadband co-axial cables are not popular.

3.17.6. Connector for Co-axial Cable

Coaxial cable is another important type of guided transmission media. It has higher bandwidth as compared to that of twisted pair cable. To connect coaxial cable to device, we need coaxial cable connector. The most common type of connector used for coaxial cables is the Bayone-Neil-Concelman or BNC connectors. Figure 3.30 shows the various types of BNC connectors. The BNC connectors are available in three different types,

(i) BNC connector
(ii) BNC-T connector
(iii) BNC terminator

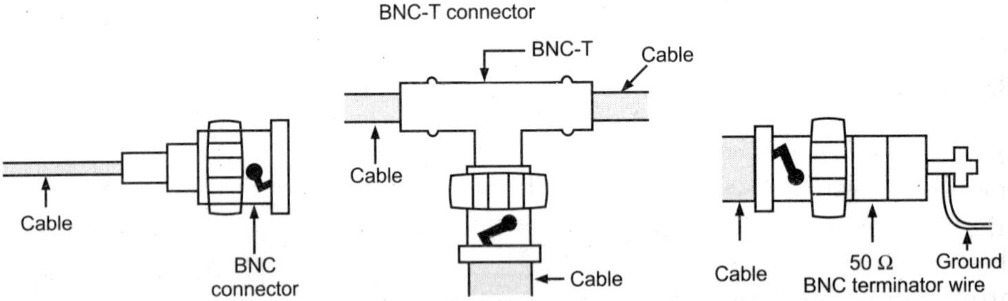

Fig. 3.30. BNC connectors of different types

The BNC connector is used to connect the end to end of the cable to a device such as a TV set. The BNC – T connector is used in Ethernet networks for branching out a cable for connection to a computer or other devices. The BNC terminator is used at the end of the cable to prevent the reflection of the signal.

3.18. OPTICAL FIBER CABLES

(RGTU, Bhopal, Sem. Exam., June, 2014)

The construction of an optical fiber cable is as shown in figure 3.31. It consists of an inner glass core surrounded by a glass cladding which has a lower refractive index. Digital signals are transmitted in the form of intensity - modulated light signal which is trapped in the glass core. Light is launched into the fiber using a light source such as a light emitting diode (LED) or laser. It is detected on the other side using a photo detector such as a phototransistor. The optical fiber cables are costlier than the other two types but they have many advantages over the other two types.

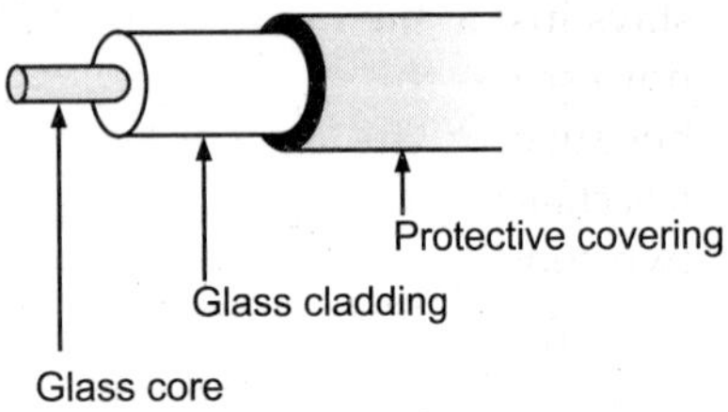

Fig. 3.31. Construction of optical fiber cable.

3.18.1. Applications

(i) Optical fiber transmission systems are widely used in the backbone of networks. Current optical fiber systems provide transmission rates from 45 Mb/s to 9.6 Gb/s using the single wavelength transmission.
(ii) The installation cost of optical fibers is higher than that for the co-axial or twisted wire cables.
(iii) Optical fibers are now used in the telephone systems.
(iv) In the Local Area Networks (LANs).

3.18.2. Light Sources for Fiber

For data transmission to take place, the sending device that is the, transmitter must be cabable of inducing data bits 0 to 1 into the light source. At the receiver, a photodiode is used to translate this light back into data bits as shown in figure 3.32.

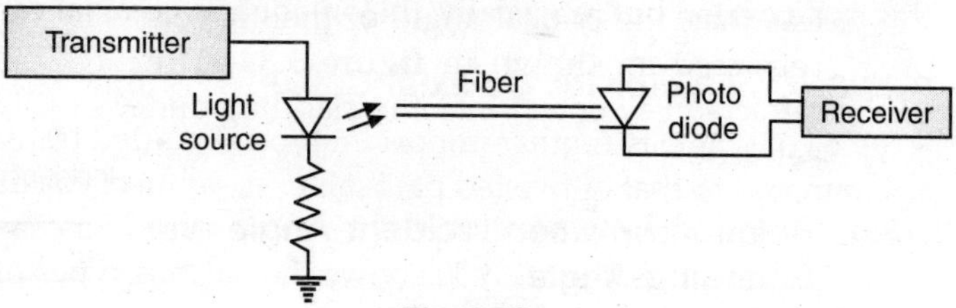

Fig. 3.32.

The two light sources that are most widely used may listed as under:

(i) LED (light emitting diode)

(ii) Injection laser diode (ILD)

The LED is cheaper but has a drawback that it provides an unfocussed light which hits the core boundaries and gets diffused. Hence, LED is preferred only for short distances. The laser diode can provide a very focussed beam that can be used for a long distance communication.

3.18.3. Working Principle of Light Propagation in a Fiber

1. Basic Concepts: The light enters into a glass fiber from one end, and gets reflected within the fiber. It follows a zigzag path along the length of the fiber as shown in figure 3.33(a). Figure 3.33(b) illustrates the principle of light travel through the optical fiber. When the light enters into a glass fiber from one end, most of it propagates along the length of the fiber and comes out from the far end.

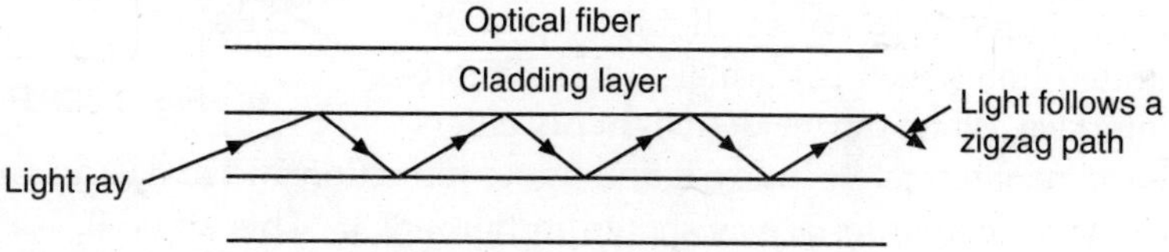

(a) Light follows a zigzag path within the optical fiber

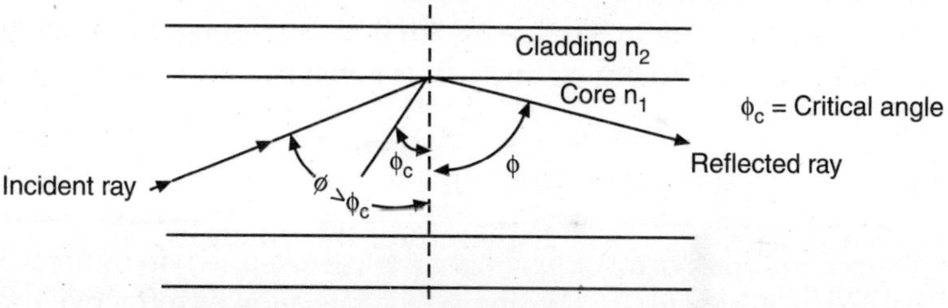

(b) Reflection at the interface of core and cladding

Fig. 3.33.

A small portion of the incident light escapes through the side walls of the fiber. The light that travels from one end to the other end of the glass fiber is said to have guided through the fiber. The light stays inside the fiber and does not escape through the walls because of the *total internal reflection* taking place inside the fiber. This total internal reflection can take place only if the following two conditions are satisfied :

(i) The glass fiber core must have a refractive index which is higher than the refractive index of the cladding around the core.

(ii) The angle of incidence of the light entering the fiber must be greater than the critical angle ϕ_c

$$\sin \phi_c = \frac{n_2}{n_1}$$

This has been shown in figure 3.33.

2. Observations from Figure 3.33(b): Few important observations from figure 3.33(b) may be listed as under:

(i) The angle of incidence (angle made by the incident ray) *i.e.*, ϕ is greater than the critical angle ϕ_c. Therefore, the incident light ray will be reflected within the core totally. The reflected ray is at same angle as that of the incident ray.

(ii) If the incident light makes an angle that is less than the critical angle ϕ_c then it gets refracted as shown in figure 3.34. The refracted ray enters into the cladding and gets lost.

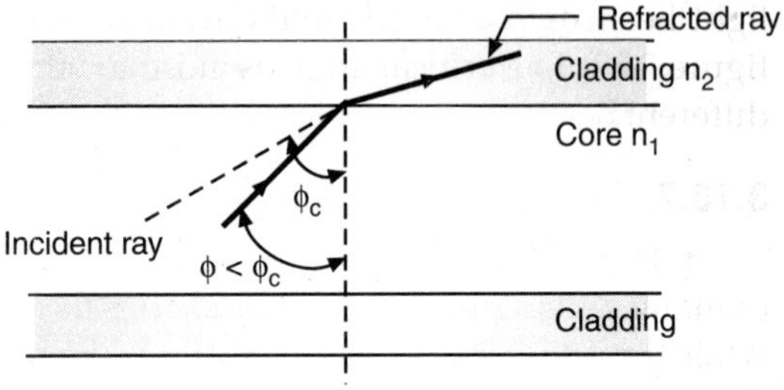

Fig. 3.34. Refraction takes place at the core cladding interface if $\phi < \phi_c$

3.18.4. Relation between Incident Angle and Emerging Angle

Let us obtain the relation between the incident angle ϕ_0 and the emerging angle θ_1 by referring to figure 3.35.

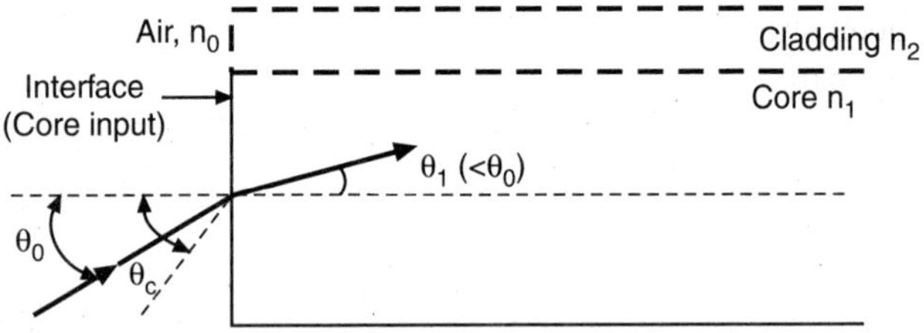

Fig. 3.35. Refraction at the interface

Let us assume that the refractive index of air is n_0 and that of the fiber core is n_1 such that $n_1 < n_0$. As shown in figure 3.35, the light ray enters the fiber core at an angle θ_0, through the air-core interface. The angle θ_0 is measured between the light ray and the dotted line which is normal to the air-core interface. When the incident light ray enters the core of refractive index, it undergoes refraction and makes an angle θ_1 with the dotted line normal to the air-core interface as shown in figure 3.35. This angle θ_1 is called as the emerging angle. The relation between the incident angle θ_0 and emerging angle θ_1 is given by *Snell's relationship* which states that,

$$n_0 \sin \theta_0 = n_1 \sin \theta_1 \qquad ...(3.5)$$

Therefore, the emerging angle θ_1 is given by,

$$\sin \theta_1 = \frac{n_0}{n_1} \sin \theta_0$$

Since, $n_0 < n_1$, $\frac{n_0}{n_1} < 1$, therefore, the emerging angle would be less than the angle of incidence θ_0.

3.18.5. Modes of Propagation

Basically, the modes refer to the number of paths followed by light rays inside the optical cable. Figure 3.36 shows different modes of operation of an optical fiber. There are two types namely single mode and multimodes. In single mode, light follows a single path through the core whereas in multimode, the light takes more than one paths through the core.

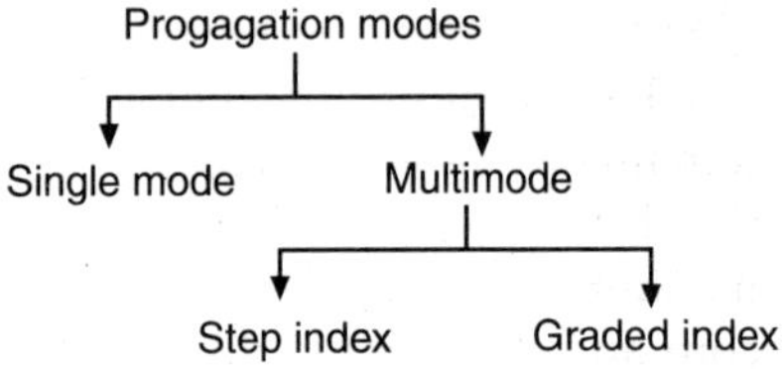

Fig. 3.36. Propagation modes in optical fibers

3.18.6. Single Mode Fibers

These are called as single mode fibers since they support on one mode of propagation (TE, TM or TEM). The optical signal travelling inside this fiber has only one group velocity. Due to single mode travelling, the amount of dispersion is less than that introduced in multimode fibers. These fibers can have either step index or graded index profile. They are high quality fibers used for wideband long haul communication and they are fabricated from doped silica to reduce internal attenuation. A highly focussed

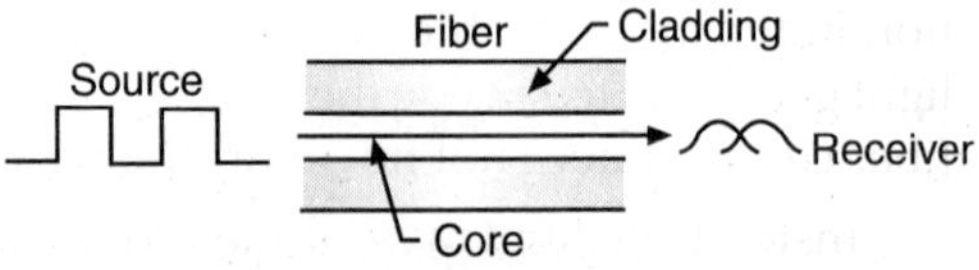

Fig. 3.37. Single mode fiber

light beam is produced by the light source. This beam travel's almost horizontally as shown in figure 3.37. The critical angle is close to 90°. As the critical angle is close to 90°, the propagation of different beams is more or less similar, the delays re negligible and the signal reconstruction is easier.

3.18.7. Multimode Fibers

1. Definition: These are called as multimode fibers since they support simultaneous propagation of many several modes. Each mode has its own group velocity and each mode follows its own path while travelling from the transmitter to the receiver. Because of the presence of more than one modes, the intermodal dispersion exists. Multimode fibers can have the step index or graded index profile and they are fabricated using the multicomponent glasses or doped silica.

2. Step Index Fibers: The construction of an optical fiber with a core and glass cladding has been shown in figure 3.38 (a). The refractive index of the core is n_1 and that of the glass cladding is n_2, with $n_1 > n_2$. Therefore, the index profile of glass clad core fiber has been shown in figure 3.38(c).

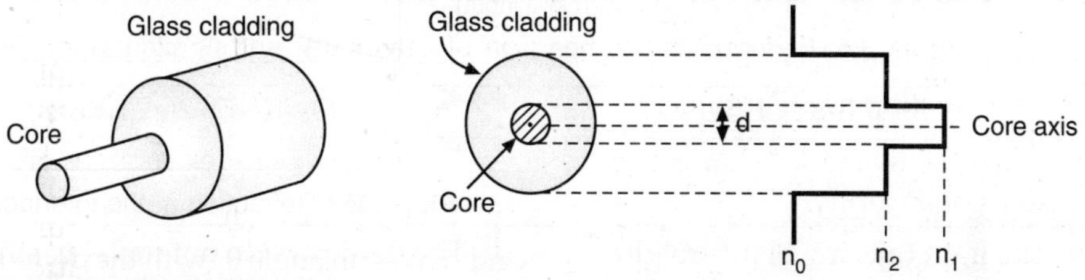

(a) Construction of a glass clad core type fiber

(b) Cross sectional view

(c) Index profile

Fig. 3.38.

Because of the sudden change in refractive index at the boundary of core and cladding, this fiber is called as **step index** fiber. Figure 3.39 illustrates the propagation of light over a step index fiber.

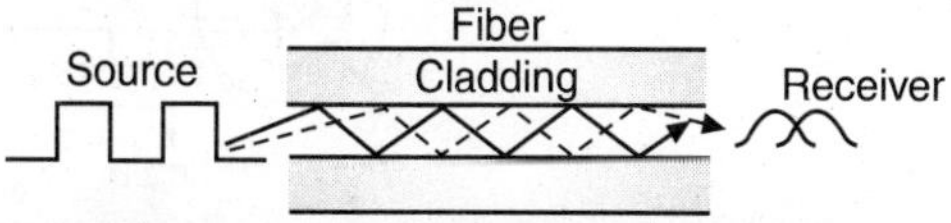

Fig. 3.39. Multimode step index fiber

Multiple beams follow different zigzag paths as shown in figure 3.39. The number of reflections that a beam undergoes, depends on the angle of incidence of that beam. Hence, at the destination, all the beams do not reach simultaneously. This leads to diffusion of signal at the receiver. The step index multimode fibers are therefore not used for long distance communications.

3. Graded Index Fibers: As shown in figure 3.40, the refractive index of the fiber core does not remain constant throughout its bulk. Instead it is maximum at the center of the core and reduces gradually towards the walls of the core. In order to get this type of index profile, the material in the fiber core is modified.

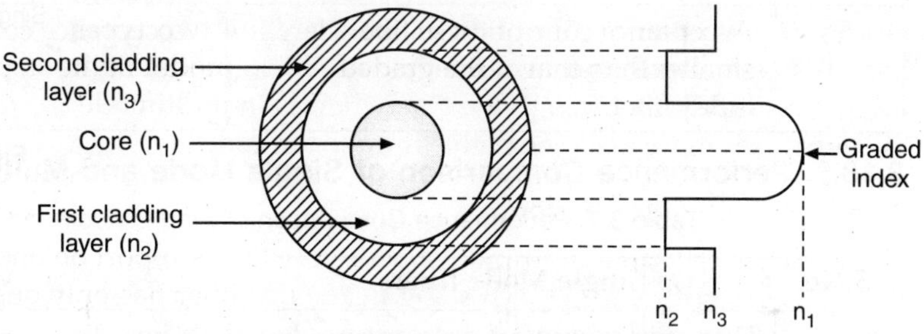

Fig. 3.40. Refractive index profile of a graded index fiber

Because of the modification in the index profile, the light gets refracted inside the fiber core and does not travel in straight line as shown in figure 3.41.

Instead, the light rays are curved towards the center of the core. These rays have been launched into the core within the acceptance cone. The acceptance cone of a graded index core is larger than that of the step index core. In graded index fibers as well, different beams result in different curves or waveforms. Table 3.6 shows the comparison of step index and graded index fibers.

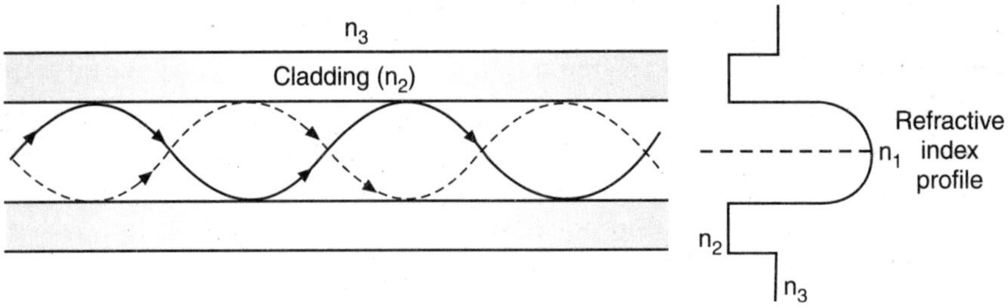

Fig. 3.41. Propagation of light in a graded index fiber

3.18.8. Performance Comparison of Step index and Graded index Fibers

Table 3.6. Performance Comparison of Step index and Graded index Fibers

S.No.	Step index fibers	Graded index fibers
1.	The refractive index changes in steps or abruptly.	The refractive index changes gradually.
2.	The light rays travel in straight line through the step index fibers.	The light rays do not travel in straight line through the graded index fibers.
3.	Index profile 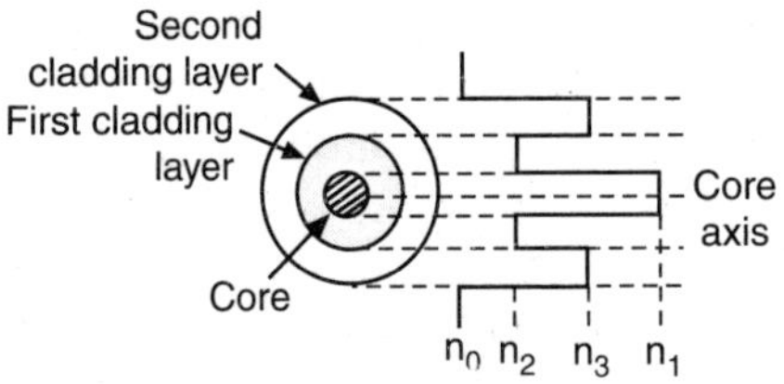 **Fig. A**	Index profile 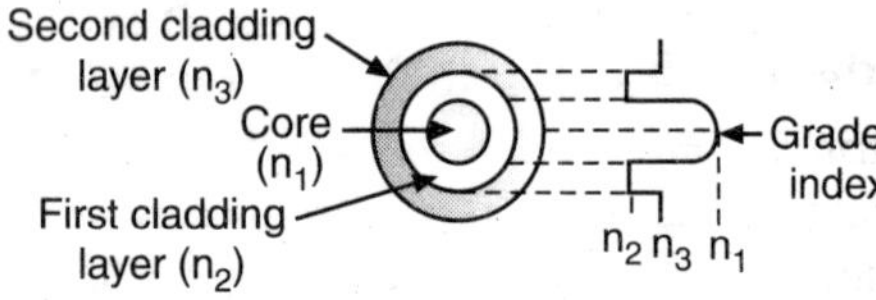 **Fig. B**
4.	The light rays travel in a straight line due to constant refractive index of the fiber throughout the bulk of the core.	The light rays do not travel in straight line due to the continuous refraction. This is due to the continuously changing refractive index throughout the core bulk.
5.	Acceptance cone of these fibers is smaller than that of the graded index fiber.	Acceptance cone of these fibers is larger than that of the step index fiber.

3.18.9. Performance Comparison of Single Mode and Multimode Fibers

Table 3.7. Performance Comparison of Single Mode and Multimode Fibers

S.No.	Single Mode fibers	Multimode fibers
1.	These fibers support only one mode of propagation (TE or TM or TEM).	These fibers support the propagation of many modes.
2.	The travelling signal inside the fiber has only one group velocity.	The different modes have different group velocities and each mode will follow its own path betwen the transmitter and receiver.

3.	The amount of dispersion introduced is less than that introduced in the multimode fibers.	The intermodal dispersion exists due to different group velocities of various modes.
4.	These fibers can have either a step index or graded index profile.	These fibers can have either step index or graded index profile.
5.	These are high quality fiber for wideband long haul transmission and are fabricated from doped silica for reducing the atte-nuation	These are fabricated using the multi-component glasses or doped silica.

3.18.10. Characteristics of Optical Fiber Cables

Fiber optic cables have the following characteristics:

(i) Fiber optic cabling can provide extremely high bandwidths in the range from 100 mbps to 2 gigabits because light has a much higher frequency than electricity.

(ii) The number of modes which a fiber optic can support does not depend on its length but on the hub or hubs that connect cables together.

(iii) Fiber optic cable has much lower attenuation and can carry signal to longer distances without using amplifiers and repeaters in between.

(iv) Fiber optic cable is not affected by EMI effects and can be used in areas where high voltages are passing by.

(v) The cost of fiber optic cable is more compared to twisted pair and co-axial.

(vi) The installation of fiber optic cables is difficult and tedious.

Important Point:

Three wavelength bands are used for fiber optic communication respectively 850 nanometer, 1300 nanometer, 1550 nanometer. Single mode fiber devices are more expensive and more difficult to install than multi-mode devices. Fiber optic cable connectors and splice (joint) attenuate the signals. Fiber optic cable supports 75 nodes in an Ethernet network. Single mode fiber optic cable are used to provide network links of several hundred kilometres in length. Fiber optic cable does not leak signals so it is immune to eves dropping (tapping of signals). Fiber optic cable does not require a ground, hence it is not affected by potential shifts in the electrical ground, nor does it produce sparks.

3.18.11. Advantages of Optical Fibers

Some of the advantages of fiber optic communication over the conventional means of communication may be listed as under :

1. Small Size and Light Weight: The size (diameter) of the optical fibers is very small (it is comparable to the diameter of human hair). Therefore, a large number of optical fibers can fit into a cable of small diameter.

2. Easy Availability and Low Cost: The material used for the manufacturing of optical fibers is *silica glass*. This material is easily available. So, the optical fibers cost lower than the cables with metallic conductors.

3. No electrical or Electromagnetic Interference: Since the transmission takes place in the form of light rays, the signal is not affected due to any electrical or electromagnetic interference.

4. Large Bandwidth: As the light rays have a very high frequency in the GHz range, the bandwidth of the optical fiber is extremely large. This allows transmission of more number of channels. Therefore, the information carrying capacity of an optical fiber is much higher than that of a co-axial cable.

5. Other Advantages: In addition to the advantages discussed earlier, the optical fiber communication has the following other advantages:

(i) No cross-talk inside the optical fiber cable.
(ii) Signal can be sent upto 100 times faster.
(iii) Intermediate amplifier are not required as the transmission losses in the fiber are low.
(iv) Ground loops are absent.
(v) Installation is easy as the fiber optic cables are flexible.
(vi) These cables are not affected by the drastic environmental conditions. Because of all these advantages, the optical fiber cable is replacing the conventional metallic conductor cable rapidly in many areas.

DO YOU KNOW?

Optical fiber transmissions systems provide the facilities for long-distance telephone communications and data communications. Regeneratorless optical fiber transmission systems are also used to interconnect telephone offices in metropolitan areas.

3.18.12. Drawbacks of Optical Fiber

Some of the disadvantages of optical communication system are :

(i) Sophisticated plants are required for manufacturing optical fibers.
(ii) The initial cost incurred is high.
(iii) Joining the optical fibers is a difficult job.

3.18.13. Fiber Optic Cable Connectors

Fiber optic cables use three types of connectors as shown in figure 3.42. The types are :

(i) Subscriber channel (SC) connector
(ii) Straight tip (ST) connector
(iii) MT–RJ connector.

The SC connector is used for cable TV. It uses a push/pull locking system.

The ST connector is used for connecting a cable to networking devices. It uses a bayonet locking system and is more reliable than SC.

MT-RJ is a new connector. It has the same size as RJ 45.

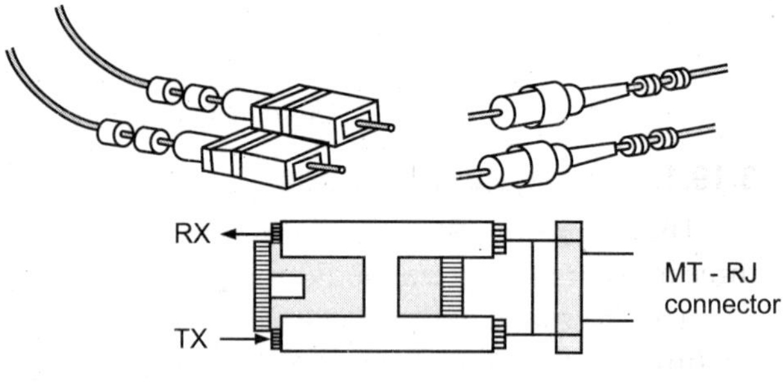

Fig. 3.42. Fiber optic cable connectors

3.18.14. Performance Comparison of Wired Media

S.No.	Twisted pair cable	Co-axial cable	Optical fiber
1.	Transmission of signals take place in the electrical form over the metallic conducting wires.	Transmission of signals takes place in the electrical form over the inner conductor of the cable.	Signal transmission takes place in an optical form over a glass fiber.
2.	Noise immunity is low. Therefore more distortion.	Higher noise immunity than the twisted pair of shielding conductor.	Highest noise immunity as the light rays are cable due to the presence unaffected by the electrical noise.
3.	Affected due to external magnetic field.	Less affected due to external magnetic field.	Not affected by the external magnetic field.
4.	Short circuit between the two conductors is possible.	short circuit between the two conductors is possible.	Short circuit is not possible.

5.	Cheapest.	Moderately expensive.	Expensive.
6.	Can support low data rates.	Moderately high data rates.	Very high data rates.
7.	Power loss due to conduction and radiation.	Power loss due to conduction.	Power loss due to absorption, scattering, dispersion and bending.
8.	Low bandwidth.	Moderately high bandwidth.	Very high bandwidth.
9.	Node capacity per segment is 2.	Node capacity per segment is 30 to 100.	Node capacity per segment is 2.
10.	Attenuation is very high.	Attenuation is low.	Attenuation is very low.
11.	Installation is easy.	Installation is fairly easy.	Installation is difficult.
12.	Electromagnetic interference (EMI) can take place.	EMI is reduced due to shielding.	EMI is not present.

3.19. UNGUIDED (WIRELESS) MEDIA *(RGTU, Bhopal, Sem. Exam., June, 2014)*

The electromagnetic spectrum used for wireless communication is shown in figure 3.43.

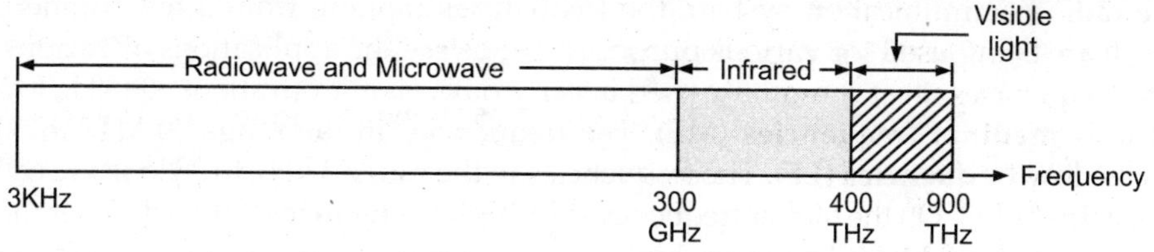

Fig. 3.43. Electromagnetic spectrum for the wireless communication

3.19.1. Propagation Methods

The unguided signals can travel from the transmitter to receiver in many different ways; the three most important methods are :

(i) Ground wave propagation (ii) Sky propagation

(iii) Space propagation or line of sight propagation.

3.19.2. Bands

The electromagnetic spectrum is divided into several subbands. Table 3.8 gives various frequency bands, corresponding types of propagation and application.

Pioneer in Communications: Jean Baptiste Joseph Fourier (1768–1830)

Jean Baptiste Joseph Fourier was born in Auxerre, Bourgogne, France, in 1768. During his youth age, Fourier studied Latin and French and, by the time he was 13, attained a strong passion for mathematics. Fourier also had deep interests in religion and began training for the priesthood in 1787. Ultimately, though, his love of mathematics, coupled with his strong aspiration to make a significant contribution to the mathematics field, led Fourier in 1789 to give up his religious studies to pursue mathematical research.

Table 3.8. Segments of the Electromagnetic Spectrum

S.No.	Name	Frequency	Wavelength
1.	Extremely low frequency (ELF)	30-300 Hz	10^7 to 10^6 m
2.	Voice frequencies (VF)	300-3000 Hz	10^6 to 10^5 m
3.	Very low frequencies (VLF)	3-30 kHz	10^5 to 10^4 m
4.	Low frequencies (LF)	30-300 kHz	10^4 to 10^3 m
5.	Medium frequencies (MF)	300 kHz - 3 MHz	10^3 to 10^2 m
6.	High frequencies (HF)	3-30 MHz	10^2 to 10 m
7.	Very high frequencies (VHF)	30-300 MHz	10 to 1 m
8.	Ultra high frequencies (UHF)	300 MHz-3GHz	1 to 10^{-1} m
9.	Super high frequencies (SHF)	3-30 GHz	1^{-1} to 10^{-2} m
10.	Extremely high frequencies (EHF)	30-300 GHz	10^{-2} to 10^{-3} m
11.	Infrared	—	0.7 to 10 μm
12	Visible light	—	0.4 μm to 0.8 μm

3.19.3. EM Spectrum and Communication Applications

In the radio communication system, the frequencies ranging from a few kilohertz to many gigahertz all are being used for various purposes. Let us see the applications of various frequency bands. The frequencies most commonly used in early days were from about 300 kHz to 3 MHz and were called as **medium frequencies (MF).** The frequencies in the range 30 kHz to 300 kHz are known as the **low frequencies (LF).** The frequencies in the range 3 kHz to 30 kHz are called as **very low frequencies (VLF).** On the higher frequency side, **high frequencies (HF)** will cover the frequency range from 3 MHz to 30 MHz. Then **very high frequencies (VHF)** from 30 MHz to 300 MHz and so on.

Table 3.9 gives the details of entire usable frequency spectrum and its applications.

Table 3.9. The Radio Frequency Spectrum

S. No.	Name	Frequency	Wavelength
1.	30 Hz-300 Hz. Extremely low frequencies ELF.	10^4 km to 10^3 km	Power transmission
2.	300 Hz-3 kHz. Voice frequencies (VF)	10^3 km to 100 km	Audio applications
3.	3 kHz-30 kHz. Very low frequencies (VLF)	100 kim to 10 km	Submarine communications, Navy, Military communications
4.	30 kHz-300 kHz. Low frequencies (LF)	10 km to 1 km. Long waves	Aeronautical and marine, navigation, these frequencies act as sub-carriers
5.	300 kHz-30 MHz Medium frequencies (MF)	1 km to 100 m. Medium waves.	AM radio broadcast, Marine and aeronautical communications
6.	3 MHz-30 MHz High frequencies (HF)	100 m to 10 m Short waves.	Shortwave transmission, Amateur and CB communication
7.	30 MHz - 300 MHz Very high frequencies (VHF)	10 m to 1 m	TV broadcasting, FM broadcasting

8.	300 MHz - 3 GHz Ultra high frequencies (UHF)	1 m to 10 cm. Microwaves.	UHF TV channels, Cellular phones, Military applications
9.	3 GHz - 30 GHz (SHF)	10^{-1} m to 10^{-2} m	Satellite communication and radar
10.	30 - 300 GHz (EHF)	10^{-2} m to 10^{-3} m	Satellites and specialized radars.

3.19.4. Infrared Signals

The EM signals having frequencies above 300 GHz are not referred as radio wave. The signal occupying the range between 0.1 mm and 700 nanometer (nm) are called infrared signals. These are used in various special kinds of communications as listed below :

(i) In astronomy to detect stars and other heavenly bodies.
(ii) For guidance in weapon systems.
(iii) TV remote control.

3.19.5. Visible Light

Light is a special type of electromagnetic radiation. It has wavelength in the range of 0.4 to 0.8 μm. Light is used for various kinds of communications. Light waves can be modulated and transmitted through the glass fibers in the optical fiber communication system. Light signals can also be transmitted through free space. Laser is a type of light, which can be easily modulated with voice, video and data information.

3.20. TYPES OF WIRELESS MEDIA

It does not use an electrical or optical conductor. In most cases, the earth atmosphere is the physical path for the data. Wireless media is used when distance or obstructions make cable media difficult. There are three main types as under:

(i) Radiowave
(ii) Microwave
(iii) Infrared.

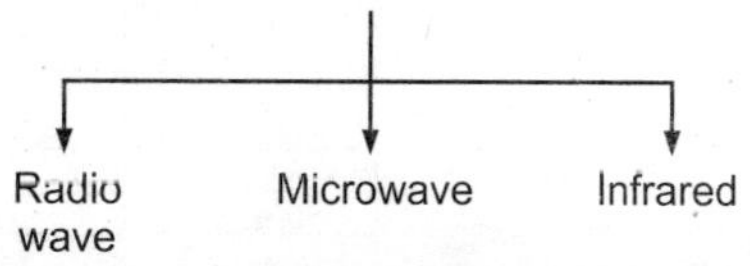

Fig. 3.44. Classification of wireless media

3.20.1. Radio Wave Transmission Systems

1. Basic Concepts: Radiowaves have frequencies between 10 kHz and 1 gigahertz. The range of electromagnetic spectrum between 10 kHz and 1 gHz is called radio frequency (RF).

2. Radiowaves include the following types :

(i) Short wave used in AM radio.
(ii) Very High Frequency (VHF) used in FM radio and TV.
(iii) Ultra High Frequency (UHF) used in TV.

The radio frequency bands are regulated and require a license from the regulatory body. Unregulated frequency bands are also present which operate at less than 1 watt transmitted power.

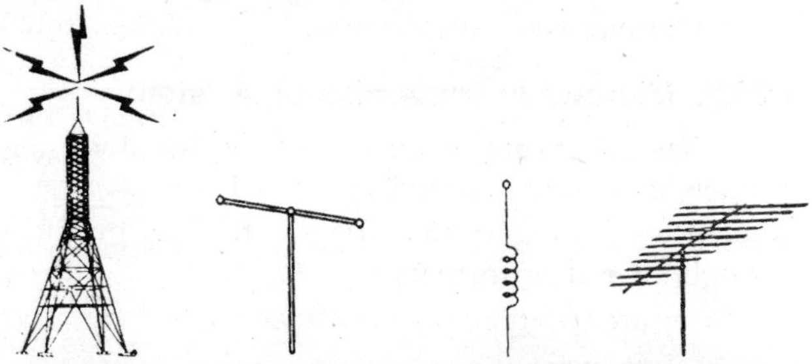

Fig. 3.45. Various types of antennas

Radiowaves can broadcast omni-directionally or directionally. Various kinds of antennas are used to broadcast these signals as shown in figure 3.45.

The power of the RF signal is determined by the antenna and transceiver.

Each range has charac-teristics that affect its use in computer network. For computer network applica-tions, radiowaves fall into three categories :

(i) Low power, single frequency
(ii) High power, single frequency
(iii) Spread - spectrum.

3. Characteristics of three Types of Radio Waves: The characteristics of the three types of radiowaves are given in table 3.10.

Table 3.10.

S.No.	Characteristics	Low power single frequency	High power single frequency	Spread spectrum
1.	Frequency range	All radio frequencies	All radio frequencies	All radio frequencies (typically 902 to 928 MHz)
2.	Bandwidth capacity	1-10 Mbps	1-10 Mbps	2-6 Mbps
3.	Attenuation	High	Low	High
4.	EMI	Poor	Poor	Fair
5.	Installation	Simple	High	Moderate
6.	Cost	Low	Higher	Moderate

The various areas of applications and the corresponding distances involved are given in table 3.11.

Table 3.11.

S.No.	System	Distance
1.	Paging	Tens of kilometres
2.	Cordless telephone	Tens of meters
3.	Cellular phone	Few hundred km
4.	Wireless LAN	100 m

4. Important Applications: Some of the important applications of radio transmission systems are as under:

(i) Cellular communication
(ii) Wireless LAN
(iii) Point to point and point to multipoint radio systems
(iv) Satellite communication.

DO YOU KNOW?

Radio transmission medium found limited applications in data communications till recently. Radio medium is electromagnetic interference prone and has limited bandwidth. Wireless local area networks have become very popular in last few years.

3.20.2. Microwave Transmission System

1. Basic Concept: It makes use of the lower gigahertz frequencies of the electromagnetic (EM) spectrum. These frequencies, are higher than the RF and they produce better throughput and performance.

There are two types of microwave data communication systems as under:

(i) Terrestrial
(ii) Satellite

2. Microwaves: Microwaves are basically electromagnetic waves having frequencies between 1 and 300 GHz. Microwaves are unidirectional. Microwave propagation is a line-of-sight propagation.

As a matter of fact, the microwave band is wide (299 GHz) and so it is possible to allot wider band. Therefore, it can support high data rates.

3. Terrestrial Microwave Systems: These systems use directional parabolic antennas to transmit and receive signals in the lower gigahertz range as shown in figure 3.46. The signals are highly focussed and the physical path must be line of sight. Relay towers are used to extend signals. Smaller terrestrial microwave systems can be used within a building. Microwave LANs operate at low power using small transmitters that communicate with omni-directional hubs. Hubs can then be connected to form an entire network.

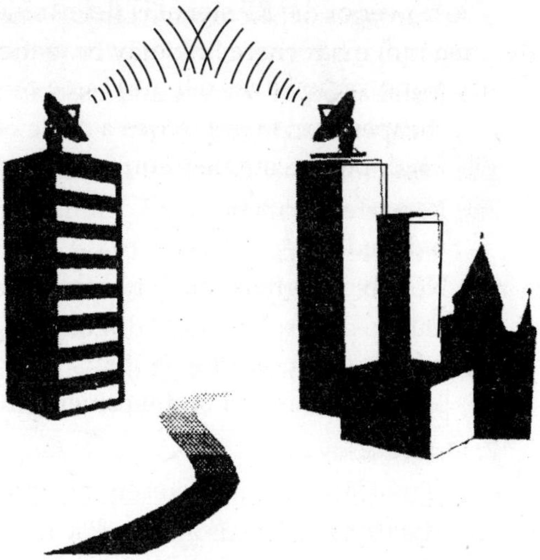

Fig. 3.46. A terrestrial microwave system

4. Characteristics of Terrestrial Microwave Systems: Terrestrial microwave systems have the following characteristics:

(i) The frequency range used is from 4 - 6 GHz and 21 to 23 GHz.

(ii) It supports a bandwidth from 1 to 10 mbps.

(iii) Attenuation is affected by frequency, signal strength, antenna size and atmospheric conditions.

(iv) The signals are affected by EMI effect, jamming and eavesdropping.

(v) Line of sight requirements make installation difficult.

(vi) Short distance systems can be inexpensive but long distance systems are relatively expensive.

5. RF link (microwave link): Long form of RF link is radio frequency link. This is actually a type of point to point wireless communication.

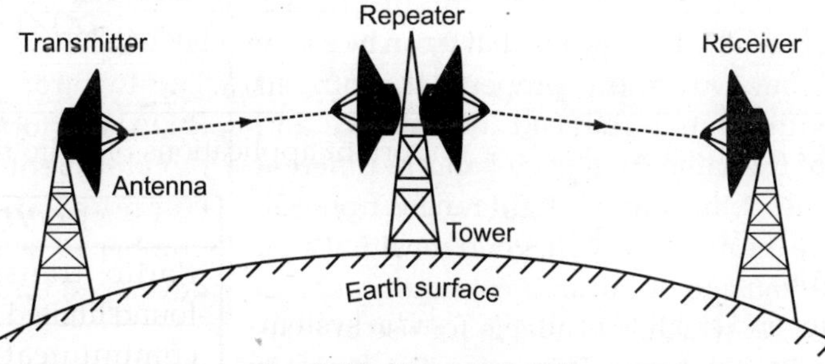

Fig. 3.47. Microwave link

The radio frequencies used for RF links are in microwave range, therefore, RF links are also called microwave links. This is shown in figure 3.47. Although many wire communication systems use copper wires or optical fiber, some just send the signal into the air. This happens when infrared, lasers, microwaves and radio are used for the transmissions of data, as they do not need any physical medium. For long distance communication, microwave radio transmission is widely used as an alternative to co-axial cable. The signal transmission takes place in the form of electromagnetic waves which have wavelengths of few centimeters. Parabolic antennas can be mounted on the towers to send a beam of waves to another antenna, tens of kilometres away. The transmitting and receiving antennas are highly directional to enable a point to point communication. This system is widely used for both telephone and television transmission. The higher the tower which holds the antenna, the greater is the range. With a 100 metre high tower, the distances of 100 km can be easily covered.

DO YOU KNOW?

The widest use of coaxial cable is for distribution of television signals in cable TV systems. Existing coaxial cable systems use the frequency range from 54 MHz to 500 MHz.

6. Advantages of Microwave Link: Some of the important advantages of microwave link may be listed as under:

(i) Installation of towers and associated equipment is cheaper than laying down a cable of 100 km length.

(ii) Less maintenance as compared to cables.

(iii) Repeaters can be used. Hence effect of noise is reduced.

(iv) No adverse effects such as cable breakage etc.

(v) Due to the use of highly directional antenna, these links do not make any interference with other communication systems.

(vi) Size of transmitter and receiver reduces due to the use of high frequency.

7. Drawbacks

(i) Signal strength at the receiving antenna reduces due to multipath reception.

(ii) The transmission is affected by the thunderstorms, and other atmospheric phenomena.

8. Applications of Microwave Transmission

(i) One-to-one communication

(ii) In cellular phones

(iii) In satellite networks

(iv) In the wireless LANs.

3.21. USE OF INFRARED LIGHT FOR TRANSMISSION

The electromagnetic waves having frequencies from 300 GHz to 400 THz (wavelengths from 1 mm to 770 nm) are known as **infrared waves**. IR uses line-of sight propagation. Infrared light is a communication medium whose properties are significantly different from those of the radio frequencies. A very important property of the infrared light is that it cannot penetrate walls. This means that it can be easily contained within a room. Due to this property, the infrared light can be used with a much reduced interference and possibility of reuse of the same frequency band in different rooms. The wavelength of the infrared light ranges from 850 nm and 900 nm, where the receivers with good sensitivity are available. Another advantage of infrared communication is the very large bandwidth which is available for use but has not been exploited to its full extent. The major disadvantage is that the sun generates radiation in the infrared band. This can cause a lot of interference with the IR communication. The infrared band can be used in development of very high speed wireless LANs in future.

DO YOU KNOW?

The Infrared Data Association (I_rDA) was formed to promote the development of Infrared light communications systems. A number of standards have been developed under its auspices.

3.21.1. Applications and Standards

A number of standards have been developed for an infrared data link (IRDA). The IRDA-C standard provides the standards for the bidirectional communications used in cordless devices such as mice; keyboards, joysticks and hand held computers. The IRDA-C standard operates at a bit rate of 75 kbits/sec and the distance range is upto 8 meters. Another standard called IRDA-D standard provides for the data rates from 115 kb/s to 4 Mb/s, with a distance range upto 1 metre. The IR data link was designed as a wireless alternative to connecting devices such as laptop to a printer. Point to point transmission allows for better data rates but devices must remain in their locations. Broadcast allows for more flexibility but with lower data rates. One advantage of infrared is that an FCC license is not required to use it. The only disadvantage of infrared signals is that they cannot penetrate walls or other objects and they are diluted by strong light sources.

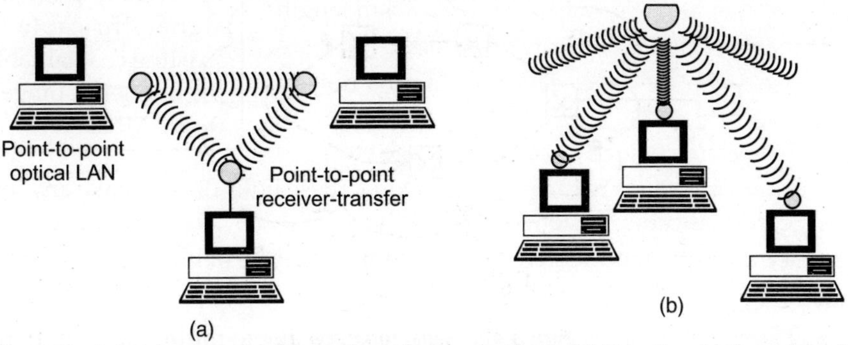

Fig. 3.48. (a) Point-to-point infrared media in a network (b) Broadcast infrared media

3.21.2. Performance Comparison of Point to Point and Broadcast Infrared System

S. No.	Characteristics	Point to Point	Broadcast
1.	Frequency range	100 GHz to 1000 terahertz	100 GHz to 1000 terahertz
2.	Bandwidth capacity	Data rates between 100 kbps to 16 mbps	Data rates less than 1 mbps.
3.	Node capacity	2 (source to destination)	More than one.
4.	Attenuation	Depends on quality of emitted light, its purity, atmospheric conditions and signal obstructions.	Depends on quality of emitted light, its purity, atmospheric conditions.
5.	EMI	Affected by intense light.	Affected by intense light.
6.	Installation	Requires precise alignment.	Fairly simple.
7.	Cost	Moderately high if laser is used.	High if laser is used.

3.21.3. Applications of Infrared

(i) Very high data rates can be supported, due to very high bandwidth (approximately 400 THz).

(ii) For communication between keyboard, mouse PCs and printers.

3.22. INTRODUCTION TO SWITCHING *(RGTU, Bhopal, Sem. Exam., June, 2014)*

A network consists of many switching devices. In order to connect multiple devices one solution could be to have a point to point connection between each pair of devices. But this increases the number of connections. The other solution could be to have a central device and connect every device to each other via the central device (Star topology). Both these methods are wasteful and impractical for very large networks. The other topologies also cannot be used.

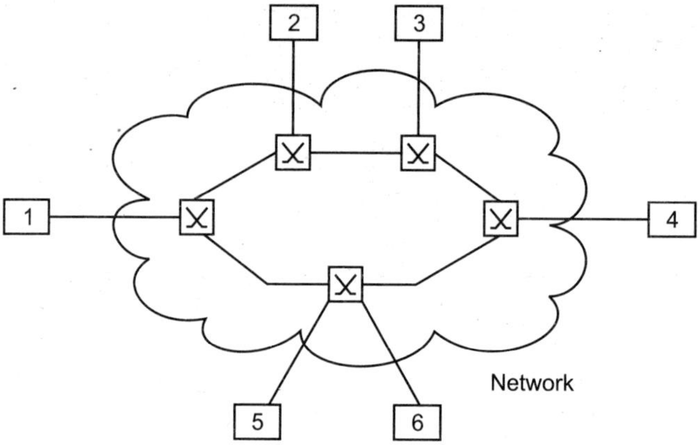

Fig. 3.49. Switched network

Hence, a better solution is **switching**. A switched network is made of a series of interconnected nodes called switches. Switch is a device that creates temporary connections between two or more devices. Figure 3.49 shows a switched network.

3.23. SWITCHED COMMUNICATION NETWORKS

As a matter of fact, the transmission of data beyond a local area is achieved by using a network of intermediate switching nodes. These switching nodes are not concerned about the contents of data being transferred but they provide a fabric to move the data from source to destination. Figure 3.50 shows a simple communication network. The devices connected to it are called as stations. The stations may be of following:

(i) Computers
(ii) Terminals
(iii) Telephones
(iv) Other communication devices

The switching devices used for the purpose of communication are called as nodes. These nodes are connected to each other in some topology by the transmission links such as co-axial cables, microwave links etc.

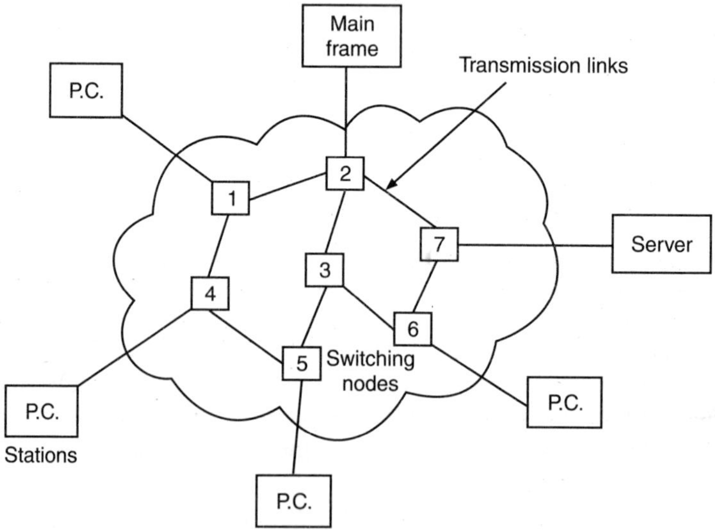

Fig. 3.50. Switched communications network

Key Point: The stations, nodes and transmission links are collectively known as the **switched communications network**.

The data emerging from the source is said to be routed to the destination. The node to node links are generally multiplexed using either FDM (Frequency Division Multiplexing) or TDM (Time Division Multiplexing).

3.24. SWITCHING METHODS *(RGTU, Bhopal, Sem. Exam., June, 2014)*

The three basic methods of switching are :

(i) Circuit switching

(ii) Packet switching

(iii) Message switching

Out of these, the circuit and packet switching are commonly used today but the message switching has been phased out in general communication but is still used in the networking applications.

Figure 3.51 shows the classification of switching methods.

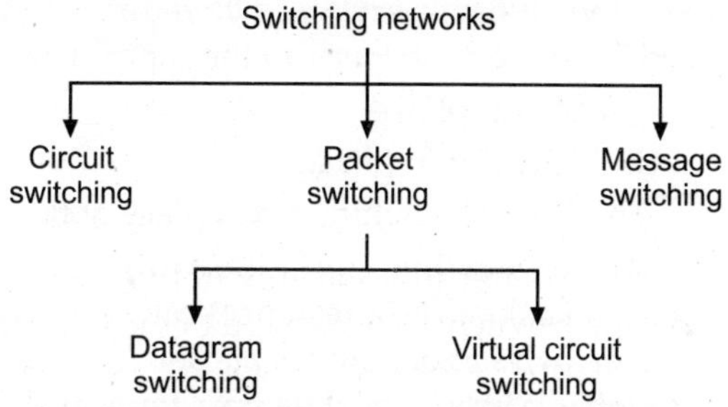

Fig. 3.51. Classification of switching methods.

Key Point: A communication network that has more than two computers must establish links between computers in order for them to be able to communicate with each other.

3.25. CIRCUIT SWITCHING NETWORKS* *(RGTU, Bhopal, Sem. Exam., June, 2014)*

Circuit switching is used in public telephone networks. It was developed to handle voice traffic but it can also handle digital data.

However handling of digital data using the circuit switching generally proves to be inefficient. Using the circuit switching, a dedicated path is established between two stations for communication.

The telephone network provide telephone service which involves the two way, real-time transmission of voice signals across a network.

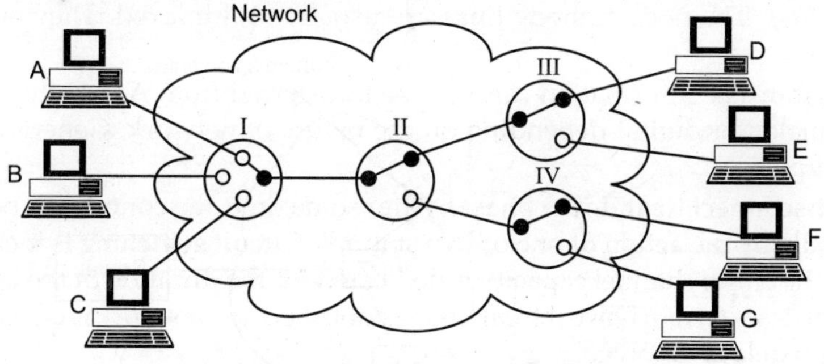

Fig. 3.52. Circuit-switched network

The network connection allows electrical current and the associated voice signal to flow between the two users. The end to end connection is maintained for the duration of the call. The telephone networks are connection oriented because they require the setting up of a connection before the actual transfer of information can take place.

* A device called a switch is used to connect stations.

The transfer mode of a network that involves setting up a dedicated end to end connection is called circuit switching. In circuit switching the routing decision is made when the path is set up across the network. After the link has been set between the sender and receiver, the information is forwarded continuously over the link. After the link has been set up no addition address information about the receiver or destination machine is required. In circuit switching a dedicated path is established between the sender and the receiver which is maintained for the entire duration of conversation, as shown in figure 3.52. In telephone systems circuit switching is used. If circuit switching is used in computer networks the sending PC first establishes a link with the receiving link. After the link is established data is transmitted from the sender to the receiver. After the data flow stops, the link is released. In figure 3.52, I, II, III and IV are called as the switching nodes. They are used to connect one user to the other. The circuit switched networks operate in three phases :

(i) Set up phase (ii) Data transfer phase

(iii) Tear down phase.

The circuit switching takes place at the physical layer.

Before starting communication in the set up phase the resources are reserved during communication. These resources can be channels, switch buffers, input/output ports. Data transferred between two stations are not packetized instead the data gets transferred continuously. No addressing is involved during the data transfer as the dedicated connection is established between the sender and receiver. The switches route the data on the basis of their occupied frequency band (FDM) or time slot (TDM).

Key Point: Circuit switching is a network switching method that physically connects two nodes to facilitate communications.

3.25.1. Three Phases

Communication via circuit switching takes place over three phases of operation as follows :

(i) Circuit establishment (ii) Data transfer

(iii) Circuit disconnect (tear down).

(i) Circuit establishment: In a circuit switching network, before any signal is transmitted, it is necessary to establish an end-to-end (station to station) link. For example, in figure 4.3, if the communication is to be between A and D, then the path from A to node I to node II to III and D has to be established first. The node to node links are usually multiplexed. They either use FDM or TDM.

(ii) Data transfer: The information can now be transferred from A to D through the network. The data can be analog or digital depending on the nature of network. Generally, all the internal connections are duplex.

(iii) Circuit disconnect (tear down phase): After some time, the connection between two users is terminated usually by the action of one or two stations. Circuit switching is inefficient in most of the applications. The entire channel capacity is dedicated for the duration of the connection, even if the data is not being transferred. Once the circuit is established, the network is effectively transparent to the users with no delays involved.

3.25.2. Efficiency*

In circuit switching, the resources remain dedicated as long as a connection is alive. Due to the allocation of resources during the entire duration of the connection, the efficiency of circuit switched networks is lower than the other two types of switching.

* In circuit switching, also called a connection-oriented circuit, a physical connection must be established for the duration of the transmission (such as in a telephone system).

3.25.3. Delay

Even though the efficiency is low, the delay in this type of networks is minimum. Figure 3.53 explains the idea of delay in the circuit switched networks. When only two switches are used. During the data transfer, the data are not delayed at each switch. The total delay is due to the time needed to create the connection, transfer data and disconnect the circuit.

The delay at the time of set up is the sum of the following four parts :

1. The propagation time of the source computer request (slope of the first gray box in figure 3.53.
2. The request signal transfer time (height of the first gray box in figure 3.53)
3. The propagation time of the acknowledgement from the destination computer (slope of the second gray box in figure 3.53).
4. The propagation time required to transfer the acknowledgement from destination computer (height of second gray box).

The delay for the data transfer phase is the sum of the following two parts :

1. The propagation delay (slope of hatched portion).
2. Time to transfer data (height of hatched portion) which can be very long.

The third component of delay is the delay corresponding to the disconnect or tear down phase. In figure 3.53, we have shown the case in which the destination computer requests disconnection because this creates the maximum delay.

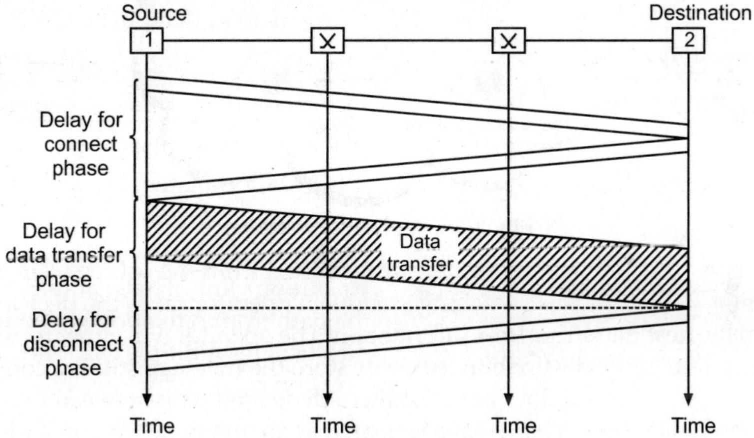

Fig. 3.53. Delay in circuit switching

Application

The circuit switching is used in the telephone networks.

3.25.4. Advantages*

(i) The major advantage of circuit switching is that the dedicated transmission channel, the computers establish provides a guaranteed data rate.

(ii) In circuit switching because of the dedicated path there is no delay in data flow.

3.25.5. Disadvantages

(i) The disadvantages of circuit switching is that since the connection is dedicated it cannot be used to transmit any other data even if the channel is free.

* Circuit switching is used for real-time communication. There is no delay or congestion in the communication link because a physical connection exists between the source and the destination.

(ii) Dedicated channels require more bandwidth.
(iii) It takes long time to establish connection.

3.25.6. Circuit Switched Technology in Telephone Networks

The telephone companies previously used the circuit switching technology for switching and routing a call. This was a physical layer technology. However, today the tendency is to use other switching techniques. For example the telephone number is used as the global address and a signalling system (called SS7) is used for set up and tear down phases.

3.26. TELEGRAPH NETWORKS AND MESSAGE SWITCHING

(RGTU, Bhopal, Sem. Exam., June, 2014)

In telegraphy, the text message is encoded using the Morse code into sequences of dots and dashes. Each dot or dash is communicated by transmitting short and long pulses of electrical current over a copper wire.

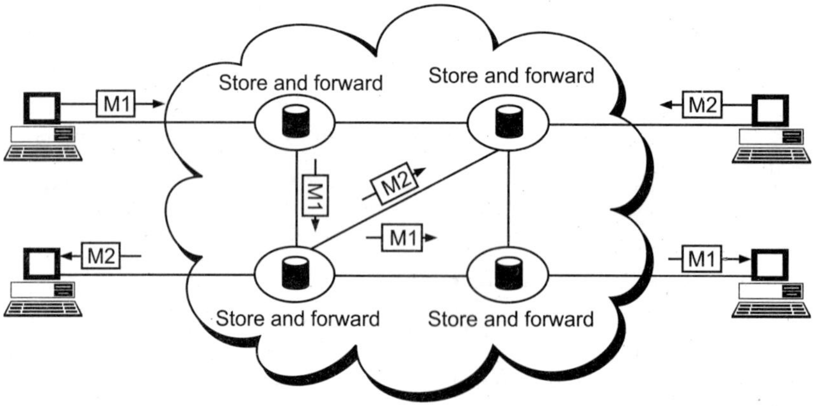

Fig. 3.54. Message switching

In telegraph networks, the text message is transmitted from the source telegraph office to the telegraph switching station. At this switching station an operator takes the decision of routing the message based on the destination address information. The operator will either forward the message if a communication line to the destination is free or store the message still the communication line becomes free. Message switching does not establish a dedicated path between two communicating devices. In message switching, each message is treated as an independent unit and includes its own destination and source address. Each complete message is then transmitted from device to device through the internetwork as shown in figure 3.54.

Each intermediate device receives the message, stores it, until the next device is ready to receive it and then forwards it to the next device. For this reason, a message switching network is sometimes called as a store and forward network. Messages switches can be programmed with information about the most efficient routes as well as information regarding neighbouring switches that can be used to forward messages to their ultimate destination.

3.26.1. Advantages

(i) It provides efficient traffic management by assigning priorities to the messages to be switched.
(ii) It reduces network traffic congestion because it is able to store message until a communication channel becomes available.
(iii) With message switching, the network devices share the data channels.
(iv) It provides asynchronous communication across time zones.

3.26.2. Disadvantages

1. The storing and forwarding introduces delay hence cannot be used for real time applications like voice and video.
2. The intermediate devices require a large storing capacity since it has to store the message unless a free path is available.

3.26.3. Delay in Message Switching*

A message switched network consists of store-and-forward switches interconnected by trunks. A single trunk is usually sufficient between a pair of switches. Multiple trunks can be provided to increase reliability. Each switch is equipped with a storage device wherein all incoming messages are temporarily stored for onward transmission. The basic operation of the store-and-forward service is similar to the telegram service. A message along with the destination address is sent from switch to switch till it reaches the destination. Let us say end system A wants to send a message to end system B as shown in figure 3.55. A sends its message along with the address of the destination and its own address to entry switch 1. The addresses are included in the header of the message. Switch 1 accepts the message and analyzes the destination address. A routing table is maintained at each node. It contains entries indicating destination nodes and the corresponding outgoing trunks from the switch. There is separate queue for each trunk.

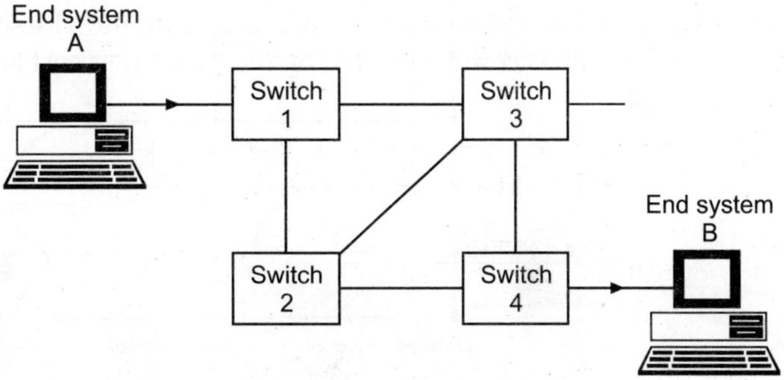

Fig. 3.55. Message switched data network

Since the destination node may be accessible via more than one route, decision to send the message to a particular next switch depends on the expected delay in its queue. Let us say, the message from A is put in the queue for node 2. The message received at switch 2 is again put in a queue of message awaiting transmission to switch 4. When its turn comes, the message is sent to switch 4 which delivers it to the destination.

Some of the basic features of store-and-forward message switching are as under :

(i) The store-and-forward service is unidirectional. After delivery of the message, the network does not send back any confirmation to the source. If end system B is required to send an acknowledgement to the message received from A, the acknowledgement is treated like any other message by the network and carries the addresses of the destination and the source.

(ii) For switch-to-switch transfer of the message, the network may employ some error control mechanism. The message may be appended with error-checking bits and if any error is detected by the receiving switch, it may request the sending switch for retransmission of the message. Therefore, the sending switch is required to keep a copy of the message till an acknowledgement is received.

(iii) Since the message is stored in a buffer at the switch at each stage of transmission each switch-to-switch transfer is an independent operation. The trunks can operate at different data rates. Even the source and destination and systems can operate at different data rates.

* In message switching, the switch needs to have a large buffer to store incoming messages from other links.

(iv) In message switching every message is treated as an independent entity by the network and therefore, destination and source addresses are repeated on each message.

(v) Delay in Delivery, figure 3.56 shows the timing diagram for routing a message through a message switched network. The message passes through the entry switch, two transit switches and finally through the exit switch to arrive at the destination.

Message delivery time is the sum of the following components :

(i) Time required to send the message to the èntry switch.

(ii) Switch delay.

(iii) Transmission time at each switch.

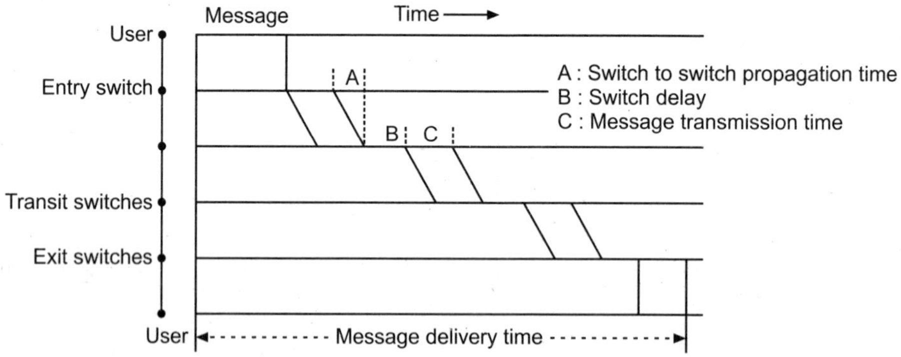

Fig. 3.56. End to end delay in message delivery

The time required to send a message to the entry switch is determined by the transmission data 1 rate and the message size. Propagation time to the entry switch is usually negligible. Switch delay is due to two factors.

(i) Message processing at each switch (time required for error checking, routing etc.)

(ii) Waiting time in the queues at each switch.

The transmission time at each switch is determined by the transmission data rate and propagation p time for transmission across the trunk. The total time required to deliver the message is the linear sum of all these components of time as they occur in a sequential manner. The delivery time varies from message to message because of random waiting times, in queues and alternate routes between the same pair of entry and exit switches. Therefore, time relationship of the messages and their sequence are not guaranteed in a message switched network. As traffic increases, there is increase in message delivery time, because the queues get longer and there may be congestion on the route.

Key Point: Message switching is a network switching achieved by storing incoming messages and sending them out as lines become available.

3.27. PACKET SWITCHING

(RGTU, Bhopal, Sem. Exam., June, 2014)

In packet switching, message are broken up into packets, each of which includes a header with source, destination and intermediate node address information. Individual packets take different routes to reach the destination. Independent routing of packets gives two advantages:

(i) Bandwidth is reduced by splitting data onto different routes in a busy circuit.

(ii) If a certain link in the network goes down during the transmission, the remaining packets can be sent through another route.

In packet switching, the packet length is restricted to a maximum length. This length is short enough to allow the switching devices to store the packet data in memory without writing any of it to disk*.

There are two methods of packet switching**

(i) Datagram packet switching

(ii) Virtual - circuit packet switching

3.27.1. Datagram Packet Switching

In this method a message is divide into a stream of packets. Each packet is separately addressed and treated as an independent unit with its own control instructors. The switching devices route each packet independently through the network with each intermediate node determining the packet's next route segment. Before transmission starts, the sequence of packets and their destinations are established by the exchange of control information between the sending terminal, the network and the receiving terminal. In packet switching, the resources are not allocated for any packet so there is no reserved bandwidth and no scheduled processing time for each packet. No dedicated connection is established between the sender and receiver. The resources are allocated on demand and on the first come first serve basis. When a switch receives a packet, it has to wait if there are any other packets being processed. This will increase the delay. The datagram packet switching is normally done at the network layer. The packet are called as **datagrams**. Datagram packet switching is shown in figure 3.57.

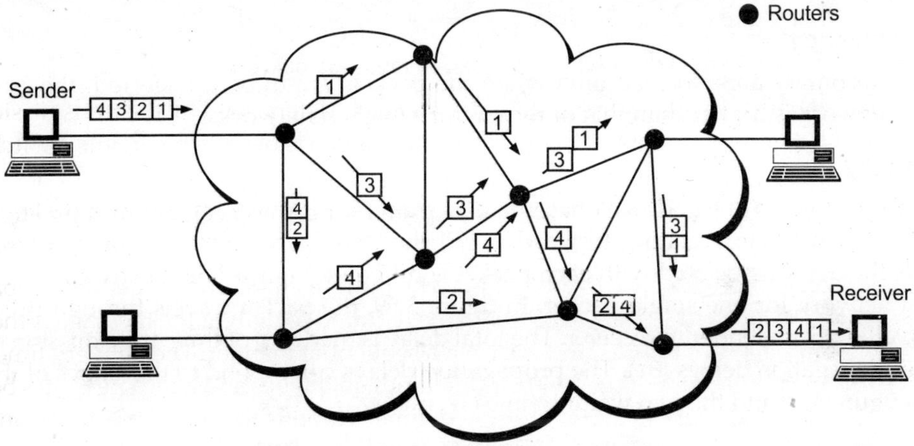

Fig. 3.57. Datagram packet switching

In this circuit, four packets are to be delivered from the sender to receiver. The switches in the datagram network are referred to as **routers.** All the four packets (datagrams) belong to the same message in this circuit however actually they can get originated from any computer. The four datagrams, as shown in figure 3.57 may travel different paths to reach the destination. Due to this the packet may arrive out of order at the destination. The delay associated with each packet will be different as a result of the different paths followed by them. The datagrams may get lost of dropped out due to lack of resources. The upper layer protocols are supposed to reorder the received datagrams or ask for the lost ones before passing them on the application. The datagram networks are called as the **connectionless** networks. This is because the switch (packet switch) does not keep any information about the connection state. There are no set up or tear down phases.

* The Internet uses packet switching. This type of service is also called as a connectionless-oriented circuit.

** A message is divided into pieces. Each piece is called a packet.

3.27.2. Routing Table

In packet switched networks, each switch or packet switch has a routing table. This table is based on the destination address. The routing tables are dynamic and are updated on periodic basis. The routing table consists of destination address and the corresponding output part over which the packet is to be forwarded as shown in figure 3.58.

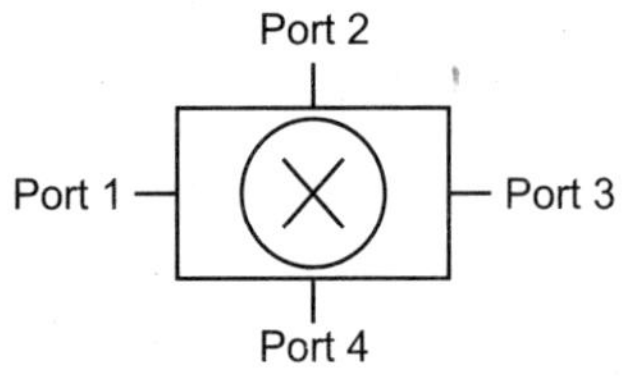

Destination address	Output port
1323	1
4360	2
9140	3
6436	4

Fig. 3.58. Router and Routing table

Destination Address

Every packet in the datagram network carries a header that contains the destination address where the packet is to be delivered and some additional information. When the router receives a packet, it examines the destination address of the packet and refers to its routing table to decide the port through which the packet is to be forwarded. For example in the routing table of figure 3.58, if the destination address on the received packet is 4360 then it will be forwarded through port 2.

3.27.3. Efficiency*

As the resources are allocated only when the packets are to be transferred, the efficiency of datagram network is **higher** than that of the circuit switched network.

3.27.4. Delay

There are no set up of tear down phases in datagram circuit switching but each packet may have to wait at a switch before getting forwarded. All the packets in a message take different paths. Therefore, the delay associated with each packet is different. Figure 3.59 illustrates the delays in a datagram network for one single packet. In figure 3.59, the packet travels through two switches while travelling from sender to receiver. The total delay is made up of three transmission times (3T) and three propagation delays (3τ). The propagation delays correspond to the slopes of the lines as shown in figure 3.59 and the two waiting times w_1 and w_2.

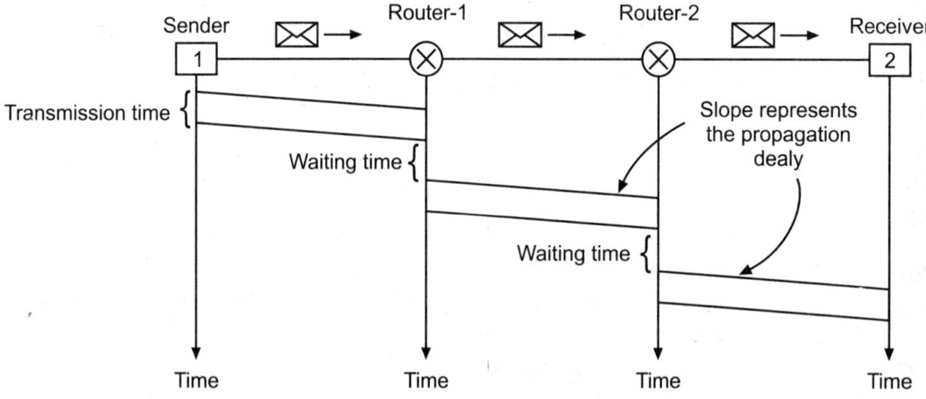

Fig. 3.59. Delay in datagram network

* Packet switching treats each packet of a message separately.

Therefore Total delay = $3T + 3\tau + w_1 + w_2$ ... (3.6)

The datagram switching is used in internet.

Key point: A packet is a small message unit that is part of a larger message.

3.27.5. Advantages of Packet Switching

(i) Greater line utilization efficiency as a single node-to-node link can be dynamically shared by many packets over time.

(ii) A packet switching network can perform data-rate conversion.

(iii) When traffic become heavy on circuit switching network, some calls are blocked. On a packet switching network, packets are still accepted, but delivery delay increases.

(iv) Priorities can be used.

(v) Each terminal in group sharing the same physical circuit may be connected to a totally different destination. This versatility is one of the major strengths of packet switching.

(vi) Unlike message switching, packet switching has a block size limit, usually 128 bytes or 256 bytes, but upto 4096 bytes on some networks, thereby reducing storage demands or buffering in the nodes, as well as delay at each node.

(vii) No single user or large data block can tie up circuit or node resources indefinitely, making it well suited for interactive traffic.

(viii) Data protection against corruption or loss, errors are corrected by retransmission.

(ix) Users can select different destinations for each virtual call, overcoming the inflexibility of point to point dedicated networks.

(x) Simultaneously calls allow PC users to access multiple windows to different remote applications.

(xi) Since many users can share transmission resources efficiently, the cost of intermittent data communication is reduced.

(xii) New calls can be added and odd ones disconnected without affecting other users.

3.27.6. Disadvantages of Packet Switching

(i) Delay :

(a) Transmission delay = length of packet divided by incoming channel rate.

(b) Variable delay due to the processing and queueing.

(ii) Overall packet delay can vary substantially (jitter)

(a) Packets may vary in length.

(b) May take different routes.

(c) May be subject to varying delays in switches.

(d) Not good for real time applications.

(iii) Header overhead reduces capacity available for user data.

(iv) More processing required at node.

Key Point: Packet switching is a network switching technique that interleaves smaller message units, called packets, and sends them to their destinations.

3.27.7. Datagram Networks in Internet

The internet uses the datagram approach to switching at the Network layer. It uses the universal addresses defined in the network layer for routing of packets from the source to destination.

3.28. HYBRID SWITCHING

(RGTU, Bhopal, Sem. Exam., June, 2014)

As a matter of fact, the hybrid forms of switching include some of the properties of packet switching, circuit switching and random access. Irrespective of the mode of switching used, the performance degrades with increase in utilization of channel. Therefore, a new term called **usable capacity** for a channel is defined.

Higher utilization of channel is preferred because that will increase the cost effectiveness. But, as the utilization crosses a certain point, the performance degrades to an unacceptable low level. Hence, the design must be created based on a uniform utilization limit.

3.29. VIRTUAL CIRCUIT PACKET SWITCHING*

It establishes a logical connection between the sending and receiving devices called virtual circuit. The sending device starts the conversation by communicating with the receiving device and agreeing as communication parameters, such as maximum message size and the network path to be taken. Once this virtual circuit is established the two devices use it for the rest of the conversation. In virtual circuit packet switching, all the packets travel through the logical connection established between the sending device and the receiving device. Virtual circuit switching has some characteristics of both circuit switched network and a datagram network. Similar to circuit switched network, there are set up and tear down phases alongwith the data transfer phase. It is possible to allocate the resources either in the set up phase similar to the circuit switched networks or as per requirement similar to the datagram networks. Similar to datagram networks, the data is sent in the form of packets. Each packet carries the address of the next switch and not the final destination address. Similar to circuit switching networks all the packets follow the same path established during the set up phase. Virtual circuit is established in the data link layer.

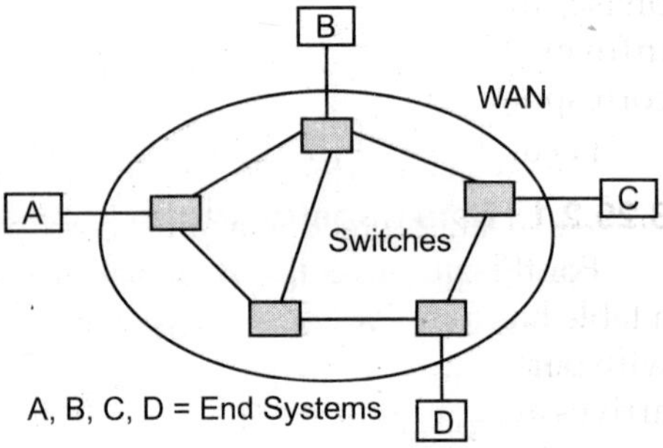

Fig. 3.60. Virtual circuit network

Figure 3.60 shows the virtual circuit network. The network consists of switches which route the traffic from source to destination.

3.29.1. Addressing

The virtual circuit networks use two types of addressing.

(i) Global addressing.

(ii) Local addressing (Virtual circuit identifiers)

(i) Global addressing: A source of destination can be a computer, router, bridge or any device which connects other networks such as LANs to the switched WAN. The source or destination must have a global address. This is a unique address. The global addressing in virtual circuit networks is used only to create a virtual circuit identifier.

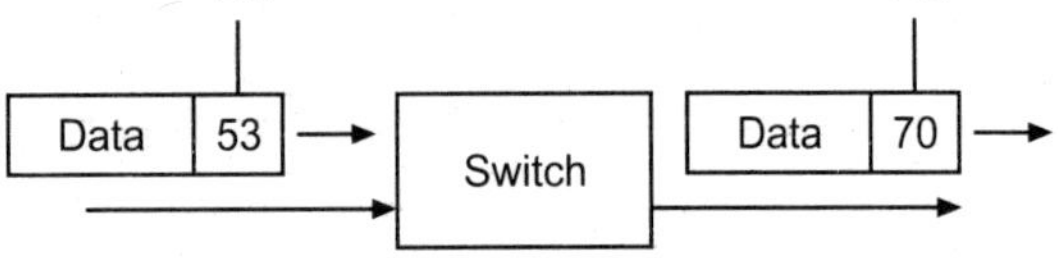

Fig. 3.61. VCI

(ii) Virtual circuit identifier: It is the identifier which is used for the actual data transfer and denoted by VCI. VCI is small number which is used by a frame between two switches. When a frame arrives at a switch, it contains one VCI as shown in figure 3.61 but when the frame levels that switch it contains another VCI.

* A virtual circuit is a type of packet switching that operates on the same concept as packet switching, but the routing of the packets is specified before transmission.

3.29.2. Three Phases of Communication

A source and destination have to undergo three phases to communicate between each other. The three phases are :

(i) Set up

(ii) Data transfer

(iii) Teardown

(i) Set up phase: In this phase the source and destination use their global addresses. This will help switches to make table entries for the connection.

(ii) Data transfer: The data transfer is the second phase in which the frames are transferred from source to destination.

(iii) Teardown: In the teardown phase, the source and destination will inform the switches to erase the corresponding entry.

Let us discuss these phases one by one.

Incoming		Outgoing	
Port	VCI	Port	VCI
1	70	3	20
1	15	2	42

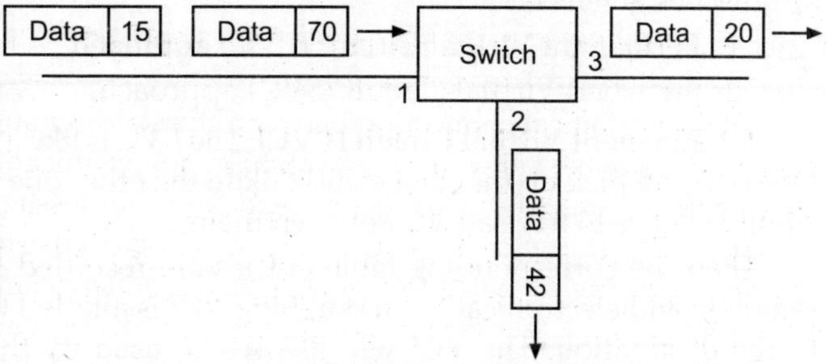

Fig. 3.62. Switch and its table

3.29.2.1. Data transfer phase

For this phase all the switches need to have a table entry for this particular virtual circuit. Such a table has four columns. Figure 3.62 shows such a switch and its table. The switch has three ports with port 1 acting as the incoming port while ports 2 and 3 as outgoing ports. A frame with VCI of 70 arrives at port 1. The switch scans its table to find port 1 and VCI 70. After finding them, the switch knows to change the VCI to 20 and send out the frame from port 3. Similarly the frame with VCI 15 is sent out on port 2 with a new VCI 42 as shown in figure 3.62.

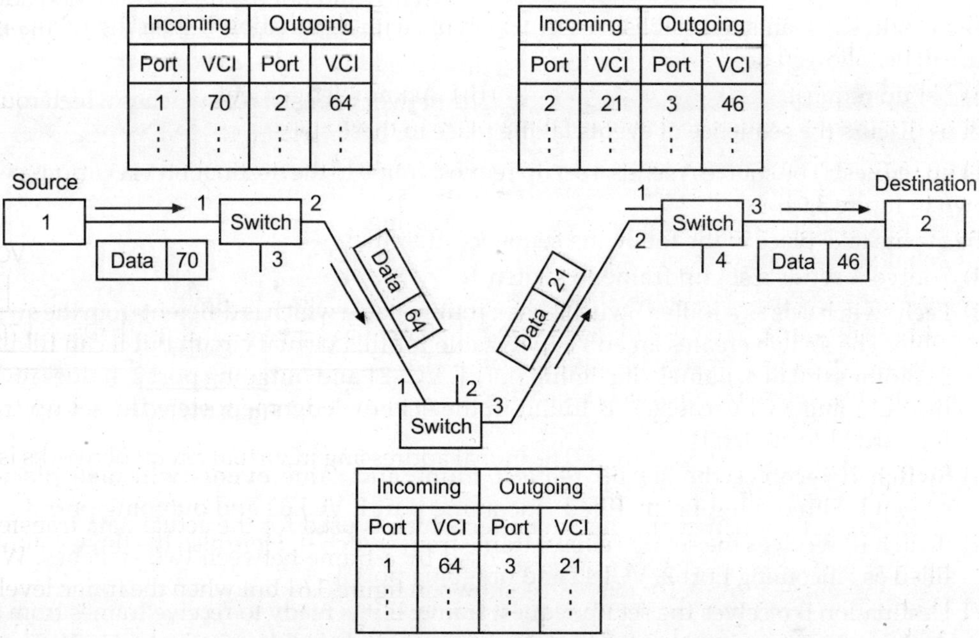

Incoming		Outgoing	
Port	VCI	Port	VCI
1	70	2	64
⋮	⋮	⋮	⋮

Incoming		Outgoing	
Port	VCI	Port	VCI
2	21	3	46
⋮	⋮	⋮	⋮

Incoming		Outgoing	
Port	VCI	Port	VCI
1	64	3	21
⋮	⋮	⋮	⋮

Fig. 3.63. Source to destination data transfer

3.29.2.2. Source to destination data transfer

Figure 3.63 shows the actual data transfer taking place between source 1 and destination 2. It shows how the VCI is changed by each switch for routing it from source to destination.

The data transfer phase is active as long as the source is sending its frames to destination. The procedure followed by each switch is same for each frame of the message being communicated. This process creates a virtual circuit which is not real to connect a source to destination.

3.29.2.3. Set up Phase

In the set up phase, a switch creates an entry for a virtual circuit by following one of the two approaches given below :

1. Permanent Virtual Circuit (PVC) approach.
2. Switched Virtual Circuit (SVC) approach.

1. Permanent Virtual Circuit (PVC): The PVC is like a leased telephone line between two parties. One part can pick up the phone and talk to the other one without dialing. A source and destination should choose to have a PVC between them.

Then the corresponding table entries are recorded for all the switches by the administrator remotely and electronically. An outgoing VCI is allotted to the source and incoming VCI is allotted to the destination. The VCI will always be used by this source to send frames to a particular destination.

Disadvantages of PVC: Following are the disadvantages of PVC

(i) PVC is costly because the connection always remains established between two parties whether it is required or not.

(ii) It needs many PVCs from a source to different destinations.

To overcome these disadvantages the other approach (SVC) is tried.

2. Switched Virtual Circuit (SVC): In SVC a temporary connection is established between the source and destination. This connection exists only when the data is to be transferred.

When source A wants to establish a virtual circuit with destination B then the following two steps are to be followed :

(i) Set up request. (ii) Acknowledgement.

Let us discuss the sequence of events taking place in these steps.

Set up request: The source A sends a set up request frame to the destination via various switches as shown in figure 3.64.

The events take place in the following sequence after that :

(i) Source A sends a set up frame to switch 1.

(ii) Each switch acts as a router. Switch I has a routing table which is different from the switching table. The switch creates an entry in its table for this virtual circuit but it can fill up only 3 columns out of 4, namely incoming port 1, VCI 21 and outgoing port 2. It does not know the outgoing VCI because it is found in the acknowledgement step The set up frame is forwarded to switch II.

(iii) Switch II receives the set up request frame and same events will take place as at switch I. Three columns are filled : incoming port I, VCI 62 and outgoing port 3.

(iv) Switch III receives the set up request frame from switch II. Here also the three columns are filled as : incoming port 2. VCI 43 and outgoing port 3.

(v) Destination B receives the set up request frame. If it is ready to receive frames from source A, then a VCI is assigned to the frames coming from A. In this case the VCI is 56. The VCI is an indication that frames come from A and not from any other source.

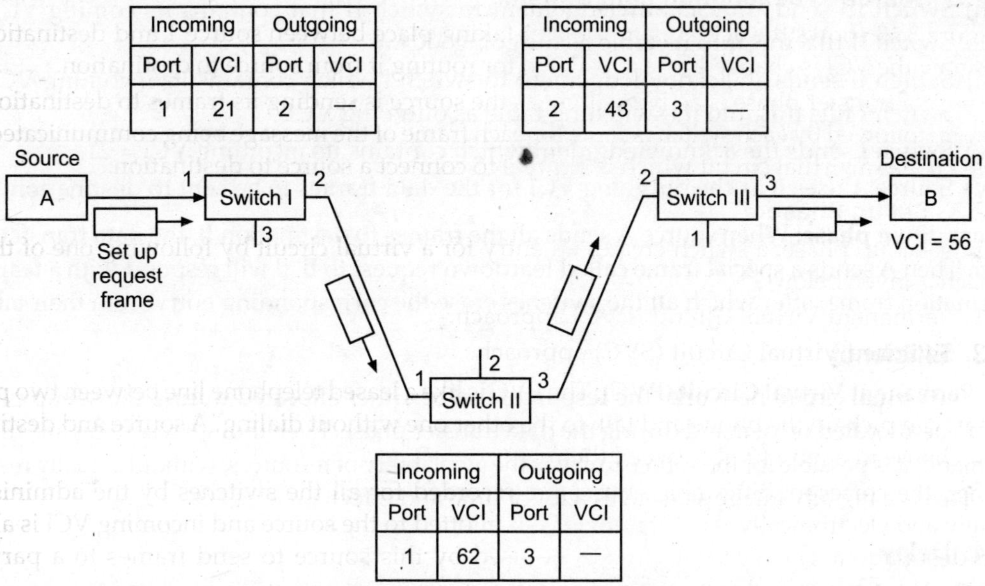

Fig. 3.64. SVC set up request

3.29.2.4. Acknowledgement

The acknowledgement frame is sent from the destination B to source A via the three switches. This completes the entries in the switching tables. The acknowledgement process is illustrated in figure 3.65.

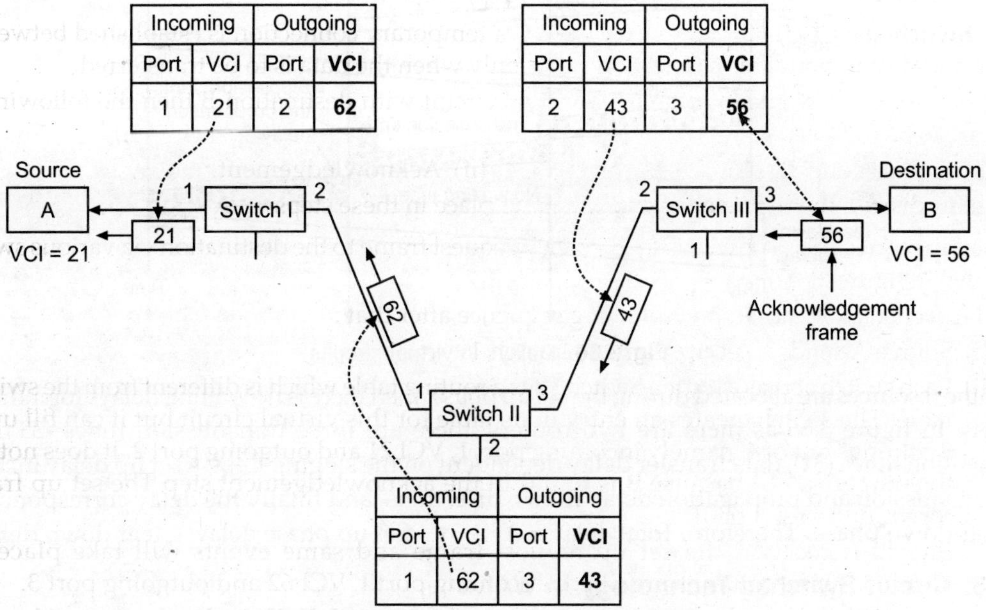

Fig. 3.65. SVC Acknowledgement

The sequence of events is as follows :

(i) Destination B sends an acknowledgment frame to switch III. The frame contains the global source and destination address and VCI 56. Switch III fills this number into the fourth column of the switching table.

(ii) Switch III sends the acknowledgement to switch II that contains its coming VCI = 43. Switch II fills this into its table as outgoing VCI.
(iii) Switch II sends an acknowledgement to switch I which contains its incoming VCI = 62. Switch I fills this into its switching table as outgoing VCI.
(iv) Switch I sends the acknowledgment which contains its incoming VCI = 21.
(v) Source A uses it as the outgoing VCI for the data frames to be sent to destination B.

Tear down phase: When source A sends all the frames to destination B, the data transfer phase is over. Then A sends a special frame called teardown request to B. B will respond with a tear down confirmation frame, after which all the switches erase the corresponding entry from their tables.

3.29.3. Efficiency

In the virtual circuit networks the resources can be either allocated during the set up phase or they can be allocated on demand during the data transfer phase. Even though resources are allocated on demand, it is possible for the source to check the availability of resources, without actually reserving them. This is a big advantage as it saves a lot of time and effect.

3.29.4. Delay

Different delays in virtual circuit networks are illustrated in figure 3.66. One component of delay is the time delay for set up phase. The other component is the time delay corresponding to the tear down phase.

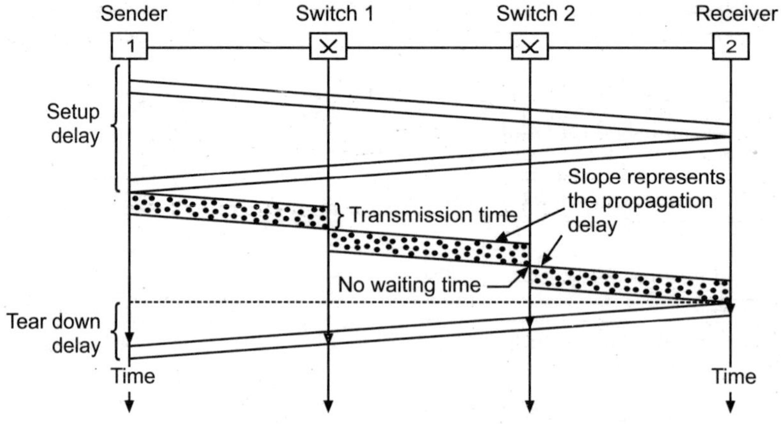

Fig. 3.66. Delays in virtual circuits

If the resources are allocated during the set up phase, then there is no waiting delays for individual packets. In figure 3.66 as there are two routers, there are three transmission times (3T), three propagation times (3τ), data transfer delay dependent on the sloping lines, a set up delay including the transmission and propagation delay in two directions, and finally the delay corresponding to the tear down phase. Therefore, Total delay = 3T + 3τ + Set up phase delay + Tear down delay.

3.29.5. Circuit Switched Technology in WANs

The virtual circuit networks are used in switched WANs such as Frame Relay and ATM networks.

3.29.6. Advantages of Virtual Circuit Packet Switching

(i) Virtual circuit packet switching uses abbreviated headers and hardware based table lookup, which allows fast processing and forwarding of packets.

(ii) In the virtual circuit packet switching resources can be allocated during connection setup.

(iii) The number of bit required in the header is much smaller than the number required to provide full destination network addresses. This reduces the wastage of transmission bandwidth.

(iv) Virtual circuit packet switching uses virtual-circuit identifier (VCI) which uses to identify connection and to specify the type of priority given to the packet by scheduler that controls the transmissions in next output port.

3.29.7. Disadvantages of Virtual Circuit Packet Switching*

(i) The switches in the network need to maintain information about the flows they are handling. The amount of required 'state' information grows very quickly with number of flows.

(ii) In the case of fault occurs in the network, all affected connections must be set up again.

(iii) Connection set up is not possible, if the switch is unable to handle the volume of traffic allowed or link utilization exceeds certain thresholds.

3.29.8. Comparison of Datagram and Virtual Circuits

S.No.	Circuit switching	Datagram packet switching	Virtual-circuit packet switching
1.	Dedicated transmission path	No dedicated path	No dedicated path
2.	Continuous transmission of data	Transmission of packets	Transmission of packets
3.	Fast enough for interactive	Fast enough for interactive	Fast enough for interactive
4.	Message are not stored	Packets may be stored until delivered	Packets stored until delivered
5.	The path is established for entire conversation	Route established for each packet	Route established for entire conversation
6.	Call setup delay; negligible transmission delay	Packet transmission delay	Call setup delay; packet transmission delay
7.	Busy signal if called party	Sender may be notified if busy packet not delivered	Sender notified if connection denial
8.	Overload may block call setup; no delay for established calls	Overload increases packet delay	Overload may block call setup; increases packet delay
9.	Electromechanical or computerized switching nodes	Small switching nodes	Small switching nodes
10.	User responsible for message loss protection	Network may be responsible for individual packets	Network may be responsible for packet sequences.
11.	Usually no speed or code conversion	Speed and code conversion	Speed and code conversion
12.	Fixed bandwidth transmission	Dynamic use of bandwidth	Dynamic use of bandwidth
13.	No overhead bits after call setup	Overhead bits in each message	Overhead bits in each packet

* A virtual circuit is a type of packet switching. In a virtual circuit, all packets of a message are transmitted in a specific path called a virtual path.

3.30. COMPARISON OF MESSAGE, CIRCUIT AND PACKET SWITCHING*

(RGTU, Bhopal, Sem. Exam., June, 2014)

S.No.	Parameter	Message switching	Circuit switching	Packet switching
1.	Application	Telegraph network for transmission of telegrams. voice signals.	Telephone network bidirectional, real time transfer of	Internet for datagram and reliable stream service between computers.
2.	End terminal	Telegraph, teletype	Telephone, modem	Computer
3.	Information type	Morse, Baudot, ASCII	Analog voice or PCM digital voice	Binary information
4.	Transmission system	Digital data over different transmission media	Analog and digital data over different transmission media	Digital data over different transmi-ssion media.
5.	Addressing scheme	Geographical	Hierarchical number-ing plan	Hierarchical address space.
6.	Routing scheme during call setup.	Manual independently.	Route selected	Each packet routed
7.	Multiplexing scheme	Character or message	Circuit multiplexing	Packet multiplexing shared media access networks.

GLOSSARY

- **Analog** Any physical device or signal that varies continuously in strength or quantity over an infinite range of voltage or current. An example may be a voltage in a circuit.
- **Analog Data** Any data type which is continuously in nature and varies in strength and quantity over time. Further, analog data may be represented as either *analog signals* or *digital signals.*
- **Attenuation** A decrease in signal strength, that occurs as the signal travels through a circuit or along a cable. The longer the cable, the greater, the attenuation. Also, the higher the frequency of the signal, greater the attenuation.
- **Bandwidth** In *analog communication*, the total capacity of a communication channel measured in hertz (Hz). It is the difference between the highest and lowest frequencies capable of being carried over a channel. The greater the bandwidth, the more signals that may be carried over a given frequency range. In *digital communication* and networking stream, basically, bandwidth is the theoretical capacity of a communication channel and is expressed in bits per second (bps), which is known as *data rate.*
- **Baseband Cable** Uses the entire bandwidth of the cable to carry a single signal.
- **Baud** A unit of signalling speed, named after the French engineer Jean maurice Emile Baudot (1845–1903). In fact, it is another term that is used to express the capacity of a channel, but is different from bits per second.
- **Baud Rate** A measure of the number of times line conditions (i.e., frequency, amplitude, voltage, or phase) change each second. At low speeds (under 300 bps), data rate (measured in bps) and baud rate are the same since signaling methods are relatively simple. As speed increases, signaling methods become more complex. Baud rate then differs from data rate because several bits are typically encoded per baud. This means that each signal can represent more than one bit of information.

* Compare the different types of switching techniques used in communication networks. Discuss their advantages and disadvantages. *(RGTU, Bhopal, Sem. Exam., June, 2014)*

- **Coaxial Cable** A type of cable which consists of a single wire conductor, surrounded by a dielectric material and two types of shielding—a foil shield and a braided shield—arranged concentrically and encashed in a PVC or Teflon outer jacket.
- **Crosstalk** Electrical interference (*i.e., noise*) that occurs when energy radiated from one wire pair of a twisted-pair wire spills over into another pair. In one type of crosstalk, called *near-end crosstalk* (NEXT), a signal on the transmit pair is so strong that it radiates to the receive pair. A serious consequence of this spilled-over radiation is that the receiving device cannot decipher the real signal.
- **Digital** Any device or signal which varies discretely in strength or quantity between two values, usually 0 and 1. 0 implies *off*; 1 implies on. Digital signals are represented as binary digits called *bits* and are *discrete*.
- **Digital Communication** Any type of communication in which data are represented in the form of binary digits.
- **Digital Data** Any data that had been converted into a binary code such as ASCII.
- **Digital Signal** An electronic signal which is represented using 0s and 1s instead of as a continuum of voltages, as is the case with *analog signals*. Analog data are usually converted to digital signals via *pulse-code modulation*; digital data are converted to digital signals using any one of several techniques, including *Manchester encoding, differential Manchester encoding*, and *NRZI*.
- **Period** The reciprocal of *frequency*. It is the amount of time it takes to complete a single cycle, *i..e*, seconds per cycle.
- **Physical Layer** The lowest layer (layer 1) of the *OSI model*. The *physical layer* translates *frames* received from the *data link layer* (layer 2) into electrical, optical, or electromagnetic signals representing 0 and 1 values, or bits. Abbreviated PHY in the documentation.
- **Shannon's Limit** A mathematical theorem which describes a model for determining the maximum data rate of a noisy analog communications channel. Shannon's limit is given by the following formula : maximum data rate (*MDR*) = $H \log_2 (1 + S/N)$, where *MDR* is given in bits per second, *H* = bandwidth in hertz, and S/N is a measure of the *signal-to-noise ratio*. It has been named for the mathematician who derived it, Claude Shannon.
- **Single-Mode Fiber** A type of fiber-optic cable with a core diameter ranging from 7 to 9 μm. In single-mode fiber, only a single ray of light, known as the *axial ray*, can pass. Thus, a light wave entering the fiber exits with very little distortion, even at very long distances and very high data rates.
- **Step-Index Multimode Fiber** A type of multimode fiber in which light pulses are guided along the cable from source to destination by reflecting off the cladding.
- **Circuit-Switched Network** A network design in which a dedicated physical circuit gateway is established between the source and destination nodes before any kind of data transmission. Further, this circuit must remain in place for the duration of a transmission.
- **Circuit-Switching** A switching technique in which a dedicated circuit path is established between two entities on demand. This path is established prior to any data transmission and is used exclusively by the connected parties until the connection is terminated. Contrast with *packet-switching*.

SUMMARY

- Media are what the message is transmitted over. In other words, a communication channel is also called a medium. Different media have different properties and used in different environments for different purpose.
- Media are roughly grouped into two classes :
  (i) Guided media; (ii) Unguided media
- Guided media is a communication medium which allows the data to get guided along it. For this, the media need to have a point to point physical connection.
- The wireless media is also called as an unguided media.
- The examples of guided media are copper wire and fiber-optics. Similarly radio and lasers through the air are examples of unguided media

- In computer networking, the medium affects nearly every aspect of communication. Most important, it determines how quickly and to whom a computer can talk and how expensive the process is.
- In the wireless media, the signal energy propagates in the form of unguided electromagnetic waves. The examples of wireless media are radio and infrared light. The wireless media is used for radio broadcasting in all the directions.
- Optical fiber transmission systems are widely used in the backbone of networks. Current optical fiber systems provide transmission rates from 45 Mb/s to 9.6 Gb/s using the single wavelength transmission. The installation cost of optical fibers is higher than that for the co-axial or twisted wire cables. Optical fibers are now used in the telephone systems.
- The unguided signals can travel from the transmitter to receiver in many different ways; the three most important methods are :
  (i) Ground wave propagation (ii) Sky propagation.
  (iii) Space propagation or line of sight propagation.
- Radio waves have frequencies between 10 kHz and 1 gigahertz. The range of electromagnetic spectrum between 10 kHz and 1 gHz is called radio frequency (RF).
- Some of the important applications of radio transmission systems are as under:
  (i) Cellular communication (ii) Wireless LAN
  (iii) Point to point and point to multipoint radio systems
  (iv) Satellite communication.
- It makes use of the lower gigahertz frequencies of the electromagnetic (EM) spectrum. These frequencies, are higher than the RF and they produce better throughput and performance.
  There are two types of microwave data communication systems which are as under:
  (i) Terrestrial
  (ii) Satellite
- Switching sends data along different routes. There are three switching techniques:
  (i) Circuit switching (ii) Message switching
  (iii) Packet switching
- A network consists of many switching devices. In order to connect multiple devices one solution could be to have a point to point connection between each pair of devices. But this increases the number of connections.
- The other solution could be to have a central device and connect every device to each other via the central device (Star topology).
- Both these methods are wasteful and impractical for very large networks. The other topologies also cannot be used.
- Switch is a device that creates temporary connections between two or more devices.
- The three basic methods of switching are :
  (i) Circuit switching; (ii) Packet switching; (iii) Message switching
  Out of these, the circuit and packet switching are commonly used today but the message switching has been phased out in general communication but is still used in the networking applications.
- Circuit switching is used in public telephone networks. It was developed to handle voice traffic but it can also handle digital data.
- The telephone network provide telephone service which involves the two way, real-time transmission of voice signals across a network.
- The network connection allows electrical current and the associated voice signal to flow between the two users. The end to end connection is maintained for the duration of the call.
- The circuit switched networks operate in three phases :
  (i) Set up phase; (ii) Data transfer phase; (iii) Tear down phase.
  The circuit switching takes place at the physical layer.
- In telegraphy, the text message is encoded using the Morse code into sequences of dots and dashes. Each dot or dash is communicated by transmitting short and long pulses of electrical current over a copper wire.

- In packet switching, the packet length is restricted to a maximum length. This length is short enough to allow the switching devices to store the packet data in memory without writing any of it to disk.
- There are two methods of packet switching :
  (i) Datagram packet switching (ii) Virtual - circuit packet switching.
- The virtual circuit networks use two types of addressing.
  (i) Global addressing. (ii) Local addressing (Virtual circuit identifiers)
- The various types of circuit switches are as follows :
  (i) Space-division switches (ii) Time-division switches
  (iii) Multistage switches (iv) Time space time switches
- Packet switch is the switch used in a packet switch. It has four different components :
  (i) Input ports (ii) Output ports
  (iii) Routing processor (iv) Switching fabric

SHORT QUESTIONS WITH ANSWERS

Q.1. What do you mean by transmission medium?

Ans. Media are what the message is transmitted over. In other words, a communication channel is also called a medium. Different media have different properties and used in different environments for different purpose. The purpose of the physical layer is to transport a raw bit stream from one computer to another.

Q.2. What do you mean by wired media?

Ans. In this type of media, the signal energy is contained and guided within a solid media. The examples of wired media are copper pair wires, coaxial cables and fiber optic cables. The wired media is used for point to point communication.

Q.3. What are the types of wired media?

Ans. The wired media can be classified as under:
(i) Co-axial cable (ii) Twisted pair cable
(iii) Optical fiber cable.

The selection of networking media depends on various factors such as cost, connectivity, width, performance in presence of noise, geographical coverage etc.

Q.4. Write the applications of twisted pair cables.

Ans. Some of the applications of twisted pair cables are as follows:
(i) In telephone lines to carry voice and data channels.
(ii) In the local loop (iii) In the DSL line (ADSL)
(iv) Local area networks such as 10 Base-T and 100 Base-T use the twisted pair cables.
(v) In the ISDN (Integrated Services Digital Network).

Q.5. Write the applications of co-axial cables.

Ans. (i) Analog telephone networks. (ii) Digital telephone network
(iii) Cable TV (iv) Traditional Ethernet LANs
(v) Digital transmission (vi) Thick and thin Ethernet

Q.6. Write few disadvantages of optical fiber.

Ans. Some of the disadvantages of optical communication system are :
(i) Sophisticated plants are required for manufacturing optical fibers.
(ii) The initial cost incurred is high. (iii) Joining the optical fibers is a difficult job.

Q.7. List the communications media.

Ans. Conductors, optical cable, and wireless communication.

Q.8. What is the application of Single Mode fiber?

Ans. Long distance communication.

Q.9. What are the signal sources for optical cable?

Ans. Laser Diode and Light Emitted Diode (LED).

Q.10. Distinguish between data rate and baud rate.

Ans. Data rate is the number of bits per second and baud rate is the number of signals per second.

Q.11. What is circuit switching networks?

Ans. Circuit switching is used in public telephone networks. It was developed to handle voice traffic but it can also handle digital data.

However, handling of digital data using the circuit switching generally proves to be inefficient. Using the circuit switching, a dedicated path is established between two stations for communication.

Q.12. What is Packet Switching?

Ans. In packet switching, message are broken up into packets, each of which includes a header with source, destination and intermediate node address information. Individual packets take different routes to reach the destination. Independent routing of packets gives two advantages:

(i) Bandwidth is reduced by splitting data onto different routes in a busy circuit.

(ii) If a certain link in the network goes down during the transmission, the remaining packets can be sent through another route.

Q.13. Write the components of packet switches.

Ans. Packet switch is the switch used in a packet switch. It has four different components :

(i) Input ports
(ii) Output ports
(iii) Routing processor
(iv) Switching fabric

Q.14. What is meant by Switching Fabrics.

Ans. During the earlier days, the switching fabric used to be the memory of a dedicated computer. But now a days packet switches with specialized mechanisms are being used. Some of them are as follows :

(i) Crossbar switch
(ii) Banyan switch
(iii) Batcher Banyan switch.

Q.15. Explain switch operation.

Ans. The switch is used to connect different segments of a LAN, in order to increase network throughput.

REVIEW QUESTIONS

1. State the types of guided media.
2. Explain about the difference between UTP and STP.
3. What is the effect of twisting the wires in UTP cables?
4. Give applications of co-axial cable.
5. What is the advantage of using shielding?
6. Compare the guided transmission media.
7. State advantages of optical fiber cable.
8. State the three ways of wireless transmission.
9. Write a short technical note on : (i) microwave communication, (ii) circuit switching networks.
10. State applications of infrared transmission.
11. Write a short technical note on Infrared transmission.
12. Compare twisted pair (UTP and STP).
13. Compare twisted pair, co-axial and fiber optic cable.
14. What is meant by switching? Explain.
15. What are switching methods? Explain.
16. What is meant by packet switching?
17. Explain about advantages and disadvantages of packet switching.
18. Explain about structure of a switch.

UNIT – II

SYLLABUS:

The Data Link layer: Design issues, error detection and correction, data link protocols-stop and wait and sliding window ARQ, utilization of ARQ techniques, example of data link protocol-HDLC.

The Medium Access Control Layer: Static and dynamic channel allocation, multiple access protocols- Pure and slotted ALOHA, CSMA, Collision free protocols, limited contention protocols, CSMA/CD (ETHERNET), fast Ethernet, Gigabit Ethernet.

The above syllabus has been covered in following chapters:

Chapter 4 : The Data Link Layer

Chapter 5 : The Medium Access Control Layer

CHAPTER 4

THE DATA LINK LAYER

Syllabus

The Data Link layer: Design issues, error detection and correction, data link protocols-stop and wait and sliding window ARQ, utilization of ARQ techniques, example of data link protocol- HDLC.

CHAPTER OUTLINE

- Introduction
- Position of Data Link Layer
- Services Provided by Data Link Layer to Network Layer
- Framing
- Error Control
- Error Detection and Correction
- Error Detection
- Two Dimensional Parity Check
- Cyclic Redundancy Check (CRC)
- Error Correction
- Linear Block Codes
- Hamming Codes
- Automatic Repeat Request (ARQ) Techniques
- Flow Control
- Error Detection and Correction
- Data Link Protocols
- Sliding Window Protocols
- Other Data Link Protocols
- Example of Data Link Protocol: High Level Data Link Control (HDLC) Protocol
- Synchronous Data Link Control (SDLC) Protocol
- The data link Layer in the Internet
- SLIP-Serial Line IP
- Point to Point Protocol (PPP)

4.1. INTRODUCTION

As a matter of fact, the physical layer deals with the transmission of signals over different transmission medias. For achieving reliable, efficient communication between two adjacent machines, the data link layer plays an important role. The layer basically deals with frame formation, flow control, error control, addressing and link management. While sending data from source to destination, the communication circuits make errors. Furthermore they have only a finite data rate, and there is non-zero propagation delay between the time a bit is sent and the time it is received. These limitations affect the efficiency of data transfer. The data link protocols used for communication take care of all these problems. Data link layer is the second layer in OSI reference model. This means that this layer is above the physical layer. The data link layer is basically concerned with sending data to the next immediate neighbour. To ensure the transmission is error-free, it performs several checks. We shall call them the duties of data link layer, as the data link layer is bound to carry out these tasks.The first requirement is to convert the streams of data into a packetized format, which is called as **framing** The second requirement is to provide some form of error-handling process so that the receiver either gets the correct data or can verify whether the data received is correct or not. The third requirement is to provide flow control so that a fast sender cannot swamp a slow receiver.

4.2. POSITION OF DATA LINK LAYER

The 4.1 shows the position of data link layer in the five layer Internet model. It is the second layer. It receives services from the physical layer and provides services to the network layer.

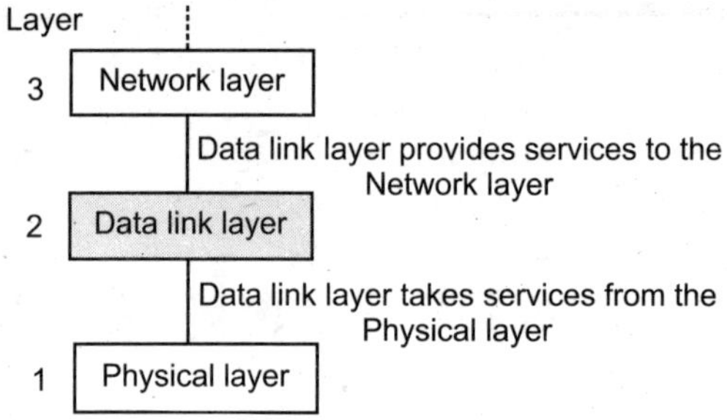

Fig. 4.1. Position of Data link layer

4.2.1. Design Issues for Data Link Layer: Functions of Data Link Layer

The data layer is supposed to carry out many specified functions. For effective data communication between two directly connected transmitting and receiving stations, the data link layer has to carry out a number of specific functions as under:

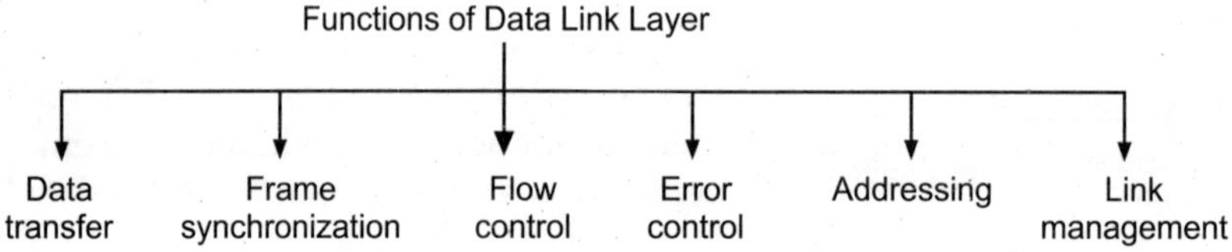

Fig. 4.2. Functions of Data link layer

1. Services provided to the network layer: A well defined service interface to the network layer. The principle service is to transfer data from the network layer on source machine to the network layer on destination machine.

2. Frame synchronization: The source machine sends data in block called frames to the destination machine. The starting and ending on each frame must be recognized by the destination machine.

3. Flow control: The source machine should not send data frames at a rate faster than the destination machine can accept them.

4. Error control:

The errors made in bits during transmission from source to destination machines must be detected and corrected.

5. Addressing: On a multipoint line, such as many machines connected together (LAN), the identity of the individual machines must be specified while transmitting the data frames.

> **DO YOU KNOW?**
>
> Data link layer can provide connection-oriented or connectionless service. In connection-oriented service, there are link establishment and link termination phases. In connectionless mode data transfer service, the data link need not be established and terminated. The data link entities are always in data transfer mode.

6. Control and data on same link: The data and control information is combined in a frame and transmitted from the source to destination machine. The destination machine must be able to recognize control information from the data being transmitted.

7. Link management: The initiation, maintenance and termination of the link between the source and destination is required for effective exchange of data. It requires coordination and cooperation among stations. Protocols or procedures are required for the link management. Let us discuss them one by one.

4.3. SERVICES PROVIDED BY DATA LINK LAYER TO NETWORK LAYER

As a matter of fact, the network layer is the layer above the network layer in the OSI model. Therefore, it is supposed to provide services to the network layer. The main service to be provided is to transfer data from the network layer on the sending machine to the network layer of the receiving machine. The virtual path followed for such a communication is shown in figure 4.3(a). It is not the actual path. The actual path followed by the data from sending machine to destination is shown in figure 4.3(b) which follows all the layers below the network layer, then the physical medium, then layers 1, 2, 3 of receiving machine.

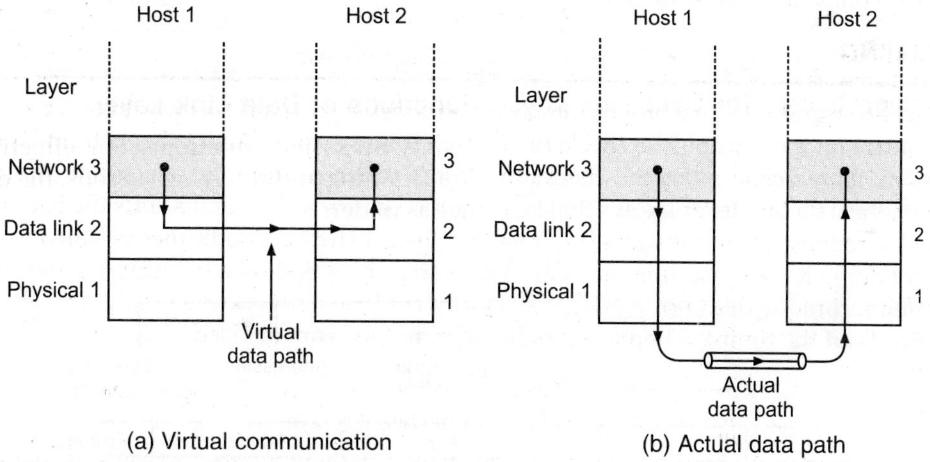

Fig. 4.3.

However it is always easier to think that the communication is taking place through the data link layers as depicted in figure 4.3(a).

4.3.1. Types of Services Provided

Data link layer can be designed to offer different types of services. Some of them may be listed as under:

(i) Unacknowledged connectionless service.
(ii) Acknowledged connectionless service.
(iii) Acknowledged connection oriented service.

4.3.1.1. Unacknowledged Connectionless Service

In this use of service, the destination machine do not send back any acknowledgement of receiving frames. It is a connectionless service. Therefore, no connection is established before communication or released after it is over. If a frame is lost due to noise, then no attempt is made to recover it. Hence, this service is suitable only if the error rate is low. Further, It is suitable for real time traffic such as speech.

4.3.1.2. Acknowledged Connectionless Service

This is the next step to improve reliability. In this type of service, there are no connections used but for each frame receive, the receiver sends an acknowledgement to the sender. If a frame is not received within specified time, then the sender will retransmit it. This service is quite suitable for communication over unreliable channels.

> **DO YOU KNOW?**
>
> The data link and transport layers perform many of the same duties. The data link layer functions in a single network, while the transport layer operates across an Internet.

4.3.1.3. Acknowledged Connection Oriented Service

This is the most sophisticated one. The source and destination machines establish a connection before transferring the data. A specific number is given to each frame being sent and the data link layer guarantees that each transmitted frame is received. All the frames are guaranteed to be received in the same order as the order of transmission. The data transfer takes place by following three distinct phases given below:

(i) Connection is established.
(ii) One or more frames are actually transmitted.
(iii) The connection is released.

4.4. FRAMING

The bits to be transmitted is first broken into discrete frames at the data link layer. In order to guarantee that the bit stream is error free, the checksum of each frame is computed. When a frame is received, the data link there, recomputes the checksum. If it is different from the checksum present in the frame, then the data link layer knows that an error has occurred. It then discards the bad frame and sends back a request for retransmission. Breaking the bit stream into frames is called as framing. One way of doing it is by inserting time gaps between frames as shown in figure 4.4. But practically this framing technique does not work satisfactorily because networks generally do not make any guarantees about the timing. Therefore, some other methods are devised.

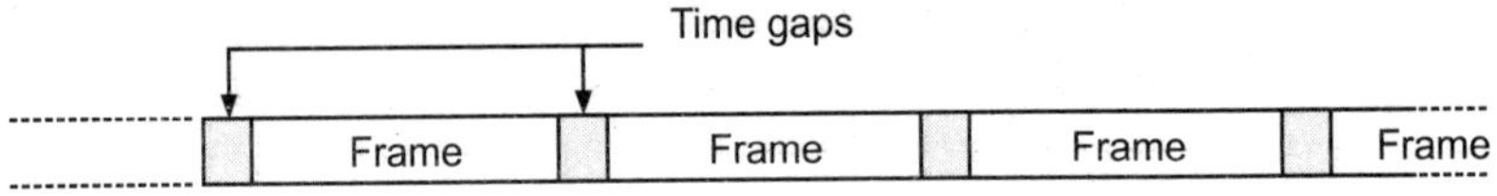

Fig. 4.4. Framing

4.4.1. Framing Methods

Following methods are used for carrying out framing:

(i) Character count
(ii) Starting and ending characters, with character stuffing.
(iii) Starting and ending flags with bit stuffing.
(iv) Physical layer coding violations.

4.4.2. Character Count

In this method, a field in the header is used to specify the number of characters in the frame. This number helps the receiver to know the number of characters in the frame following this count. The character count method is illustrated in figure 4.5.

The two frames shown in figure 4.5 are of 3 and 8 characters respectively. The disadvantage of this method is that, an error can change the character count. If the wrong character count number is received then the receiver will get out of synchronization and will be unable to locate the start of next frame. The character count method is rarely used in practice.

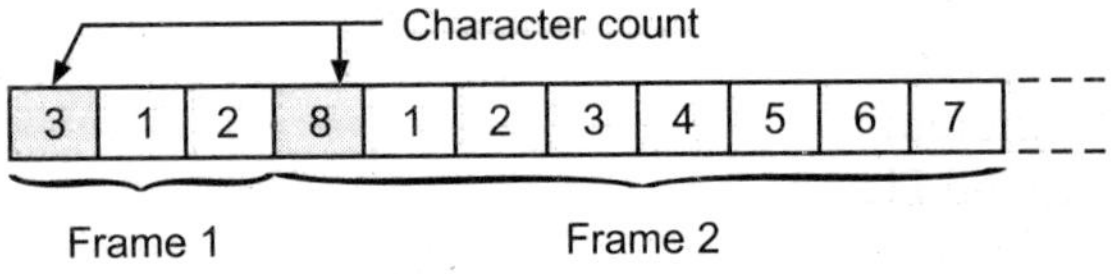

Fig. 4.5. Character count method

4.4.2.1. Starting and Ending Character with Character Stuffing

The problem of character count method is solved here by using a starting character before the starting of each frame and an ending character at the end of each frame. Each frame is preceded by the transmission of ASCII character sequence DLE STX (DLE stand for data link escape and STX is start of Text). After each frame, the ASCII character sequence DLE ETX is transmitted. Here, DLE stands for data Link Escape and ETX stands for END of Text. Hence, if the receiver loses the synchronization, it just has to search for the DLE STX or DLE ETX characters to return back on track. This is shown in figure 4.6.

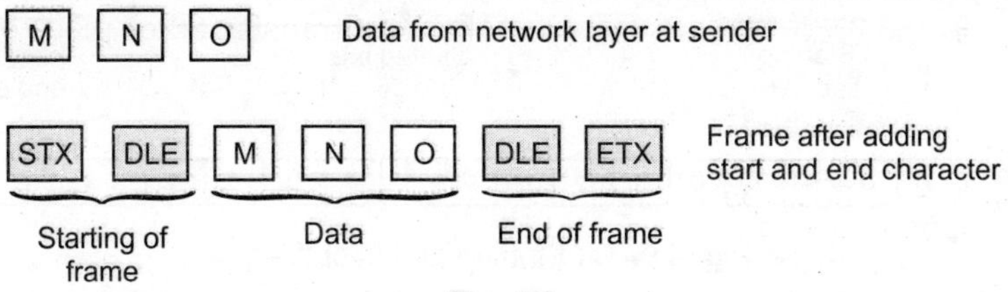

Fig. 4.6.

4.4.2.2. Character Stuffing

1. Definition: The problem with this system is that the characters DLE STX or DLE ETX can be a part of data as well. If so, they will be misinterpreted by the receiver as start or end of frame. This problem is solved by using a technique called **character stuffing**, which is as follows :

The data link layer at the sending end inserts an ASCII DLE character just before each accidental DLE character in the data. The data link layer at the receiving end will remove these DLE characters before handing over the data to the network layer. Thus, the framing DLE STX or DLE ETX can be distinguished from the one in data because DLEs in the data are always doubled. This is called character stuffing and it is shown in figure 4.7. It may be noted that at the receiving end, the destuffing is essential.

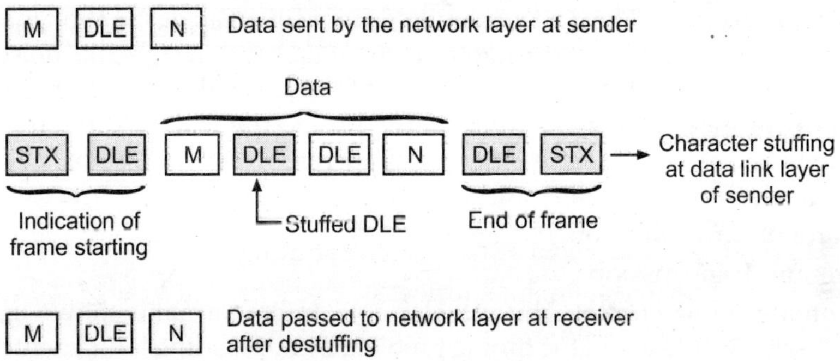

Fig. 4.7. Character Stuffing

2. Drawback: The main drawback of this framing method is that we have to use the 8 bit characters and ASCII code. This problem can be overcome by using the next framing technique.

4.4.2.3. Starting and Ending Flags, with Bit Stuffing

1. Definition: This technique allows the frames to contain an arbitrary number of bits and codes different from ASCII code. At the beginning and end of each frame, a specific bit pattern 0111 1110 called **Flag byte** is transmitted. Since there are six consecutive 1s in this byte, a technique called **bit stuffing** which is similar to character stuffing is used.

2. Bit Stuffing: Whenever the sender data link layer detects the presence of five consecutive ones in the data, it automatically stuffs a 0 bit into the outgoing bit stream. This is called bit stuffing and it is illustrated in figure 4.8.

3. De-stuffing: When a receiver detects presence of five consecutive ones in the received bit stream, it automatically detects the 0 bit following the five ones. This is called de-stuffing. It is shown in figure 4.8. Due to bit stuffing, the possible problem if the data contains the flag byte pattern (0111 1110) is eliminated.

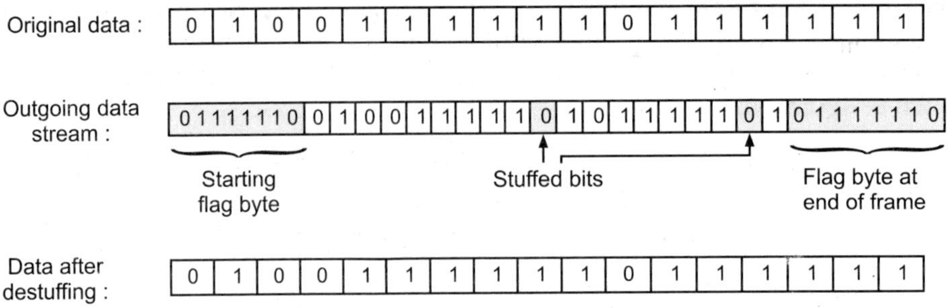

Fig. 4.8. Bit Stuffing and destuffing

4.4.2.4. Physical Layer Coding Violations

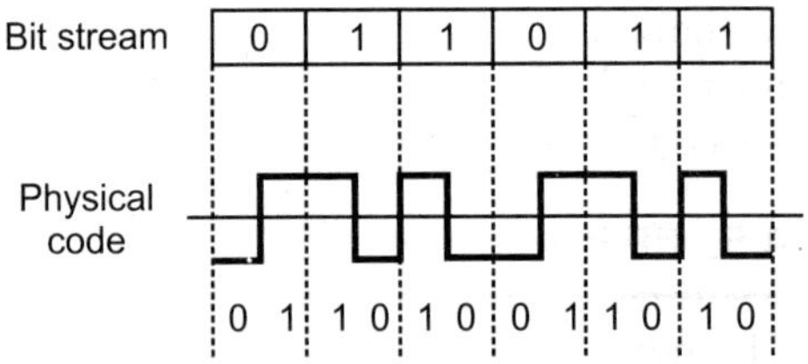

Fig. 4.9.

1. Basic Concept: This method of framing is applicable only to the networks in which the encoding on the physical medium contains some redundancy. Some LANs encode each bit of data using two physical bits. The Manchester coding is generally used. Normally, a 1 bit is encoded into a 10 pair and a 0 bit is encoded into a 01 pair as shown in figure 4.9. This use of invalid physical code is a part of 802 LAN standards.

2. Which Method of Framing is Used Practically?: Many data link protocols use the combination of the character count technique with one of the other techniques so as to have an extra safety.

3. Frame synchronization: The data transmitted from source to destination machine is serial in nature. Due to errors occurring in bit during transmission, due to factors like noise and others, the start of bit and end of bit or start of frame and end of frame may not be recognized by the receiver properly. The receiver may loose synchronization with the transmitter if the transmitter sends a long stream of bits and if no steps are taken to synchronize the transmitter and receiver. Serial transmission occurs in one of the following ways :

(i) Asynchronous transmission

(ii) Synchronous transmission.

4. Asynchronous Frame Format: The asynchronous frame format is shown in figure 4.10. The strategy with this scheme is to avoid the timing problem by not sending long, uninterrupted streams of bits. In this format, data is transmitted one character at a time, where each character is five to eight bits in length.

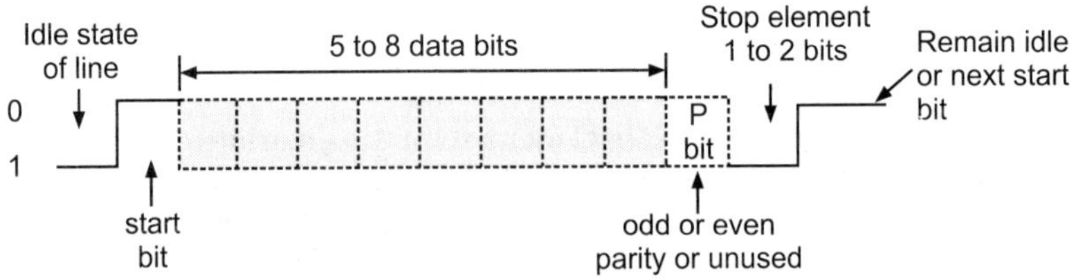

Fig. 4.10. Asynchronous frame format

Timing or synchronization must only be maintained within each character, the receiver has the opportunity to resynchronize at the beginning of each new character. For synchronization start and stop bits are added at the beginning and end of the character. The receiving machine resynchronizes at the beginning of each new byte. When the receiver detects a start bit, it sets a timer and begins counting bits as they come in. After n bits, the receiver looks for a stop bit. As soon as, it detects the stop bit, it ignores any received pulses until it detects the next start bit.

5. Synchronous Frame Format: In synchronous frame format, the bit stream is combined into longer frames which may contain multiple bytes without start and stop bits as shown in figure 4.11.

8 bit flag	Control field	Data field	Control field	8 bit flag

Fig. 4.11. Synchronous frame format

To prevent timing drift between the transmitter and receiver, their clocks must be synchronized. One of the ways for synchronization is to provide a separate clock line between transmitter and receiver. The other way to provide synchronization is to include the clocking information in the data signal. As shown in the figure 4.11, the frame starts with a preamble called a flag which is eight bits long. The same flag is used as a postamble. The receiver looks for the occurrence of the flag pattern to signal the start of the frame. This is followed by some number of control fields then a data field, more control fields and finally the flag is repeated. The advantage of synchronous transmission is its speed because it uses lesser number of overhead bit than asynchronous frames.

4.5. ERROR CONTROL

The next problem to be dealt with is to make sure that all frames are eventually delivered to the network layer at the destination, in proper order. Generally, the receiver sends back some feedback (positive or negative) to convey the information about whether it has received a frame or not. A positive acknowledgement (feedback) indicates a successful and error free delivery of a frame. Whereas a negative acknowledgement means that something has gone wrong and that the frame needs to be retransmitted. Due to the presence of noise burst, a frame may vanish completely. Therefore, the receiver does not receive anything and it does not react at all (no acknowledgement). This problem is overcome by introducing a timer in the data link layer. Its function is as follows.

4.5.1. Function of a Timer

As soon as a sender transmits a frame, it also starts the data link timer. The timer timing is set by taking into account the factors such as the required for the frame to reach the destination, processing time at the destination and the time required for the acknowledgement to return back. Normally, the frame is received correctly and the acknowledgement will return back to the sender before the timer runs out. This shows that a frame has been received and the timer is cancelled. But if a frame is lost or acknowledgement is lost, then the timer will go off. This will alert the sender that there is some problem The solution to this problem is that the sender retransmits the same frame. But, when a frame is transmitted multiple times, there is a possibility that the receiver will accept the same frame two or more times and pass it to the network layer more than once. To avoid this, each outgoing frame is assigned a distinct sequence number. This will help the receiver to distinguish retransmission.

4.6. ERROR DETECTION AND CORRECTION

When transmission of digital signals takes place between two systems such as computers as shown in figure 4.12, the signal get contaminated due to the addition of Noise to it. The noise can introduce an error in the binary bits travelling from one system to the other. This means that a 0 may change to 1 or a 1 may change to 0.

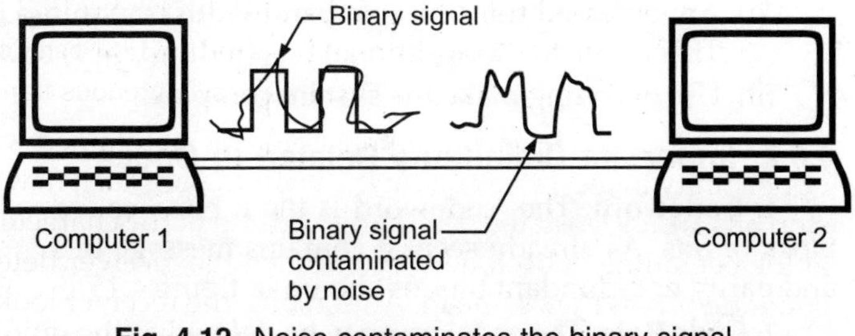

Fig. 4.12. Noise contaminates the binary signal

These error can become a serious threat to the accuracy of the digital system. Therefore, it is necessary to detect and correct the errors.

4.6.1. Types of Errors

Basically, whenever the received data differs from the transmitted data, an error is said to have occurred. Errors can occur due to various reasons. The errors introduced in the data bits during their transmission can be categorized as :

(i) Content errors

(ii) Flow integrity errors.

The content errors are nothing but errors in the content of a message e.g. a 0 may be received as 1 or vice versa. Such errors are introduced due to noise added into the data signal during its transmission. Flow integrity errors means missing blocks of data. It is possible that a data block may be lost in the network as it has been delivered to a wrong destination. Depending on the number of bits in error we can classify into two types as under

(i) Single bit error and

(ii) Burst errors.

1. Single bit error: The term signal bit error suggest that only one bit in the given data unit such as byte is in error. This means that only one bit will change from 1 to 0 or 0 to 1, as shown in figure 4.13.

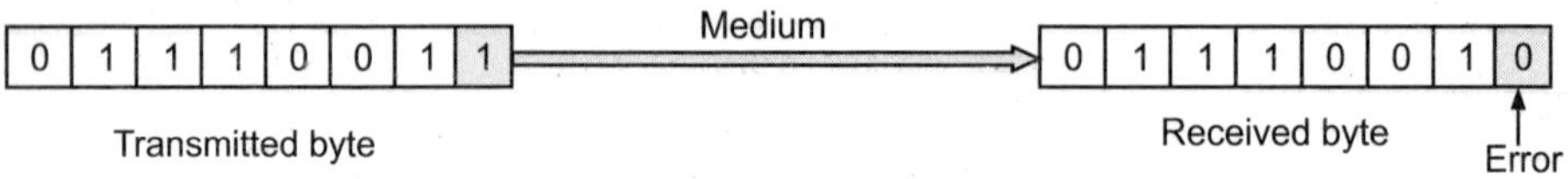

Fig. 4.13. Single bit error

2. Burst errors: If two or more bits from a data unit such as a byte change from 1 to 0 or from 0 to 1 then burst errors are said to have occurred. The length of the burst is measured from the first corrupted bit to the last corrupted bit. Some of the bits in between may not have been corrupted. Burst errors are illustrated in figure 4.14.

DO YOU KNOW?

Transmission errors are usually detected at the physical layer of the OSI model.

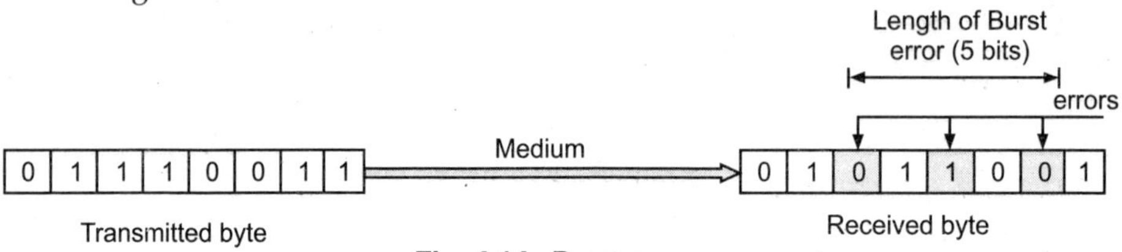

Fig. 4.14. Burst errors

Drawbacks of Coding

Some of the drawbacks of the coding technique may be listed as under :

(i) An increased transmission bandwidth is required in order to transmit the encoded signal. This is due to the additional bits (redundancy) added by the encoder.

(ii) Use of coding make the system complex.

4.6.2. Important Definitions Related to Codes

1. Codeword: The codeword is the n bit encoded block of bits. As already seen, it contains message bits and parity or redundant bits, as shown in figure 4.15

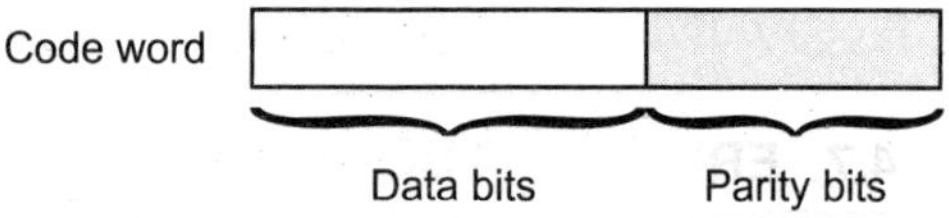

Fig. 4.15. Structure of a transmitted codeword

2. Code Rate: The code rate may be defined as the ratio of the number of message bits (k) to the total number of bits (n) in a codeword.

This means that

$$\text{Code rate (r)} = \frac{k}{n} \quad ...(4.1)$$

3. Hamming Weight of a Codeword [w(x)]: The Hamming weight of a codeword x may be defined as the number of non zero elements in the codeword. Hamming weight of a code vector (codeword) is the distance between that codeword and an all zero code vector (a code having all elements equal to zero).

4. Code Efficiency: The code efficiency is defined as the ratio of message bits to the number of transmitted bits per block. This means that

$$\text{code efficiency} = \text{Code rate} = \frac{k}{n} \quad ...(4.2)$$

5. Hamming Distance: Let us consider two code vectors (or codewords) having the same number of elements. The *Hamming Distance* or simply distance between the two codewords may be defined as the number of locations in which their respective elements differ. For example, let us consider the two codewords given below :

Code word No. 1 :	1	1	0	1	0	1	0	0
	↕				↕		↕	
Code word No. 2 :	0	1	0	1	1	1	1	0

It may be noted that the bits 2, 4 and 8 are different from each other. Hence, Hamming distance is 4.

EXAMPLE 4.1. Find the Hamming weight of the following code vector :

x = 1 1 0 1 0 1 0 0

Solution : As the number of non-zero elements in the above codeword is 3, therefore, the Hamming weight w(x) = 4.

6. Minimum Distance d_{min} : The minimum distance d_{min} of a linear block code may be defined as the smallest Hamming distance between any pair of code vectors in the code. Therefore, the minimum distance is same as the smallest Hamming weight of difference between any pair of code vectors. It can be proved that the minimum distance of a linear block code is the smallest hamming weight of the non-zero vectors in the code.

7. Role of d_{min} in error detection and correction: The error detection is always possible when the number of transmission errors in a codeword is less than the minimum distance d_{min} because then the erroneous word is not a valid codeword. But when the number of errors equals or exceeds d_{min}, the erroneous codeword may correspond to another valid codeword and errors cannot be detected. The error detection and correction capabilities of a coding technique depend on the minimum distance as shown in the Table 4.1

Table 4.1. Role of d_{min} for detection and correction of errors

Detect upto s errors per word.	$d_{min} \geq (s + 1)$
Correct upto t errors per word.	$d_{min} \geq (2t + 1)$
Correct upto t errors and detect s > t errors per word.	$d_{min} \geq (t + s + 1)$

4.7. ERROR DETECTION*

When a codeword is transmitted, one or more number of transmitted bits will be reversed (0 to 1 or

* Errors are inevitable; we have to live with them. The idea is to communicate correctly despite errors.

vice versa) due to transmission impairments. Thus, errors will be introduced. It is possible for the receiver to detect these errors if the received codeword (corrupted) is not one of the valid codewords. When the errors are introduced, the distance between the transmitted and received codewords will be equal to the number of errors as illustrated in figure 4.16

Hence to detect the errors at the receiver, the valid codewords should be separated by a distance of more than 1. Otherwise, the incorrect received codewords will also become some other valid codewords and the error detection will be impossible. The number of errors that can be detected depends on the distance between any two valid codewords.

Transmitted code word	1 0 1 0 1 1 0 0	1 1 1 0 1 0 1 1	0 0 1 0 0 1 0 1
Received code word	1 (1) 1 0 1 1 0 0 (error)	(0) 1 1 (1) 1 0 1 1 (error)	(1) 0 1 (1) 0 (0) 0 1 (error)
Number of errors	1	2	3
Distance	1	2	3

Fig. 4.16.

4.7.1. Error Detection Methods

Some of the most important error detection methods are as under :

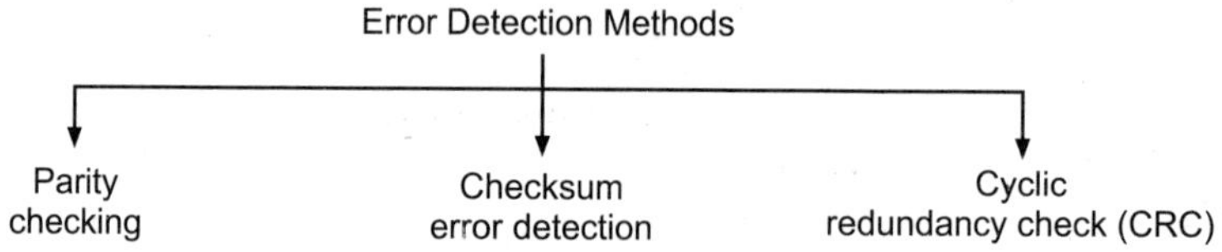

Before thinking of correcting the errors introduced in the data bits it is necessary to first detect them. Some of the popular error detection methods are as under :

(i) Parity Checking

(ii) Checksum error detection

(iii) Cyclic redundancy check (CRC).

For error detection and correction, it is necessary to add some check bit to a block of data bits. These check bits are also known as **redundant bits** because they do not carry any useful information.

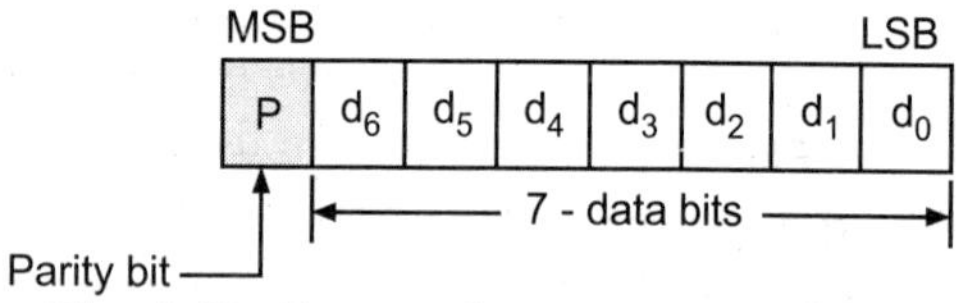

Fig. 4.17. Format of a transmitted word with parity bit

4.7.2. Parity

1. Definition: The simplest technique for detecting errors is to add an extra bit known as parity bit to each word being transmitted. As shown in figure 4.17, generally, the MSB of an 8 - bit word is used as the parity bit and the remaining 7 bits are used as data or message bits.

> **DO YOU KNOW?**
>
> The method used to handle error control depends on the method used for flow control.

The parity of the 8 - bit transmitted word can be either even parity or odd parity. **Even parity** means the number of 1's in the given word including the parity bit should be even (2, 4, 6..). **Odd Parity** means the number of 1's in the given word including the parity bit should be odd (1, 3, 5..).

2. Use of Parity bit to decide parity: The Parity bit can be set to 0 to 1 depending on the type of parity required. For odd parity, this bit is set to 1 or 0 at the transmitter such that the number of 1 bits in the entire word is odd. For even parity, this bit is set to 1 or 0 such that the number of 1 bits in the entire word is even. This has been illustrated in figure 4.18.

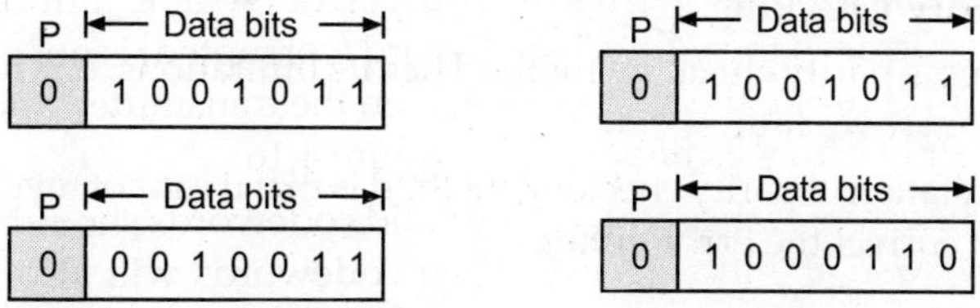

(a) Inclusion of a parity bit to obtain an even parity

(b) Inclusion of a parity bit to obtain the odd parity

Fig. 4.18.

3. How does parity checking fail to detect errors?: This parity at the receiver can detect the presence of an error if the parity of the received signal is different from the expected parity. This means that if it is known that the parity of the transmitted signal is always going to be even and if the received signal has an odd parity then the receiver can conclude that the received signal is not correct. This has been shown in figure 4.19. When a single error or an odd number of errors occur during transmission the parity of the codeword changes. Parity of the received codeword is checked at the receiver and change in parity indicates that error is present in the received word. This has been shown in figure 4.19. If presence of error is detected then the receiver will ignore the received byte and request for the retransmission of the same byte to the transmitter.

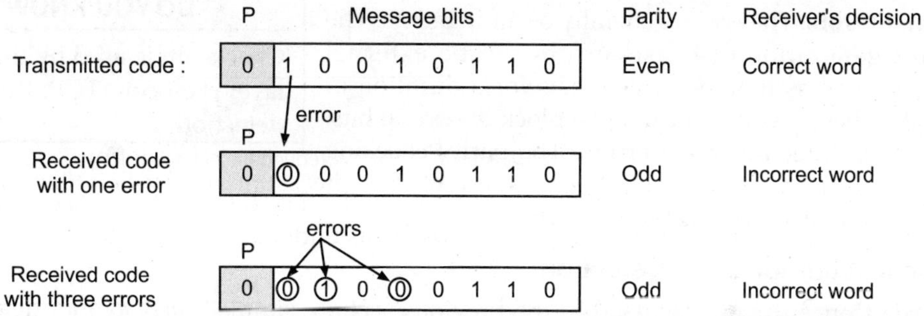

Fig. 4.19. The receiver detects the presence of error if the number of errors is odd i.e. 1, 3, 5...

If the number of errors introduced in the transmitted code is two or any even number, then the number of the received codeword will not change . It will still remain even as shown in figure 4.20 and the number will fail to detect the presence of errors.

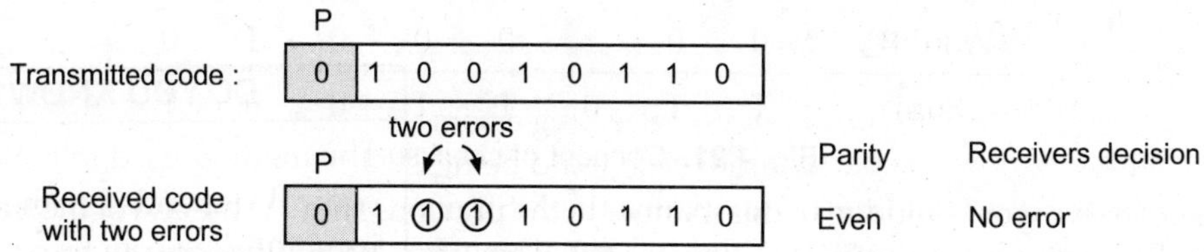

Fig. 4.20. The receiver cannot detect the presence of error if the number of errors is even i.e. 2, 4, 6....

4. Conclusions

(i) Double or any even number of errors in the received word will not change the parity. Therefore, even number of errors will be unnoticed.

(ii) If one or odd number of errors occur then the parity of the received word will be different from the parity of transmitted signal. Thus, error is noticed. However, this error cannot be corrected.

5. Limitations of Parity Checking

(i) Thus, the simple parity checking method has its limitations. It is not suitable for detection of multiple errors (two, four, six etc.)

(ii) The other limitation of parity checking method is that it cannot reveal the location of erroneous bit. It cannot correct the error either.

EXAMPLE 4.2. Write the ASCII code of word HOLE using even parity.

Solution : Table 4.2 shows the ASCII code of word HOLE using even parity.

Table 4.2.

	P	7	6	5	4	3	2	1
H	**0**	1	0	0	1	0	0	0
O	**1**	1	0	0	1	1	1	1
L	**1**	1	0	0	1	1	0	0
E	**1**	1	0	0	0	1	0	1

It may be noted that the parity bits are selected in order to obtain an even parity for each row (i.e., for each letter).

DO YOU KNOW?

Checksum is used by the higher-layer protocols (TCP/IP) for error detection.

6. Burst Errors: The errors generally occur in bursts. The reason for generation of bursts errors can be external interference such as lighting which lasts for a duration of several bits. Therefore, it will corrupt a block of several bits. Such errors are called as burst errors. The parity checking methods is not useful in detecting the burst errors. The checksum and cyclic redundancy check (CRC) methods can detect the burst errors.

4.7.3. Checksum for Error Detection

1. Basic Concepts: As discussed in the previous section, simple parity cannot detect two or even number of errors within the same word. One way to overcome this problem is to use a sort of two dimensional parity. As each word is transmitted, it is added to the previously sent word and the sum is retained at the transmitter as shown in figure 4.21.

Word A :	1	0	1	1	0	1	1	1
+								
Word B :	0	0	1	0	0	0	1	0
Sum	1	1	0	1	1	0	0	1

Fig. 4.21. Concept of checksum

Each successive word is added in this manner to the previous sum. At the end of the transmission, the sum (called a checksum) upto that time is sent. The errors normally occur in burst. The parity check method is not useful in detecting the errors under such conditions. The checksum error detection method can be used successfully in detecing such errors. In this method a checksum is transmitted along with every block of data bytes. In this method, an eight bit accumulator is used to add 8 bit bytes of a block of data to find the *checksum byte*. The carries of the MSB are ignored while finding out the checksum byte.

2. Generation of Checksum: The generation of checksum can be clear with the help of the following example.

EXAMPLE 4.3. Find the checksum of the following message.

10110001, 10101011, 00110101, 10100001

Solution: It may be noted that the carries of MSB have been ignored while writing the checksum byte.

Ignored

Carries	(10)	1	0	1	1	1	0	
Data bytes		1	0	1	1	0	0	1
		1	0	1	0	1	1	1
		+ 0	0	1	1	0	0	1
		+ 1	0	1	0	0	0	1
Checksum byte		+ 0	0	1	1	0	1	0

3. How to detect error using the checksum byte? : After transmitting a block of data bytes (say 8-data bytes) the *checksum* byte is also transmitted. The checksum byte is regenerated at the receiver separately by adding the received bytes. The regenerated checksum byte is then compared with the transmitted one. If both are identical then there is no error. If they are different then the errors are present in the block of received data bytes. Sometimes, the 2's complement of the checksum is transmitted instead of the checksum itself. The receiver will accumulate all the bytes including the 2's complement of the checksum. If there is no error, the contents of the accumulator should be zero after accumulation of the 2's complement of the checksum byte.

4. Advantage of the Checksum Method: The advantage of this method over the simple parity checking method is that the data bits are *mixed up* due to the 8 bits addition. Therefore, checksum represents the overall data block. In checksum, therefore, there is 255 to 1 chance of detecting random errors.

4.8. TWO DIMENSIONAL PARITY CHECK

1. Basic Concepts: When a large number of binary words are being transmitted or received in succession, the resulting collection of bits is considered as a block of data, with rows and columns as shown in figure 4.22. The parity bits are produced for each row and column of such block of data. The two sets of parity bits so generated are known as :

(i) Longitudinal Redundancy Check (LRC) bits

(ii) Vertical Redundancy Check (VRC) bits.

Characters		C	O	M	P	U	T	E	R	
7 bit ASCII codes (Message bits)	b_1	1	1	1	0	1	0	1	0	1
	b_2	1	1	0	0	0	0	0	1	1
	b_3	0	1	1	0	1	1	1	0	1
	b_4	0	1	1	0	0	0	0	0	0
	b_5	0	0	0	1	1	1	0	1	0
	b_6	0	0	0	0	0	0	0	0	0
	b_7	1	1	1	1	1	1	1	1	0
VRC bits (even parity) →		1	1	0	0	0	1	1	1	1

These bits will make the parity of each column even

LRC bits (even parity) → These bits will make the parity of each row even

Fig. 4.22. Vertical and longitudinal parity check bits

The LRC bits indicate the parity of rows and VRC bits indicate the parity of columns as shown in figure 4.22.

(i) The Vertical Redundancy Check (VRC) Bits: As shown in figure 4.22, the VRC bits are parity bits associated with the ASCII code of each character. Each VRC bit will make the parity of its corresponding column *an even parity*. For example, let us consider column 1 corresponding to character C. The ASCII code for the character C is :

Character	C	
b_1	1	
b_2	1	← Column-1 of the data block
b_3	0	
b_4	0	
b_5	0	
b_6	0	
b_7	1	
VRC bit →	1	← VRC bit = 1 to make the parity of first column even

DO YOU KNOW?

Redundancy is the concept of sending extra bits for use in error detection.

Therefore, the 8th bit which is a VRC bit is made 1 to make the parity even.

(ii) The Longitudinal Redundancy Check (LRC) Bits: The LRC bits are parity bits associated with the rows of the data block of figure 4.22. Each LRC bit will make the parity of the corresponding row, an even parity. For example, let us consider row 1 of figure 4.22.

Row 1 :	b_1	1	1	1	0	1	0	1	0	1	← LRC bit to make parity even

2. How to Locate the Bit in Error? : Even a single error in any bit will result in a noncorrect LRC in one of the rows and an incorrect VRC in one of the columns. The bit which is common to the row and column is the bit in error. However, there is still a limitation on the Block parity code, which is that, multiple errors in rows and columns can be only detected but they cannot be corrected. This is because, it is not possible to locate the bits which are in error. This will be clear with the help of the following example.

EXAMPLE 4.4. The following bit stream is encoded using VRC, LRC and even parity. Locate and correct the error if it is present.

11000011 11110011 10110010 00001010
00101010 00101011 10100011 01001011
11100001

Solution :

(1) Figure 4.23 shows the received data block alongwith the LRC and VRC bits.

(2) It may be noted that the parity bits corresponding to row 1 and column 5 indicate wrong parity. Therefore, the fifth bit in the first row (encircled bit) is incorrect.

Thus, using VRC and LRC, it is possible to locate and correct the bits in error.

	byte 1	byte 2	byte 3		bit in error ↓				LRC bits (even parity)	
b_1	1	1	1	0	(0)	0	1	0	1	← Wrong parity
b_2	1	1	0	0	0	0	0	1	1	
b_3	0	1	1	0	1	1	1	0	1	
b_4	0	1	1	0	0	0	0	0	0	
b_5	0	0	0	1	1	1	0	1	0	
b_6	0	0	0	0	0	0	0	0	0	
b_7	1	1	1	1	1	1	1	1	0	
VRC bits (even parity) →	1	1	0	0	0	1	1	1	1	
					↑ Wrong parity					

Data → block (b_1 to b_7)

First bit of the fifth byte is in error

Fig. 4.23.

4.9. CYCLIC REDUNDANCY CHECK (CRC)*

Cycle redundancy check (CRC) is a type of polynomial code in which a bit string is represented in the form of polynomials with coefficients of 0 and 1 only. Polynomial arithmetic uses a modulo-2 arithmetic i.e. addition and subtraction are identical to EXOR. For CRC code, the sender and receiver should agree upon a generator polynomial G(x). A codeword can be generated for a given data word (message) polynomial M(x) with the help of long division. This technique is more powerful than the parity check and checksum error detection.

> **DO YOU KNOW?**
>
> CRC, the most powerful of the redundancy checking techniques, is based on binary division.

CRC is based on binary division. A sequence of redundant bits called CRC or CRC remainder is appended at the end of a data unit such as byte. The resulting data unit after adding CRC remainder becomes exactly divisible by another predetermined binary number. At the receiver, this data unit is divided by the same binary number. There is no error if this division does not yield any remainder. But, a non-zero remainder indicates presence of errors in the received data unit. Such an erroneous data unit is then rejected.

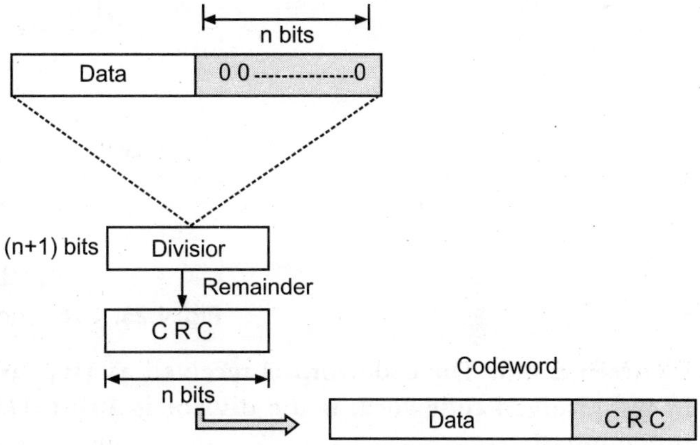

Fig. 4.24. CRC generator

4.9.1. Procedure to Obtain CRC

The redundancy bits used by CRC are derived by following the procedure given below:

(i) Divide the data unit by a predetermined divisor.

(ii) Obtain the remainder. It is the CRC.

4.9.2. Requirements of CRC

A CRC will be valid if and only if it satisfies the following requirements :

(i) It should have exactly one less bit than divisor.

(ii) Appending the CRC to the end of the data unit should result in the bit sequence which is exactly divisible by the divisor.

4.9.3. CRC Generator

Figure 4.24 shows the CRC generator. The procedure for CRC generation may be listed in the form of following points:

(i) Append a string of n 0s to the data unit where n is 1 less than the number of bits in the predecided divisor (n + 1 bit long).

(ii) Divide the newly generated data unit in step (i) by the divisor. This is a binary division.

(iii) The remainder obtained after the division in step (ii) is the n bit CRC.

(iv) This CRC will replace the n 0s appended to the data unit in step (i), to get the codeword to be transmitted as shown in figure 4.24.

* Checksum and CRC are two popular error detection mechanisms and Hamming code is a popular error correction technique.

4.9.4. CRC Checker

Figure 4.25 shows the CRC checker. The codeword received at the receiver consists of data and CRC. The receiver treats it as one unit and divides it by the same $(n + 1)$ bit divisor which was used at the transmitter. The remainder of this division is then checked. If the remainder is zero, then the received codeword is error free and hence should be accepted. But, a non-zero remainder indicates presence of errors hence the corresponding codeword should be rejected.

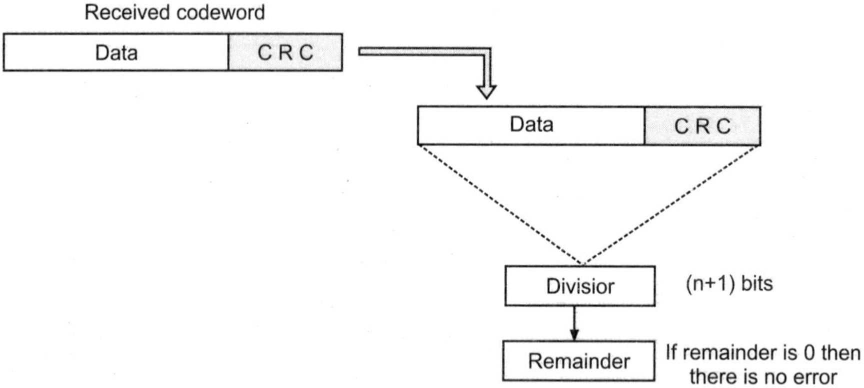

Fig. 4.25. CRC checker

EXAMPLE 4.5. The codeword is received as 1100 1001 01011. Check whether there are errors in the received codeword, if the divisor is 10101. (The divisor corresponds to the generator polynomial).

Solution : As we know that the codeword is formed by adding the dividend and the remainder. This codeword will have an important property that it will be completely divisible by the divisor. Thus, at the receiver, we have to divide the received codeword by, the same divisor and check for the remainder. If there is no remainder then there are no errors. But, if there is remainder after division, then there are errors in the received codeword. Let us use this technique and find if there are errors.

Data word : 1100 1001 01011

Divisor : 10101

```
                          1 1 1 1 1 0 0 0 0 1
Divisor ──→  1 0 1 0 1 ) 1 1 0 0 1 0 0 1 0 1 0 1 1   ← Received codeword
                        ⊕1 0 1 0 1
                         0 1 1 0 0 0
                          ⊕ 1 0 1 0 1
MOD-2                       0 1 1 0 1 0
additions                 ⊕ 1 0 1 0 1
                              0 1 1 1 1 1
                            ⊕ 1 0 1 0 1
                                0 1 0 1 0 0
                              ⊕ 1 0 1 0 1
                                  0 0 0 0 1 1 0 1 1
                                          1 0 1 0 1
                                          0 1 1 1 0   ← Remainder
```

The non-zero remainder shows that there are errors in the received codeword.

Generation of CRC Code: The generation of CRC code is clear after solving the following example.

EXAMPLE 4.6. Generate the CRC code for the data word of 1100 10101. The divisor is 10101.

Solution : Given that data word : 110010101

Divisor : 10101

The number of data bits = m = 9

The number of bits in the codeword = N

Dividend = Data word number of zeros.

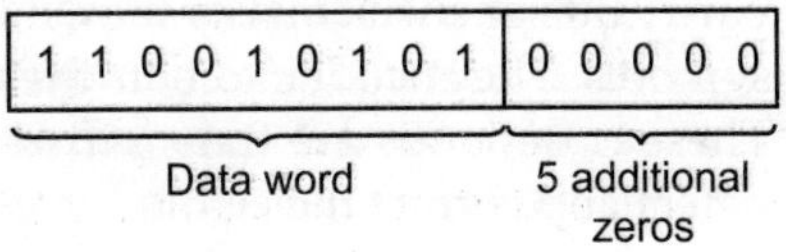

Now, let us carry out the division as under :

```
                         1 1 1 1 1 0 0 1 1 1 1        ← Quotient
Divisor →  1 0 1 0 1 ) 1 1 0 0 1 0 1 0 1 0 0 0 0 0 ← Dividend
                       1 0 1 0 1
                       ⊕ ⊕ ⊕ ⊕ ⊕
                       0 1 1 0 0 0
                         1 0 1 0 1
                         0 1 1 0 1 1
                           1 0 1 0 1
                           0 1 1 1 0 0
                             1 0 1 0 1
                               1 0 0 1 1
                               1 0 1 0 1
                                 1 1 0 0 0
                                 1 0 1 0 1
                                   1 1 0 1 0
                                   1 0 1 0 1
                                     1 1 1 1 0

                                     1 0 1 0 1
                                     0 0 1 1 0  ← Remainder
```

Codeword: In CRC, the required codeword is obtained by writing the data word followed by the remainder. Thus, we have

	11001010100000
	110
Codeword =	11001010100110

Undetected Errors in CRC

(i) CRC cannot detect all types of errors.

(ii) The probability of error detection and the types of detectable errors depends on the choice of divisor.

4.10. ERROR CORRECTION

Now, let us discuss two completely different approaches for the error control. They are as under:

1. Automatic request for retransmission (ARQ)
2. Forward error correction (FEC)

1. The ARQ Technique : In the ARQ system, the receiver can request for the retransmission of the complete or a part of message if it finds some error in the received message. This requires an additional channel called *feedback channel* to send the receiver's request for retransmission.

2. The FEC Technique : In the FEC technique, there is no such feedback path and request for retransmission. Most of the coding techniques described in this chapter are related to FEC.

4.10.1. Error Correction Techniques

In the error correction techniques, codes are generated at transmitter by adding a group of parity bits or check bits as shown in figure 4.26. The source generates the data in the form of binary symbols. The encoder accepts these bits and adds the check bits to them to produce the codewords. These codewords are transmitted towards the receiver. The check bits are used by the decoder to detect and correct the errors.

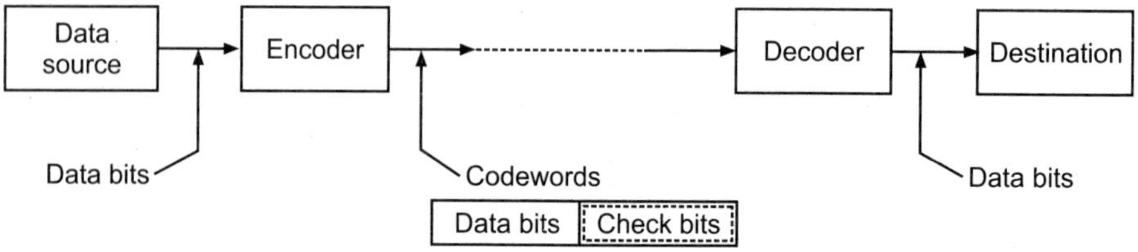

Fig. 4.26. Error correction technique

The encoder of figure 4.26 adds the check bits to the data bits, according to the prescribed rule. This rule will be dependent on the type of code being used. The decoder separates out the data and check bits. It uses the parity bits to detect and correct errors if they are present in the received codewords. The data bits are then applied to the destination.

4.10.2. FEC (Forward Error Correction)

In FEC, the receiver searches for the most likely correct codeword. When an error is detected, the distance between the received invalid codeword and all the possible valid codewords is measured. The nearest valid codeword (the one having minimum distance) is the most likely correct version of the received codeword as shown in figure 4.27.

In figure 4.27, the valid codeword 1 has the minimum distance (1), hence, it is the most likely correct codeword.

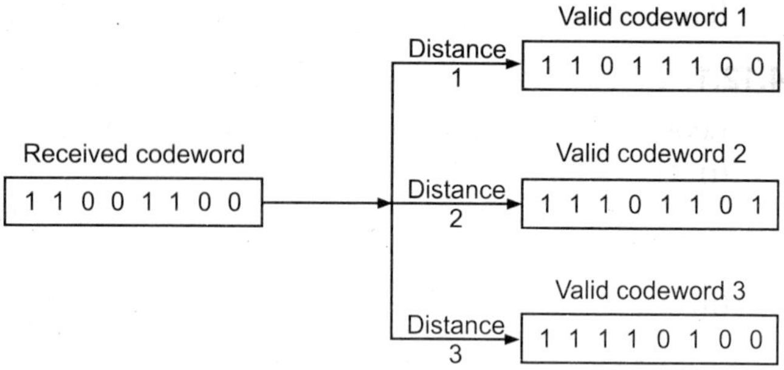

Fig. 4.27. Concept of FEC

4.10.3 Error Correction Techniques

Some of the FEC techniques may be listed as under :

(i) Linear block codes.

(ii) Convolutional coding

(iii) Hamming codes

(iv) Cyclic codes.

4.11. LINEAR BLOCK CODES

1. Basic Concept: The generation of block codes is illustrated in figure 4.28. To generate an (n, k) block code, the channel encoder accepts the information in successive k bit blocks. At the end of each such block (of k message bits), it adds (n – k) parity bits as shown in figure 4.28. As these bits do not contain any information, they are called

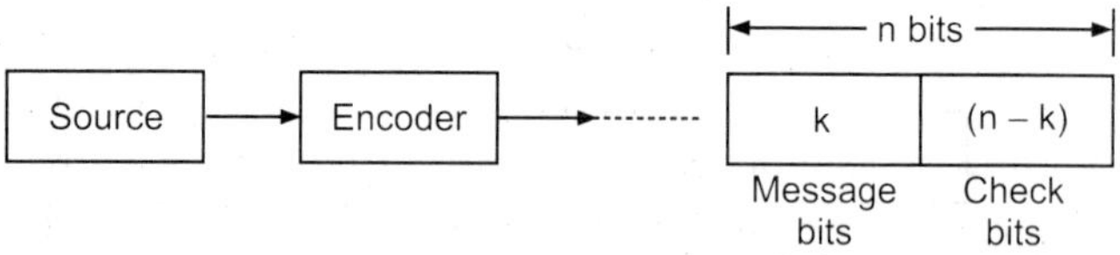

Fig. 4.28. Generation of an n bit linear block code

as *redundant* bits. It is important to note that the (n – k) parity bits are related algebraically to the k message bits. The n bit codeword is thus produced as shown in figure 4.28.

2. Why are these codes called linear codes ?: These codes have an important property that any two codewords of a linear code can be added in modulo-2 adder to produce a third codeword in the code. Non-linear codes do not exhibit such a property. All the practically used codes are linear codes.

3. Codeword Structure: The codeword structure of a linear block code is as shown in figure 4.29.

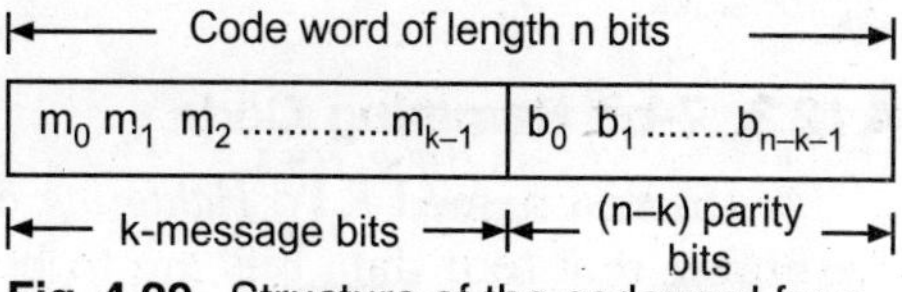

Fig. 4.29. Structure of the codeword for a linear block code

4.12. HAMMING CODES

Hamming codes are linear block codes. The family of (n, k) Hamming codes for m ≥ 3 is defined by the following equations:

Block length : $n = 2^m - 1$

Number of message bits : $k = 2^m - m - 1$ ... (4.3)

Number of parity bits : $(n - k) = m$

where, minimum number of parity bits is 3.

The minimum distance $d_{min} = 3$.

$$\text{The code rate or code efficiency} = \frac{k}{n} = \frac{2^m - m - 1}{2^m - 1} = 1 - \frac{m}{2^m - 1} \quad ...(4.4)$$

If m >> 1, then, code rate $r \cong 1$.

4.12.1. Error Detection and Correction Capabilities of Hamming Code

Looking at the table 4.1 for the minimum distance, $d_{min} = 3$, we have

(i) The number of errors that can be detected per word = 2

Since $d_{min} \geq (s + 1)$, so, $3 \geq s + 1$

so, $s \leq 2$

(ii) The number of errors that can be corrected per word = 1

Since $d_{min} \geq (2t + 1)$, so, $3 \geq (2t + 1)$

so, $t \leq 2$

Therefore, with $d_{min} = 3$, it is possible to detect upto 2 errors and it is possible to correct upto only 1 error.

4.12.2. Hamming Code Structure

Hamming code is basically a linear block code named after its inventor. It is an error correcting code. The parity bits are inserted in between the data bits as shown in figure 4.30. The 7-bits Hamming code is used commonly, but the concept can be extended to any number of bits.

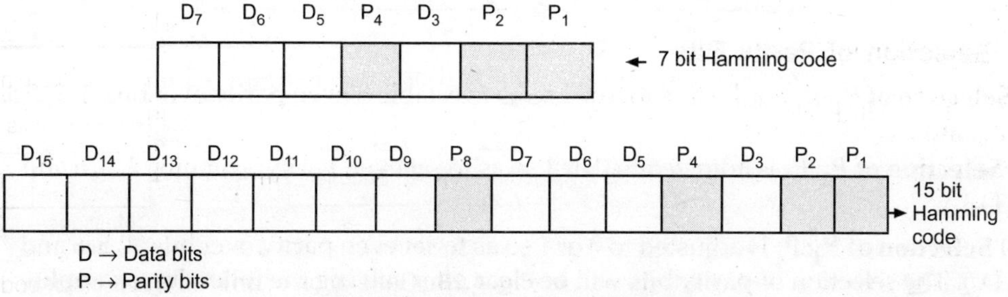

Fig. 4.30. Hamming codewords

Here, it may be noted that the parity bits are inserted at each 2^n bit where n = 0, 1, 2, 3 .... Thus, P_1 is at $2^0 = 1$, i.e., at first bit, P_2 is at $2^2 = 2P_4$ is as 2^2 and P_8 is at 2^3 as shown in figure 4.30.

DO YOU KNOW?

The Hamming code is a single-bit error corection method using redundant bits. The number of bits is a function of the length of the data bits.

4.12.3. 7-bit Hamming Code

A scientist named *R.W. Hamming* developed a coding system which was easy to implement. Assuming that four data bits are to be transmitted, he suggested a codeword pattern shown in figure 4.31.

D_7	D_6	D_5	P_4	D_3	P_2	P_1

$D_7\ D_6\ D_5\ D_3 \rightarrow$ Data bits

$P_4\ P_2\ P_1 \rightarrow$ Parity bits

Fig. 4.31. Codeword pattern for Hamming code

The D bits in figure 4.31 are data bits, whereas, P bits are parity bits. The parity bits P_1, P_2, P_4 are adjusted in a particular way as explained below.

Minimum number of parity bits

Table 4.3 gives a listing of minimum number of parity bits needed for various ranges of m information bits.

Table 4.3. Number of parity bits to be used

Number of information bits	Number of parity bits
2 to 4	3
5 to 11	4
12 to 26	5
27 to 57	6
58 to 120	7

Deciding the values of parity bits

Table 4.4 indicates which bit positions are associated with each parity bit in order to establish required parity (even or odd) over the selected bits positions

Table 4.4.

Parity Bit	Bits to be checked
P_1	1, 3, 5, 7, 9, 11, 13, 15, ...
P_2	2, 3, 6, 7, 10, 11, 14, 15, ...
P_4	4, 5, 6, 7, 12, 13, 14, 15, ...
P_8	8, 9, 10, 11, 12, 13, 14, 15, ...

4.12.4. Selection of Parity Bits

(i) Selection of P_1: P_1 is adjusted to 0 or 1 so as to establish even parity over bits, 1, 3, 5 and 7 i.e., P_1, D_3, D_5 and D_7.

(ii) Selection of P_2: P_2 is adjusted to 0 or 1 so as to set even parity over bits, 2, 3, 6 and 7 (P_2, D_3, D_6 and D_7).

(iii) Selection of P_4: P_4 is adjusted to 0 or 1 so as to set even parity over bits, 4, 5, 6 and 7 (P_4, D_5, D_6 and D_7). The selection of parity bits will be clear after solving the following example.

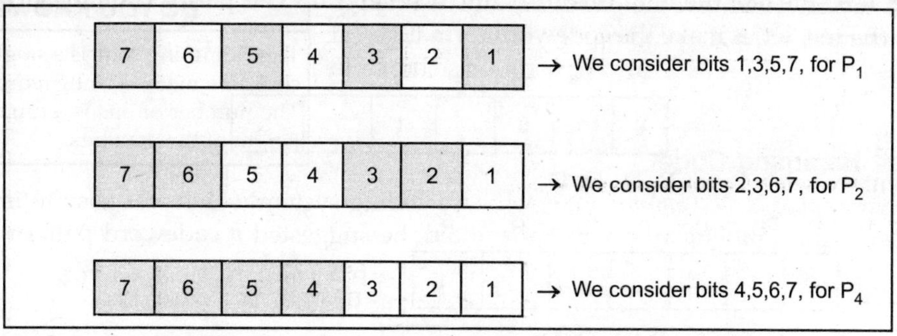

EXAMPLE 4.7. A bit word 1011 is to be transmitted. Construct the even parity seven-bit Hamming code for this data.

Solution : (i) First, let us make the codeword format.

The seven bit Hamming code format is shown below. Given bit word = 1011

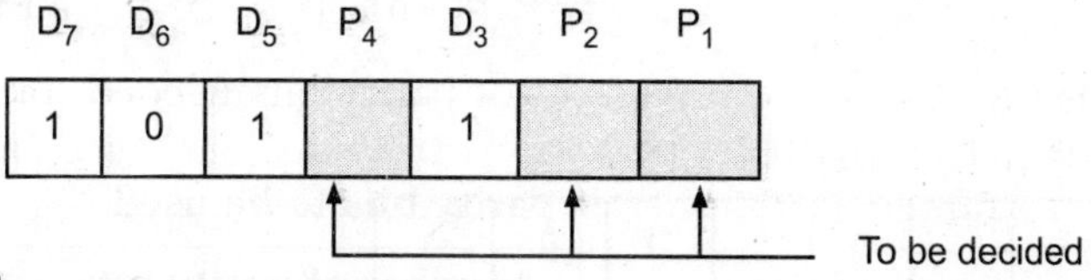

(ii) Then, we decide P_1.

P_1 sets the parity of bits P_1, D_3, D_5 and D_7. As $D_7\ D_5\ D_3 = 111$, we have to set $P_1 = 1$ in order to have the even parity.

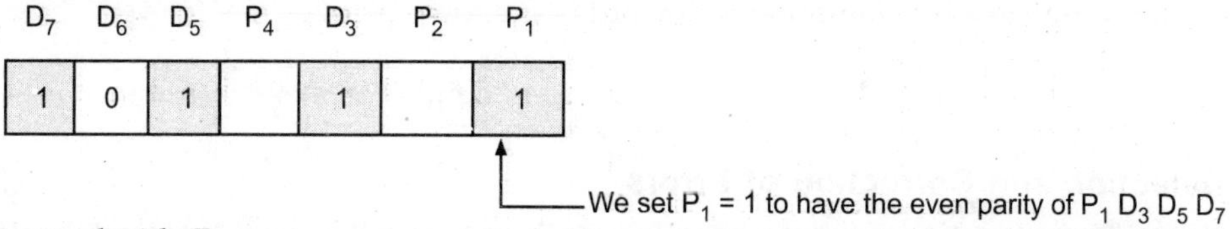

(iii) Then, we decide P_2.

P_2 is set to have the even parity of P_2 D_3 D_6 and D_7. But, $D_3\ D_6\ D_7 = 101$, hence, let us set $P_2 = 0$.

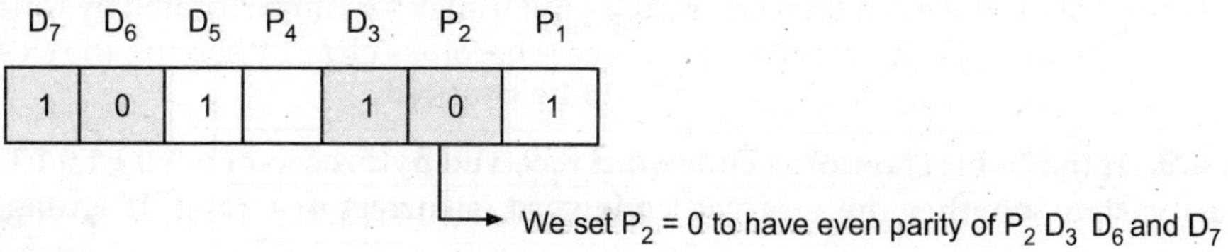

(iv) We decide P_4.

P_4 is set to have the even parity of P_4 D_5 D_6 and D_7. But, $D_5\ D_6\ D_7 = 101$, hence, let us set $P_4 = 0$.

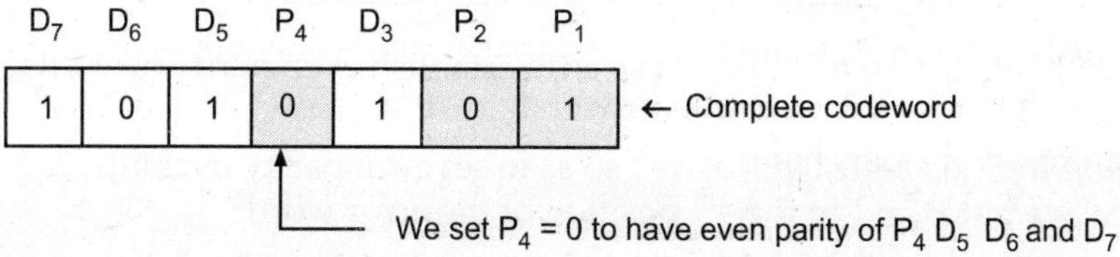

EXAMPLE 4.8. Encode the data bits 0101 into a seven bit even parity Hamming code.

Solution : (i) First, let us make the codeword format.

D_7	D_6	D_5	P_4	D_3	P_2	P_1
0	1	0		1		

(ii) Then, we select P_1 for $P_1\ D_3\ D_5\ D_7$.

D_7	D_6	D_5	P_4	D_3	P_2	P_1
0	1	0		1		1

→ $P_1 = 1$ for $P_1\ D_3\ D_5\ D_7 = 1\ 1\ 0\ 0$

(iii) Then, we select P_2 for $P_2\ D_3\ D_6\ D_7$.

D_7	D_6	D_5	P_4	D_3	P_2	P_1
0	1	0		1	0	1

→ $P_2 = 0$ for $P_2\ D_3\ D_6\ D_7 = 0\ 1\ 1\ 0$

(iv) Then, we select P_4.

D_7	D_6	D_5	P_4	D_3	D_2	D_1
0	1	0	1	1	0	1

Set $P_4 = 1$ to have $P_4\ D_5\ D_6\ D_7 = 1\ 0\ 1\ 0$

Hence, the complete 7-bit Hamming codeword is as shown below :

0	1	0	1	1	0	1

← Complete codeword

4.12.5. Detection and Correction of Errors

The Hamming coded data is now transmitted. At the receiver, it is decoded to get the data back. The bits (1, 3, 5, 7), (2, 3, 6, 7) and (4, 5, 6, 7) are checked for even parity. If all the 4 - bit groups mentioned above possess the even parity then the received codeword is correct i.e. it does not contain errors. But, if the parity is not even then error exists. Such an error can be located by forming a three bit number out of the three parity checks. This process becomes clear by solving the example given below.

EXAMPLE 4.9. If the 7 - bit Hamming codeword received by a receiver is 1 0 1 1 0 1 1. Assuming the even parity, state whether the received codeword is correct or wrong. If wrong, locate the bit in error.

Solution : Received codeword :

D_7	D_6	D_5	P_4	D_3	P_2	P_1
1	0	1	1	0	1	1

(i) First, we analyze bits 4, 5, 6 and 7.

$P_4\ D_5\ D_6\ D_7 = 1\ 1\ 0\ 1 \rightarrow$ Odd parity.

Therefore, error exists here.

Hence, we put $P_4 = 1$ in the 4th position of the error word.

(ii) Then, we analyze bits 2, 3, 6 and 7.

Therefore, $P_2\ D_3\ D_6\ D_7 = 1\ 0\ 0\ 1 \rightarrow$ Even parity so no error.

Hence, we put $P_2 = 0$ in the 2nd position of the error word.

(iii) Next, we check the bits 1, 3, 5, 7.

Therefore, $P_1 D_3 D_5 D_7 = 1\,0\,1\,1 \rightarrow$ Odd parity, hence error exists.

Hence, we put $P_1 = 1$ in the 1st position of the error word.

(iv) Next, we write the error word.

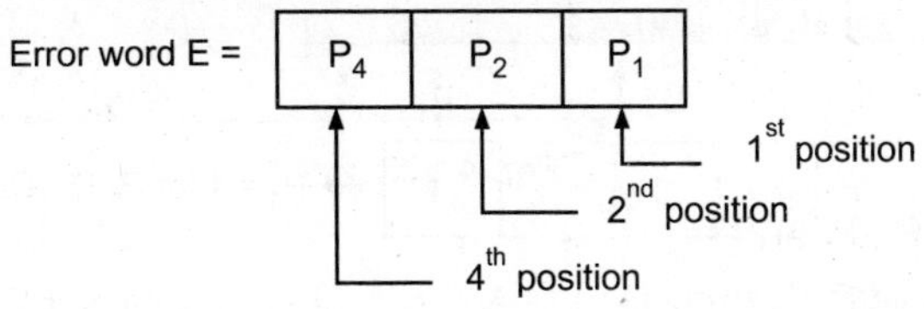

Substituting the values of P_4, P_2 and P_1 obtained in steps 1, 2 and 3, we obtain

E =		
1	0	1

$E = (5)_{10}$

Hence, bit 5 of the transmitted codeword is in error.

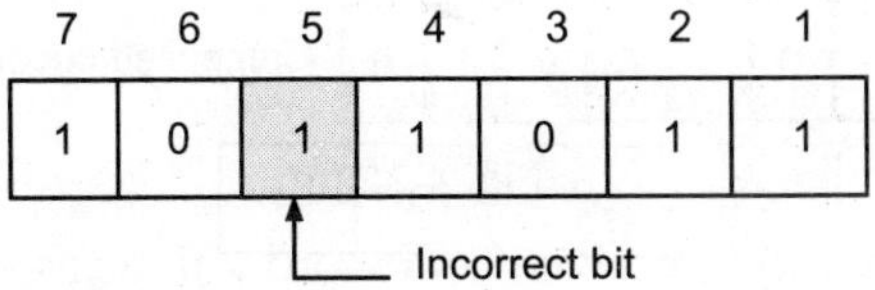

(v) Finally, we correct the error.

We invert the incorrect bit to obtain the correct codeword as under :

Correct codeword = [1 0 0 1 0 1 1]

EXAMPLE 4.10. A seven bit hamming code is received as 1 1 1 0 1 0 1. What will be the correct code ?

Solution :

7	6	5	4	3	2	1	
1	1	1	0	1	0	1	← Received codeword

1. Check bits, 4, 5, 6, 7 → odd parity, hence error ∴ $P_4 = 1$
2. Check bits, 2, 3, 6, 7 → odd parity, hence error ∴ $P_2 = 1$
3. Check bits, 1, 3, 5, 7 → even parity, hence no error ∴ $P_1 = 0$

∴ Error word E =

1	1	0
P_4	P_2	P_1

4. Decimal equivalent of $E = 1\,1\,0 = (6)_{10}$

Therefore, 6th bit in the received codeword is incorrect. Hence, we invert it.

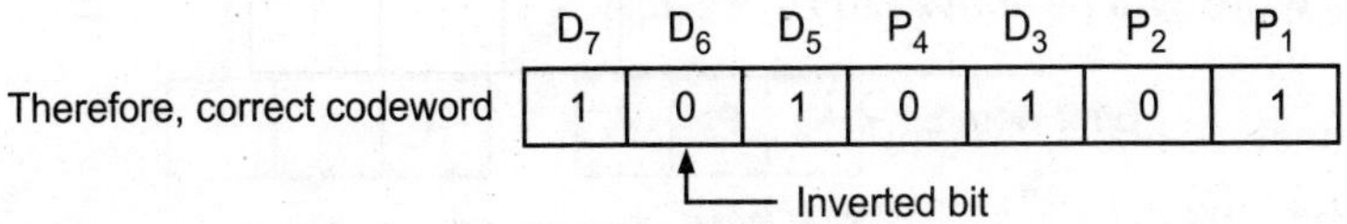

EXAMPLE 4.11. Write the steps to generate the hamming code. Prepare Hamming code for bit pattern 10101010. Suppose while transmitting, error occurs in 7th bit, write the bit pattern at the receiver. Using Hamming code, explain how will you detect and correct the error.

Solution: Let us obtain the 7-bit Hamming code for the bit pattern 1010.

(i) First, let us make the codeword format.

The seven bit Hamming codeword format is shown in figure 4.32(a)

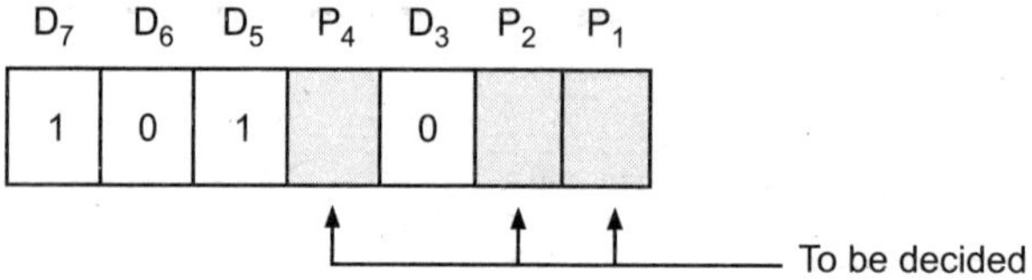

Fig. 4.32. (a)

(ii) Then, we decide P_1, P_2 and P_4.

- P_1 sets the parity of P_1, D_3 and D_5, D_7. As $D_7 D_5 D_3 = 110$, we set $P_1 = 0$, so as to have the even symmetry.
- P_2 is set to have the even parity of P_2, D_3, D_6 and D_7. But, $D_3 D_6 D_7 = 001$. Hence, we set $P_2 = 1$.
- P_4 is set to have the even parity of $P_4 D_5$ D_6 and D_7. But, D_5 D_6 $D_7 = 101$, so, we set $P_4 = 0$.
- The codeword is shown in figure 4.32 (b).

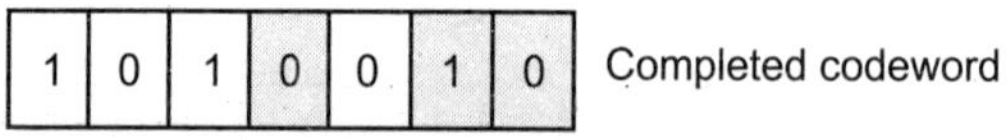

Fig. 4.32 (b)

The 7th bit is in error. Hence, the received codeword will be as shown in figure 4.33

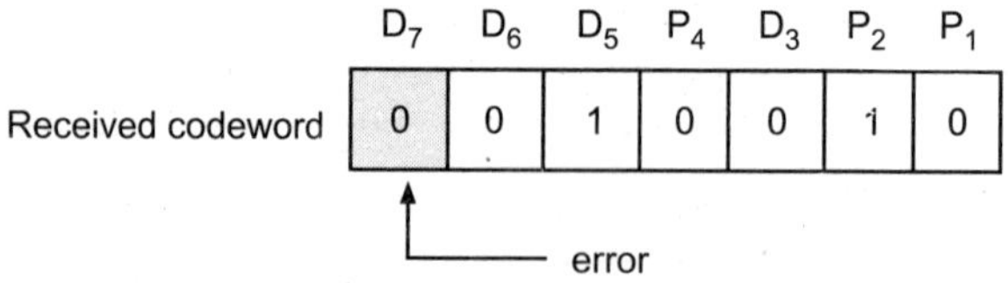

Fig. 4.33.

(iii) Finally, we carry out Error Detection and Correction.

(a) First, we analyze bits 4, 5, 6, 7.

P_2 D_5 D_6 $D_7 = 0100 \rightarrow$ odd parity so error exists here.

Hence, we put $P_4 = 1$ in the 4th position of error word.

(b) Then, we analyze bits 2, 3, 6, 7.

P_2 D_3 D_6 $D_7 = 1000 \rightarrow$ odd parity so error exists here.

Hence, we put $P_2 = 1$ in the 2nd position of error word.

(c) Next, we check the bits 1, 3, 5, 7.

P_1 D_3 D_5 $D_7 = 0010 \rightarrow$ odd parity so error exists.

Hence, we put $P_1 = 1$ in the 1st position of error word.

(d) The error word will be shown in figure 4.34.

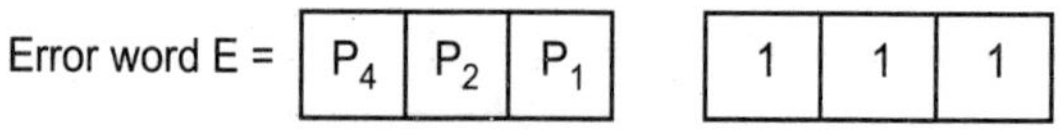

Fig. 4.34. Error word

The decimal equivalent of error word is $(7)_{10}$. Hence, bit 7 in the received codeword is in error.

EXAMPLE 4.12. Write the steps to compute the checksum in CRC code. Calculate CRC for the frame 110101011 and the generator polynomial = $x^4 + x + 1$ and write the transmitted frame.

Solution : The generator polynomial actually acts as the divisor in the process of CRC generation.

Therefore, Data word : 110101011

Divisor : $x^4 + 0\,x^3 + 0\,x^2 + x + 1 = 10011$

1 1 0 1 0 1 0 1 1	0 0 0 0 0
← Data word →	← 5 additional zeros →

The number of data bits = m = 9

The number of bits in the codeword = N

Dividend = Data word + (N – m) number of zeros.

Now, let us carry out the division as under :

```
                  1 1 0 0 0 0 1 1 1 1
         10011 ) 1 1 0 1 0 1 0 1 1 0 0 0 0 0
                 1 0 0 1 1
Mod 2 addition → ⊕ ⊕ ⊕ ⊕ ⊕
                 ---------------
                   0 1 0 0 1 1
                     1 0 0 1 1
                   ----------------
                     0 0 0 0 0 0 1 1 0 0
                               1 0 0 1 1
                               -----------
                               1 1 1 1 1
                               1 0 0 1 1
                             -------------
                             0 1 1 0 0 0
                               1 0 0 1 1
                               -------------
                               0 1 0 1 1 0
                                 1 0 0 1 1
                                 -------------
                                 0 0 1 0 1 0
                                     1 0 0 1 1
                                     ---------
                                     1 1 0 0 1   ← Remainder
```

Codeword : The codeword is given by

Codeword =	1 1 0 1 0 1 0 1 1	0 1 1 1 1
	← Data word →	← Remainder →

Internationally used CRC Polynomials

The three polynomials which are used internationally are as under:

$$\text{CRC 12} = x^{12} + x^{11} + x^3 + x^2 + x^1 + 1$$

$$\text{CRC 16} = x^{16} + x^{15} + x^2 + 1$$

$$\text{CRC – CCITT} = x^{16} + x^{12} + x^5 + 1$$

EXAMPLE 4.13. If the frame is 110101011 and generator is $x^4 + x + 1$ what would be the transmitted frame.

Solution : Given that data word : 1101011011

Generator : $x^4 + x + 1 = x^4 + 0\,x^3 + 0\,x^2 + x + 1 = 10011$

(i) First, we add five zeros at the end of the data word.

We add five zeros (equal to number of generator bits) at the end of data word to get the dividend as under :

Dividend	1 1 0 1 0 1 1 0 1 1	0 0 0 0 0
	← Data word →	← 5 additional zeros →

(ii) Next, we carryout the long division as under :

```
                            1 1 0 0 0 0 1 0 1 0 1
Generator → 10011 ) 1 1 0 1 0 1 1 0 1 1 0 0 0 0 0  ← Dividend
                    1 0 0 1 1
                    ⊕ ⊕ ⊕ ⊕ ⊕
                    ---------
                    0 1 0 0 1 1
                      1 0 0 1 1
                      -----------
                      0 0 0 0 0 1 0 1 1 0
                                1 0 0 1 1
                                ---------
                                0 0 1 0 1 0 0
                                    1 0 0 1 1
                                    ---------
                                    0 0 1 1 1 0 0
                                        1 0 0 1 1
                                        ---------
                                        0 1 1 1 1  ← Remainder
```

(iii) Finally, we write the transmitted frame

The transmitted frame is obtained by writing the data word followed by the remainder.

```
Therefore, we have    1 1 0 1 0 1 1 0 1 1 0 0 0 0 0  ← Dividend
                    +                     0 1 1 1 1  ← Remainder
                      -----------------------------
Transmitted frame →   1 1 0 1 0 1 1 0 1 1 0 1 1 1 1
```

Hence, the transmitted codeword is as under :

```
Codeword = | 1 1 0 1 0 1 0 1 1 | 1 1 0 0 1 |
           |←—— Data word ——→|←—————→|
                                 Remainder
```

EXAMPLE 4.14. What is remainder obtained by dividing $x^7 + x^5 + 1$ by the generator polynomial $x^3 + 1$?

Solution : Dividend : $x^7 + x^5 + 1 = x^7 + 0x^6 + x^5 + 0x^4 + 0x^3 + 0x^2 + 0x + 1$

$= 10100001$

Divisor : $x^3 + 1 = x^3 + 0x^2 + 0x + 1 = 1001$

The long division can be carried out as under :

```
          1 0 1 1 0
1001 ) 1 0 1 0 0 0 0 1
       1 0 0 1
       -------
       0 0 1 1 0 0
           1 0 0 1
           -------
           0 1 0 1 0
             1 0 0 1
             -------
             0 0 1 1 1  ← Remainder
```

The remainder will be 00111 = $x^2 + x + 1$ in the polynomial form.

EXAMPLE 4.15. A bit stream 10011101 is transmitted using the standard CRC method. The generator polynomial is $x^3 + 1$. Show the actual bit string transmitted. Suppose the third bit from left is inverted during transmission. Show that this error is detected at the receiver end.

Solution : Given that data word (Bit string) : 10011101

Generator polynomial : $x^3 + 1 = x^3 + 0x^2 + 0x + 1 = 1001$

(i) First, we obtain the dividend.

Dividend = Data word + 4 zeros

Therefore, the dividend is as under :

Dividend →	1 0 0 1 1 1 0 1	0 0 0 0
	Data word	4 additional zeros

(ii) Next, we carryout the division.

```
         1 0 0 0 1 1 0 0 1
1 0 0 1 ) 1 0 0 1 1 1 0 1 0 0 0 0
         ⊕1 0 0 1
         ---------
          0 0 0 0 1 1 0 1
                  1 0 0 1
                  -------
                  0 1 0 0 0
                    1 0 0 1
                    -------
                    0 0 0 1 0 0 0
                          1 0 0 1
                          -------
                          0 0 0 1 ← Remainder
```

(iii) Then, we obtain the actually transmitted bit stream.

The transmitted word is obtained by writing the data word followed by the remainder as under :

Transmitted word →	1 0 0 1 1 1 0 1	0 0 0 1
	Data word	Remainder

(iv) Finally, we carryout error detection and write the erroneous received word.

The received word = 1 0 (1) 1 1 1 0 1 0 0 0 1

(the circled 1 is marked: Error)

At the receiver, this word is divided by the same divider used at the transmitter *i.e.*, 1001.

```
          1 0 1 0 1 0 0 0 0
1 0 0 1 ) 1 0 1 1 1 1 0 1 0 0 0 1 ← Received word
          1 0 0 1
          -------
          0 0 1 0 1 1
              1 0 0 1
              -------
              0 0 1 0 0 1
                  1 0 0 1
                  -------
                  0 0 0 0 0 0 1 ← Remainder
```

A non zero remainder indicates that there is an error in the received codeword.

EXAMPLE 4.16. A bit string 0111101111101111110, needs to be transmitted at the data link layer. What is the string actually transmitted after bit stuffing ?

Solution : The original bit stream and the stream after bit stuffing are shown in figure 4.34.

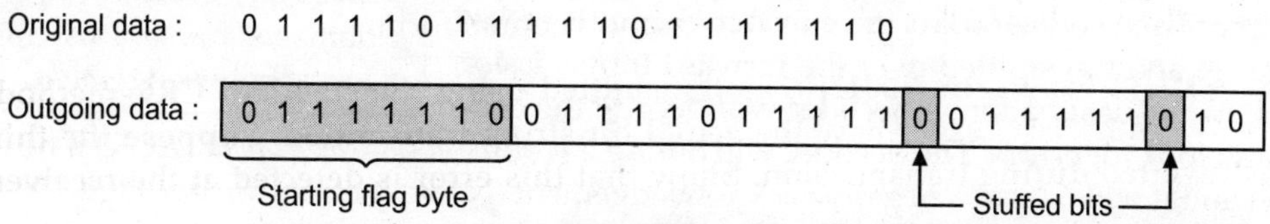

Fig. 4.35.

EXAMPLE 4.17. Apply bit stuffing to the following :

0 1 1 0 1 1 1 1 1 1 1 1 1 1 1 1 1 1 1 1 0 0 1 0

Solution : The outgoing data after bit stuffing is shown in figure 4.36

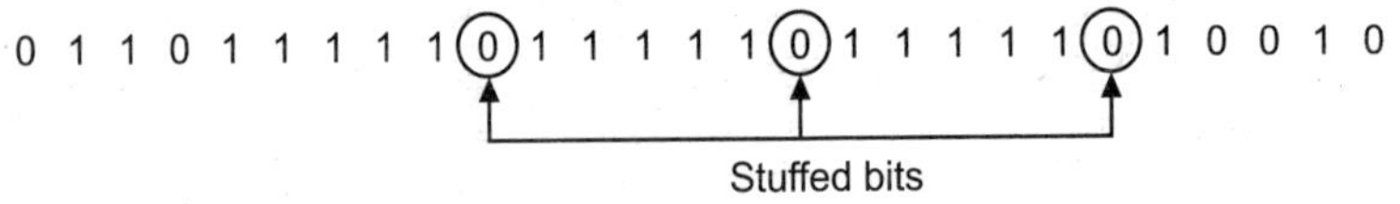

Fig. 4.36.

4.13. AUTOMATIC REPEAT REQUEST (ARQ) TECHNIQUES

(RGTU, Bhopal, Sem. Exam., June, 2014)

There are two basic systems of error detection and correction. The first on being the forward error correction (FEC) system and the second one is the automatic repeat request (ARQ) system. In the ARQ system of error control, when an error is detected, a request is made for the retransmission of that signal. Therefore, a feedback channel is required for sending the request for retransmission. The ARQ system differ from the FEC systems in three important respects. They are as under:

(i) In ARQ system, less number of check bits (parity bits) are required to be sent. This will increase the (k/n) ratio for an (n, k) block code if transmitted using the ARQ system.

(ii) A return transmission path and additional hardware in order to implement repeat transmission of codewords will be required.

(iii) The bit rate of forward transmission must make allowance for the backward repeat transmission.

4.13.1. Basic ARQ System

The block diagram of the basic ARQ system is as shown in figure 4.37.

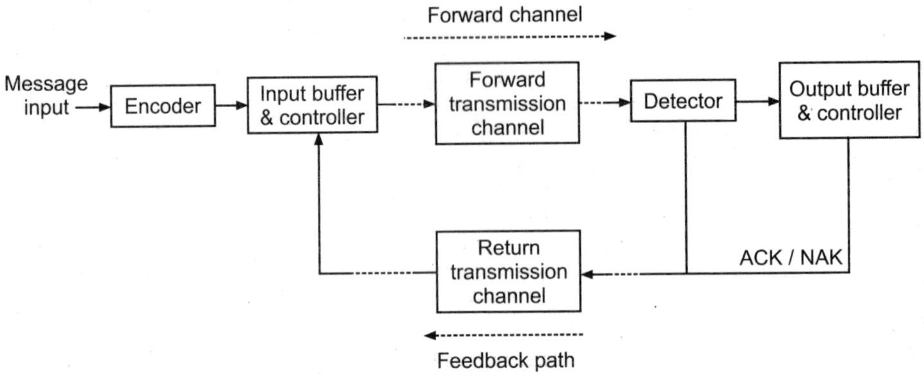

Fig. 4.37. Block diagram of the basic ARQ system

4.13.1.1. Working Operation of ARQ system

> **DO YOU KNOW?**
>
> Three types of errors require ARQ: a damaged frame, a lost frame, and a lost acknowledgement.

The encoder produces codewords for each message signal at its input. Each codeword at the encoder output is stored temporarily and transmitted over the forward transmission channel. At the destination, a decoder will decode the code words and look for errors. The decoder will output a *positive acknowledgement* (ACK) if no errors are detected and it will output a negative acknowledgement (NAK) If errors are detected. On receiving a negative acknowledgment (NAK) signal via the return transmission path, the *controller* will retransmit the appropriate word from the words stored by the input buffer. A particular word may be retransmitted only once or it may be retransmitted twice or more number of times. The output controller and buffer on the receiver side assemble the output bit stream from the codewords accepted by the decoder.

4.13.1.2. Error Probability on the Return Path

The bit rate of the return transmission which involves the return transmission of ACK/NAK signal is low as compared to the bit rate of the forward transmission. Therefore, the error probability of the return transmission is negligibly small.

4.13.1.3. Types of ARQ System

The three types of ARQ systems may be listed as under :

(i) Stop and-wait ARQ system

(ii) Go back n ARQ system, and

(iii) Selective repeat ARQ system.

Let us discuss these systems one by one.

Important Point: *Error control in the data link layer is based upon the principle of request for automatic retransmission (ARQ) of the missing, lost or damage frames.*

4.14. FLOW CONTROL

Flow control is another important design issue related to the data link layer. The problem to be handled is what to do with the sender which systematically wants to transmit frames at a faster rate than the capacity of the receiver. This happens when the sender is using a faster computer than the receiver. This will completely swamp the receiver. The receiver will keep losing some of the frames simply because they are arriving too quickly. The solution to this problem is to introduce the **flow control.** The flow control will control the rate of frame transmission to a value which can be handled by the receiver. It requires some kind of a feedback mechanism from the receiver to the sender.

Let us discuss some flow control techniques based on this principle.

It is a set of procedures that tells the sender how much data it can transmit before it must wait for an acknowledgement from the receiver; otherwise there will be overflow of data. The data flow should not be so fast that the receiver is overwhelmed. The speed of processing any receiving device is limited and it also has a limited amount of memory storage space, for storing the incoming data. There has to be some system, for reverse communication from the receiver to transmitter. The receiver can tell the transmitter about adjusting the data flow rate to suit its speed or even stop temporarily. As the rate of processing is generally slower than the rate of transmission, each receiver has a block of memory called **Buffer.** The buffer is reserved for storing incoming data until they are processed. If the buffer beings to fill up, the receiver must be able to tell the sender to stop transmission until it is once again able to receive. Similarly, the transmitter also has a buffer for storing the bits if the transmission is stopped.

DO YOU KNOW?

Flow control at the transport level is handled by a three walled sliding window.

Important Point: *Flow control may be defined as a set of procedures which are used for restricting the amount of data a transmitter can send before waiting for acknowledgement.*

4.15. ERROR DETECTION AND CORRECTION

Errors are introduced due to noise. Errors on some medium tend to come in burst or singly. If the errors come in burst, only one or two blocks of data will be affected. This is the advantage of burst errors. The disadvantage of burst errors is that they are difficult to detect and correct as compared to isolated errors. Error control includes error detection as well as error correction. Due to error control mechanism, it is possible to inform the sender about any missed or damaged transmitted frames. A request can be made for repeat transmission of such frames. In data link layer the error control corresponds to detection of an error and request for repeat transmission. This process of error control is known as the **automatic repeat request** (ARQ).

Error Detecting Codes

Some error detecting techniques are :

1. Use of parity bits.
2. Use of block parity.
3. Use of checksum.

Error Correcting Codes

Error correcting codes are used for correction of the detected errors.

Typically, used error correcting codes are :

1. Hamming codes
2. Cyclic redundancy code.

Three internationally accepted polynomials in cyclic redundancy code are CRC-12, CRC-16 and CRC-CCITT.

4.16. DATA LINK PROTOCOLS

In this article, let us discuss some elementary data link layer protocols.

4.16.1. An Unrestricted Simplex Protocols

This protocol is the simplest possible protocol. The transmission of data takes place in only one direction. Therefore, it is a simplex (unidirectional) protocol. It is assumed that the network layers of sender and receiver are always ready. It is also assumed that the processing time can be ignored and infinite buffer space is available. The communication channel is imagined to be noise free so it does not damage or lose any frame. All this is highly unrealistic. This protocol is also called as *utopia*. This protocol consists of two distinct procedures, namely a sender and a receiver. The sender runs in the data link layer of the sender machine whereas the receiver runs in the data link layer of the receiver machine. No sequence numbers or acknowledgements are used.

4.16.2. A Simplex Stop and Wait Protocol

The most unrealistic restriction in the previous protocol is the assumption that the receiving network layer can process the data with zero processing time. In this protocol it is assumed that a finite processing time is essential. However, like the first protocol, the communication channel is assumed to be noise free and the communication is simplex i.e only in one direction. This protocol deals with an important problem *i.e.* how to prevent the sender from flooding the receiver with data faster than its processing speed. In this protocol, a small dummy frame is sent back from the receiver to the transmitter to indicate that it can send the next frame. The transmitter sends one frame and then waits for the dummy frame called acknowledgement. Once the acknowledgement is received, it sends the next frame. Hence, the name **stop-and-wait.** The best thing about this protocol is that the incoming frame is always an acknowledgement.

4.16.3. A Simplex Protocol for Noisy Channel

This is the third protocol in which we go one step ahead and assume that the communication channel is noisy and can introduce errors in the data travelling over it. Frames may either be damaged or lost completely. In this protocol, the sender waits for a positive acknowledgement before advancing to the next data item. Therefore, it is called as PAR (Positive acknowledgement with retransmission) or automatic repeat request (ARQ). It may be noted that due to retransmission, there is always a possibility of duplication of frames at the receiver. To avoid this, the sender puts a sequence number in the leader of each frame it sends. The receiver can check the sequence number of each arriving frame to check the possible duplication. If a frame is duplicated then receiver will discard it. The operation can be divided into following two modes :

1. Normal operation; and
2. Time out.

DO YOU KNOW?

Data link protocols can be classified as synchronous or asynchronous.

1. Normal operation: After transmitting one packet, the sender waits for an *acknowledgement (ACK)* from the receiver before transmitting the next one. In this way, the sender can recognize that the previous packet is transmitted successfully and we could say *stop and wait* guarantees reliable transfer between nodes. To support this feature, the sender keeps a record of each packet it sends. Also, to avoid confusion caused by delayed or duplicate ACKs, *stop and wait* sends each packet with unique sequence numbers and receives that numbers in each ACKs.

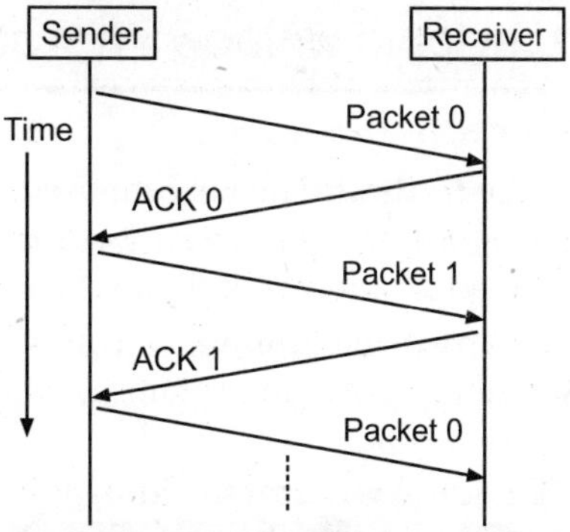

Fig. 4.38. Positive acknowledgement with retransmission

2. Time Out: If the sender does not receive **ACK** for previous sent packet after a certain period of time, the sender *times out* and *retransmit* that packet again. There are two cases when the sender does not receive ACK; one is when the ACK is lost and the other is when the frame itself is not transmitted. To support this feature, the sender keeps timer for each packet. We have already discussed that a timer is introduced in the data link layer.

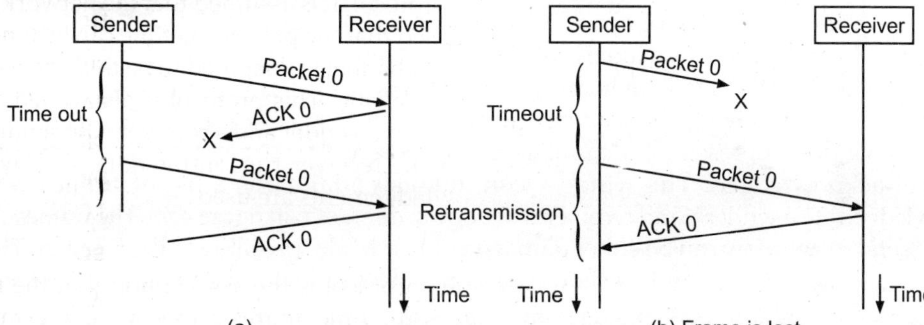

Fig. 4.39. Timeout and retransmission

4.16.4. Piggybacking

1. Basic Concepts: In all the practical situations, the transmission of data needs to be bi-directional. This is called as full-duplex transmission. One way of achieving full duplex transmission is to have two separate channels one for forward data transfer and the other for reverse data transfer (for acknowledgements). But, this will waste the bandwidth of the reverse channel almost entirely. A better solution would be to use each channel (forward and reverse) to transmit frames both ways, with both channels having the same capacity. Let A and B be the users. Then the data frames from A and B are intermixed with the acknowledgments from A to B. One more improvement can be made. When a data frame arrives, the receiver waits, does not send the control frame (acknowledgement) back immediately. The receiver waits until its network layer passes in the next data packet. The acknowledgement is then attached to this outgoing data frame. This technique in which the outgoing acknowledgement is delayed temporarily is called as **piggybacking.**

2. Advantage of piggybacking: The major advantage of piggybacking is better use of available channel bandwidth.

3. Drawbacks

(i) The drawback of piggybacking is the additional system complexity.

(ii) If the data link layer waits too log before transmitting acknowledgement, then retransmission of frame will take place.

DO YOU KNOW?
Retransmission of data is initiated by automatic repeat request (ARQ).

4.17. SLIDING WINDOW PROTOCOLS

(RGTU, Bhopal, Sem. Exam., June, 2014)

1. Definition: The next three protocols are more reburst and bi-directional protocols. All these protocols belong to a special class of protocols called **Sliding Window Protocols.** They show a different performance in terms of their efficiency, complexity and buffer requirements.

2. Sequence Number: One of the important features of all the sliding window protocols is that each outbound frame contains a sequence number, ranging from 0 to $2^n - 1$. The value of n can be arbitrary.

3. Sliding windows: Sliding windows refers to imaginary boxes at the transmitter and receiver. This window holds the frames at either end and provides the upper limit on the number of frames that can be transmitted before requiring an acknowledgement. Thus, in short, we can say that at any instant of time, the sender maintains a set of sequence numbers corresponding to the frames permitted to send. These frames which are being permitted to send are said to be falling in the **Sending Window.** The receiver also maintains a receiver window. It corresponds to the set of frames permitted to accept.

4. Significance of sender's and receiver's window: The sender's window and receiver's window need not be of same size. They may be of same size or they can grow or shrink as the frames are sent or received. The sequence numbers within the sender's window represents the number of frames sent but not yet acknowledged. The window continuously maintains a list of frames which are unacknowledged. The sender's and receiver's windows are shown in figure 4.40. The frames currently within the senders window can be either damaged or lost while travelling to the receiver. Therefore, all the frames in the senders window should be stored in memory for their possible retransmission.

DO YOU KNOW?
Three types of errors require ARQ: a damaged frame, a lost frame, and a lost acknowledgement.

The receiving window corresponds to the frames the receiver may accept. Any frame that falls out of the receiving window is discarded. At the beginning of the transmission of frames from the transmitter, the receiver window contains n – 1 spaces for frames. The sliding window of the receiver shrinks from left when frames of data are received. The sliding window of the receiver expands to right when acknowledgements are sent. The receiver window therefore represents not the number of frames received but the number of frames that may still be received before an ACK must be sent. If frames 0 through 3 (*i.e.* 4) have arrived but have not been acknowledged, the window in the receiver contains 7 – 4 = 3 frame spaces. As each ACK is sent out, the receiving window expands to include as many new spaces as newly acknowledged frames. The window expands to include a number of new frame spaces (5 – 2 = 3) equal to the number of most recently acknowledged frame (say 5) minus the number of the previously acknowledged frame (say 2). An example of the sliding window method is shown in the figure 4.40.

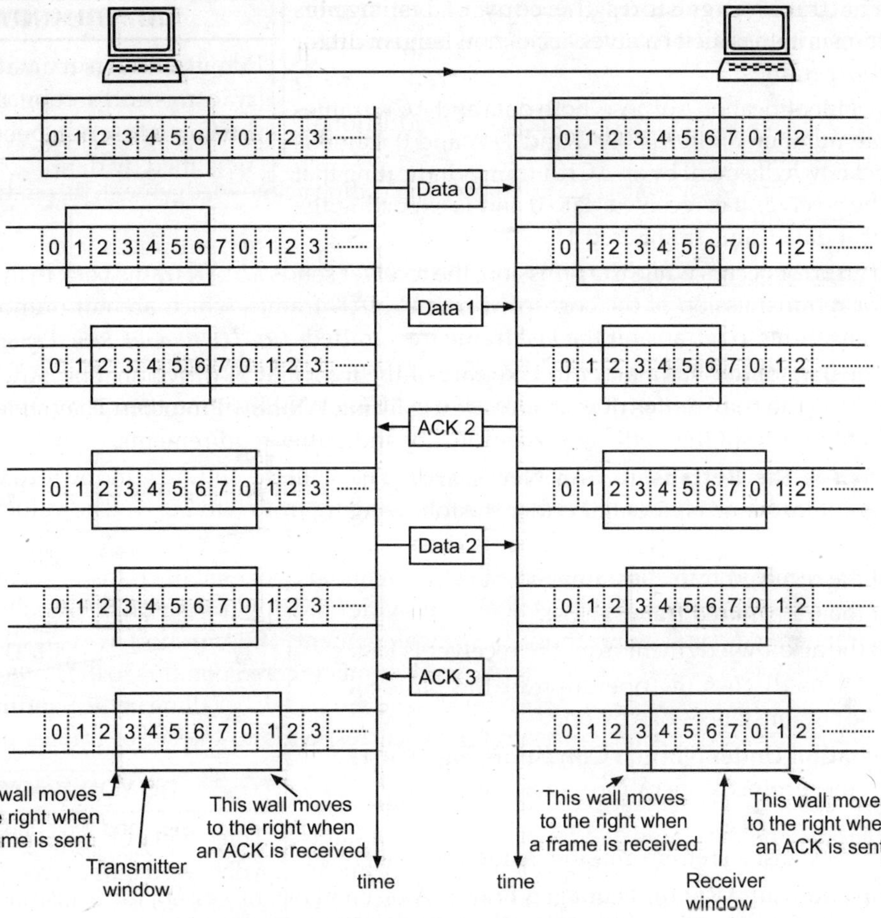

Fig. 4.40. Sliding window flow control

4.17.1. Stop and Wait ARQ: A One Bit Sliding Window Protocol

1. Basic Concepts: This protocol is called one bit protocol because the maximum window size here i.e. n is equal to 1. It uses the stop-and-wait technique. The sender sends one frame and waits to get its acknowledgement. Only after receiving the acknowledgement, it transmits the next frames. Because of this, one bit sliding window protocol is also called as **Stop and Wait protocol**. The sequence of events taking place when a frame is transmitted and received is as under:

(i) The data link layer of the sending machine fetches the first packet from its network layer.

(ii) It builds the frame for it and sends it to receiver.

(iii) The receiver data link layer checks the received frame for duplication.

(iv) If ok, it then passes the frame to its network layer.

2. The Operation of Protocol: The operation of this protocol is based on the ARQ (automatic repeat request) principle. So the sliding window protocols are also called as ARQ protocols. In this method, the transmitter transmits one frame of data and waits for an acknowledgement from the receiver. If it receives a positive acknowledgement (ACK) it transmits the next frame. If it receives a negative acknowledgement (NAK), it retransmits the same frame.

3. Features Added for Retransmission: For retransmission, four features are added to the basic flow control mechanism.

DO YOU KNOW?

Flow control is regulation of data transmission so that the receiver buffer does not become overwhelmed by data.

(i) The transmitter stores the copy of last frames transmitted until it receives an acknowledgement for those frames.

(ii) For identification purpose, both data and ACK frames are numbered alternately 0 and 1. A and 0 frame is acknowledgement by an ACK 1 frames indicating that the receiver has received data 0 and is expecting the next data frame numbered 1.

(iii) If an error occurs while transmission, the receiver sends a NAK frame back to the transmitter for retransmission of the corrupted frame. NAK frames which are not numbered tell the transmitter to retransmit the last frame transmitted.

(iv) The transmitter has a timer to take care of the frame ACK which are lost. After a specified time if the transmitter does not receive a ACK or NAK frame then it retransmits the last frame.

4. When is the Retransmission Necessary?: The retransmission of frame is essential under the following events :

(i) If the received frame is damaged.

(ii) If the transmitted frame is lost.

(iii) If the acknowledgment from the receiver is lost.

Now, let us discuss the operation of the protocol under these circumstances one by one.

5. Operation Under Normal Condition: Figure 4.41 illustrates the protocol operation when everything is normal.

No frame is lost so retransmission is not necessary.

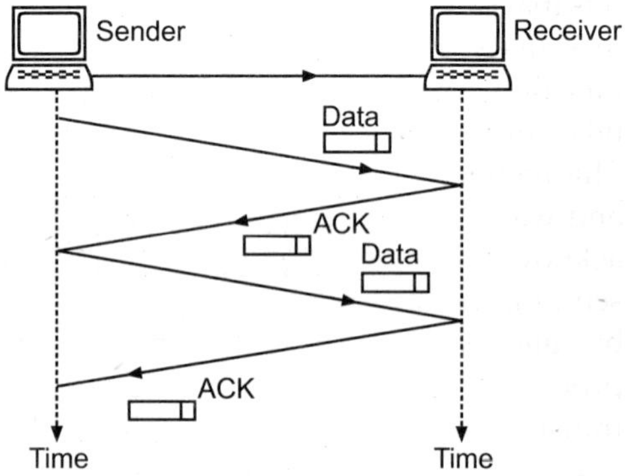

Fig. 4.41. Stop and Wait under normal condition

6. Stop and wait ARQ for Damaged Frame: As seen in figure 4.42 (*a*), the transmitter transmits data frame numbered 0. The receiver returns an ACK 1 indicating that the data frame numbered 0 is received without any error. The process goes on in this way, but if an error occurs the receiver sends a NAK requesting retransmission of the corrupted data frame.

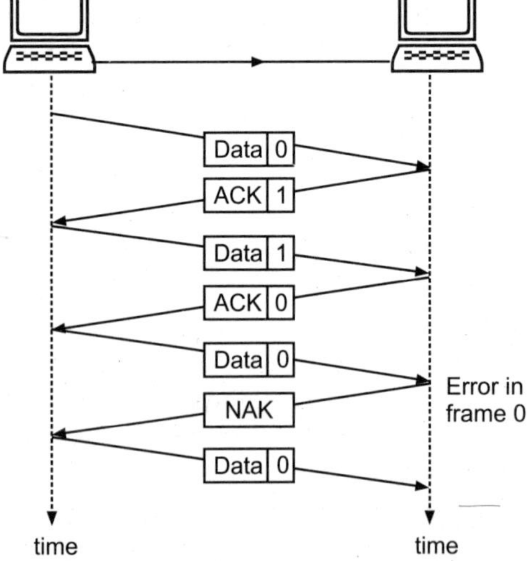

Fig. 4.42. (a) Stop and wait ARQ damaged frame

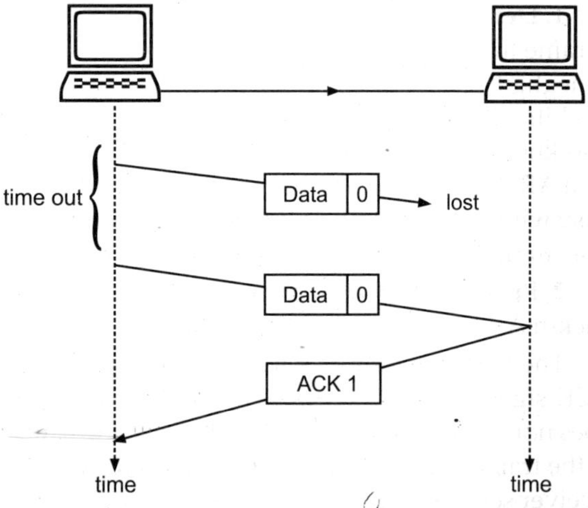

Fig. 4.42. (b) Stop and Wait ARQ, lost data frame

7. Stop and Wait ARQ for Lost Data Frame: Figure 4.42 (b) shows that if a data frame is lost and if the transmitter does not receive any type of acknowledgment from the receiver with a specified time it retransmits the same frame again.

8. Stop and Wait ARQ for Lost Acknowledgement: The figure 4.43 shows that if the acknowledgement sent by the receiver is lost, the transmitter retransmits the same data frame after its timer goes off. Stop and wait ARQ protocol becomes inefficient when the propagation delay is much greater than the time to transmit a frame e.g. let us assume that we are transmitting frames that are 800 bits long over a channel that has a speed of 1 Mbps and let us also assume that the time taken for transmission of the frame and its acknowledgement is 30 ms. The number of bits that can be transmitted over this channel in 30 ms is equal to $30 \times 10^{-3} \times 1 \times 10^{6} = 30{,}000$ bits. But in the stop-and-wait ARQ, only 800 bits can be transmitted in this time period. This inefficiency is due to the fact that in stop and wait ARQ, the transmitter waits, for an acknowledgement from the receiver before sending the next frame. The product of the bit rate and the delay that elapses before an action can take place is called the Delay-bandwidth product. The delay-bandwidth product helps in measuring the lost opportunity in terms of transmitted bits.

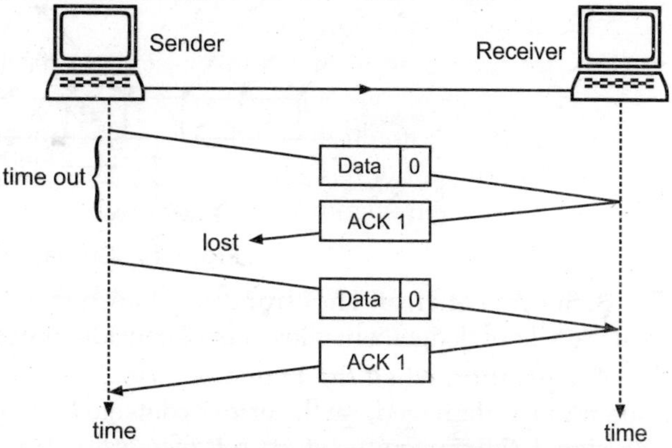

Fig. 4.43. Stop and wait ARQ, lost ACK frame

Important Point: *Stop and wait ARQ was used in IBM's Binary Synchronous Communications (Bisync) Protocol. It is also used in Xmodem, a popular file transfer protocol for modem.*

9. Drawback of stop and wait protocol: Problem with stop and wait protocol is that it is very inefficient. At any one moment, only one frame is in transition. The sender will have to wait at least one round trip time before sending next. The waiting can be long for a slow network such as satellite link.

4.17.2. A Protocol Using Go Back n

(RGTU, Bhopal, Sem. Exam., June, 2014)

1. Definition: In stop and wait protocol, it was assumed that the transmission time required for a frame to arrive at the receiver plus the transmission time for the acknowledgement to come back is negligible. But in some practical situations, this assumption is not correct. In the systems like satellite system, the round trip time can be as long as 500 ms (propagation delay). This protocol is also known as **Go-Back-n ARQ.** It is a method used to overcome the inefficiency of the stop and wait ARQ by allowing the transmitter to continue sending enough frames so that the channel is kept busy while the transmitter waits for acknowledgements. In this method, if one frame is damaged or lost, all frames are sent since the last frame acknowledged are retransmitted.

2. Principle of Go-Back-n ARQ: Let us consider figure 4.44 to understand the principle of Go-Back-n ARQ.

The major difference between this and the previous system is that the sender does not wait for ACK signal for the transmission of next frame. It transmits the frames continuously as long as it does not receive the NAK signal. NAK is the negative acknowledgement signal sent by the receiver to the transmitter. When the receiver detects an error in the third frame as shown in figure 4.44 the receiver sends a NAK signal back to sender. But this signal takes some time to reach the transmitter. By that time, the transmitter has transmitted frames upto frame 7. On the reception of the NAK signal, the transmitter will retransmit all the frames from 3 onwards. The receiver discard, all the frames it has received after 3 i.e. 3 to 7. It will then receive all the frames that are retransmitted by the transmitter.

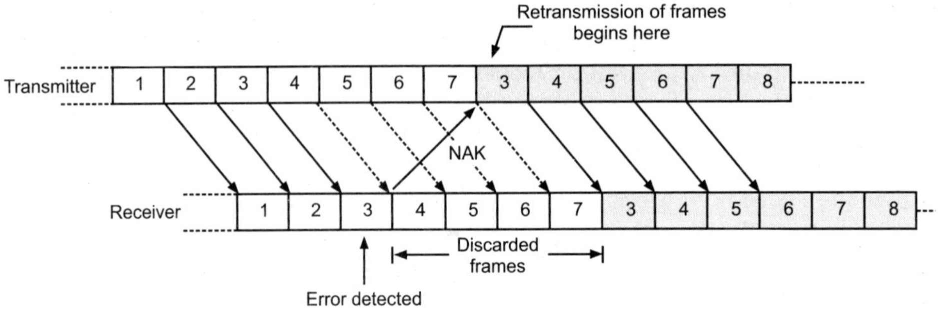

Fig. 4.44. Go Back n ARQ system

3. Sources of error: The error can get introduced, if the transmitted frames are damaged or lost or if the acknowledgement is lost. Let us consider the operation of this protocol under these conditions.

4. Operation when the Frame is Lost: This condition is illustrated in figure 4.45. The second data frame is damaged, so the error is detected and receiver send NAK-2 signal back. On receiving this signal, the transmitter starts retransmission from frame 2. All the frames received after frame 2 are discarded by the receiver.

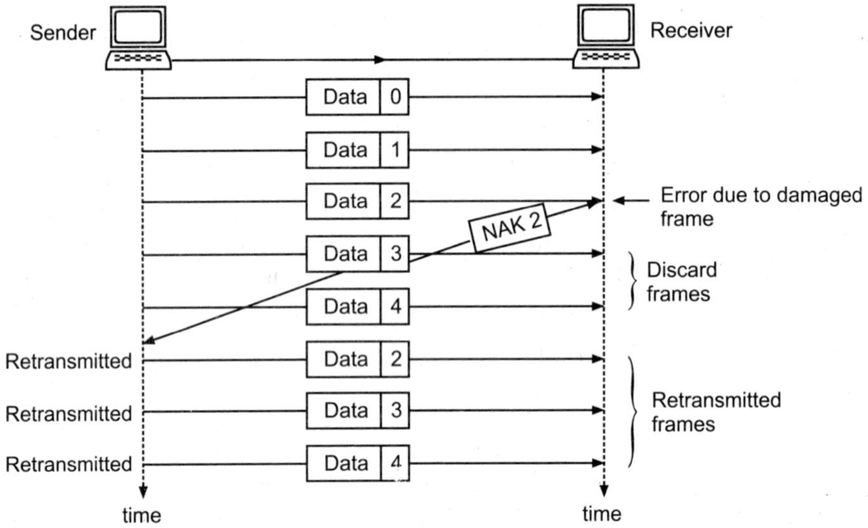

Fig. 4.45. Go-Back-n Damaged Data Frame

5. Operation when a Frame is Lost: As shown in figure 4.46 (a), the case of lost frame is also treated in the same manner as that of the damaged frame.

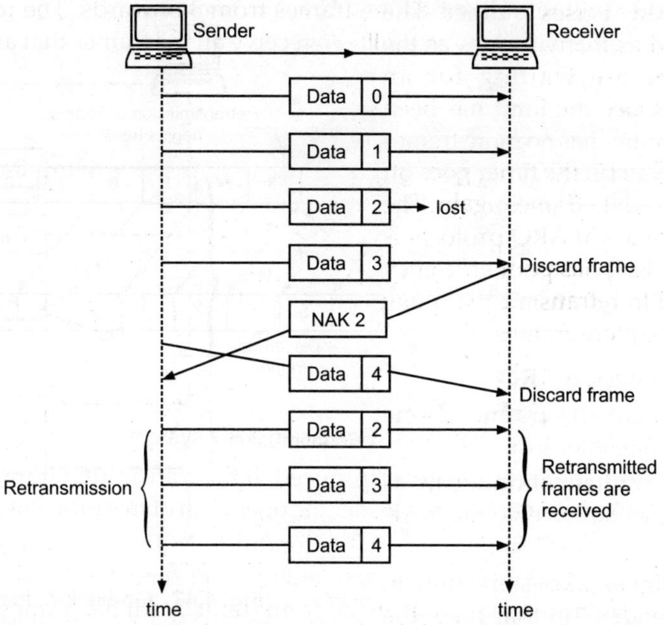

Fig. 4.46. (a) Go-Back-n, Lost Data Frame

The receiver, if it does not receive a particular data frame it sends a NAK to the transmitter and the transmitter retransmits all the frames sent since the last frame acknowledged.

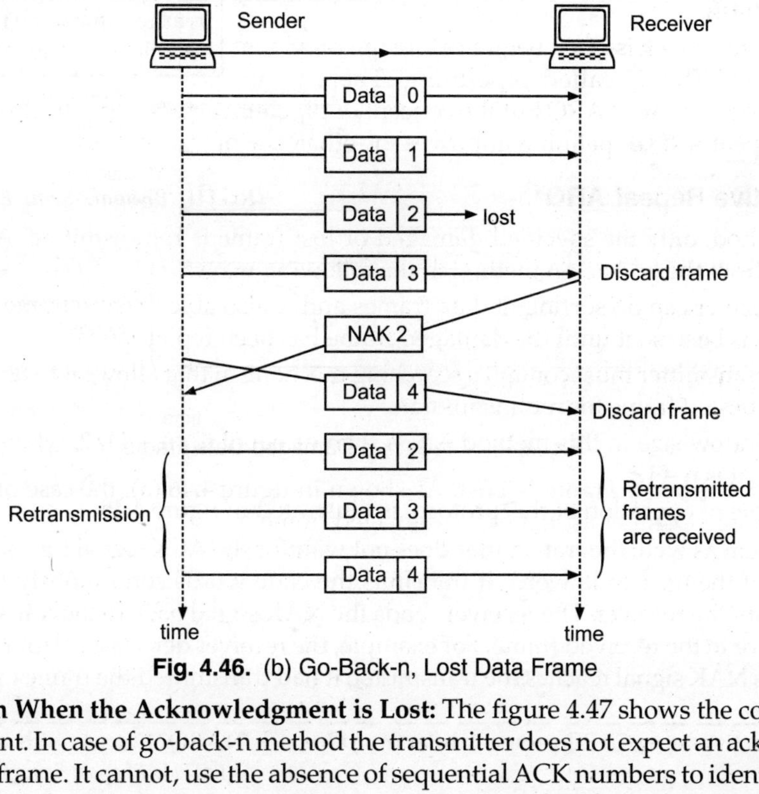

Fig. 4.46. (b) Go-Back-n, Lost Data Frame

6. Operation When the Acknowledgment is Lost: The figure 4.47 shows the condition for lost acknowledgement. In case of go-back-n method the transmitter does not expect an acknowledgement after every data frame. It cannot, use the absence of sequential ACK numbers to identify lost ACK or

NAK frames, instead it uses a timer. The transmitter can send as many frames as the window allows before waiting for an acknowledgement. Once the limit has been reached or the transmitter has no more frames to transmit it must wait till the timer goes off and retransmit all the data frames again. The disadvantage of Go-back-n ARQ protocol is that in noisy channels, it has poor efficiency because of the need to retransmit the frame in error and all subsequent frames.

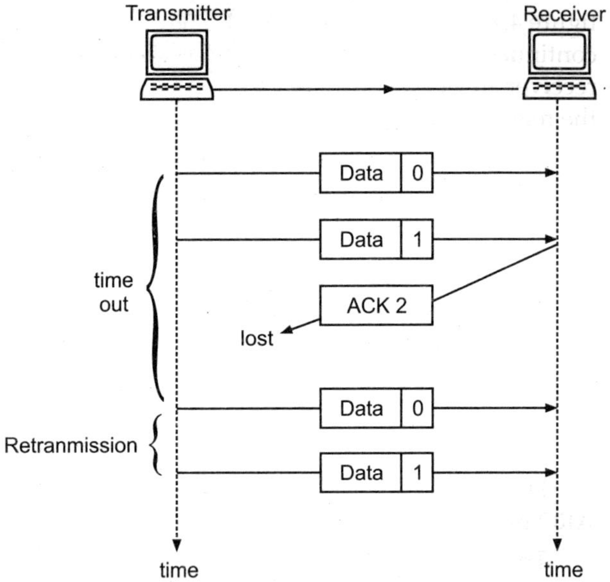

Fig. 4.47. Go-back-n, lost ACK frame

7. Drawbacks of Go back n ARQ

(i) It transmits all the frames if one frame is damaged or lost.

(ii) It transmits frames continuously as long as it does not receive the NAK signal.

(iii) The NAK signal takes some time to reach the sender. Till that time, the sender has already sent some frames. All those will be retransmitted after receiving the NAK.

(iv) The error can get introduced if the NAK is lost

> **DO YOU KNOW?**
>
> In go-back-n ARQ, retransmission begins wtih the last unacknowledged frame even if subsequent frames have arrived corectly. Duplicate frames are discarded.

4.17.3. Pipelining

In networking, a task is often begun before the previous task is complete. This is called pipelining. There is no pipelining in stop-and -wait ARQ but the concept of pipelining does apply to Go-Back-n ARQ and the selective repeat ARQ. Pipelining improves the efficiency of transmission.

4.17.4. Selective Repeat ARQ* *(RGTU, Bhopal, Sem. Exam., June, 2014)*

In this method, only the specified damaged or lost frame is retransmitted. A selective repeat system differs from the go-back-n method in the following ways :

(i) The receiver can do sorting of data frames and is also able to store frames received after a NAK has been sent until the damaged frame has been replaced.

(ii) The retransmitter must contain a searching mechanism that allows it to find and select only the requested frame for retransmission.

(iii) The window size in this method is less than or equal to (n + 1)/2, whereas in case of go-back-n, it is n – 1.

The principle of operation of this protocol is illustrated in figure 4.48.

In this system as well, the transmitter does not wait for the ACK signal for the transmission of the next code word. It transmits the code words continuously till it receives the NAK signal from the receiver. The receiver sends the NAK signal back to the transmitter as soon as it detects an error in the received frame. For example, the receiver detects an error in the third frame. By the time this NAK signal reaches the transmitter, it had transmitted the frame upto 7 as shown in

* How does selective-repeat-ARQ protocol differ from Go-back-N ARQ? Discuss with suitable diagrams. *(RGTU, Bhopal, Sem. Exam., June, 2014)*

figure 4.48. On reception of NAK signal, the transmitter will retransmit only the frame-3 and then continues with the sequence 8, 9... as shown in figure 4.48. The frames 4, 5, 6 and 7 received by the receiver are not discarded by the receiver. The receiver receives the retransmitted frames in between the regular frames. Therefore, the receiver will have to maintain the frames sequentially.

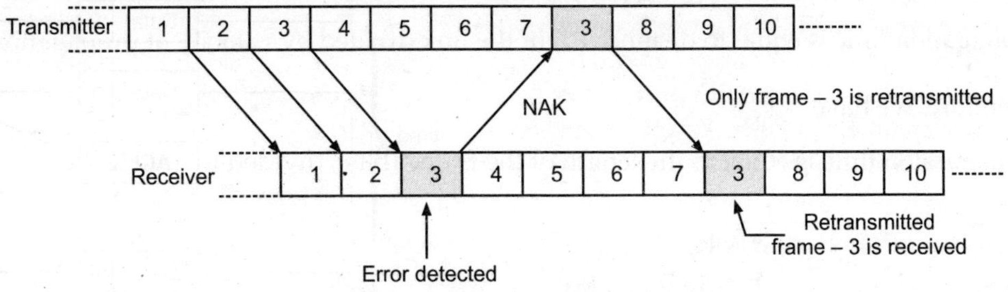

Fig. 4.48. Selective repeat ARQ system

Hence the selective repeat ARQ is the most efficient but the most complex protocol, of all the ARQ protocols.

Thus in selective repeat ARQ, only the frame which is damaged or lost is retransmitted by the transmitter. The lost ACK or NAK frames are treated in the same manner as the go-back-n method. When the transmitter reaches either the capacity of its window [(n + 1)/2] or the end of its transmission it sets a timer. If no acknowledgement arrives in the time allowed, all the frames that remain unacknowledged retransmitted. The disadvantage of this method is that because of the complexity of sorting and storage required by the receiver and the extra logic required by the transmitter to select frames for retransmission, the system becomes more expensive. The advantage of this system is that it gives the best throughput efficiency. This is due to the use of pipelining in selective repeat ARQ.

DO YOU KNOW?

In selective-repeat ARQ, only the unacknowledged frame is retransmitted.

4.17.5. Utilization of ARQ Techniques

The throughput efficiency is the measure of the performance of an ARQ protocol. For any channel a certain bandwidth and bit error rate are specified. For such a channel there will be an optimum operating condition that will support for the maximum *Net Data Throughout (NDT)*. NDT indicates the number of usable character detected at the receiver. It indicates the number of correct bits detected in a specified period of time. This is done by distinguishing between the total number of bits received (including the parity bits) and the number of correct bits. Throughput efficiency is defined as

$$\eta = \frac{t_f}{t_f + 2t_p}$$

where t_f = Transmission time required to transmit a frame

t_p = Propagation time required to reach destination for a transmitted bit

N = Frame size (bits)

R = Data rate

Let us consider A is a sender and B is a receiver. Then, the assumption are as under:

Assumptions

Receiver sends an immediate acknowledgement on the reception of the data frame. Size of acknowledgement frame is very small. Flow is unidirectional. Sender receives the acknowledgement after $t_f + t_p + t_p$ time. It can send data immediately after receiving acknowledgement. If t_f and t_p are constant, t_p/t_f is also constant.

Let $A = \frac{t_p}{t_f}$

Therefore, $\eta = \frac{1}{(1+2A)}$

Propagation time is equal to distance (d) of the link divided by velocity of propagation (v).

Therefore, we have $t_p = \frac{d}{v}$

Transmission time is equal to the length of the frame (bits), divided by rate R.

Hence, $t_f = \frac{L}{R}$

or $A = \frac{d/v}{L/R} = \frac{Rd}{Lv}$

EXAMPLE 4.18. Calculate the throughput of stop-and-wait flow control mechanism if the frame size is 4800 bits, bit rate is 9600 bps and distance between device is 200 km. Speed of propagation over the transmission is 200,000 km/s.

Solution : We have $t_f = \frac{\text{Frame size}}{\text{Bit rate}} = \frac{4800}{9600} = 0.5 \text{ sec}$

$$t_p = \frac{2000}{200000} = 0.01 \text{ sec}$$

We know that $A = \frac{t_p}{t_f} = \frac{0.01}{0.5} = 0.02$

Since, $\eta = \frac{1}{(1+2A)} = \frac{1}{(1+2\times 0.02)} = 0.96$

or $\% \eta = 96\%$ **Ans.**

EXAMPLE 4.19. A channel has a bit rate of 4 kbps and propagation delay of 20 msec. For what range of frame sizes does stop and wait gives an efficiency of at least 50 percent.

Solution : Given that Bit rate = 4 kbps,

Propagation delay $t_p = 20$ ms,

Efficiency $\eta \geq 50\%$ i.e $0.5 \leq \eta \leq 1$

We have to find the range of frame size.

(i) First, we calculate value of t_f.

We know that $\eta = \frac{t_f}{t_f + 2t_p}$

For $\eta = 0.5$, we get, $0.5 = \frac{t_f}{t_f + (2\times 20\times 10^{-3})}$

or $0.5\, t_f + 20 \times 10^{-3} = t_f$

or $t_f = 40 \times 10^{-3}$ sec.

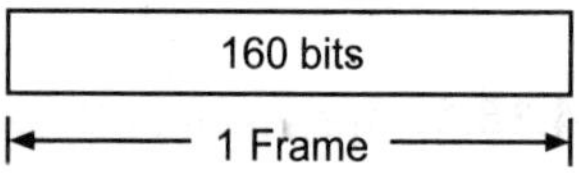

Fig. 4.49. 1 Frame size

Here, it may be noted that t_f = Transmission time for 1 frame

(ii) Next, we calculate the frame size.

R = Data rate = 4 kbps = 4000 bps

Therefore, $N = R \times t_f$

where N = frame size

Therefore, $N = 40 \times 10^{-3} \times 4 \times 10^{3} = 160$ bits **Ans.**

EXAMPLE 4.20. A channel has a bit rate of 4.8 Kbits/sec and a propagation delay of 20 msec. For what range of a frame size does stop and wait protocols gives an efficiency of 50%.

Solution : If the channel capacity is B bits/sec, the frame is L bits and the round trip propagation time T seconds, the time required to transmit a single frame is L/B sec. After the last bit of a data frame has been sent, there is delay of at least T/2 for the acknowledgment to come back , for a total delay of T. In stop and wait, the line is busy for T/2 and idle for T, giving an efficiency of L/(L+BT).

Given that

Bit rate (B) =4.8 k bits/sec, propagation delay (T) = 20 msec

Efficiency 50%, frame size (L) = ?

$$\text{Efficiency} = \frac{L}{(L+BT)}$$

$$0.5 = \frac{L}{(L+4.8\times10^{3}\times20\times10^{-3})} = \frac{L}{(L+96)}$$

or $$0.5(L + 96) = L$$

$$0.5\,L + 48 = L$$

or $$L = \frac{48}{0.5} = 96 \text{ bits}$$

Improvement in the throughput efficiency

If the data signalling rate (R) is increased, then the time taken to transmit each block (B/R) will be reduced. However, as delay remains unchanged, the throughput efficiency will decrease. To compensate for this, it will be necessary to use longer blocks for higher data rates (R). Longer blocks however will have a greater probability of error, therefore an optimum block length must be obtained for any particular system. Throughput efficiency also depends on the type of system used. For a half duplex system the transmission efficiency is very poor. An alternative method which gives greater efficiency is to use a continuous mode of transmission instead of block by block transmission. In this system, the data blocks are transmitted without interruption unless a negative acknowledgement signal (NAK) is received by the transmitting end. When NAK is transmitted back to the transmitter it will retransmit the error block. The continuous transmission method avoids the dead time but needs more storage or buffering.

4.18. OTHER DATA LINK PROTOCOLS

A data link protocol is a set of rules or specifications used to implement the data link layer. Data link protocols are divided into two subgroups :

(i) Asynchronous protocols - treat each character in a bit stream independently

(ii) Synchronous protocols - take the whole bit stream and divides it into characters of equal size.

4.18.1. Asynchronous Protocols

These are generally used in modems. They use the start stop bit format and are slow in implementation . The different types of asynchronous protocols are as under:

(1) X MODEM

(2) Y MODEM

(3) Z MODEM

(4) BLAST

(5) KERMIT

1. X MODEM: It is a file transfer protocol for telephone line communications between PC's. It is a half - duplex stop and wait ARQ protocol. The frame format is as shown in figure 4.50.

As seen in figure 4.50, the first field is a one byte start of header (SOH). The second field is a two bytes header. The first byte header contains the sequence number which carries he frame number and the second byte header is used to check the validity of the sequence number. The data field is 128 bytes. The last field in the CRC used for the error detection and correction. This protocol uses the ACK and NAK frames for acknowledgement which is already discussed. Besides this, the transmitter can receive a cancel signal (CAN) for stopping (aborting) the transmission.

⟸ direction of transmission	SOH	Header	Data 128 bytes	CRC
	1 byte	2 bytes		

Fig. 4.50. Frame format for X MODEM

2. Y MODEM : This protocol is similar to the X MODEM protocol with the following major differences :

(i) Multiple files can be sent simultaneously.

(ii) The data fields is of 1024 bytes.

(iii) ITU-T CRC 16 is used for error checking.

(iv) Two cancel signals (CAN) are used for stopping (aborting) the transmission.

> **DO YOU KNOW?**
>
> All bit-oriented protocols are related to high-level data link control (HDLC).

3. Z MODEM: This protocol is the combination of X MODEM and Y MODEM.

4. Blocked Asynchronous Transmission (BLAST): It is a full duplex protocol. It uses sliding window flow control. It allows the transfer of data and binary files.

5. Kermit: This is a terminal emulation program as well as a file transfer protocol which is similar to the X MODEM, with the transmitter waiting for a NAK before it starts transmission. It allows the transmission of control characters as text using two steps :

(i) The control character which is used as text is transformed to a printable character by adding a fixed number to its ASCII code representation.

(ii) The # character is added to the front of the transformed character.

When the receiver comes across a # character, it knows that this must be dropped and that the next character is a control character. If the transmitter wants to send a # character it will send two of them.

4.18.2. Synchronous Protocols

These protocols are used for LAN, MAN and WAN because of their high speed. These protocols are divided into two classes :

(1) Character- Oriented Protocol: These are also called byte-oriented protocols. In this protocol, the frame or packet is interpreted as a series of characters. The control information is inserted into the data stream either as separate control frames or as additions to existing data frames. IBM's Binary Synchronous Communication (BSC) is an example of this protocol.

(2) Bit- Oriented Protocol : In this protocol, the frame or packet is interpreted as a series of bits. Bit - oriented protocols can pack more information into shorter frames and avoid the transparency problem of character - oriented protocol. **High Level Data Link Control (HDLC)**, Synchronous Data Link Control (SDLC) are examples of bit-oriented protocols.

4.18.3. Binary Synchronous Control (BSC) Protocol

It supports half-duplex transmission using stop and wait ARQ flow control and error control. It is used for point to point and multipoint configurations. The BSC frame has data frame and control frame. The data frame contains the message information along with control information applicable to that frame. The control frames are used to exchange information between communicating devices for examples to establish initial connection, for flow and error control and to disconnect the devices at the end of the session. The data frame is shown in figure 4.51.

Fig. 4.51. BSC data frame

As shown in figure 4.51, the frame consists of the following characters along with the data:

1. SYN – The first two characters are the synchronization characters. These characters alert the receiver to the arrival of a new frame and provide a bit pattern used by the receiving device to synchronize its timing with that of the sending device.
2. SOH – It is the start of header character. It comes after SYN and before the STX character. The header includes the address of the receiving and transmitting device and the identifying number of the frame (0 or 1).
3. STX – Start of text. This character signals to the receiver that the control information is ending and the next byte will be data.
4. ETX – End of text. It indicates the transition between text and more control characters.
5. BCC – Block check count . It is included for error detection ,.It can be one -character longitudinal redundancy check (LRC) or two character cyclic redundancy check CRC. A multi -block multi frame transmission is shown in the figure 4.52. As shown in the figure 4.52, the frame consists of the following extra fields.

(a) ITB – Intermediate text block. It indicates the end of intermediate block.

(b) ETB – End of transmission block. It indicates the end of multi-block transmission.

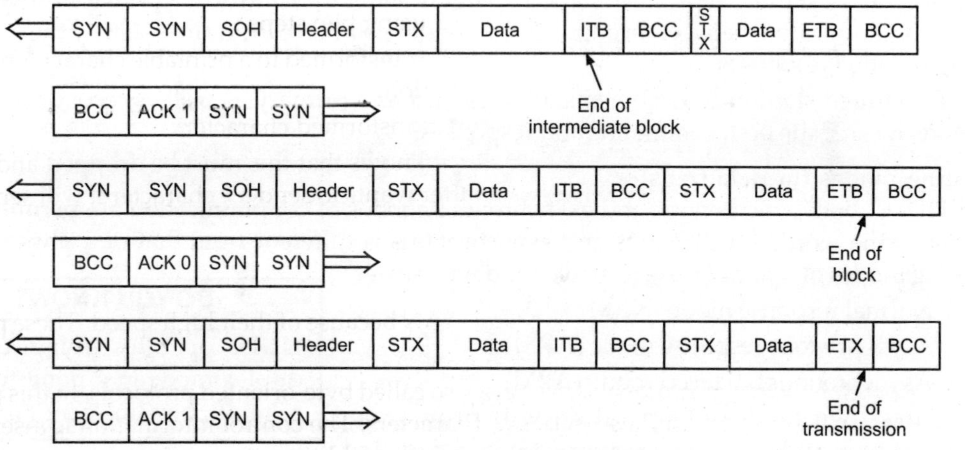

Fig. 4.52. Multi- frame transmission

The control frame is used by one device to send commands to or get information from another device. It consists of control characters and no data. The control frame is used for establishing connections, maintaining flow and error control and terminating connection. The control frame format is shown in figure 4.53.

Fig. 4.53. BSC control frame

If the text field of a transmission includes an eight bit pattern that locks like a BSC control character the receiver interprets it as one destroying the sense of message. Confusion between control information and data is called a lack of data transparency. The problem is overcomed by a process called byte stuffing . As shown in figure 4.54, a data link character (DLE) is added and after the character which may cause the confusion.

SYN	SYN	DLE	STX	Transparent text	DLE	ETX	BCC

Fig. 4.54. Byte stuffing

4.19. EXAMPLE OF DATA LINK PROTOCOL: HIGH LEVEL DATA LINK CONTROL (HDLC) PROTOCOL

(RGTU, Bhopal, Sem. Exam., June, 2014)

The high level data link control (HDLC) protocol was developed by ISO. It is the most widely excepted data link protocol. It offers a high level of flexibility, adaptability, reliability and efficiency of operation . HDLC is a bit oriented data link control protocol, and it satisfies a variety of data control requirements. To make HDLC protocol applicable to various possible network configurations, three types of stations have been defined :

1. Primary station
2. Secondary station
3. Combined station

1. Primary station : A primary station looks after the data link management. In the event of communication between the primary and secondary stations, the primary station has a responsibility of connecting and disconnecting the data link . The frames sent by a primary station are called commands.

2. Secondary station: A primary station operates under the control of a primary station. In the event of communication between primary and secondary stations, the frames sent by the secondary station are called responses.

3. Combined station: A combined station can act as a primary as well as secondary stations. Therefore, it can issue both commands and responses.

Operating modes for data transfer

In HDLC, both synchronous and asynchronous modes of communication are permitted. The meaning of the words synchronous and asynchronous is different from that of a physical layer. Following modes of operation are possible for data transfer :

1. Normal response mode (NRM)
2. Asynchronous response mode (ARM)
3. Asynchronous balanced mode (ABM)

The first two modes of operation will provide an unbalanced type of data transfer between one primary and the secondary stations whereas the third one provides a balanced type of data transfer.

> **DO YOU KNOW?**
>
> HDLC operates in half-or full-duplex mode in a point-to-point or multipoint link configuration.

1. Normal Response Mode (NRM): This mode is suitable for point to point and point to multipoint configuration. Here, the primary station will control the overall data link management. It is a synchronous mode of communication.

2. Asynchronous Response Mode (ARM): This mode is used for communication between primary and secondary stations. As the name indicates it is an asynchronous mode of communication. In ARM, the secondary station can transmit response (frame) without taking permission from the primary station. This is not allowed in NRM. Therefore, NRM is a more disciplined mode than ARM. The responsibility of link management function still lies with the primary station.

3. Asynchronous Balanced Mode (ABM): This mode is applicable to the point to point communication between two combined station. As both these stations are combined stations, they are capable of link management functions. As the communication is asynchronous, one station can transmit a frame without permission from the other station. In this mode, information frames can be transmitted in full duplex manner.

4.19.1. Frame Structure in HDLC

(RGTU, Bhopal, Sem. Exam., June, 2014)

1. Basic Concept: In the discussion of ARQ, we have observed that the functionality of a protocol depends on the control fields that are used in the header. The format of the HDLC frame is defined so that it can accommodate the various data transfer modes. The HDLC uses two different frame format as shown in figure 4.55(a) and 4.55(b). If you compare them, then it will be clear that except the information field both the frames are identical to each other. The frame is transmitted from left to right with the lowest order bit transmitted first.

2. Flag field: The flag is a unique 8-bit word pattern (01111110). It is used to identify the start and end of each frame as shown in figure 4.55 (a). It is also used to fill the idle time between consecutive frames.

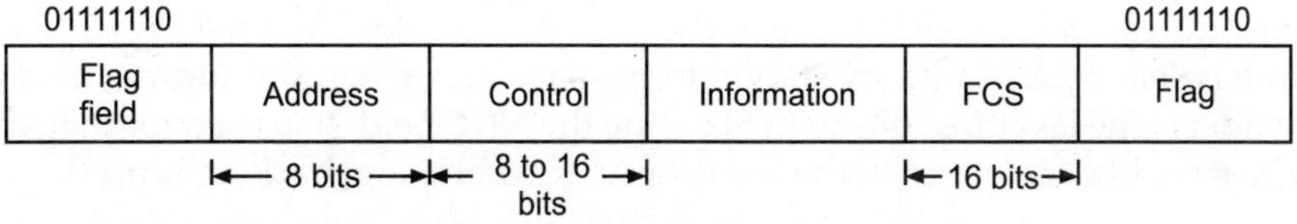

Fig. 4.55. (a) Information transfer frame

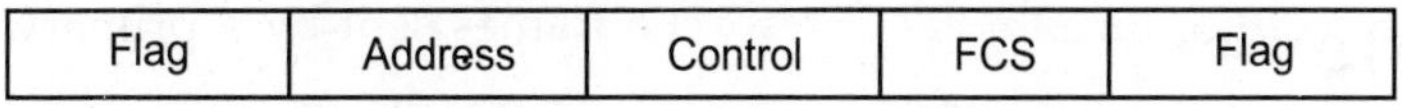

Fig. 4.55. (b) Supervisory and unnumbered frames

3. Address field: The address field consists of the address of secondary station irrespective of whether a frame is being transmitted by primary or secondary station. Address field consists of 8 bits hence it is capable of addressing 256 addresses.

4. Control field: The control field usually consists of 8 bits but the number of bits can be extended to 16. It carries the sequence number of the frame, acknowledgements , request for transmission and other control commands and responses.

5. Information field: The field size of the information field is variable and it can consists of any number of bits. It consists of the user's data bits and it is completely transparent.

6. Frame Check Sequence (FCS) Field: This is a 16 bit field which is used for detection of errors in the address , control and information field. It is nothing else but a 16 bit CRC code for error detection.

> **DO YOU KNOW?**
>
> HDLC handles data transparency by adding a 0 whenever there are five consecutive 1s following a 0. This is called bit stuffing.

4.19.2. Frame Types in HDLC

There are three types of frames defined in HDLC as under:

1. The I-frame or information frame.
2. The S-frame or supervisory frame.
3. The U- frame or the unnumbered frame.

1. The I-frame: Figure 4.56 shows the format of the information frame or I-frame.

It is supposed to carry the user data from the network layer. It is also possible to include the flow and error control information which is also called piggybacking.

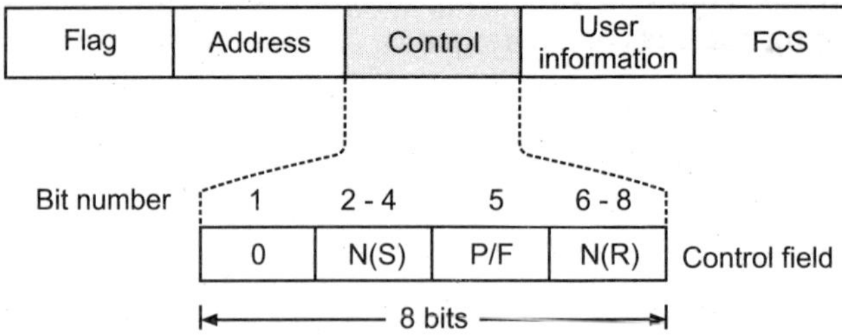

Fig. 4.56. I - frame format

Explanation: Let us concentrate on the control field of the I-frame. As shown in the figure 4.56, if the first bit in the control field is 0, it is identified as an information frame (I-frame). The next three bits (2 to 4) are called N(S) and they are used to define the sequence number of the frame. Since there are only 3 bits , we can define a sequence number between 0 and 7 only. The value of N(S) field corresponds to the value of control variable S as discussed for the three ARQ mechanisms. The next bit (5th) is the poll/final (P/F) bit. It can have two possible values 0 or 1 out of which it has meaning only when P/F = 1. When P/F = 1, it means poll when a frame is sent by primary station to secondary (*i.e.* when the address fields contains the address of the receiver). When P/F = 0 , it means final when a frame is sent by a secondary station to primary .(i.e. when the address field contains the address of sender). The last three bits (6 to 8) define the N(R) field. It is used for piggybacking. The 3 bits in the N(R) field correspond to the value of ACK when piggybacking is used.

2. The S-Frame: Figure 4.57 shows the format of S frames or supervisory frames. An S- frame does not contain any information field. These frames are used for flow and error control when piggybacking is not possible.

Let us discuss the control field of the S - frame. A 10 in the first two bits of the control fields identifies it as a supervisory frame as shown in figure 4.57. The next two bits define the code field marked SS. There are four types of supervisory frames corresponding to the four possible value of the S bits in the control field.

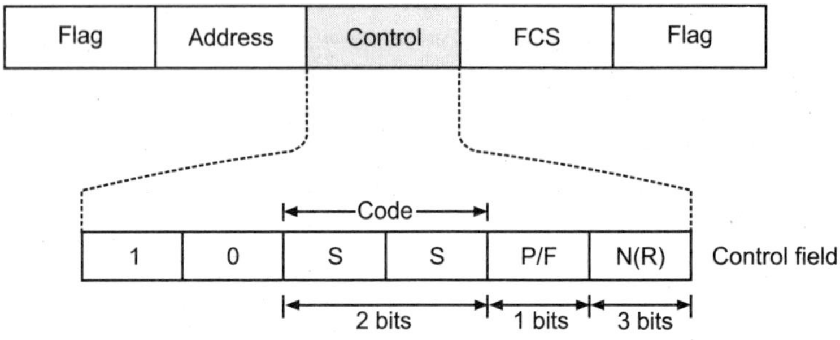

Fig. 4.57. S- frame format

(i) SS = 00 → corresponds to receive ready (RR) frames which are used to acknowledge frames when no I frames are available to piggyback the acknowledgement.

(ii) SS = 01 → corresponds to reject (REJ) frames which are used by the receiver to send a NAK when error has occurred.

(iii) SS = 10 → corresponds to a Receive Not Ready (NRN) frame and it is used for flow control.

(iv) SS = 11 → corresponds to a Selective Repeat Frame which indicates to the transmitter that it should retransmit the frame indicated in the N(R) subfield.

The fifth bit in the control field is P/F bit as discussed earlier, and the next 3 bits called N(R) correspond to the ACK or NAK value.

3. U - Frames: The format of U-frame, *i.e.*, the unnumbered frame is shown in figure 4.58. These frames are used for exchanging the session management and control information between the communicating devices.

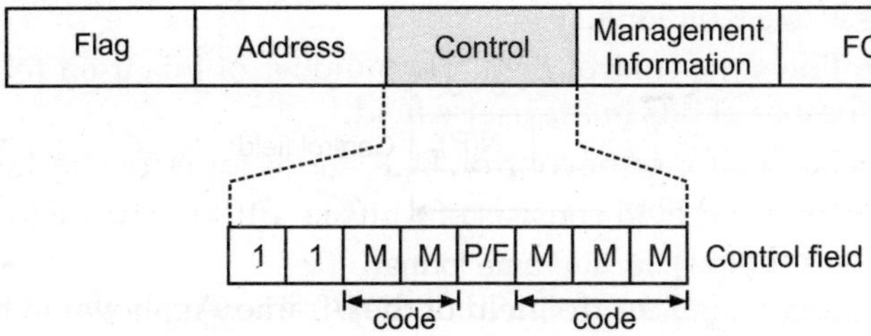

Fig. 4.58. Format of U-frame

A 11 in the first two bits of the control field identities an unnumbered frame as shown in figure 4.58. The information field in U-frame is used for carrying the system management information. It does not carry the user data. The U - frame code bits (M bits in figure 4.58) are divided into two sections. Two bits before P/F bit and three bits after the P/F bits. These five codes bits can create upto 32 different types of U - frames. The unnumbered frame types are used for function such as initialization , status reporting and resetting. The information frame and supervisory frames implement the error and flow control functions of the data link layer. The combination of the I - frame and supervisory frames allows HDLC to implement stop and wait , Go back n and selective repeat ARQ.

4.19.3. Transparency in HDLC

The data field of HDLC frame is capable of carrying text and non-text information. But a problem is introduced for some message types during the transmission. If the data field of an HDLC frame contains the pattern 01111110 which is reserved for the flag field , then the receiver will treat that sequence as the end flag. Naturally the remaining bits are interpreted as the bits from next frame. This is called as lack of data transparency.

4.19.4. Bit Stuffing

Bit stuffing is used to overcome the lack of data transparency. In HDLC, transparency is achieved by ensuring that the unique flag sequence (01111110) does not occur in the address, control, information and FCS fields. At the transmitter, an extra '0' bit is inserted after five consecutive 1's occurring anywhere after the opening flag and before the closing flag. At the receiver the extra '0' bit following five consecutive 1 is deleted. This technique is called *zero stuffing* or bit stuffing. The bit stuffing is not done for three operating conditions. First is when the bit sequence is really a flag , second is when the transmission is being aborted, and third is when the channel is idle.

4.20. SYNCHRONOUS DATA LINK CONTROL (SDLC) PROTOCOL

Synchronous Data Link Control (SDLC) protocol is used by IBM. This is not a standard but it is widely used. SDLC is actually a subset of HDLC but it includes several minor additional features. In the previous section, we have studied HDLC. Since SDLC is a subset of HDLC they will be similar. Figure 4.59 shows the frame structure for SDLC.

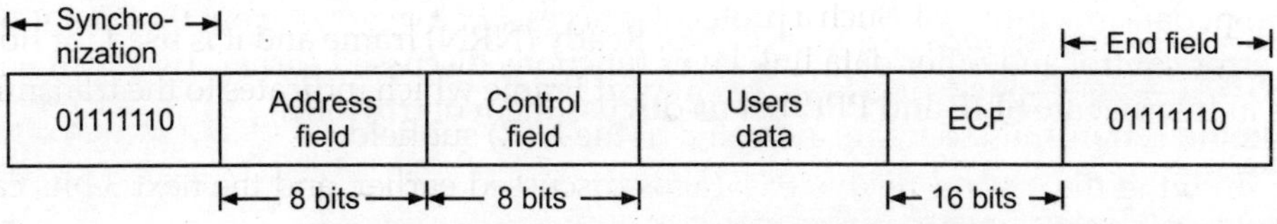

Fig. 4.59. The frame format of SDLC protocol

The important points about the frame format of SDLC may be described as under :

(i) A group of 8 bits i.e (01111110) in the beginning of the frame is used for synchronization.

(ii) The group of synchronizing bits is followed by an address field which is 8 bit long and then a control field which is also a 8 bit long.

(iii) The user's data field follows the control field. The number of bits used for this field can vary. We can use any number of bits in the user's field.

(iv) The ECF field of 16 bits is used for error control. ECF stands for error checking field.

(v) The last field in frame called end field consists of 8 bits i.e., (0111 1110) , which are same as those in the synchronization field.

There are primary and secondary stations defined for SDLC. They have the same meaning as explained in the previous section for HDLC. Most of the things explained for HDLC are also applicable to the SDLC as well.

4.20.1. Why is CRC in Data Link Protocols in Trailer and not in Header ?

Note that for all the data link protocols discussed so far, The CRC field that contains the checksum for error detection and correction, always appears in the trailer i.e at the end of the frame and not in header. The CRC is obtained by adding all the bits being transmitted, and appended to the outgoing stream as soon as the last bit is transmitted. If we want CRC to be in the header i.e. at the beginning of the frame, then the CRC has to be calculated by making a pass over the frame before transmission. This would require each byte to be handled twice, once for computing CRC and then for transmission. But, if CRC is put in the trailer, then each byte will have to be handled only once.

4.21. THE DATA LINK LAYER IN THE INTERNET

The Internet consists of individual machines plus the infrastructure required for the communication between them. Internet is basically a wide area network which is built up from point to point leased lines. So, let us examine the data link layer protocol used on point to point lines in the Internet. In case of the Internet , millions of individuals have the home connections to the Internet using modems and dial up telephone lines. Generally, the user's home PC calls up an Internet provider (IP). But in some situations, the home PC functions as a character oriented terminal logged into the Internet service providers time sharing system. In this mode, the user can type commands and run programs but the services such as WWW are not available . This way of working is called having a shell account. There is one more alternative. The home PC can call an Internet service provider's router and then act as a full fledged Internet host. This is illustrated in figure 4.60.

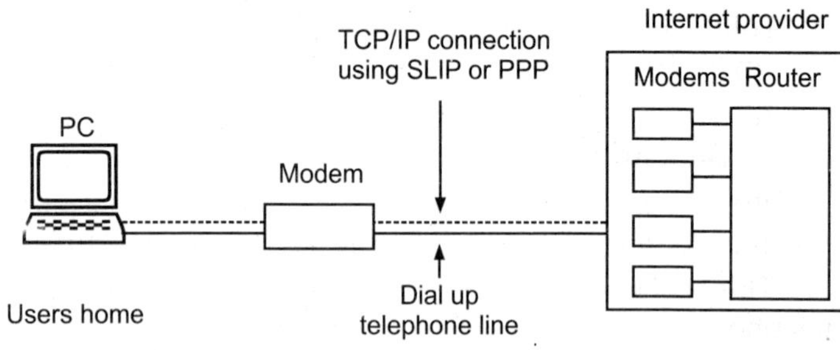

Fig. 4.60. Home PC acting as an Internet host

With this type of approach , it is possible to get all the Internet facilities including WWW. For the router-router leased line connection or dial up host router approach, it is necessary to use some point to point data link protocol. Such a protocol is necessary for carrying out the operations such as framing, error control and other data link layer functions discussed earlier. Two such widely used protocols in Internet are SLIP and PPP. Let us discuss them one by one.

4.22. SLIP-SERIAL LINE IP

This protocol was devised in 1984 , to connect a workstation to the Internet over a dial-up line using a modem. This protocol is very simple The workstation sends raw IP packets over the line with a flag byte at the end for framing. If the flag format appears in the data, then a two byte sequence (OXDB, OXDC) is sent in its place. If OXDB occurs in the flag byte, then it is also stuffed. In some SLIP implementation,s a flag byte is attached at the front and back of each IP packet sent.

Problem with SLIP Protocol

SLIP is a simple and widely used protocol. But it has some serious problems. They are as under :

(i) It does not have any error detection or correction facility.

(ii) SLIP supports only IP. So, it cannot be used over the networks which do not use IP.

(iii) It is necessary that both the communicating sides must know the other's IP address in advance. It is not possible to dynamically assign the address during the set up.

(iv) SLIP does not provide any authentication. So neither party knows whom it is talking to.

(v) It is not an approved Internet standard. So many versions exist which makes networking difficult.

4.23. POINT TO POINT PROTOCOL (PPP)

One of the most common protocols for point to point protocol PPP. This protocol is used by millions of Internet users to connect their home computers to the server of an Internet service provider. Most of these users have a traditional modem and they are connected to the Internet through a telephone line or a TV cable. The PPP is used for controlling and managing the data transfer.

4.23.1. Services Provided by PPP

Following are some of the services provided by PPP:

(i) Defines the format of the frames.

(ii) It defines how the link between two devices is to be established and how the data exchange should take place.

(iii) It decides the encapsulation of network layer data into the data link frame.

(iv) It defines the way in which the two devices can authenticate each other.

This protocol was designed for users who wanted to connect their computer system through a telephone line to the computer of an Internet provider to access Internet. The PPP protocol can operate over a full duplex point to point transmission link as well as over asynchronous links, bit synchronous links or Asymmetric digital subscriber line (ADSL) and Synchronous Optical Network (SONET). The PPP was an improvement over the Serial Line Internet Protocol (SLIP).

4.23.2. Frame Format of PPP

The PPP protocol uses an HDLC like frame format to include datagrams over points to point links as shown in figure 4.61.

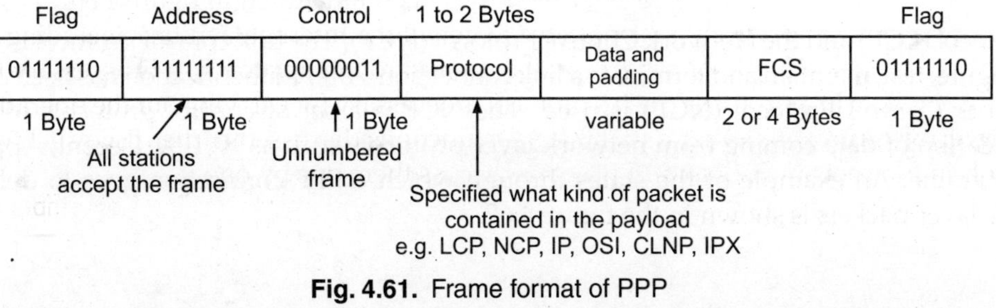

Fig. 4.61. Frame format of PPP

The description of various fields is as under:

1. Flag: The PPP frame always begins and ends with the standard HDLC flag i.e. 01111110.

2. Address: Since PPP is used for a point to point connection, it uses the broadcast address of HDLC i.e 11111111, to avoid a data link address in the protocol. All 1's in the address field indicates that all stations are to accept the frame.

3. Control: This field uses the format of the U-frame in HDLC. The value is 00000011 to show that the frame does not contain any sequence numbers and that there is no flow or error control.

4. Protocol: It defines what is being carried in the data field i.e user data or other information.

5. Data field: It carries the user data or other information.

6. FCS (Frame Check Field): This Field is simply a 2 or 4 byte CRC. It can use the CCITT 16 or CCITT 32 generator polynomial. The PPP protocol provides many useful capabilities with the help of a Link Control.

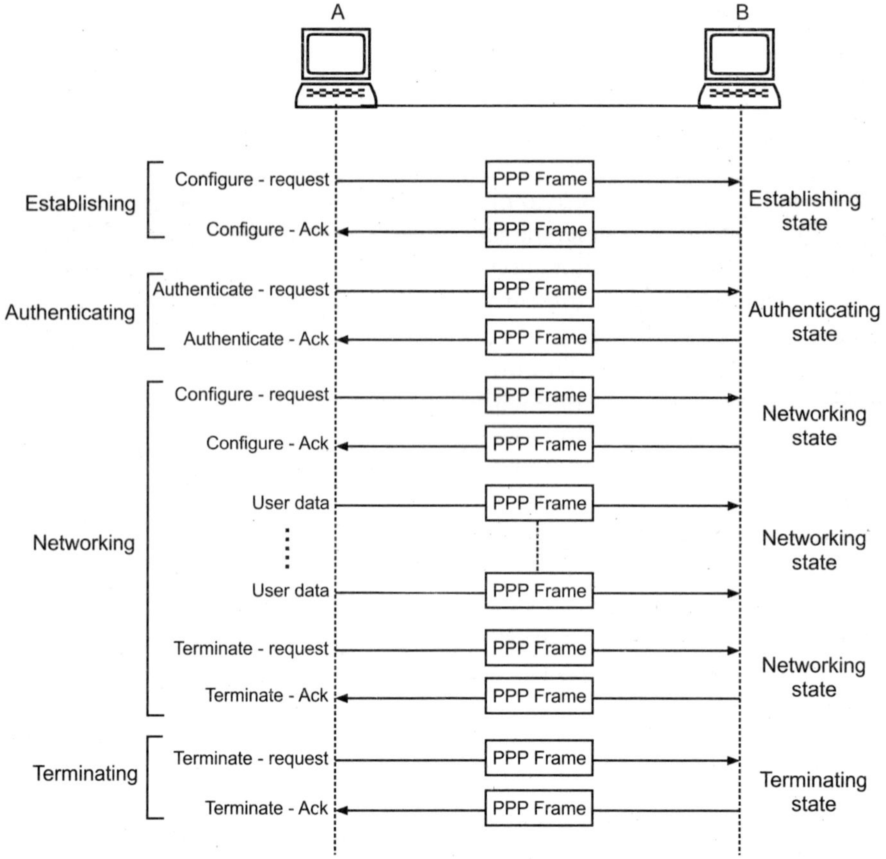

Fig. 4.62. States of PPP connection

Protocol (LCP) and the Network Control Protocol (NCP). The link control protocol is used to set up configure, test, maintain and terminate a link connection. After authentication has been completed, a Network Control Protocol (NCP) is used. The NCP is a set of control protocols to allow the encapsulation of data coming from network layer protocols such as IP, IPX, Decent, Apple Talk in the PPP frame. An example of the states through which a PPP connection goes to deliver some network layer packets is shown in the figure 4.62.

4.23.3. Transition States

The PPP connection goes through different phases as shown in the transition state diagram of figure 4.63.

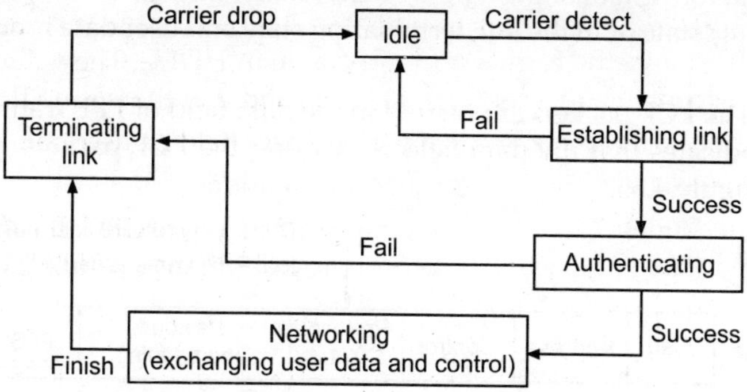

Fig. 4.63. Flow diagram for PPP connection

Figures 4.62 and 4.63 show the following steps :

1. Idle: It means that the link is not being used. There is no active carrier and the line is quiet.

2. Establishing: When one of the end machine starts, the communication, the connection goes into the establishing state. In this state, the user sends the configure request packet to negotiate the options for establishing the link. If negotiation is successful, then the system goes to the authenticating state. The LCP packets are used for this purpose.

3. Authenticating: The user sends the authenticate request packet and includes the user name and password. After it receives the configure-Ack packet, the authentication process is over.

4. Networking: When a connection reaches this state, the exchange of user control and data packets can be started. The connection remains in this state until one of the end points want to terminate the connection.

5. Terminating: The user sends the terminate packet to terminate the link. With the reception of the terminate, Ack packet, the link is terminated.

4.23.4. PPP Stack

PPP is a data link protocol. But it uses stack of other protocols in order to establish the link , to authenticate the users and to carry the network layer data. PPP uses three sets of protocols namely :

(i) Link Control Protocol (LCP)

(ii) Authentication Protocols

(iii) Network Control Protocols (NCP)

Figure 4.64 shows the protocol stack for PPP. It shows that any moment, a PPP packet can carry the packets related to one of the protocols mentioned above.

Flag	Address	Control	Protocol	Packet from one of the protocol stack	FCS	Flag

Fig. 4.64. Protocol stack

The value of the protocol field defines the protocol stack.

4.23.5. Link Control Protocol (LCP)

1. Responsibility: One of the protocols in the protocol stack is Link Control Protocol (LCP). The responsibility of LCP are as under :

(i) To Establish Links (ii) To maintain the establish links
(iii) To configure the links and (iv) Termination of the links

It also provides negotiation mechanisms. It may be noted that both the users should agree on the various options before establishing a link. Hence, when the PPP is carrying the LCP packet, it is in the link establishing state or in the link terminating state and user data is not carried during these states.

2. LCP packet: The LCP packets are carried in the data field of PPP frame. The value $C021_{16}$ in the protocol field indicates that the data field is carrying the LCP packets. The format of an LCP packet is shown in figure 4.65.

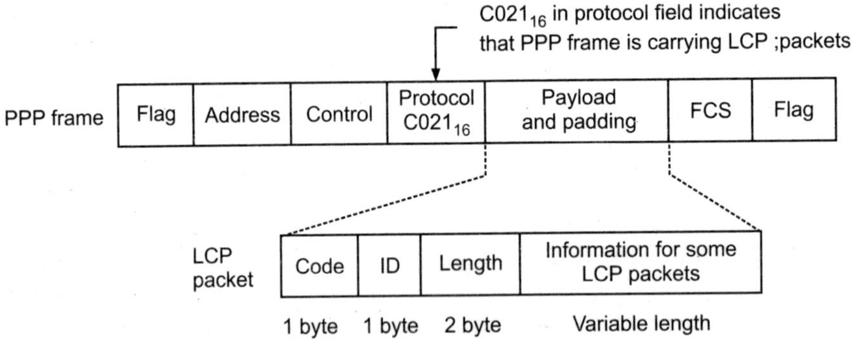

Fig. 4.65. LCP packet

Various fields in LCP packet are as under :

(i) Code: It is a one byte length field which defines the type of LCP packet.

(ii) ID: It is a one byte length field which holds a value used to match a request with the reply.

(iii) Length: It is a two byte long field which defines the length of entire LCP packet.

(iv) Information: It consists extra information needed for some LCP packets.

3. Types of LCP Packets: The LCP packets are broadly classified into three types as under :

(i) Configuration packets
(ii) Link termination packets
(iii) Link Monitoring and debugging packets

4.23.6. Authentication Protocols

Meaning of authentication is to validate the identity of a user who needs to access the set of resources. These are two protocols created by PPP for authentication purpose namely:

(i) Password Authentication Protocol (PAP)
(ii) Challenge Handshake Authentication Protocol (CHAP)

These protocols are used during authentication state and during this state, no user data are exchanged.

4.23.7. Network Control Protocol (NCP)

The next step after link establishment and authentication is getting connected to the network layer. For this state, PPP uses another protocol called Network Control Protocol (NCP). NCP is a set of control protocols which facilitates the encapsulation of data coming from network layer protocols into the PPP frame.

4.23.8. Difference Between SLIP and PPP

S.No.	SLIP	PPP
1.	Error detection and correction is not possible.	Error detection and correction is possible.
2.	SLIP supports only IP	IP and other protocols are supported.
3.	Does not provide any authentication	Provides authentication and security.
4.	SLIP is not an approved Internet standard	PPP is an approved Internet standard
5.	IP address is assigned statically to the user.	Assignment of IP address is done dynamically.

But if it finds 'δ' correct headers in a row, then it moves into the SYNCH state to start the normal operation. If 'α' HECs in a row are incorrect, then the TC sublayer concludes that it has lost synchronization and returns to the HUNT state again.

Few Solved Examples

EXAMPLE 4.21. Calculate CRC for a 10 bit sequence 1010011110. The generator polynomial is $x^3 + x + 1$.

Solution: Given that

Data word : 1010011110

Generator : $x^3 + x + 1 = 1011$

First, we add four zeroes (equal to the number of generator bits) at the end of the data word to get the dividend as under :

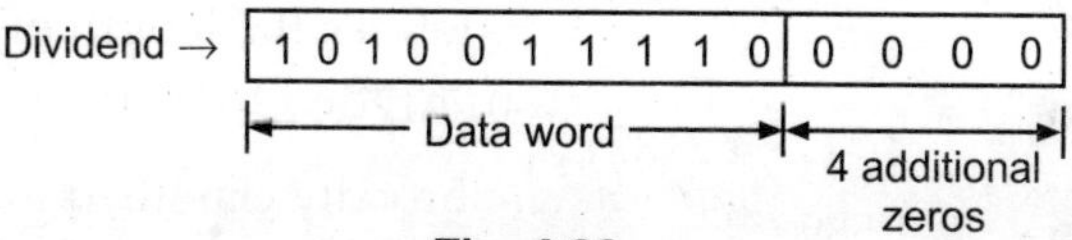

Fig. 4.66.

Then we perform the long division as under :

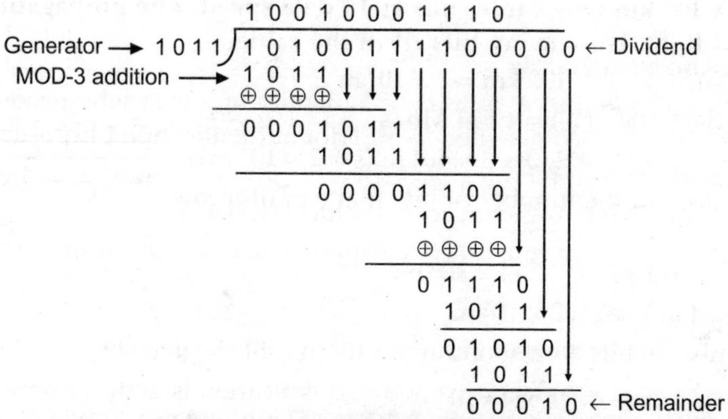

Fig. 4.67.

EXAMPLE 4.22. Given an error free 64 kbps satellite channel which is used to send 512 byte data frames in one direction with very short acknowledgements coming back the other way. What will be the maximum throughput for window sizes of 1, 7, 15 and 127.

Solution: Given that

$$\text{Data rate} = R = 64 \text{ kbps} = 64 \times 10^3 \text{ bps.}$$
$$\text{Frame size} = N = 512 \text{ bytes} = 512 \times 8 \text{ bits}$$
$$\text{Window sizes} = 1, 7, 15 \text{ and } 127.$$

We have to determine the maximum throughput.

(i) First, let us calculate t_f.

We know that the transmission time for 1 frame is given by,

$$t_f = \frac{\text{Frame size}}{\text{Bit rate}} = \frac{N}{R} = \frac{512 \times 8}{64 \times 10^3}$$

Therefore, $t_f = 64 \times 10^{-3}$ sec

(ii) Then, we evaluate the value of A as under :

$$A = \frac{t_p}{t_f}$$

But t_p = propagation delay = 270 ms for satellite channel

Therefore, $$A = \frac{270 \times 10^{-3}}{64 \times 10^{-3}} = 4.2187.$$

(iii) Finally, let us evaluate the value of maximum throughput.

We know that $$\eta_{max} = \frac{W}{1+2A}$$

where W = Window size

(a) For W = 1, $$\eta_{max} = \frac{1}{1+(2 \times 4.2187)} = 0.1059$$

(b) For W = 7, $$\eta_{max} = \frac{7}{1+(2 \times 4.2187)} = 0.7417$$

(c) For W = 15, $\eta_{max} = 1.589$

(d) For W = 127, $\eta_{max} = 13.459$ **Ans.**

EXAMPLE 4.23. A 100 km long cable runs at T1 data speed. The propagation speed in cable is 2/3 of the speed of light. How many bits fit in the cable.

Solution: Given that, $L = 100 \text{ km} = 1 \times 10^5$ m,

We know that data rate of T1 = 1.544 Mb/s

Also, Speed $v = 2/3 \times 3 \times 10^8 \text{ m/s} = 2 \times 10^8 \text{ m/s}$.

Now, we have to find the number of bits that can fit in the cable.

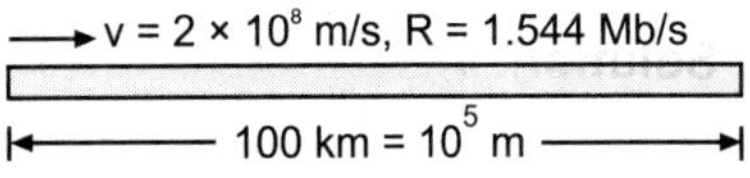

Fig. 4.68.

Number of bits in 1 sec = 1.544×10^6 bits

Distance covered in 1 sec = 2×10^8 m

Therefore, number of bits corresponding to 10^5 m cable is given by,

$$x = \frac{1.544 \times 10^6}{2 \times 10^8} \times 10^5 = 772 \text{ bits.}$$ **Ans.**

EXAMPLE 4.24. Sketch the Manchester encoding for the given bit stream : 0001110101.

Solution : The required waveforms have been shown in figure 4.69.

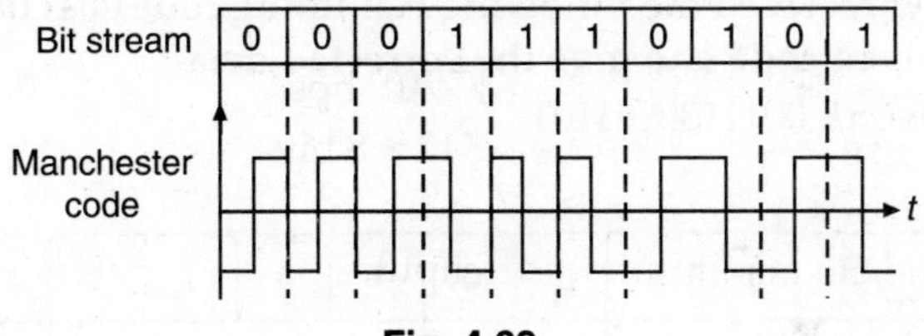

Fig. 4.69.

EXAMPLE 4.25. Given the use of 1000 bit frames on a 1 Mbps satellite channel. What will be the maximum Link utilization for

(i) Stop and wait ARQ.

(ii) Continuous ARQ with Window size 7.

(iii) Continuous ARQ with Window size of 127.

Solution: Given that frame size = 1000 bits, Bit rate = 1 Mbps

We have to determine link utilization.

(i) For stop and wait ARQ.

We know that $t_f = \dfrac{\text{Frame size}}{\text{Bit rate}} = \dfrac{1000}{1 \times 10^6} = 1 \times 10^{-3}$ s = 1 ms.

Also, t_p = 270 ms propagation delay for a satellite channel

Therefore, $A = \dfrac{t_p}{t_f} = \dfrac{270}{1} = 270$

We have, $\eta = \dfrac{1}{1+2A} = \dfrac{1}{1+2(270)} = 1.848 \times 10^{-3} = 0.1848\%$. **Ans.**

(ii) For continuous ARQ with *W* = 7.

We know that $\eta = \dfrac{W}{1+2A} = \dfrac{7}{1+(2 \times 270)} = 1.2936\%$. **Ans.**

(iii) Continuous ARQ with *W* = 127.

We know that $\eta = \dfrac{127}{1+(2 \times 270)} = 23.4696\%$. **Ans.**

EXAMPLE 4.26. Sketch the Manchester and Differential Manchester encoding for the following given bit stream ;

0 0 0 1 1 1 0 1 0 1

Solution: The required waveforms have been shown in figure 4.70.

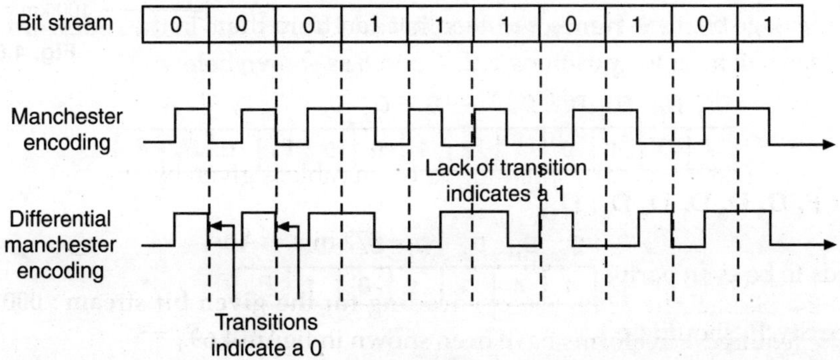

Fig. 4.70.

EXAMPLE 4.27. A receiver received the following Hamming code 0011100101101 with odd parity. Find the error in the received code and give the corrected data.

Solution: Received codeword: 0011100101101

Type of parity: odd

0	0	1	1	1	0	0	1	0	1	1	0	1
					P_8				P_4		P_2	P_1

1. Check bits 8, 9, 10, 11, 12, 13

Here, it is odd parity, so no error.

Thus, $P_8 = 0$

2. Check bits 4, 5, 6, 7, 12, 13

The bits are 0 0 0 1 0 1 → even parity, so, error exists.

Thus, we make $P_4 = 1$

3. Check bits 2, 3, 6, 7, 10, 11

The bits are 1 1 0 1 1 0 → even parity, so , error exists.

Thus, we make $P_2 = 1$.

4. Check bits 1, 3, 5, 7, 9, 11, 13

The bits are 0110011 → even parity, so, error exists.

Thus, we make $P_1 = 1$

5. Write the error word

Error word = E =

P_8	P_4	P_2	P_1
0	1	1	1

6. Obtain correct codeword

$$E = 0111 = (7)_{10}$$

Hence, the seventh bit in the received codeword is incorrect so, we invert it from 0 to 1. Hence, the correct codeword is,

0 0 1 1 1 0 (1) 1 0 1 1 0 1

Inverted seventh bit

Limitation of Hamming Code

The main limitation of Hamming code is that it can detect and correct only one error.

EXAMPLE 4.28. Generate the Hamming Code for the data 111011001 with even parity.

Solution: Data number: 111011001

1. Number of message bits is 9. Hence, we need to add 4 parity bits in the codeword. The parity bits will be at the positions 1, 2, 4, and 8 as shown below:

D_{13}	D_{12}	D_{11}	D_{10}	D_9		D_7	D_6	D_5		D_3		
1	1	1	0	1	P_6	1	0	0	P_4	1	P_2	P_1

2. Select P_1 for $P_1\ D_3\ D_5\ D_7\ D_9\ D_{11}\ D_{13}$

Parity needs to be even parity

D_{13}	D_{11}	D_9	D_7	D_5	D_3	
1	1	1	1	0	1	P_1

For even parity, P_1 should be 1. Thus, $P_1 = 1$

3. Select P_2

To select P_2, we have to consider the bits in positions 2, 3, 6, 7 10 and 11.

Thus, $1\,0\,1\,0\,1\,P_2 \rightarrow P_2 = 1$ Hence, $P_2 = 1$

4. Select P_4

For P_4, we have to consider the bits in the following positions 4, 5, 6, 7, 12, 13 and select the value of P_4 for even parity.

Thus, $1\,1\,1\,0\,0\,P_4$, Hence, $P_4 = 1$

5. Select P_8

To select P_8, we have to consider the bit in the following positions 8, 9, 10, 11, 12, 13 and select P_8 for even parity.

Thus, $1\,1\,1\,0\,1\,P_8$ Hence, $P_8 = 0$

Thus, the codeword is as follows:

						P_8				P_4		P_2	P_1
Codeword	1	1	1	0	1	0	1	0	0	1	1	1	1

EXAMPLE 4.29. Encode the following 8 bit data using Hamming code.

(1) 1010 1111 – Even parity

(2) 0111 1100 – Odd parity

Solution: As the number of data bits is 8 i.e., between 5 and 11, we have to use 4 parity bits.

Part I: For data word 1010 1111

1. Codeward structure

Figure 4.71 shows the codeword structure. Here, P_8, P_4, P_2 and P_1 are four parity bits. Remaining are eight data bits.

D_{12}	D_{11}	D_{10}	D_9	P_8	D_7	D_6	D_5	P_4	D_3	P_2	P_1
1	0	1	0		1	1	1		1		

Fig. 4.71. Codeword structure

Step 2: Select P_1 for $P_1\,D_3\,D_5\,D_7\,D_9\,D_{11}$

D_{11}	D_9	D_7	D_5	D_3	P_1
0	0	1	1	1	

For even parity, $P_1 = 1$.

3. Select P_2

To select P_2, we have to consider the bits in the positions 2, 3, 6, 7, 10 and 11.

They are $0\,1\,1\,1\,1\,P_2$.

For even parity, $P_2 = 0$.

4. Select P_4

For P_4, we have to consider the bits in the positions 4, 5, 6, 7, 12.

They are $1\,1\,1\,1\,P_4$.

For even parity, $P_4 = 0$.

5. Select P_8

For P_8, we consider the bits in the positions 8, 9, 10, 11,12.

They are $1\,0\,1\,0\,P_8$.

For even parity, $P_8 = 0$.

6. Write the codeword

				P_8				P_4		P_2	P_1
1	0	1	0	0	1	1	1	0	1	0	1

Fig. 4.72. Codeword

Similarly, we can solve part II of the problem.

EXAMPLE 4.30. Calculate link utilization efficiency for stop-and-wait protocol, if bit rate = 19.2 kbps, frame size = 960 bits and propagation time = 0.06 sec. for window size = 3 and 7.

Solution: Given: Bit rate R = 19.2 kbps = 19.2×10^3 bps, Frame size N = 960 bits

Propagation time t_p = 0.06 sec. Window size W = 3 and 7.

To find: Link utilization efficiency (η)

1. Calculate t_f

Tranamission time for 1 frame is t_f, is given by,

$$t_f = \frac{\text{Frame size (N)}}{\text{Bit rate (R)}} = \frac{960}{19.2 \times 10^3} = 0.05 \text{ sec.}$$

2. Calculate A

We have $$A = \frac{t_p}{t_f} = \frac{0.06}{0.05} = 1.2$$

3. Calculate efficiency

We have $$\eta = \frac{W}{1+2A}$$ Where, W = window size

Hence, $$\eta = \frac{3}{1+(2 \times 1.2)} = 0.8823$$ **Ans.**

EXAMPLE 4.31. A channel with 10 kbps bit rate and propagation delay of 10 msec, what should be the frame size to obtain efficiency of at least 50% for stop and wait ARQ.

Soluton: Given Bit rate : 10 kbps, Propagation delay = 10 msec

$$0.5 \le \eta \le 1$$

To find: Frame size

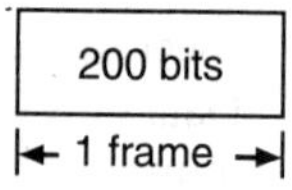

Fig. 4.73.

1. Calculate value of t_f

We have $$\eta = \frac{t_f}{t_f + 2t_p}$$ where, t_f = time for one frame

Thus, $$0.5 = \frac{t_f}{t_f + (2 \times 10 \times 10^{-3})}$$

or, $$t_f = 0.02 \text{ sec} = 20 \text{ msec}$$

2. Calculate the frame size (N)

$$R = 10 \text{ kbps} = 10000 \text{ bps}$$

Hence, $$N = R \times t_f = 10 \times 10^3 \times 20 \times 10^{-3} = 200 \text{ bits}$$ **Ans.**

EXAMPLE 4.32. Using 5 bit sequence numbers, what is the maximum size of the send and receiver window for

(i) Stop-and-wait ARQ.

(ii) Go-back-N ARQ.

(iii) Selective-repeat ARQ.

Solution: The concept of sender sliding window is used in order to hold the outstanding frames until they are acknowledged. That means it is imagined that all the frames stored in a buffer and outstanding frames are enclosed in a window.

1. Stop and wait ARQ

There are no outstanding frames at the sending end.

Hence, the size of sending window is zero. The size of receive window is always 1.

2. Go-back-N ARQ

The maximum size of send window with an m bit sequence number is 2^{m-1}. Hence, for a 5 bit sequence number (m = 5), the maximum send window size is $2^5 - 1 = 31$.

The maximum receive window size is always 1.

3. Selective repeat ARQ

The maximum send and receive window size is $2^m/2$.

Hence, for m = 5, the window size is $2^5/2 = 16$.

GLOSSARY

- **Bit-Stuffing** A data link process that is used to guarantee the uniqueness of a flag bit so that user data may not be interpreted as a given start-stop bit pattern.
- **Bit-Time** A unit of measure equal to 0.1 µs. Hence, a one-bit transmission requires 0.1 µs. Transmitting a 64-byte Ethernet/802.3 frame needs 512 bit-times, or 51.2 µs.
- **Codeword** A data frame which comprises both user data and *redundancy bits*, which are used for *error control*.
- **CRC Checksum** The result of a polynomial division which uses a predetermined *generator polynomial* as the divisor.
- **CRC Error** It is an invalid *CRC* checksum.
- **Cyclic Redundancy Check (CRC)** An *error detection* technique that constructs a polynomial whose terms' coefficients are the value of each of the bits of a data frame. This polynomial is divided by a predetermined *generator polynomial*. The remainder of this division, known as CRC *checksum*, is then assigned to a frame's checksum field. The most common CRC used in most LAN protocols is called CRC 32, a 32-bit checksum.
- **Data Link Layer** The second layer (layer 2) of the OSI model. The data link layer regulates and formats transmission of information from software on a node to the network cabling facilities. In fact, this layer is partitioned into two sublayers; the *logical link control* sublayer (LLC), which provides framing, flow control, and error control; and the *media access control* sublayer (MAC), that specifies the manner in which nodes access a shared medium.
- **Error Control** The process of guaranteeing reliable deliver of data. Error control may be provided through *error detection* or *error correction*.
- **Error Correction** The process in which a destination node, upon detecting a data transmission error, has sufficient information to correct the error autonomously. In fact, error correction implies *error detection.*
- **Error Detection** The process in which a destination node detects a data transmission error and requests a retransmission from the sending node. Error detection is also known as error correction through retransmission.
- **Flow Control** A process which controls the rate at which data messages are exchanged between two nodes. Flow control provides a mechanism to ensure that a sending node does not overwhelm a receiving node during data transmission.
- **Fragmenting** A process in which a *packet* is broken into smaller units to accommodate the maximum transmission unit a physical network is capable of supporting. Fragmented packets are transmitted to the destination separately and then reassembled at the destination node before they are passed to the higher

levels. In *IP*, reassembly of a *datagram* occurs at the destination node and not at any of the intermediary nodes the packet traverses.

- **Frame** A specially formatted sequence of bits which incorporates both data and control information.
- **Frame Relay** A public *WAN* packet-switching protocol which provides LAN-to-LAN connectivity. Its name implies what it does, namely, relays frames across a network between two sites. Frame relay was originally part of the *ISDN* standard.
- **Framing** A *data link layer* process that partitions a bit stream into discrete units or blocks of data called *frames.*
- **HDSL** A modified HDSL designed and packaged for corporate clients.
- **High bit-rate Digital Subscriber Line (HDSL)** A DSL variant which provides symmetrical service at T1 rates over two pairs of UTP and E1 rates over three pairs of UTP. Telephone service is not supported. Few applications include connecting PBXs and serving as an alternative to T1/E1. HDSL is suitable for campus network and *ISPs.*
- **Post Office Protocol (POP)** A TCP/IP application protocol that is used for transferring e-mail from a server to a desktop PC client.
- **Redundancy Bits** Extra bits incorporated into a data frame which provide error-correction information. A data set composed of both user data and redundancy bits is known as a *codeword*. Also called *check bits.*
- **Sliding Window Protocol** A *flow-control* strategy which enables a sender to transmit frames continuously without having to wait for acknowledgments to these frames from the receiver. The sliding windows concept is implemented by requiring the sender to sequentially number each data frame it sends and by having the sender and receiver maintain information about the number of frames they can respectively send or receiver.
- **Stop-and-Wait Protocol** A *flow-control* strategy which requires the sender to transmit one fame and then wait for the receiver to acknowledge receipt of this frame before any additional frames are transmitted.

SUMMARY

- The data layer is supposed to carry out many specified functions. For effective data communication between two directly connected transmitting and receiving stations, the data link layer has to carry out a number of specific functions.
- The data and control information is combined in a frame and transmitted from the source to destination machine. The destination machine must be able to recognize control information from the data being transmitted.
- Data link layer can be designed to offer different types of services. Some of them are as under:
  - (i) Unacknowledged connectionless service.
  - (ii) Acknowledged connectionless service.
  - (iii) Acknowledged connection oriented service.
- Following methods are used for carrying out framing:
  - (i) Character count
  - (ii) Starting and ending characters, with character stuffing.
  - (iii) Starting and ending flags with bit stuffing.
  - (iv) Physical layer coding violations.
- Many data link protocols use the combination of the character count technique with one of the other techniques so as to have an extra safety.
- The errors introduced in the data bits during their transmission can be categorized as :
  - (i) Content errors
  - (ii) Flow integrity errors.
- The codeword is the n bit encoded block of bits. it contains message bits and parity or redundant bits.
- When a codeword is transmitted, one or more number of transmitted bits will be reversed (0 to 1 or vice versa) due to transmission impairments. Thus, errors will be introduced. It is possible for the receiver to detect these errors if the received codeword (corrupted) is not one of the valid codewords.

- The redundancy bits used by CRC are derived by following the procedure given below:
  (i) Divide the data unit by a predetermined divisor.
  (ii) Obtain the remainder. It is the CRC.
- There are two basic systems of error detection and correction. The first on being the forward error correction (FEC) system and the second one is the automatic repeat request (ARQ) system. In the ARQ system of error control, when an error is detected, a request is made for the retransmission of that signal.
- The bit rate of the return transmission which involves the return transmission of ACK/NAK signal is low as compared to the bit rate of the forward transmission. Therefore, the error probability of the return transmission is negligibly small.
- The three types of ARQ systems are as under :
  (i) Stop and-wait ARQ system (ii) Go back n ARQ system, and
  (iii) Selective repeat ARQ system.

SHORT QUESTIONS WITH ANSWERS

Q.1. Write the types of Serviced Provided by a Data Link Layer.

Ans. Data link layer can be designed to offer different types of services. Some of them are as under:
(i) Unacknowledged connectionless service.
(ii) Acknowledged connectionless service.
(iii) Acknowledged connection oriented service.

Q.2. What are Framing Methods?

Ans. Following methods are used for carrying out framing:
(i) Character count
(ii) Starting and ending characters, with character stuffing.
(iii) Starting and ending flags with bit stuffing.
(iv) Physical layer coding violations.

Q.3. Write the Types of Errors.

Ans. The errors introduced in the data bits during their transmission can be categorized as:
(i) Content errors (ii) Flow integrity errors.

Q.4. Write the drawbacks of Coding.

Ans. Some of the drawbacks of the coding technique may be listed as under :
(i) An increased transmission bandwidth is required in order to transmit the encoded signal. This is due to the additional bits (redundancy) added by the encoder.
(ii) Use of coding make the system complex.

Q.5. What does SDLC stand for?

Ans. Synchronous Data Link Control.

Q.6. What is the application of HDLC?

Ans. High Level Data Link Control is used to define the frame format of the Data Link Layer.

Q.7. Which layer performs error detection?

Ans. Data link layer.

Q.8. What protocol is used to access the Internet by a modem?

Ans. Point-to-Point Protocol (PPP) or Slip.

Q.9. What does PPP stand for?

Ans. Point-to-Point Protocol.

Q.10. What are the functions of Link Control Protocol?

Ans. Establish connection, configure connection parameters, and negotiate link parameters such as authentication method and MRU.

REVIEW QUESTIONS

1. State the various design issues for the data link layer.
2. State and explain the various services provided to the network layer.
3. What are different framing methods ?
4. Explain about character stuffing.
5. What is bit stuffing ?
6. Explain about the function of timer.
7. Write a short technical note on error control.
8. Explain about the simplex protocol for noisy channel
9. What is piggybacking ?
10. Explain about the stop and wait protocol.
11. State drawbacks of stop and wait protocols.
12. Explain about Go back n protocol.
13. What is pipelining ?
14. Write a short technical note on Selective repeat ARQ.
15. Define throughput efficiency and explain how it can be increased?
16. Draw and explain the frame structure of HDLC.
17. State and explain various frame types in HDLC.
18. Explain about transparency and bit stuffing in HDLC.
19. Write short technical note on (a) SDLC, (b) PPP, (c) HDLC protocol, (d) sliding window protocols.
20. Explain about the frame format of PPP.
21. Explain about the flow diagram for PPP connection.
22. Explain about role of minimum distance in error correction and detection.
23. Which are the error detection methods ?
24. Give the limitations of parity checking.
25. What are the advantages of checksum method ?
26. Calculate CRC for a 10-bit sequence 1010011110. The generator polynomial is $x^3 + x + 1$
27. Briefly explain the sliding window protocols.

Chapter 5

The Medium Access Control Layer

Syllabus

The Medium Access Control Layer: Static and dynamic channel allocation, multiple access protocols-Pure and slotted ALOHA, CSMA, Collision free protocols, limited contention protocols, CSMA/CD (ETHERNET), fast Ethernet, Gigabit Ethernet.

Chapter Outline

- Introduction
- MAC and LLC sublayers
- The Channel Allocation Schemes: Static and dynamic channel allocation
- Multiple Access Protocols: Multiple Access Communications
- Multiple Access ALOHA System
- Comparison Between Pure and Slotted ALOHA
- Carrier Sense Multiple Access (CSMA)
- Collision-Free Protocols
- Controlled Access
- Channelization
- Introduction to LANs
- Components of LANs
- Promiscuous Mode
- IEEE 802 Standards
- 802.3 IEEE Standard: Ethernet
- Medium Access Control (MAC)
- Logical Link Control (LLC)/ IEEE 802.2
- Local Area Network Technologies
- Traditional Ethernet
- Physical Layer Implementations of Traditional Ethernet
- Fast Ethernet
- Gigabit Ethernet
- Token Bus : IEEE 802.4

5.1. INTRODUCTION

As discussed earlier, the networks can be divided into two categories as under :

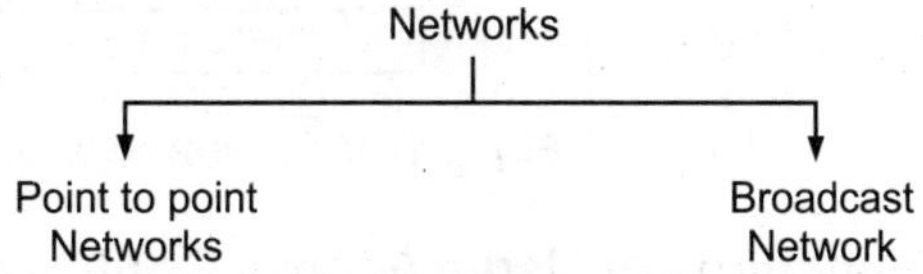

In this chapter, we shall discuss about the broadcast networks and their protocols. The broadcast channels are also called multi-access channels or random access channels. In the broadcast networks, the most important point is the criteria to determine who is allowed to use the channel when more than one users want to use it. A protocol is used to make this decision. Such a protocol, belongs to a sublayer of data link layer called the MAC (medium access control) sublayer. The MAC sublayer is very important in LANs because it is a broadcast network.

5.2. MAC AND LLC SUBLAYERS

Figure 5.1 shows the layered OSI model (partial) to show the position of MAC and LLC sublayers. We shall discuss the broadcast protocols corresponding to the lower layers (1 and 2) of the OSI model as shown in figure 5.1.

Figure 5.1 relates the LAN protocols with the OSI architecture. This architecture was developed by IEEE 802 committee and it has been accepted as LAN standard.

It is called IEEE 802 reference model. Let us discuss this model layer by layer.

Pioneer in Communication: Radia Perlman (1951)

Radia Perlman's work has had a great impact upon the field of networking, and she has been considered as specialist in areas of bridging and routing, as well as sabotage-proof networks. Best known as the inventor of spanning tree algorithm used in bridges and switches, Perlman is also the inventor of several important algorithms that make modern link-state routing protocols robust and quite efficient, including self-stabilizing flooding of routing updates and designated routers.

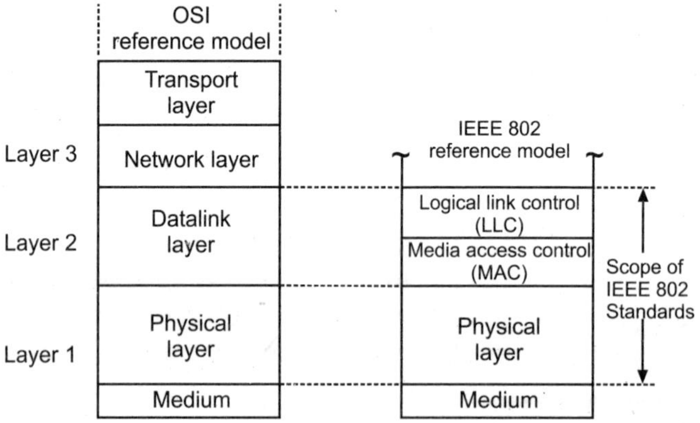

Fig. 5.1. IEEE 802 protocol layers compared to OSI model

5.2.1. Functions of Media Access Control Sublayer

(i) To perform the control of access to media.

(ii) It performs the unique addressing to stations directly connected to LAN.

(iii) Detection of errors.

DO YOU KNOW?

The MAC sublayer coordinates the data link tasks within a specific LAN.

5.2.2. Functions of Logical Link Control (LLC) Sublayer

(i) Error Recovery

(ii) It performs the flow control operation

(iii) User addressing.

5.3. THE CHANNEL ALLOCATION SCHEMES: STATIC AND DYNAMIC CHANNEL ALLOCATION

In a broadcast network, the single broadcast channel is to be allocated to one transmitting user at a time. This is called as **channel allocation**. There are two different schemes used for channel allocation as follows.

5.3.1. Static Channel Allocation in LANs and MANs

The traditional way of allocating a single channel, such as a telephone channel among many users is FDM. The Frequency Division Multiplexing (FDM) and Time Division Multiplexing (TDM) are the examples of static channel allocation. In these methods, either a fixed frequency or a fixed time slot is allotted to each user. The problem in these methods is that if all the N number of users are not using the channel the channel bandwidth is wasted and if there are more than N users who want to use the channel they cannot do so for the lack of bandwidth. To see the poor performance of static channel, let us consider an example for FDM system where the mean time delay (T) for a channel of capacity C bps, with an arrival rate of λ grames/sec. Each frame having a length drawn from an exponential probability density function with mean $1/\mu$ bits/frame is given as,

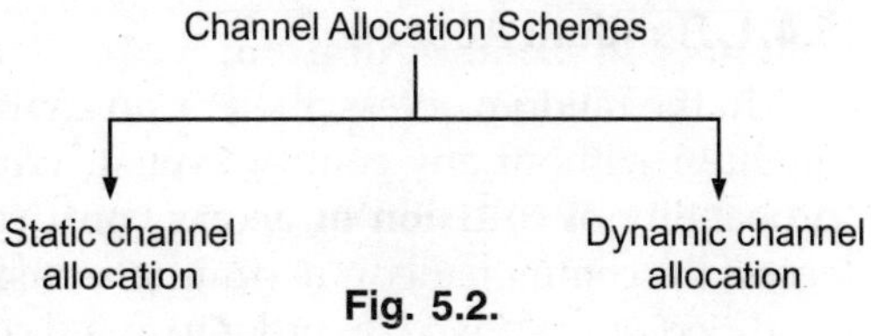

Fig. 5.2.

$$T = \frac{1}{\mu C - \lambda}$$

If the single channel is divided into N independent subchannels, the above equation is modified as follows:

$$T_{FDM} = \frac{1}{\mu(C/N) - (\lambda/N)} = \frac{1}{\mu C - \lambda}$$

$$T_{FDM} = NT$$

DO YOU KNOW?

The MAC sublayer is manufacturer-specific and dependent on the LAN type.

Important Point:

From the above equation, it is obvious that the mean delay using FDM is worse. The static channel allocation has a poor performance with bursty traffic and hence generally dynamic channel allocation is used, for computer networks where the traffic is of bursty nature.

5.3.2. Dynamic Channel Allocation

In this method, no fixed frequency or fixed time slot is allotted to the user. The user can use the single channel as per his requirement. Following assumptions are made for the implementation of this method:

(i) Station model – Consists of N independent stations each with a program or user that generates frames for transmission.

(ii) Single channel – A single channel is available for all communication.

(iii) Collision – If frames are transmitted at the same time by two or more stations, there is an overlap in time and the resulting signal is garbled. This event is called **collision**.

(iv) Continuous or slotted time – There is no master clock dividing time into discrete time intervals or time is divided into discrete intervals.

(v) Carrier or No carrier sense – Stations sense the channel before transmission or they directly transmit without sensing the channel.

5.4. MULTIPLE ACCESS PROTOCOLS: MULTIPLE ACCESS COMMUNICATIONS

When a number of stations (users) use a common link of communication system we have to use a multiple access protocol in order to coordinate the access to the common link. The three techniques used to deal with the multiple access problem are as follows:

(i) Random Access

(ii) Controlled Access

(iii) Channelization.

Let us discuss them one by one.

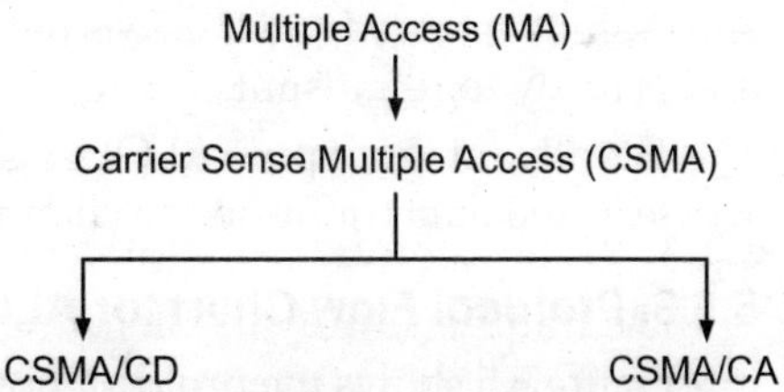

Fig. 5.3. Evolution of random access methods

5.4.1. Random Access

In the random access, there is no control station. Each station has the right to use the common medium without any control over it. With increase in number of stations, there is an increased probability of **collision or access conflict.** The collisions occur when more than one user try to access the common medium simultaneously. As a result of such collisions, some frames can be either modified or destroyed. In order to avoid collisions, we have to set up a procedure. The evolution of the random access methods is shown in figure. 5.3.

5.4.2. Evolution of Random Access Methods

The first method, known as ALOHA used a simple procedure called multiple access (MA). It was improved to develop the carrier sense multiple access (CSMA).

The CSMA further evolved into two methods namely CSMA/CD (CSMA with collision detection) and CSMA/CA(CSMA with collision avoidance) which avoids the collisions.

5.5. MULTIPLE ACCESS ALOHA SYSTEM *(RGTU, Bhopal, Sem. Exam., June, 2014)*

5.5.1. ALOHA System

Systems in which multiple users share a common channel in a way that can lead to conflicts are widely known as contention systems. The ALOHA system is a contention protocol which was developed at the University of Hawaii in the early 1970's by Norman Abramson and his colleagues.

DO YOU KNOW?

Broadband networks can combine serveral local area networks with voice and video on the same cable. They are more complex to configure than baseband LANs.

The ALOHA system has two versions :

(i) Pure ALOHA – does not require global time synchronisation.

(ii) Slotted ALOHA – requires time synchronisation.

5.5.2. Pure ALOHA

It works on a very simple principle. Essentially, it allows for any station to broadcast at any time. If two signals collide, each station simply waits a random time and tries again. Collisions are easily detected. As shown in the figure. 5.4, when the central station receives a frame it sends an acknowledgement on a different frequency.

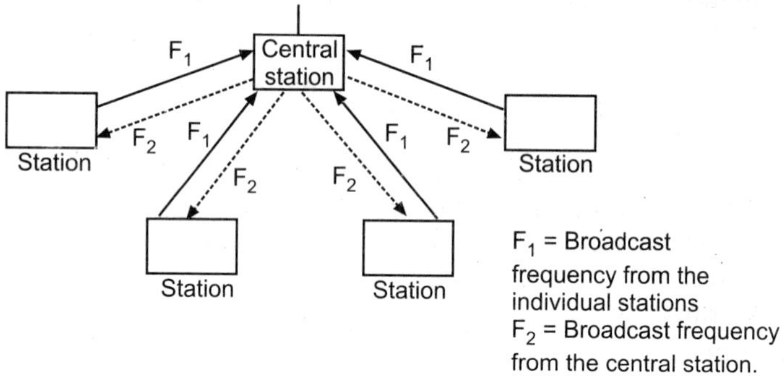

Fig. 5.4. Pure ALOHA System

If a user station receives an acknowledgement it assumes that the transmitted frame was successfully received and if it does not get an acknowledgement it assumes that collision had occurred and is ready to retransmit.

The advantage of pure ALOHA is its simplicity in implementation but its performance becomes worse as the data traffic on the channel increases.

5.5.3. Protocol Flow Chart for ALOHA

Figure 5.5 shows the protocol flowchart for ALOHA.

5.5.3.1. Explanation of Flow Chart

A station which has a frame ready, will send then it waits for some time. If it receives the acknowledgement then the transmission is successful. Otherwise the station uses a backoff strategy, and sends the packet again. After many times, if there is no acknowledgement then the station aborts the idea of transmission.

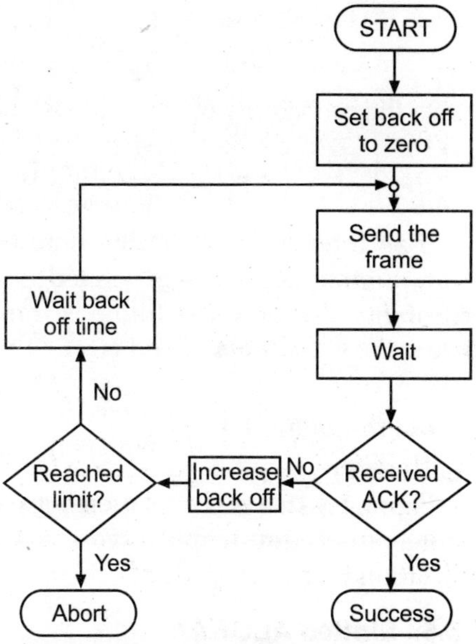

Fig. 5.5. Protocol flow chart for ALOHA

5.5.3.2. Contention System

Systems in which multiple users share a common channel in a way that can lead to conflict are known as the contention systems. Whenever two frames try to occupy the channel at the same time, there is bound to be a collision and both will be garbled. Retransmission is essential for all the destroyed frames.

5.5.4. Efficiency of an ALOHA Channel

Efficiency is that fraction of all transmitted frames which escape collisions. Let us consider infinite number of interactive users at their computers (stations). **Each user is either** typing or waiting. Initially all of them are in the typing state. When a line is typed, the user stops and waits. The station then transmits a frame containing this line and checks the channel to confirm the success. If it is successful then the user starts typing again, otherwise the user waits and its frame is retransmitted over and over again till it succeeds.

Frame time

Let the frame time denote the amount of time required to transmit the standard fixed length frame. Note that

$$\text{Frame time} = \frac{\text{Frame length}}{\text{Bit rate}}$$

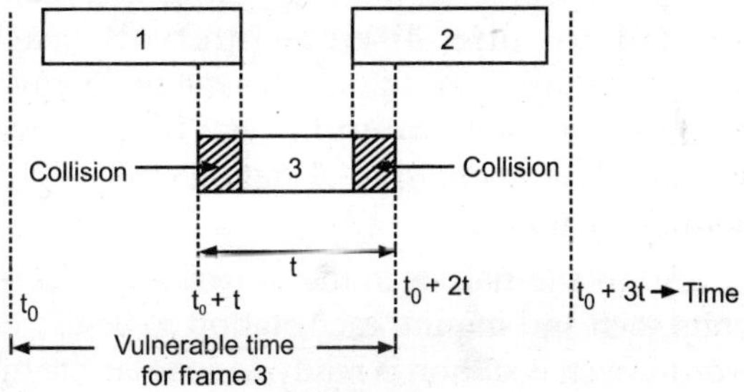

Fig. 5.6.

We assume that infinite number of users generate new frames according to the Poisson's distribution with mean N frames per frame time. If $N > 1$, then users are generating frames at a rate higher than that can be handled by the channel. So, every frame will face collision. Hence, $0 < N < 1$. Let there be k transmission attempts (including retransmissions) per frame time. The probability of k transmissions per frame time is also Poisson. Let the mean of number of transmissions be G per frame time. So, $G \geq N$. At low load $N \cong 0$ there will be less number of collisions so less number of retransmission and $G \cong N$. With increase in load there are many collisions so $G > N$. For all the loads, the throughput is given by,

$$S = GP_0$$

where P_0 = Probability that a frame dos not suffer a collision.

Now, let us consider figure 5.6. What is the condition for frame 3 in figure 5.6 to arrive without collision? Let t = time required to send a frame. If frame 1 is generated any where between t_0 to $(t_0 + t)$ then it will collide with frame 3. Similarly, any frame (2) generated between $(t_0 + t)$ and $(t_0 + 2t)$ also collides with frame 3.

The probability that k frames are generated during a given frame time is given by the Poisson's distribution.

$$P[k] = \frac{G^k e^{-G}}{k!}$$

so the probability of zero frames i.e. k = 0 is

$$P_0 = \frac{G^0 e^{-G}}{0!} = e^{-G}$$

If an interval is two frame time long, the mean number of frames generated is 2G. The probability that no other frame is transmitted during the vulnerable period is,

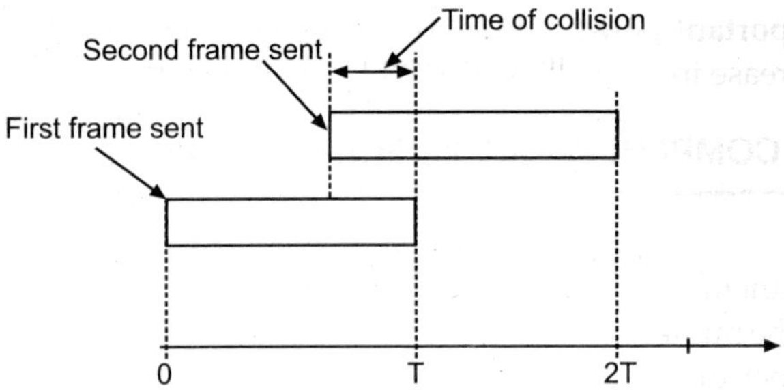

Fig. 5.7. Transmission using pure ALOHO

$$P_0 = e^{-2G}$$

But throughput $S = GP_0$

Therefore, $S = Ge^{-2G}$

Figure 5.9 shows the relation between the offered traffic G and the throughput S. It shows that the maximum throughput occurs at G = 0.5 and S_{max} = 0.184. Hence, the best possible channel utilization is on 18.4 percent.

5.5.5. Slotted ALOHA*

(RGTU, Bhopal, Sem. Exam., June, 2014)

To overcome the disadvantage of the pure ALOHA system, Robert published a method for doubling the capacity of traffic on the channel. In this method, time is divided up into discrete intervals, each interval corresponding to one frame. This method requires the users to agree on slot boundaries. In this method, for synchronisation one special station emits a pip at the start of each interval,iuat arbitrary times can waste time upto 2T.

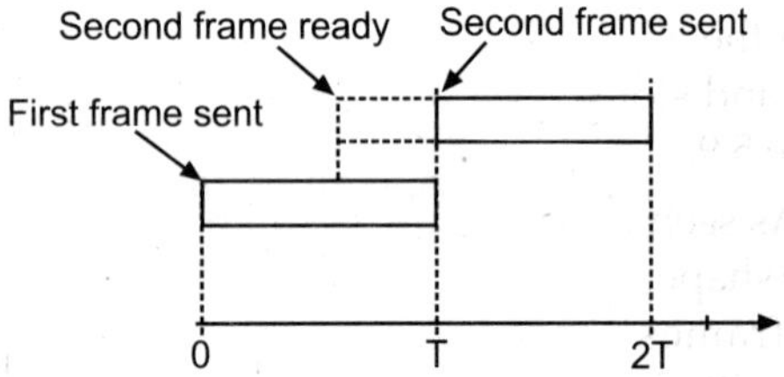

Fig. 5.8. Transmission using slotted ALOHA

As an alternative, in the slotted ALOHA method, the time is divided into intervals (slots) of T units each and require each station to begin each transmission at the beginning of a slot. In other words, even if station is ready to send in the middle of a slot, it must wait until the start of the next one as shown in figure 5.8. In this method a collision occurs when both stations become ready in the same slot. Slotted ALOHA is thus a discrete time system whereas pure ALOHA is a continuous time system. The Vulnerable period has been reduced to half that of pure ALOHA, the throughput for slotted ALOHA is,

$$S = Ge^{-G}$$

The maximum throughput corresponds to G = 1 and it is given by $S_{max} = 1/e = 0.368$ as figure 5.9. So, for a slotted ALOHA with G =1 the probability of success is 37%. The probability of empty slots is,

$$P(k) = \frac{G^0 e^{-G}}{k!}$$

For G = 1 and k = 0, we get P(k = 0) = 0.368 and the probability of collisions is 24%. The probability of transmission requiring exactly k attempts (i.e. k – 1 collisions followed by one success) is given by,

$$P_k = e^{-G} (1-e^{-G})^{k-1}$$

And the expected number of transmission E per carriage return typed is

$$E = e^G$$

* Prove that the throughput of Network using slotted ALOHA can be given as $S = Ge^{-G}$, where, G is the load and S is the throughput.

Important point: As E depends exponentially on G, with a small increase in G, there is a large increase in E and drastic fall in performance.

5.6. COMPARISON BETWEEN PURE AND SLOTTED ALOHA*

(RGTU, Bhopal, Sem. Exam., June, 2014)

A mathematical model can be created for the relationship between the number of frames transmitted and the number of frames transmitted successfully. Let G represent the traffic measured as the average number of frames generated per slot. Let S be the success rate measured as the average number of frames sent successfully per slot. The relationship between G and S for both pure and slotted ALOHA is given as follows :

Pure ALOHA → $S = Ge^{-2G}$

Slotted ALOHA → $S = Ge^{-G}$

Where e is the mathematical constant = 2.718. From the above equation, a success rate curve for pure and slotted ALOHA can be plotted as shown figure 5.9.

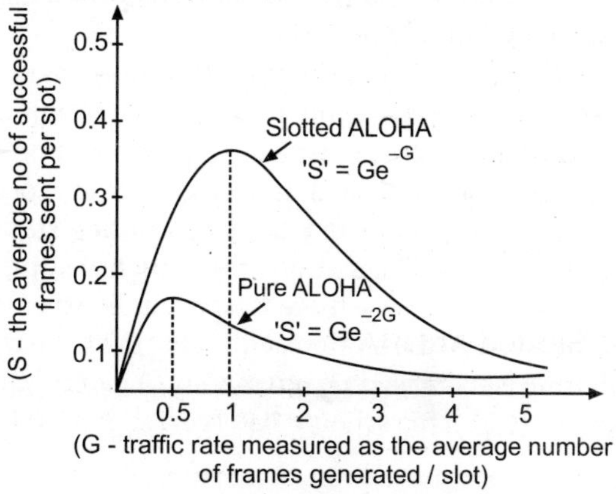

Fig. 5.9. Comparison of pure and slotted ALOHA

As seen in the figure 5.9, both graphs have the same shape. If G is small so is S, which means that if few frames are generated few frames will be transmitted successfully. As G increases so does S but upto a certain point. As G continues to increase S approaches to 0 which means that if more frames are generated there will be more collisions and the success rate falls to 0. Similarly, for pure ALOHA, the maximum occurs at G = 0.5 for which S = 1/2e = 0.184 which means the rate of successful transmissions is approximately 18.4%. As seen from the graph, the maximum for slotted ALOHA occurs at G = 1 for which S = 1/e = 0.368. In other words, the rate of successful transmissions is approximately 0.368 frames per slot time or 37% of the time will be spent on successful transmission. Hence, the slotted ALOHA has a double throughput efficiency than the pure ALOHA system. The maximum utilization achievable using CSMA can be increased much beyond that obtainable using ALOHA or slotted ALOHA. The maximum utilization is dependent on length of the frame and on the propagation time. With increase in the length of the frame or reduction in the propagation time, the utilization gets improved.

5.7. CARRIER SENSE MULTIPLE ACCESS (CSMA) *(RGTU, Bhopal, Sem. Exam., June, 2014)*

The CSMA protocol operates on the principle of carrier sensing. In this protocol, a station listens to see the presence of transmission (carrier) on the cable and decides to act accordingly.

(i) Non-Persistent CSMA: In this scheme, if a station wants to transmit a frame and it finds that the channel is busy (some other station is transmitting) then it has to wait for fixed interval of time. After this time, it again checks the status of the channel and if the channel is free transmits.

(ii) 1-Persistent CSMA: In this scheme, the station which wants to transmit, continuously monitors the channel until it is idle and then transmits immediately. The disadvantage of this strategy is that if two stations transmit simultaneously a collision takes place. Retransmission starts only after finding the channel free.

* Derive an expression for the throughput of a slotted ALOHA system and compare this with pure ALOHA system. *(RGTU, Bhopal, Sem. Exam., June, 2014)*

(iii) p-Persistent CSMA: The possibility of such collisions and retransmissions is reduced in the p-persistent CSMA. In this scheme, all the waiting stations are not allowed to transmit simultaneously as soon as the channel becomes idle. A station is assumed to be transmitting with a probability p. For example, if p = 1/6 and if 6 stations are waiting then on an average, only one station can transmit while others will wait.

5.7.1. Carrier Sense Multiple Access/Collision Detection (CSMA/CD)*

(RGTU, Bhopal, Sem. Exam., June, 2014)

The CSMA/CD specifications have been standardized by IEEE 802.3 standard. It is a very widely used MAC protocol.

(i) Media Access Control: The problem in CSMA explained earlier is that a transmitting station continues to transmit its frame even though a collision occurs. The channel time is unnecessarily wasted due to this. In CSMA/CD, if a station receives other transmissions when it is transmitting, then a collision can be detected as soon as it occurs and the transmission time can be saved. As soon as a collision is detected, the transmitting stations release a jam signal. The jam signal will alert the other stations. The stations then are not supposed to transmit immediately after the collision has occurred. Otherwise there is a possibility that the same frames would collide again. After some back off delay time the stations will retry the transmission. If again the collision takes place then the back off time is increased progressively. A careful design can achieve efficiencies of more than 90% using CSMA/CD. This scheme has been shown in figure 5.10.

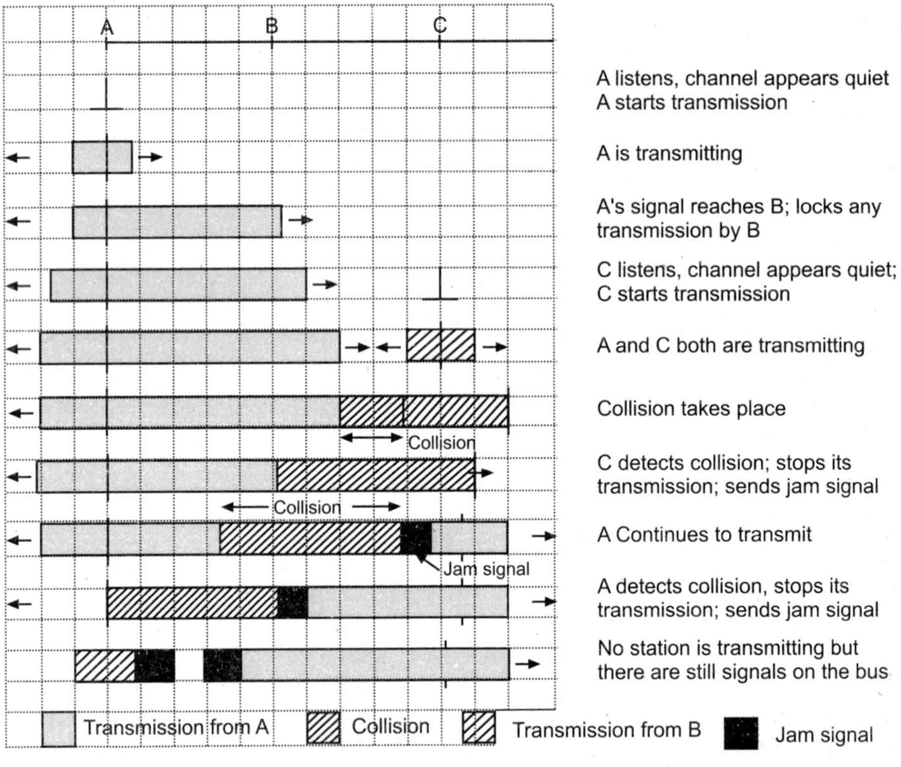

Fig. 5.10. CSMA/CD scheme

* What is CSMA scheme? Discuss non-persistent, 1-persistent and p-persistent CSMA with a diagram.
(RGTU, Bhopal, Sem. Exam., June, 2014)

(ii) Frame format of CSMA/CD: The frame format as per IEEE 802.3 standard is as shown in figure 5.11.

Field sizes →	7	1	2 to 6	2 to 6	2	46 to 1500		4 bytes
	PR	SFD	DA	SA	L	DATA	PAD	FCS

PR : Preamble

SFD : Start Frame Delimiter

DA : Destination Address

SA : Source Address

L : Length of data field

FCS : Frame check sequence

Fig. 5.11. Frame format of IEEE 802.3 CSMA/CD frame

The frame consists of following fields :

Preamble (PR) : Preamble is a seven byte long pattern which is used for bit synchronization.

Start Frame Delimiter (SFD) : It is a one byte long, unique bit pattern which is used to mark the start of the frame.

Destination Address (DA) : This field is 2 to 6 bytes (octet) long and contains the destination address.

Source Address (SA) : This field is 2 to 6 bytes (octet) long and contains the source address.

Length (L) : This field is 2 byte long. It indicates the number of bytes (octets) in the data field.

Data field : It can have 46 to 1500 bytes if the address field has 6 bytes.

Frame Check Sequence (FCS) : This is 4 octet long and contains CRC code, for error detection.

The frame size is variable from 64 to 1518 bytes. The data field can be from 46 to 1500 bytes. If the number of data bytes is less than 46 then PAD field is introduced to make a frame of minimum size.

(iii) Frame Format of Ethernet CSMA/CD Frame: Ethernet is another system of specifications developed to standardize the specifications of CSMA/CD. Ethernet specifications are almost same as those of IEEE 802.3. However, the Ethernet CDMA/CD frame format is slightly different than that of IEEE 802.3. The Ethernet frame structure is as shown in figure 5.12. The Type field which is 2 octet long identifies the higher level protocol type associated with the frame.

Field size →	8	6	6	2	46-1500	4 Octet
	PR	DA	SA	TYPE	DATA	FCS

PR : Preamble

DA : Destination address

SA : Source address

FCS : Frame check sequence

Fig. 5.12. Format of Ethernet CSMA/CD format

5.7.2. CSMA/CD Procedure*

(RGTU, Bhopal, Sem. Exam., June, 2014)

Figure 5.13 shows a flow chart for the CSMA/CD protocol.

Explanation

The station that has a ready frame sets the back off parameter to zero. Then it senses the line using one of the persistent strategies. It then sends the frame. If there is no collision for a period corresponding to one complete frame, then the transmission is successful. Otherwise the station sends the jam signal to inform the other stations about the collision. The station then increments the

* Distinguish between ALOHA and CSMA protocols in detail.

back off time and waits for a random back off time and sends the frame again. If the back off has reached its limit then the station aborts the transmission. CSMA/CD is used for the traditional Ethernet. CSMA/CD is an important protocol. IEEE 802.3 (Ethernet) is an example of CSMA/CD. It is an international standard. The MAC sublayer protocol does not guarantee reliable delivery. Even in absence of collision, the receiver may not have copied the frame correctly.

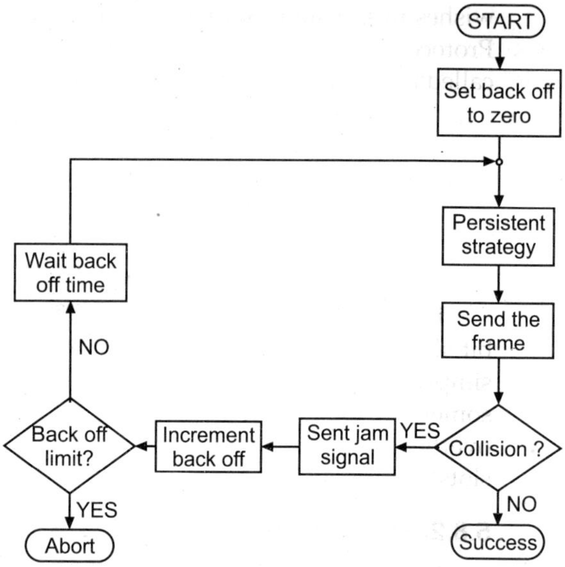

Fig. 5.13. CSMA/CD procedure

5.7.3. CSMA Protocol with Collision Avoidance (CSMA/CA)

This is the CSMA protocol with collision avoidance. Figure 5.14 shows the flow chart explaining the principle of CSMA/CA.

The station ready to transmit, senses the line by using one of the persistent strategies. As soon as it finds the line to be idle, the station waits for an IFG (Interframe gap) amount of time. It then waits for some random time and sends the frame. After sending the frame, it sets a timer and waits for the acknowledgement from the receiver. If the acknowledgement is received before expiry of the timer, then the transmission is successful. But if the transmitting station does not receive the expected acknowledgement before the timer expiry then it increments the back off parameter, waits for the back off time and resenses the line.

5.8. COLLISION-FREE PROTOCOLS

As we have seen that almost collisions can be avoided in CSMA/CD, they can still occur during the contention period. The collision during contention period adversely affects the systems performance, this happens when the cable is long and length of packets are short. This problem becomes serious as fiber optic networks come into use. Here, we shall discuss some protocols that resolve the collisions during the contention period.

5.8.1. Bit-map Protocol

Bit -map protocol is collision-free protocol. In bit-map method, each contention period consists of exactly N slots. If any station has to send frame, then it transmits a 1 bit in the respective slot. For example, if station 2 has a frame to send, it transmits a 1 bit during the second slot. In general, station 1 may announce the fact that it has a frame to send by inserting a 1 bit into slot i. In this way, each station has complete knowledge of which stations

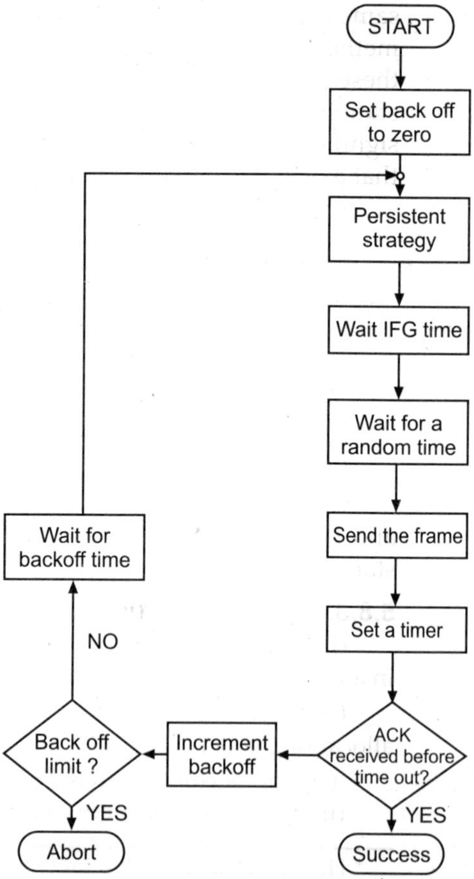

Fig. 5.14. CSMA/CA procedure

wishes to transmit. Since everyone agrees on who goes next, there will never be any collisions. Protocols like this in which the desire to transmit is broadcast before the actual transmissions are called reservation protocols.

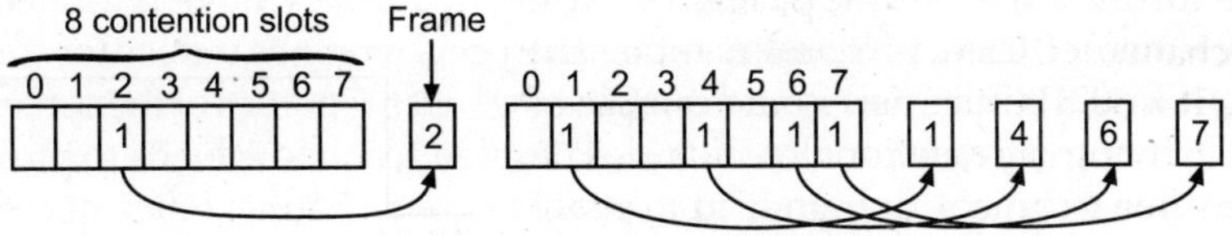

Fig. 5.15. A bit-map protocol

For analyzing the performance of this protocol, we will measure time in units of the contention bit slot, with data frame consisting of d time units. Under low load conditions, the bit map will simply be repeated over and over, for lack of data frames. At high-load, when all the stations have something to send all the time, the N bit contention period is prorated over N frames, yielding an overhead of only 1 bit per frame. Generally, high numbered stations have to wait half a scan (N/2 bit slots) before starting to transmit, low-numbered stations have to wait on an average 1.5 N slots.

5.8.2. Binary Countdown

Binary countdown protocol is used to overcome the overhead 1 bit per station. In binary countdown binary station addresses are used. A station wanting to use the channel broadcasts its address as a binary bit string, starting with the high-order bit. All addresses are assumed to be of same length. Here, we will see the example to illustrate the working of binary countdown. In this method, different station addresses are ORed together to decide the priority of transmitting. If these stations, 0001, 1001, 1100, 1011 all are trying to size the channel for transmission. All the stations at first broadcast their most significant address bit i.e. 0, 1, 1, 1 respectively. The most significant bits are ORed together. Station 0001 see the 1 MSB in other station addresses and knows that a higher numbered station is competing for the channel, so it gives up for the current round.

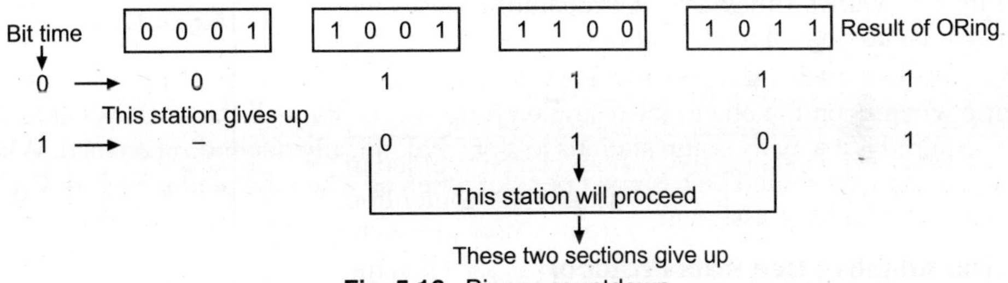

Fig. 5.16. Binary countdown

Other three stations 1001, 1100, 1011 continue. The next bit is 1 at station 1100, so station 1011 and 1001 give up. Then station 110 starts transmitting a frame, after which another bidding cycle starts.

5.8.3. Limited Contention Protocols*

(i) Meaning of contention system: The systems in which multiple users share a common channel in a way that can lead to conflicts are known as **contention systems**.

(ii) Contention protocols: Till now we have considered two different techniques for the channel allocation, namely:

(i) Contention (such as CSMA)

(ii) Collision-free methods.

* How do contention-free protocols differ from contention oriented protocols ? Discuss any one protocol of contention free category ?

The performance of these techniques can be judged based on two performance parameters, namely delay and efficiency. As the load of the channel increases, contention becomes increasingly less attractive, because the overhead associated with channel arbitration becomes greater. Just the reverse is true for the collision-free protocols. At low load, they have high delay, but as the load increases, the channel efficiency improves rather than gets worse as it does for contention protocols.

Obviously, it would be nice if we could combine the best properties of the contention and collision-free protocols, arriving at a new protocol that used contention at low loads to provide low delay; but used a collision-free technique at high load to provide good channel efficiency. Such protocols, are called **limited contention protocols.** These protocols are a combination of contention and collision-free protocols provide good channel efficiency at high loads. The contention protocols like CSMA/CD are symmetric in nature i.e. each station attempts to acquire the channel with some probability P, with all stations using the same P. In case of limited contention protocols, the overall performance is improved by assigning different probabilities to different stations.

In case of symmetric protocols for small number of stations, the chance of success are good but it becomes worse as the number of stations increase. In case of limited contention protocols which are asymmetric in nature, the probability of some station acquiring the channel can be increased only by decreasing the amount of competition. In this method the stations are first divided up into groups. Only the members of group 0 are permitted to compete for slot 0. If one of them succeeds it acquires the channel and transmits its frame. If the slow lies empty or if there is a collision, the members of group 1 compete for slot 1 and so on. Thus, by making an appropriate division of stations into groups, the amount of contention for each slot can be reduced.

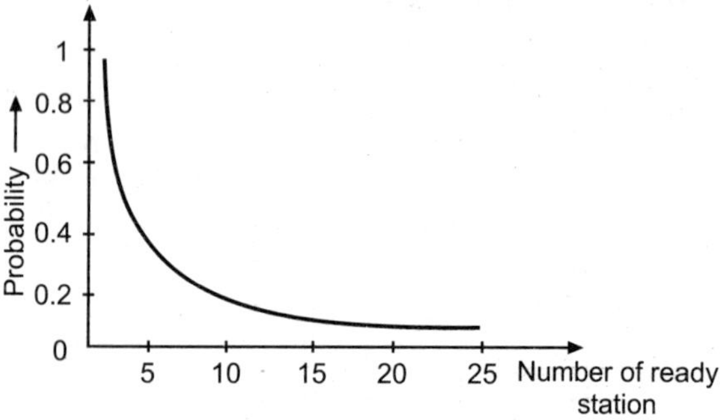

Fig. 5.17.

Figure 5.17 shows the graph of probability plotted against number of ready stations.

The important question is how to assign stations to slots? At one extreme, we have one member per group whereas on the other side a single group will contain all the stations (slotted ALOHA). What is required is a way to assign stations to slots dynamically depending on load. When the load is low, many stations should be assigned per slot whereas when the load is high few (or even one) station per slot should be assigned.

5.8.4. The Adaptive Tree Walk Protocol

The assignment of stations to the slots can be done with the help of a simple algorithm called adaptive tree walk protocol. It is imagined that the stations are leaves of a binary tree as shown in figure 5.18. The tree for eight stations is shown in figure 5.18. As shown in the figure 5.18 in the first contention slot which is slot 0 if a successful frame transmission occurs all stations are allowed to compete for the channel. If there is a collision then during slot 1 only those stations falling under node 2 in the tree may compete. If one of the stations acquires the channel, the slot following the frame is reserved for stations falling under node 3. On the other hand, if there is a collision under node 2 during slot 1 then during slot 2, it is the turn of stations falling under node 4 to compete for the channel.

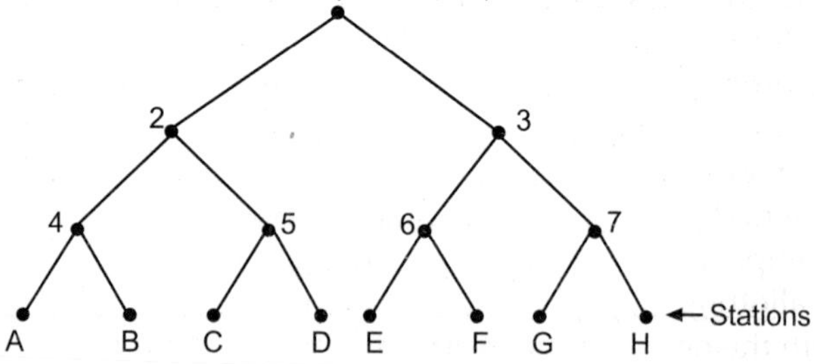

Fig. 5.18. Adaptive tree walk protocol

If a collision occurs, the search continues recursively with the nodes, left and right children. If a bit slot is idle or if there is only one station that transmits in it, then, the searching of its node can stop, because all ready stations have been located.

5.9. CONTROLLED ACCESS

In the previous section, we have considered the random access approach for sharing a transmission medium. They are simpler to implement and are useful in handling the light traffic. In this section we will discuss the scheduling approaches to the medium access control. There are two main approaches in the scheduling approach as follows :

1. Reservation system
2. Polling system
3. Token passing ring networks.

5.9.1 Reservation Systems

The principle of reservation system can be understood from figure 5.19. In this system each station transmits a single packet at the full rate R bps. The transmission from the stations can be organized into frames of variable length. Before each frame a reserved slot (interval) is being transmitted.

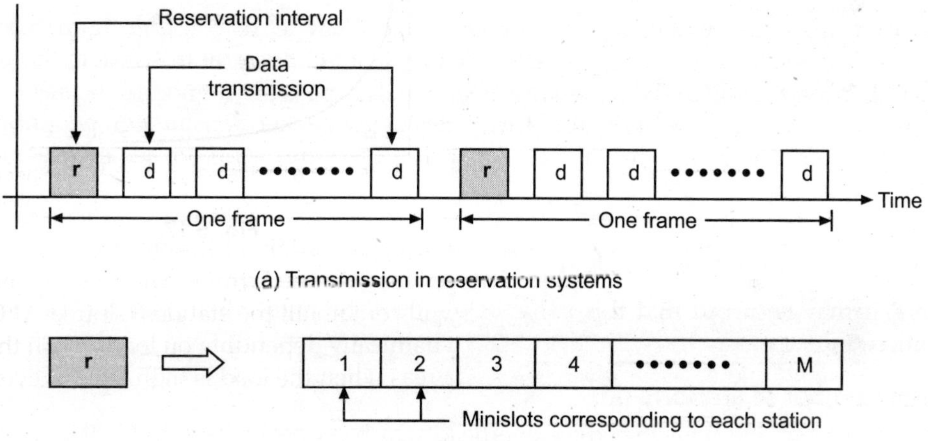

(a) Transmission in reservation systems

(b) Details of reservation interval

Fig. 5.19. Basic reservation system

Figure 5.19 (b) shows the details of the reservation interval r. The reservation interval consists of M minislots one per station. Stations use their minislots to indicate that they have a packet to transmit in the corresponding frame. The station that wants to transmit packet by broadcasting their reservation bit during the appropriate minislot. By listening to the reservation interval, the stations can determine the order of packet transmissions in the corresponding frame. The frame length corresponding to the number of stations which have a packet to transmit. If the length of the packet is variable, then it can be handled if the reservation message includes packet length information. This reservation system described above is called as the basic reservation system. The basic reservation system can be improved by using the time division multiplexing scheme by using the time slots which are idle by allotting them to the other stations. The operation of the basic reservation system can be explained with the help of figure 5.20.

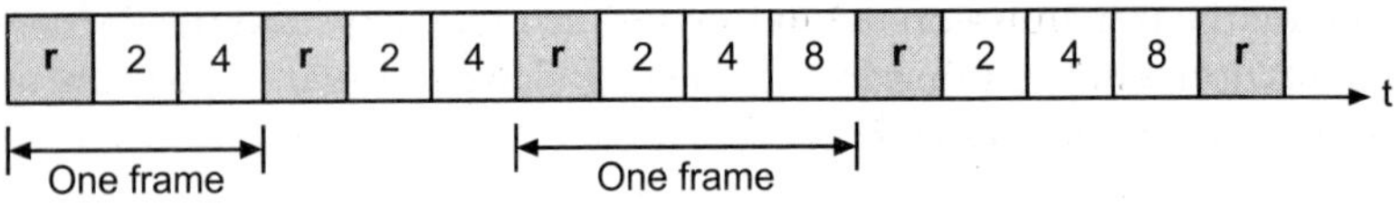

(a) Negligible propagation delay

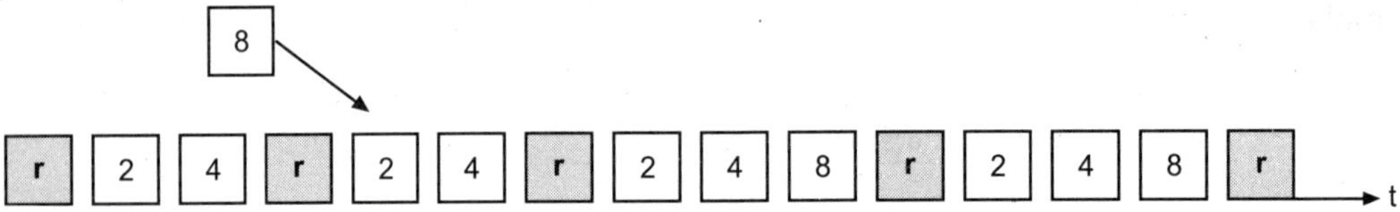

(b) Non negligible propagation delay

Fig. 5.20. Operation of reservation system with negligible and non-negligible delays

Let us consider figure 5.20 (a) which shows a system with negligible propagation delay. In the initial portion only the stations 2 and 4 transmit their packets. But in the middle portion, the station also wants to transmit its packet. Hence, the frame gets expanded from two slots to three slots. The maximum throughput from this system can be attained when all the stations are busy.

The corresponding maximum throughput is given by :

$$\rho_{max} = \frac{1}{1+v} \text{ for one packet reservation/minislot.}$$

If v << 1, then the value or ρ_{max} can be very high. Now let us consider figure 5.20 (b) which shows a reservation system with non-negligible propagation delay. In this system the stations will transmit their reservation in the same way as before. It is possible to modify the basic reservation system so that stations can reserve more than one slot per packet transmission per minislot. Let us assume that a minislot can reserve say upto k packets. Then the maximum achievable throughput is given by,

$$\rho_{max} = \frac{1}{1+(v/k)} \text{ for k packet reservation/minislot.}$$

Here, it may be noted that this value of ρ_{max} will be higher than that for the single packet reservation/minislot.

Effect of number of stations (M)

The reservation intervals introduce overhead which is proportional to M. This means that the reservation interval becomes M × v. As the number of stations (M) become very large, this overhead will become significant. Then this becomes a serious problem. This problem can be sorted out by not allocating a minislot to each station and then instead making stations contend for a reservation minislot by using a random access technique such as ALOHA or slotted ALOHA.

5.9.2. Polling

Now, let us consider polling system shown in figure 5.21. In this system, the stations take turns accessing the medium. At any given time only one of the stations will transmit into the medium.

When a station does transmitting then some mechanism is used to pass the right of transmission to the another station. There are different ways of passing the right of transmission from one station to the other station figure 5.21 (a) shows M stations communicate with a central controller. The outbound line carries the information from the central computer to the M users whereas the shared inbound line will carry the information from users to the central computer. Thus, the inbound line is a shared medium that requires a medium access control in order to co-ordinate the transmissions from the user to the host computer. The host computer acts as a central controller. It issues control messages for co-ordinating the transmissions from the stations. The central controller sends a polling message to a particular station. That station sends its message on the inbound line. Once this process is over, the station gives a go-ahead message. It is possible that the central controller may poll the stations in a round robin fashion or may do it according to some pre-determined order. Figure 5.21 (b) shows another system where it is possible to use polling. The central controller of this system can make use in radio transmission. Figure 5.22 shows the sequence of polling message.

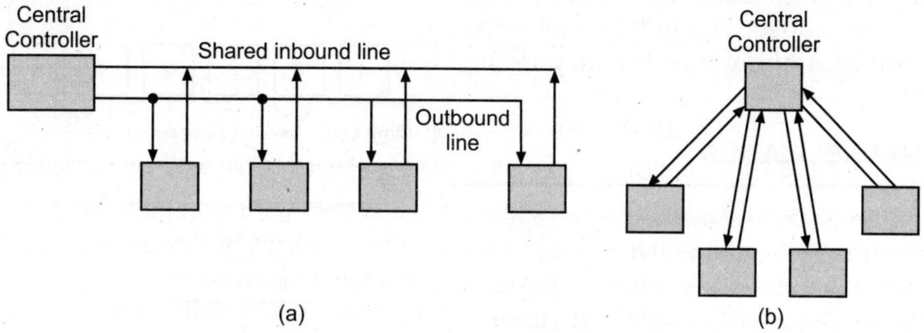

Fig. 5.21. Examples of polling systems.

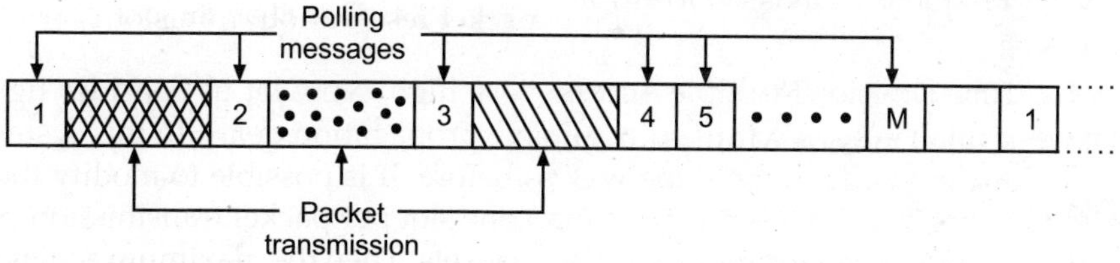

Fig. 5.22. Polling messages and transmission in a polling system

Station 1 is being polled first. While the polling message propagates and is received and until station 1 begins transmission a certain time called walk time will elapse. The next period is occupied by the transmission from station 1. This period will then be followed by the walk time corresponding to station 2. This process will continue until all the M stations are polled. The walk time can be considered to be an overhead in the polling system. The total walk time 'τ' is the sum of walk time corresponding to each station.

5.9.3. Token Passing

Token is a special frame which is used to authorize a particular station for transmission. In the token passing method, a station is authorized to send its data receives the token.

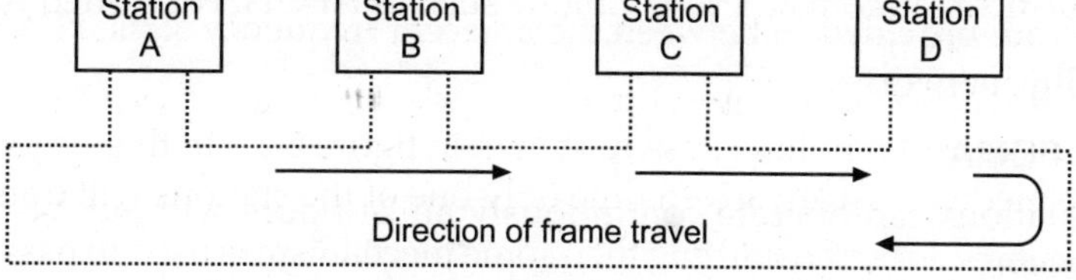

Fig. 5.23. Token passing network

Each station has a predecessor and successor. The frames come from the predecessor and go to the successor as shown in figure 5.23. When no data is being transmitted and the line is idle, the token circulates around the ring. If a station is ready to send data, then it will wait for the token. The station will capture the circulating token and transmit one or more frames. This station will keep sending the frames as long as it has frames to send or the allotted time is not complete. If then passes this token to the next station in the ring. This is the simplest possible token passing technique in which all the stations have equal priority or right to send. In the practical system, the other features such as priority and reservation are added. Figure 5.24 shows the flow chart explaining the token passing procedure.

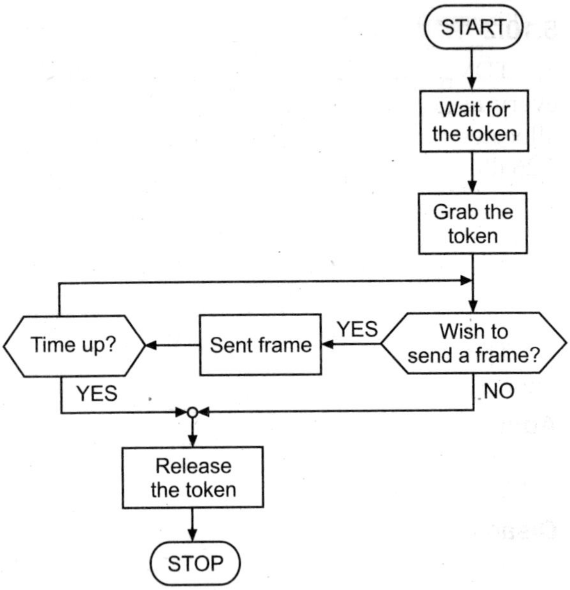

Fig. 5.24. Flow chart for token passing technique.

5.10. CHANNELIZATION

This is a multiple access method in which the total bandwidth of the common link is shared in the frequency domain, the domain or through codes. Depending on the method of sharing, there are three channelization techniques.

(i) FDMA : Frequency Division Multiple Access

(ii) TDMA : Time Division Multiple Access

(iii) CDMA : Code Division Multiple Access

5.10.1. FDMA

In the frequency division multiple access, the available channel (medium) bandwidth is shared by all the station. So each station uses its allocated frequency band to send its data. Each band is thus reserved for a specific station. The concept of FDMA is illustrated in figure 5.25. FDMA is a data link layer protocol which uses FDM at the physical layer. The station 1 uses frequency band f_0 to f_1, station-2 uses the band f_2 to f_3 etc., as shown in figure 5.25.

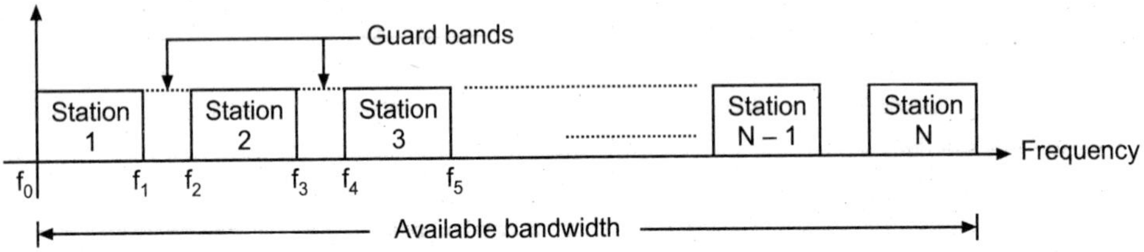

Fig. 5.25. Concept of FDMA

Guard bands are provided in between the adjacent frequency slots. FDMA is used in cellular phones and satellite networks.

Advantages of FDMA

(i) All the stations can operate continuously all 24 hours without having to wait for their turn to come.

(ii) No synchronization is necessary.

Disadvantages of FDMA

(i) Each channel or station can use only a part of the total bandwidth.

(ii) In spite of guard bands being provided, there is some adjacent channel interference present.

(iii) Due to the non-linearity of devices used, the intermodulation products are generated.

5.10.2 TDMA

TDMA stands for time division multiple access. In TDMA, the entire bandwidth can be used by every user (station) but not simultaneously. A station can use the entire bandwidth only for the allocated time. Thus, each channel is allocated a time slot during which it can send its data. Figure 5.26 illustrates the concept of TDMA.

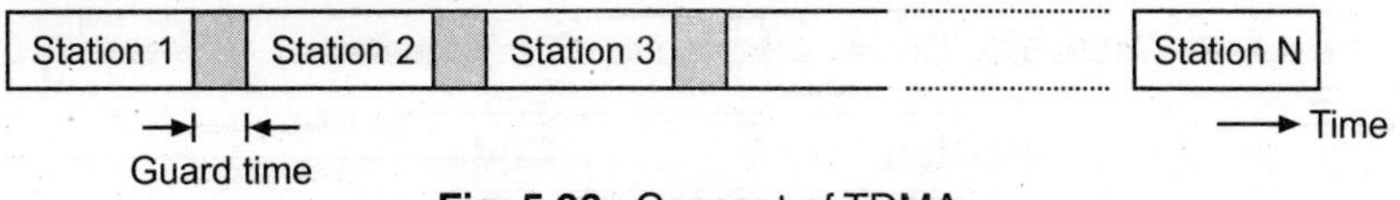

Fig. 5.26. Concept of TDMA

TDMA is a data link layer protocol which uses TDM at the physical layer. TDMA is used cellular phones and satellite networks.

Advantages of TDMA

Since only one station is present at any given time, the generation of intermodulation products will not take place.

Disadvantages of TDMA

TDMA needs synchronization which makes it more complicated as compared to FDMA.

5.10.3. Code Division Multiple Access (CDMA)

An alternative to FDMA and TDMA is an another system called code division multiple access (CDMA). In CDMA, more than one user is allowed to share a channel or subchannel with the help of direct-sequence spread spectrum signals. In CDMA, each user is given a unique code sequence or signature sequence. This sequence allows the user to spread the information signal across the assigned frequency band. At the receiver the signal is recovered by using the same code sequence. At the receiver, the signals received from various users are separated by checking the cross-correlation of the received signal with each possible user signature sequence. In CDMA, the users access the channel in a random manner. Hence the signals transmitted by multiple users will completely overlap both in time and in frequency. The CDMA signals are spread in frequency. Therefore, the demodula-tion and separation of these signals at the receiver can be achieved by using the pseudorandom code sequence. CDMA is sometimes also called as spread spectrum multiple access (SSMA). In CDMA, the bandwidth as well as time of the channel is being shared by the users. It is necessary to introduce the guard times and guard bands as shown in figure 5.27. CDMA does not need any synchronization but the code sequences or signature waveforms are required.

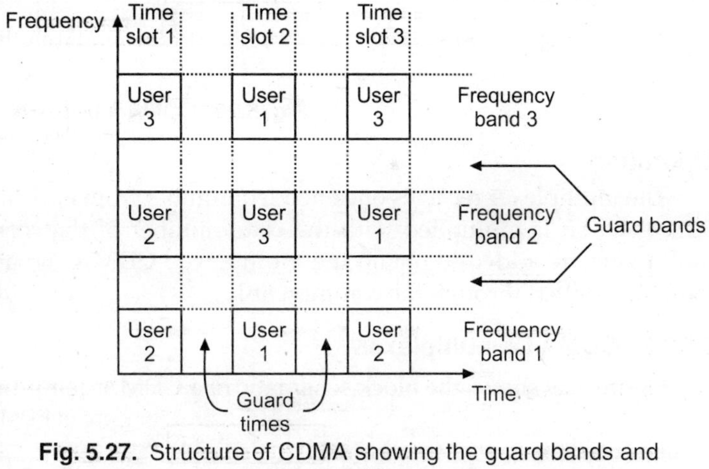

Fig. 5.27. Structure of CDMA showing the guard bands and the guard times

5.10.4. Encoding Rules for CDMA

In CDMA, each station is assigned a code which is simply a sequence of numbers called chips. If a station wants to send a 0 bit then, it sends a – 1, if it wants to send a 1 bit then it sends a + 1. If the station does not want to transmit i.e., wants to remain idle. It is represented by a 0. These are called as the *encoding rules* for CDMA.

5.10.5. CDMA Multiplexer

Let A, B and C be the three stations in a CDMA system. Let the unique codes assigned to them be as under :

Table 5.1. Codes assigned to different stations.

Station	Code
A	+ 1, – 1, – 1, + 1
B	+ 1, + 1, + 1, + 1
C	+ 1, – 1, + 1, – 1

Let us consider figure 5.28 which shows a CDMA multiplexer. We assume that station A wants to transmits a 0 bit station B wants to remain silent and station C wants to transmit a 1.

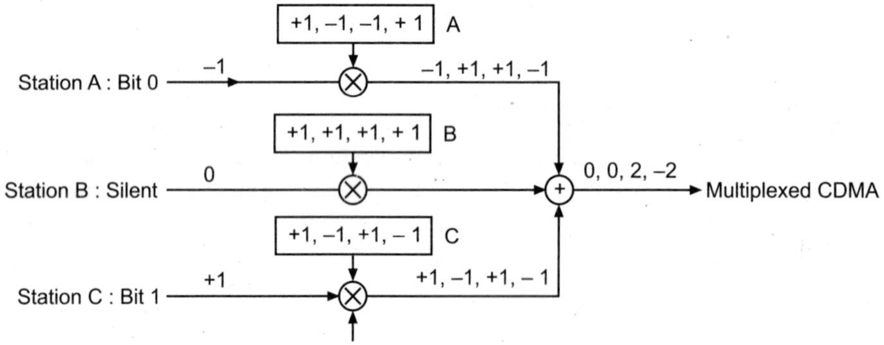

Fig. 5.28. CDMA multiplexer

Operation

The multiplexer receives one encoded number from each station (– 1, 0, + 1) as shown in figure 5.28. Each bit is multiplied with the code number of the corresponding station. The outputs of multipliers are added to obtain the multiplexed CDMA signal. The multiplexed CDMA signal is then transmitted through the common link.

5.10.6. CDMA Demultiplexer

Figure 5.29 shows the block schematic of a CDMA demultiplexer.

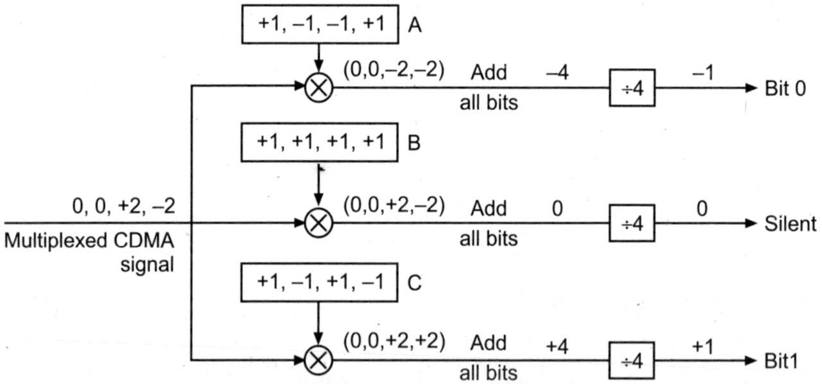

Fig. 5.29. CDMA Demultiplexer

Operation

The multiplexed CDMA signal is applied to all the multiplexers. It is multiplied with the code numbers assigned to the stations A, B and C. The bits available at the output of each multiplier are added together. This addition will always be either – 4, 0 or + 4. The result of addition at the output of each multiplier is divided by 4 to obtain the originally transmitted bits.

Orthogonal sequences

The chip sequences i.e., the codes assigned to various stations are not selected randomly. They are chosen carefully. The codes used in the multiplexed and demultiplexer are called as orthogonal sequences.

5.10.7. Performance Comparison of FDMA, TDMA and CDMA

S. No.	FDMA	TDMA	CDMA
1.	Overall bandwidth is shared among many stations.	Time sharing takes place.	Sharing of bandwidth and time both takes place.
2.	Due to non-linearity of devices inter modulation products are generated due to interference between adjacent channels.	Due to incorrect synchronization, there can be an interference between the adjacent time slots.	Both type of interferences will be present.
3.	Synchronization is not necessary.	Synchronization is essential.	Synchronization is not necessary.
4.	Code word is not required.	Code word is not required.	Codewords are required.
5.	Guard bands between adjacent channels are necessary.	Guard times between adjacent time slots are necessary.	Guard bands and Guard times both are necessary.

5.11. INTRODUCTION TO LANs*

It is an obvious fact that the development of LANs was motivated by the requirement to share resources and information among workstations in a department or works group. Also, the requirement for LANs are different than those in a wide area or in a public network. The short distances between computers imply that low-cost, high-speed, reliable communications is possible. The emphasis on low cost implies a broadcast network approach that does not use equipment that switches information between stations. Instead, the stations cooperate by executing a MAC protocol that minimizes the incidence of collisions generals. Electronic Engineers (IEEE), which has been accredited in the area of the LANs the American National Standards Institute (ANSI). The set of standards includes the CSMA-CD (Ethernet) LAN and the token-passing ring LAN. It also includes the definition of the logical link control, which places LANs within the data link layer of the OSI reference model.

DO YOU KNOW?

Local-area networks (LANs) are small data networks that can be arranged in star, ring, or bus configurations.

5.12. COMPONENTS OF LANs

(i) Basic Concept: Figure 5.30 shows the components of LANs. A number of computers and network devices, such as printers, are interconnected by a shared transmission medium, typically a

* Describe the various components and topologies for LANs. Also discuss the various design issues related with LAN

cabling system, which is arranged in a bus, ring, or star topology. The cabling system may use twisted-pair cable, coaxial cable, or optical fiber transmission media. In some cases, the cabling system is replaced by wireless transmission based on radio or infrared signals. The Ethernet bus topology using co-axial cable is shown in figure 5.30 (a). LAN standards define physical layer protocols that specify the physical properties of the cabling or wireless system for example, connectors and maximum cable lengths, as well as the digital transmission system, for example, modulation, line code, and transmission speed.

(ii) Network Interface Card (NIC): The computers and network devices are connected to the cabling system through a network interface card (NIC) or LAN adapter card as shown in figure 5.30 (b). For desktop computers, the NIC is inserted into an expansion slot or built into the system. Laptop computers typically use the smaller PCMCIA card, which is inserted into a slot that can also be used by a modem or other device.

(iii) Functions of NIC: The NIC card coordinates the transfer of information between the computer and the network. The NIC card transfers information in parallel format to and from main memory (RAM) in the computer. On the other hand, the NIC card transfers information in serial formation to and from the network, so parallel-to-serial conversion is one of the NIC's functions. The speed of the network and the computer are not matched, so the NIC card must also buffer data.

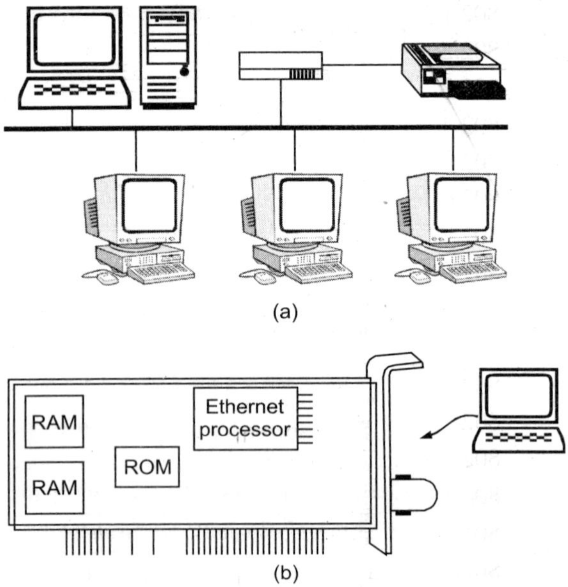

Fig. 5.30. (a) Typical LAN structure and (b) Network interface card

The NIC card has a port that meets the connector and transmission specifications of physical layer standards. The NIC card includes reads-only memory (ROM) containing firmware that allows the NIC to implement the MAC protocol of a LAN standard. This process involves taking network layer packets, encapsulating them inside MAC frames, and transferring the frames by using the MAC protocol, as well as receiving MAC frames and delivering the network layer packets to the computer.

> **DO YOU KNOW?**
>
> The simplest way to configure a small LAN is to use peer-to-peer networking, but larger networks use a dedicated server, with the other nodes operating as clients sharing files on the server.

Each NIC card is assigned a *unique* MAC *or physical address* that is burned into the ROM. Typically, the first three bytes of the address specify the NIC vendor, and the remaining bytes specify a unique number for that vendor. The NIC card contains hardware that allows it to recognize its physical address, as well as the broadcast address. The hardware can also be set to recognize multicast addresses that direct frames to groups of stations.

5.13. PROMISCUOUS MODE

The NIC card can also be set to run in *promiscuous* mode where it listens to all transmissions. This mode is used by system administrators to troubleshoot the information that can facilitate unauthorized access to computers in the LAN.

5.14. IEEE 802 STANDARDS

(RGTU, Bhopal, Sem. Exam., June, 2014)

The Institution of Electrical and Electronics Engineers (IEEE) has developed the layered architecture and other standards of LAN, under their project 802 set up in 1980. The IEEE 802 standards are as follows :

802.1	Architecture, Management and Internetworking
802.2	Logical Link Control (LLC)
802.3	Carrier Sense Multiple Access/Collision Detect (CSMA/CD)
802.4	Token Bus
802.5	Token Ring
802.6	Metropolitan Area Networks (MANs)
802.7	Bandpass Technical Advisory Group
802.8	Fiber Optic Technical Advisory Group
802.9	Integrated Data and Voice Network
802.10	Security Working Group
802.11	Wireless LAN working Group
802.12	Demand Priority Working Group
802.13	Not used
802.14	Cable Modem Working Group
802.15	Wireless Personal Area Networking Group
802.16	Broadband Wireless Access Study Group.

DO YOU KNOW?

Project 802 divides the data link layer into sublayers:

(i) Logical link control (LLC)

(ii) Medium access control (MAC)

In LANs, all the stations share the common cable (i.e., media). Therefore, IEEE adopted three mechanisms of media access control, namely :

(i) Carrier sense multiple access/collision detection (CSMA/CD)

(ii) Token bus and

(iii) Token ring.

Thus, there are three protocols for the MAC sublayer. The IEEE standards 802.3 (CSMA/CD), 802.4 (Token bus), 802.5 (Token ring) are associated with these protocols as shown in figure 5.31. The physical layer protocols do the job of signal encoding, data rate control and interfacing to the transmission medium. The Logical link control layer (LLC) specifications are given in IEEE 802.2.

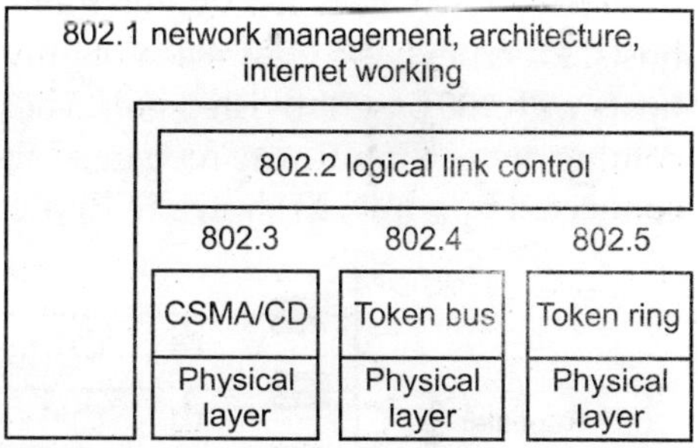

Fig. 5.31. IEEE LAN and related standards

5.15. 802.3 IEEE STANDARD: ETHERNET*

The Ethernet is most successful local area networking technology. It is working example of the more general carrier sense multiple access with collision detect (CSMA/CD). The Ethernet is a multi-access network, meaning that a set of nodes send and receive frames over a shared link you can, therefore, think of an Ethernet or being like a bus that has multiple stations plugged into it. The **carrier sense** in CSMA/CD means that a node listens as it transmits and can therefore detect when a frame it is transmitting has collided with a frame transmitted by another node. The Ethernet has

* Ethernet is a local area network protocol developed jointly by Xerox, Intel, and Digital Equipment Corporation (DEC) at the Xerox Palo Alto Research Center (PARC) in the mid-1970s.

its root in an early packet radio network, called ALOHA. Like the ALOHA, the problem faced by the Ethernet is how to mediate access to a shared medium fairly and efficiently. In ALOHA, the medium was the atmosphere, while in Ethernet the medium is coax-cable. In the earliest days, 10-Mbps Ethernet was used, but now it has been extended to include a 100-Mbps version called **Fast Ethernet** and a 1000-Mbps version called **gigabit Ethernet.** Here, we will discuss 10 Mbps Ethernet, since it is typically used in multiple-access mode. Both 100-Mbps and 1000-Mbps Ethernets are designed to be used in full duplex, point-to-point configurations.

DO YOU KNOW?

The purpose of the IEEE's project 802 is to setup standards so that LAN equipment manufactured by different companies is compatible.

Let us see some physical properties of Ethernet. An Ethernet segment is implemented in a coaxial cable of upto 500 m. A transceiver, which is a small device directly attached to the tap, detects when the line is idle and drives the signal when the host is transmitting. Tap must be at least 2.5 m apart. Transceiver also receives incoming signals. It is in turn, connected to an Ethernet adapter, which is plugged into the host. All the power of Ethernet is in adapter as shown in figure 5.32.

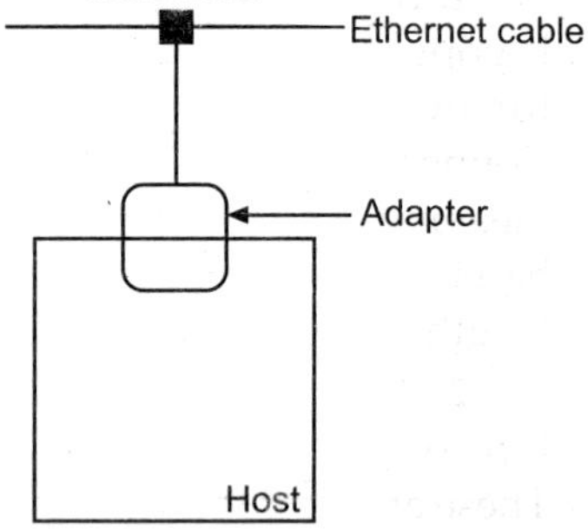

Fig. 5.32. Ethernet transceiver and adapter

Multiple Ethernet segment can be joined together by repeaters as shown in figure 5.33(a). A repeater is a device that forwards digital signals.

Key Point: The name Ethernet was derived from the old Electromagnetic theoretical substance called luminiferousether, which was formerly believed to be the invisible universal element that bound together the entire universe and all its associated parts. Thus, an "ether"net is a network that connects all components attached to the 'net'.

Here, it may be noted that no more than four repeaters may be positioned between any pair of hosts. Ethernet has a total reach of only 2500 m and it is limited to supporting a maximum of 1024 hosts with 100 base T, twisted pair. The common configuration has several point-to-point segments coming out of a multi-way repeater, called a hub, multiple 100-Mbps Ethernet segments can also be connected by a hub, as shown in figure. 5.33 (b).

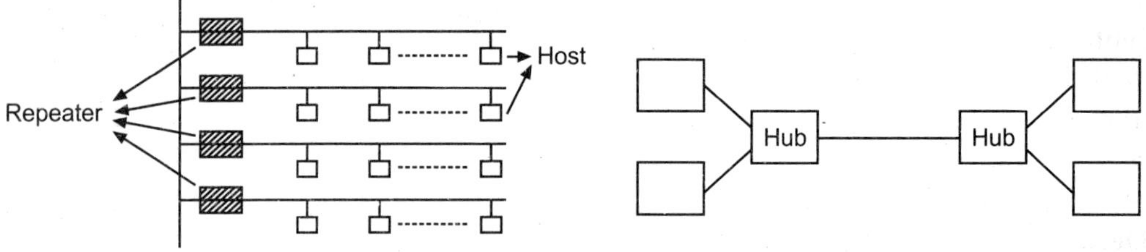

(a) Ethernet repeater (b) Ethernet hub

Fig. 5.33.

5.15.1. 802.3 Cabling

The name Ethernet refers to the cable. Four types of cabling are commonly used for Ethernet as follows :

1. 10 Base 5
2. 10 Base 2
3. 10 Base – T
4. 10 Base – F

The other specifications of these cables are as given in Table 5.2.

Table 5.2.

Name	Cable	Maximum segment	Nodes/segment
10 Base 5	Thick coax	500 m	100
10 Base 2	Thin coax	200 m	30
10 Base T	Twisted pair	100 m	1024
10 Base F	Fiber optics	2000 m	1024

1. 10 Base 5 (Thick Ethernet): The 10 Base 5 cabling is popularly known as thick Ethernet. It has a marking for every 2.5 meters to show where the taps go. Connections to it are made using vampire taps. The meaning of 10 Base 5 is that it operates at 10 Mbps, uses baseband signalling and can have maximum segment length of 500 m.

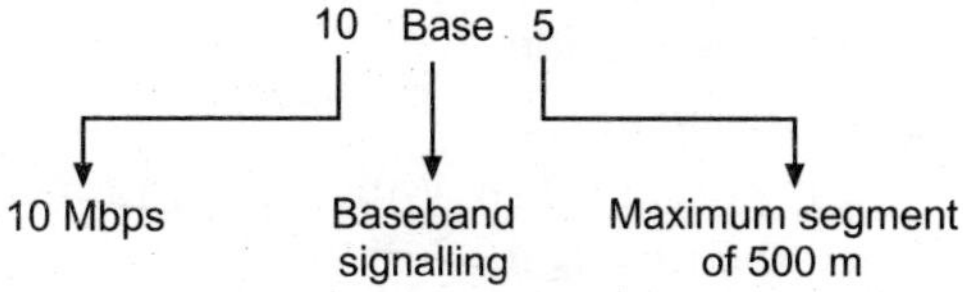

Fig. 5.34. (a) Meaning of 10 base 5

2. 10 Base 2 (Thin Ethernet): This is the second type of cable, and is called 10 Base 2 or thin Ethernet. These are easier to use and more reliable. The standard BNC connectors are used for making connections. 10 Base 2 is cheaper and easier to install. But maximum segment is only 200 m and it can handle only 30 machines per cable segment.

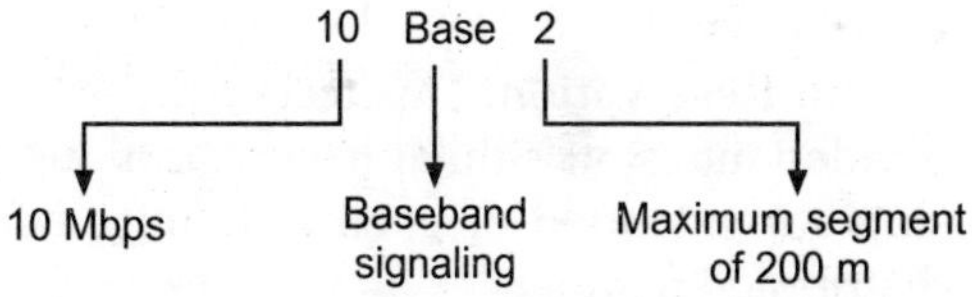

Fig. 5.34. (b) Meaning of 10 Base 2

3. 10 Base – T cabling: This uses the twisted telephone cable pair. The maximum segment length is 100 meters and it is possible to connect 1024 machines per cable segment.

5.16. MEDIUM ACCESS CONTROL (MAC)

The MAC is a protocol which controls the access to the transmission medium for an orderly and efficient use of the transmission capacity of the network.

Such a control can be exercised in two different ways :

(i) Centralized control

(ii) Decentralized control

> **DO YOU KNOW?**
>
> LAN communication speeds have recently been increased by increasing the data rate possible with all popular LAN configurations. Speed increases can also be achieved by dividing a network into segments.

(i) **Centralized Control**

In the centralized control, the controller has the authority to grant access to the network. A station who wishes to transmit its data on the network has to wait till a permission is received from the controller.

(ii) **Decentralized Control**

In a decentralized network, the stations collectively perform the medium access control function to determine the order in which the stations transmit.

5.16.1. Advantages of Centralized Control

(i) It can exercise greater control.

(ii) It can use simple access logic at each station.

(iii) It avoids the problems of distributed co-ordination.

5.16.2. Drawbacks of Centralized Control

(i) It creates a single point of failure. If this point fails, then it causes the entire network to fail.

(ii) It may act as a bottleneck to reduce the performance.

The other parameter in a Medium Access Control (MAC) is dependent on the topology and the trade off between the factors such as cost, performance and complexity. The access control techniques can be of two types namely synchronous type or asynchronous type. The synchronous technique includes the FDM and TDM techniques. The asynchronous technique can be further divided into three categories as follows :

(a) Round robin

(b) Reservation

(c) Contention.

(a) Round robin: With round robin, each station in LAN is given the opportunity to transmit. When the turn comes, the station may transmit or may not. After some time, the opportunity is given to the next station. The control of sequence in which the opportunity is given can be centralized or distributed. Polling is an example of centralized control. This technique is very efficient if many stations have data to transmit over an extended period of time. Otherwise this technique does not prove to be efficient and the other techniques should be preferred.

(b) Reservation: This technique is suitable for the stream traffic. The total time on medium is divided into slots similar to synchronous TDM. A station that wants to transmit reserves the future slots for extended or even indefinite period. The reservation may be made in a centralized on distributed manner.

(c) Contention: This technique is useful for the bursty type traffic. No control is exercised to determine which station should transmit at the given instance of time. All the stations contend for time in a very rough and tumble manner. These techniques are necessarily of distributed nature. The advantage is that these techniques are simple to implement and efficient as well. But their performance tends to degrade as the load becomes heavier. Round robin and contention techniques are the most preferred techniques in almost all types of LAN products. Table 5.3 shows the MAC protocols which are defined in the LAN standards.

Table 5.3. Standardized medium access control techniques

Round robin	Token bus (IEEE 802.4) Polling (IEEER 802.11)	Token ring (IEEE 802.5, FDDI)	Request/priority (IEEE 802.12)
Reservation	DQDB(IEEE 802.6)		
Contention	CSMA/CD(IEEE 802.3) CSMA (IEEE 802.11)		CSMA/CD (IEEE 802.3)

5.16.3. MAC Frame Format

The MAC layer receives a block of data from the LLC layer. Its job is to perform all the operations related to medium access and transmission of data. The MAC implements these functions with the help of protocol data unit at its layer. A general format of a MAC frame is shown in figure 5.35. The exact format of MAC frame differs slightly from various MAC protocols. The MAC sublayer is lower layer (Below LLC sublayer) of the data link layer of the layered architecture of LAN. The three basic functions of MAC sublayer are as under:

(i) Media access control

(ii) Error detection

(iii) Station addressing

1. MAC Service: The MAC sublayer provides connection-less mode service for the transfer of LLC PDUs.

2. MAC Frame: As mentioned earlier, the MAC frame consists of a header, data field and a trailer as shown in figure 5.35.

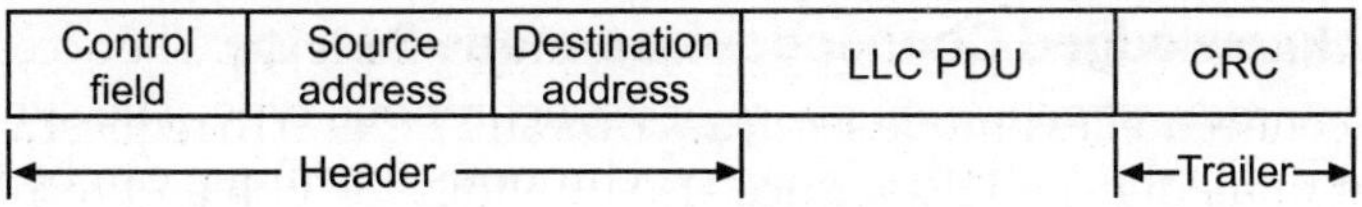

Fig. 5.35. MAC frame format

3. MAC control: The header specifies source and destination station addresses. It contains a control field as well the control field for media access control and a frame delimiter to identify start of the frame.

4. Destination MAC address: It specifies the physical destination attachment point of the device to which this frame is to be delivered.

5. Source MAC address: It specifies the physical attachment point of the source which generates this frame.

6. LLC PDU: The data field contains LLC PDU. The LLC contains the data from the next higher layer.

7. CRC: This is the cyclic redundancy check field. It is also called the Frame Check Sequence (FCS) field. This is an error detecting code. The MAC layer does the function of error detection but it does not correct them. The frame which contains error is just discarded. The LLC sublayer then notices the missing frame and requests for the retransmission of the same.

5.17. LOGICAL LINK CONTROL (LLC)/ IEEE 802.2

As we have mentioned earlier, the LLC sublayer performs the following functions :

(i) Error control

(ii) Flow control

(iii) Sequencing and user addressing functions.

It provides services to the network layer and receives services from the MAC sublayer to carry out the functions mentioned above.

5.17.1. LLC Service

The LLC sublayer provides the following three types of services to the network layer entity :

Type 1 : Unacknowledged connectionless-mode service

Type 2 : Connection-mode service

Type 3 : Acknowledged connectionless-mode service.

5.17.2. Type 1: Unacknowledged Connectionless-mode Service

As the name suggests, there is no acknowledgement of delivery or non-delivery of data units. Therefore, this is a non-reliable data link service. It just sends or receives the data units without bothering about the error control flow control or sequencing at the data link layer. Then these functions are to be performed by the higher layers.

5.17.3. Type 2 : Connection-mode Service

In this service, three functions are involved which are :

(i) To establish a connection

(ii) Data transfer

(iii) To disconnect the connection.

Flow control, error control and sequencing are the basic features of this service. This is a more reliable service as compared to the type 1 service.

5.17.4. Type 3 : Acknowledged Connectionless-mode Service

Acknowledged connectionless-mode service allows an LLC user to request an acknowledgement of data transmission immediately. This service also provides the poll and response service Polling means requesting. A user can request a protocol data unit (PDU) from a remote station.

5.17.5. LLC Protocol

As shown in figure 5.36, the LLC appends header to the protocol data unit (PDU) which is obtained from the network layer. Such an LLC PDU has been shown in figure 5.36.

The LLC PDU shown in figure 5.36 contains the addresses of the source data link layer service access point (SSAP), the destination data link layer service access point (DASP), the control field and information filed.

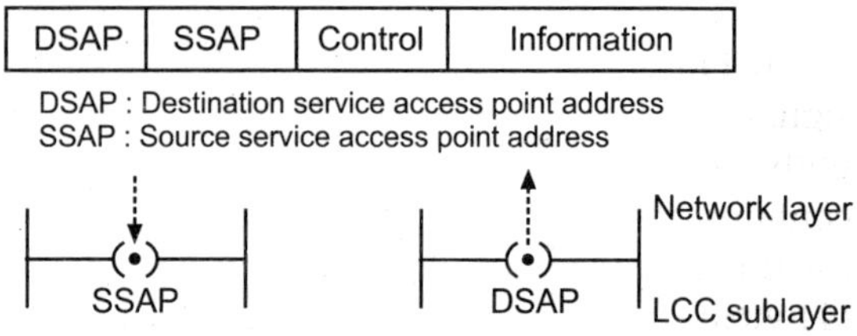

Fig. 5.36. Format of LLC PDU

5.18. LOCAL AREA NETWORK TECHNOLOGIES

There are various technologies available in the LAN market but the most popular of them is Ethernet. In the following sections, let us discuss three generations of Ethernet.

(i) Traditional Ethernet (10 Mbps)

(ii) Fast Ethernet (100 Mbps)

(iii) Gigabit Ethernet (1000 Mbps).

Traditional Ethernet was created in 1976 and has a data rate of 10 Mbps. The fast Ethernet is its next version and has a data rate of 100 Mbps. The Gigabit Ethernet operates at the data rate of 1000 Mbps.

5.18.1. Computer Connected to Internet via LAN

When a computer is connected to Internet via LAN, it needs all the five layers of the internet model. The three upper layers (network, transport and application) are common to all the LANs. The data link layer is divided into two sublayers namely the logical link control (LLC) and the medium access control sublayer (MAC). The LLC sublayer was originally designed to be the same so that all the LANs become interoperable. This means that the LANs differ only in their MAC sublayer and their physical layer. As far as the different types of Ethernets are concerned, the MAC sublayer is slightly different but the physical sublayer is quite different.

5.19. TRADITIONAL ETHERNET

The traditional Ethernet is the oldest version of Ethernet created in 1976 which is designed to operate at 10 Mbps. The access to the network by a device is through the CSMA/CD and the media are shared between all the stations.

DO YOU KNOW?

Ethernet is the most common example of a CSMA/CD network, while the IBM token-ring system is a popular token-passing network.

Medium access control sublayer

The MAC layer governs the operation of the access method (CSMA/CD). It receives the data from the upper layer, frames it and passes it to the PLS sublayer for encoding. The access method used is 1-peristent CSMA/CD.

5.19.1. Traditional Ethernet Frame

Figure 5.37 shows frame format of traditional Ethernet.

Preamble	SFD	Destination Address	Source Address	Length PDU	Data and padding	CRC
7 bytes	1 byte	6 bytes	6 bytes	2 bytes	0-46 bytes	4 bytes

Fig. 5.37. Traditional Ethernet Frame

1. Frame format: The 64–bit (8 bytes) preamble allows the receiver to synchronize with the signal, it is a sequence of alternating 0's and 1's. Both the source and destination hosts are identified with a 48-bit (6 bytes) address. The packet type field serves as the de-multiplexing key. Each frame contains upto 1500 bytes of data minimally, a frame must contain at least 46 bytes of data, the reason for this is that the frame must be long enough to detect a collision. Each frame includes 32 bit (4 bytes) checksum. The Ethernet is a bit-oriented framing protocol. An Ethernet frame has 14-byte header, two 6-bytes addresses and 2-byte type field.

The sending adapter attaches the preamble, CRC and postamble before transmitting and the receiving adapter removes them. First, an Ethernet is extremely easy to administer and maintain. There are no switches, which can fall, no routing or bath tables that have to be kept up-to-date. We can add new host easily to this network second, it is inexpensive, cable is cheap, only network adapter is little costly.

Destination address	Source address	Length PDU	Data and padding	CRC
6 bytes	6 bytes	2 bytes	46 bytes	4 bytes
64 bytes				

(a) Minimum length frame

Destination address	Source address	Length PDU	Data and padding	CRC
6 bytes	6 bytes	2 bytes	1500 bytes	4 bytes
1518 bytes				

(b) Maximum length frame

Fig. 5.38. Minimum and maximum length frame formats of traditional Ethernet

2. Start Frame Delimiter (SFD): This is the second field in the Ethernet frame and it is of 1 byte length. The byte stored at this field is 10101011. This field signals the beginning of the frame. The SDF is used to communicate to the station that is the last chance for synchronization. The last two bits 11 alert the receiver that the next field in the frame is the destination address.

5.19.2. Frame Length

There is a restriction imposed on the minimum and maximum length of the frame. The minimum frame length is 512 bits or 64 bytes and the maximum frame length is 12,144 bits or 1518 bytes. The format of the minimum length frame is shown in figure 5.38(a) and that of the maximum length frame is shown in figure 5.38(b). The restriction on the minimum length is to ensure correct operation of CSMA/CD, whereas the restriction on the maximum length is just historical.

5.19.3. Addressing

There can be various types of stations connected on an Ethernet network such as PC on workstation or printer. Each station has its own network interface card (NIC) which fits inside the station to provide the 6 byte physical address to the station. Figure 5.39 shows a 6-bytes Ethernet address in the hexadecimal notation.

04 – 02 – 01 – 06 – 1C – 5B

Fig. 5.39. Ethernet address

5.19.4. Types of Addresses

A source address is only unicast address that means the frame comes from only one station. The destination address can be of three types,

(i) Unicast

(ii) Multicast

(iii) Broadcast

Figure 5.40 shows how to differentiate between the unicast address and multicast address.

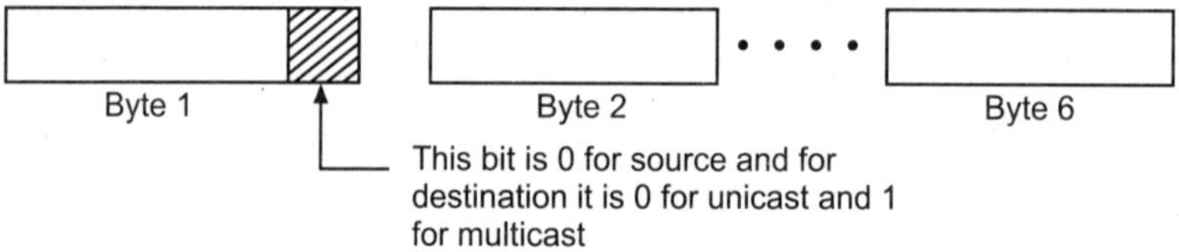

Fig. 5.40. Difference between unicast and multicast address

(i) Unicast destination address: This type of address defines only one recipient, the relation between the sender and the receiver is one-to-one.

(ii) Multicast destination address: This type of address defines a group of addresses so the relation between the sender and the receiver is one-to-many.

(iii) Broadcast address: This type of destination address is a special case of multicast address. The recipients are all the stations connected on the network.

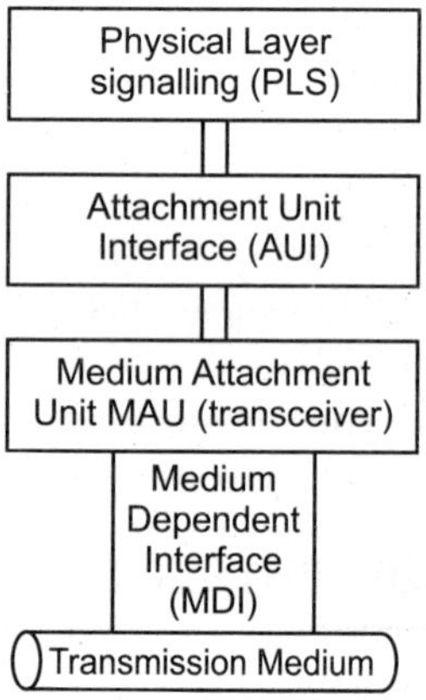

Fig. 5.41. Physical layer for traditional Ethernet

5.19.5. Physical Layer for Traditional Ethernet

The physical layer for the traditional Ethernet is shown in figure 5.41. Let us discuss the blocks one by one.

5.19.5.1. Physical Layer Signalling (PLS)

The PLS is a sublayer which is used for encoding and decoding data. The type of encoding used in the traditional Ethernet is the Manchester encoding and the data rate is 10 Mbps. Figure 5.42 Illustrates the functions of PLS sublayer.

The data from MAC sublayer is encoded using the Manchester encoder. The encoder data is passed on to the transceiver. The data from transceiver is decoded by the Manchester decoder and applied to the MAC sublayer.

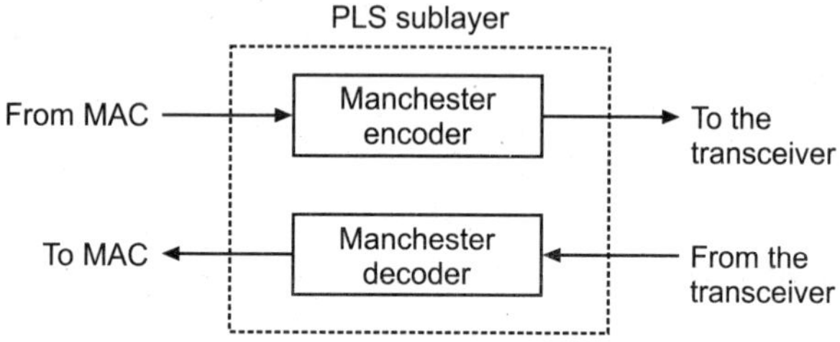

Fig. 5.42. PLS sublayer functions

5.19.5.2. Attachment Unit Interface (AUI)

AUI is a specification which is used for defining the interface between PLS and MAU. It was designed to first implement the Ethernet with thick co-axial cable.

5.19.5.3. MAU (Transceiver)

The MAU (medium attachment unit) is also called as a transceiver. It is medium dependent. It produces appropriate signals for each type of medium. Different types of mediums used in traditional Ethernet are coaxial cable, twisted pair cable, fiber optic cable. There is an MAU for each type of medium. The transceiver is a combination of transmitter and receiver. It performs the task of transmission as well as reception. It transmits the signal over the medium, it receives the signals over the medium and it also detects the collisions. All these operations have been illustrated in figure 5.43.

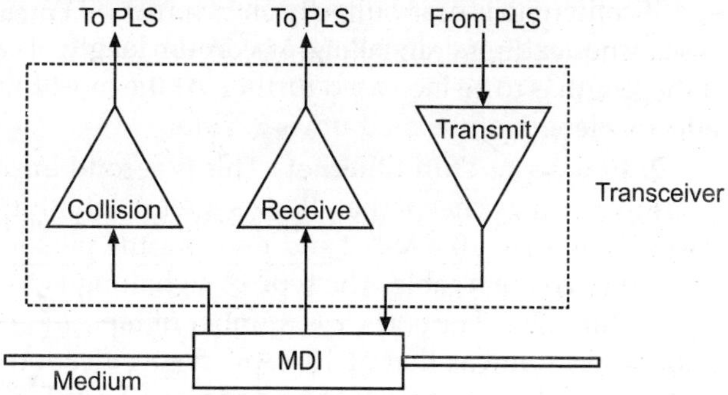

Fig. 5.43. MAU (transceiver)

5.19.5.4. Medium Dependent Interface (MDI)

In order to establish the connection between the transceiver and the medium, we need a medium dependent interface (MDI). It is a piece of hardware used for connecting the transceiver to the medium.

5.19.6. Physical Properties of Ethernet

Let us see some physical properties of Ethernet. An Ethernet segment is implemented on a coaxial cable of upto 500 m. A transceiver, which is a small device directly attached to the tap, detects when the line is idle and drives the signal when the host is transmitting. Tap must be at least 2.5 m apart. Transceiver also receives incoming signals. It is in turn, connected to an Ethernet adapter, which is plugged into the host. All the power of Ethernet is in adapter. Multiple Ethernet segments can be joined together by repeaters. A repeater is a device that forwards digital signals. Note that, no more than four repeaters may be positioned between any pair of hosts. Ethernet has a total reach of only 2500 m and it is limited to supporting a maximum of 1024 hosts with 100 base T, twisted pair.

The common configurations have several point-to-point segments coming out of a multi-way repeater, called a hub, multiple 100-Mbps Ethernet segments can also be connected by a hub.

5.20 PHYSICAL LAYER IMPLEMENTATIONS OF TRADITIONAL ETHERNET

The standard has defined four different implementations for the baseband (digital) 10 Mbps Ethernet as shown is figure 5.44.

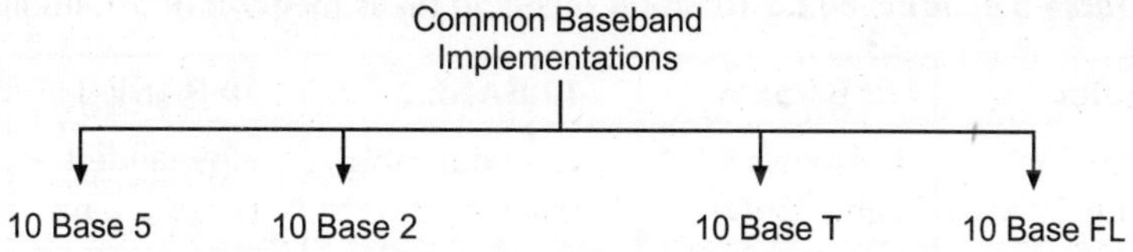

Fig. 5.44. Categories of traditional Ethernet

5.20.1. IEEE 802.3 10 Mbps Specifications (Ethernet)

IEEE 802.3 committee defines alternative physical configurations. Various defined options are as follows :

1. 10 BASE 5
2. 10 BASE 2
3. 10 BASE – T (T stands for twisted pair)
4. 10 BASE – FL (F stands for optical fiber)

All the four options stated above are for the 10 Mbps Ethernet. Table 5.4 summarizes these options.

1. 10 Base 5 : Thick Ethernet: The first implementation of the traditional Ethernet is called 10 Base 5 or thick Ethernet or thicknet. This was the first Ethernet technology. The name thicknet is due to the use of thick coaxial cable. The thicknet uses the bus topology. It is the original 802.3 medium specification and is based directly on Ethernet. 50 W co-axial cable is used. Type of signalling technique is Manchester digital signalling. Maximum length of cable segment is 500 m. We have to use repeaters if the length is to be increased further. At the most four repeaters are allowed to be used. Hence, the effective length of the medium is 2.5 km.

2. 10 Base 2 : Thin Ethernet : This is second implementation of the traditional Ethernet, and it is also known as cheapernet. It uses a relatively thin coaxial cable and bus topology. This is a low cost system than 10 BASE 2 and used for the personal computer LANs. This specification as well uses 50 W co-axial cable. The type of signalling is Manchester digital signalling. This specification uses a thin cable, supports less number of taps and specified for a distance of 185 metres only. The data rate is same as that of 10 BASE 5 specification hence it is possible to combine them. The 10 BASE 2 should not be used to bridge two 10 BASE 5 segments.

3. 10 Base-T : Twisted pair Ethernet : This is the third implementation of traditional Ethernet. It uses a physical start topology. The twisted pair cable is used instead of coaxial cable. This uses the unshielded twisted pair as medium. The type of signalling is Manchester digital signalling. The maximum segment length is reduced to only 100m. The advantage of this type is that the twisted pair wire is easily available in any building. This specification defines the star-shaped topology. As an alternative, an optical fiber link can be used. Then the maximum length becomes 500 m.

4. 10 Base FL : Fiber Link Ethernet : This is the fourth implementation of traditional Ethernet. It uses a star topology for connecting stations to a hub. The transceiver is connected to the hub by using two pairs of fiber optic cables. This standard contains three specifications as follows :

1. 10 BASE FP (P for passive)
2. 10 BASE FL (L for link)
3. 10 BASE FB (B for backbone)

All these specifications use a pair of optical fibers for each transmission link.

Manchester coding is used. Each Manchester signal element is converted into light (OFF for 0 and ON for 1). Hence, the frequency of 10 Mbps of the Manchester bit stream actually needs 20 Mbps on the fiber.

Table 5.4 . IEEE 802.3 10 Mbps physical layer medium alternatives

S. No.	Parameter	10 BASE 5	10 BASE 2	10 BASE-T	10 BASE-FP
1.	Transmission medium	Co-axial cable (50Ω)	Co-axial cable (50Ω)	Unshielded twisted pair	Optical fiber pair
2.	Signalling technique	Baseband (Manchester)	Baseband (Manchester)	Baseband (Manchester)	Manchester/ on-off
3.	Topology used	Bus	Bus	Star	Star
4.	Maximum segment length	500 m	185 m	100 m	100 m
5.	Nodes per segment	100	30	–	33
6.	Cable diameter	10	5	0.4 to 0.6	62.5/125 μm

It may be noted that all the physical layer medium alternatives mentioned in Table 5.4 are for the data rates of 10 Mbps.

5.20.2. 802.3 Encoding

None of the versions of 802.3 uses straight binary encoding because it leads to ambiguities.

Instead the bits to be sent are encoded using the Manchester encoding of different Manchester encoding.

(i) Manchester Encoding: Figure 5.45 illustrates the Manchester encoding for a binary stream of bits.

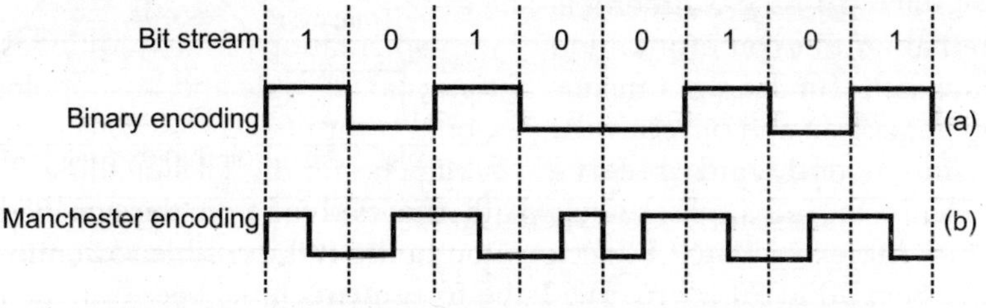

Fig. 5.45. (a) Binary encoding (b) Manchester encoding

This ensures that every bit period has a transition in the middle, making it easy for the receiver to synchronize with the sender. The disadvantage with this scheme is that it requires twice as much bandwidth as straight binary encoding, because the pulses are half the width as shown in figure 5.45.

(ii) Differential Manchester Encoding : This is the variation of the basic Manchester encoding. In this, l bit is indicated by absence of a transition at the start of the interval. A 0 bit is indicated by presence of a transition at the start of the interval as shown in figure 5.46. The advantage of 10 Base T is that it is easy to maintain.

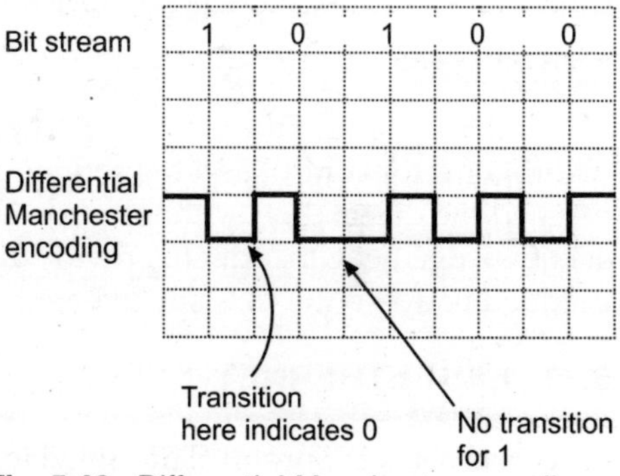

Fig. 5.46. Differential Manchester encoding

5.20.3. Different Ways of Wiring

Different ways of wiring up a building have been shown in figure 5.47.

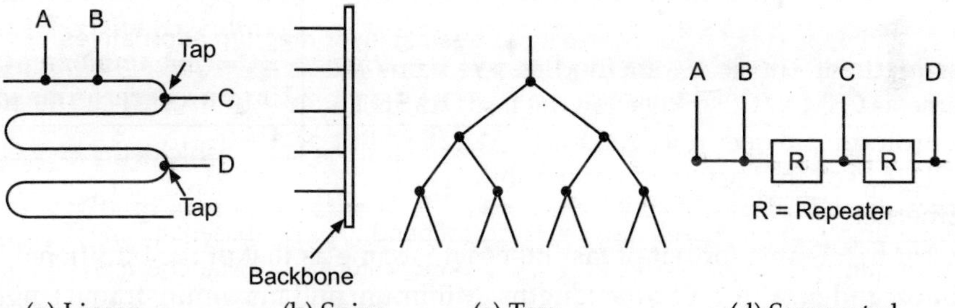

(a) Linear (b) Spine (c) Tree (d) Segmented

Fig. 5.47. Ways of cabling

The different types of cabling are as under:

(i) Linear (ii) Spine

(iii) Tree (iv) Segmented.

The type of cabling is selected depending on the application. Each version of 802.3 has a maximum length per segment (e.g. 500 m for 10 Base-5). If networks are larger, then multiple cables are to be used connected by repeaters as shown in figure 5.47 (d).

5.20.4. Bridged Ethernet

We can divide a LAN using bridges. Bridges have two effects on an Ethernet LAN. They are as follows:

(i) They increase the bandwidth.

(ii) They separate the collision domains.

5.20.5. Full Duplex Ethernet

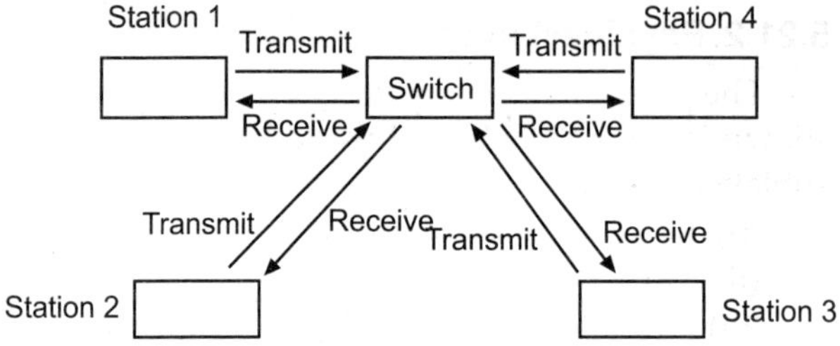

Fig. 5.48 Full duplex switched Ethernet

The 10 Base 5 and 10 Base 2 Ethernets have a serious limitation. The communication on them is always half duplex. This means that a station can either transmit or receive at a time. It cannot send and receive simultaneously. Hence, the full duplex switched Ethernet evolved from the switched Ethernet. Figure 5.48 shows the full duplex switched Ethernet. Due to the full duplex mode, the capacity of each domain increases from 10 to 20 Mbps. We have to use two links between each station and the switch, one to send and other to receive. The full duplex switched Ethernet eliminates the need of CSMA/CD because there is no more need of carrier sensing.

5.20.6. MAC

The traditional Ethernet is a connectionless protocol at the MAC sublayer. There is no flow control or error control and the sender does not get any information about whether the frame has reached the destination without error. When the receiver receives the frame, it does not send any acknowledgement. In order to provide the flow and error control, a new sublayer called MAC control is added between the LLC sublayer and MAC sublayer.

5.21. FAST ETHERNET

Fast Ethernet is the protocol designed to work upto 100 Mbps. The traditional Ethernet can operate only upto 10 Mbps. Hence for higher data rates, fast Ethernet has been developed.

1. MAC sublayer: In the evolution of Ethernet, care has been taken to keep the MAC sublayer untouched. So, MAC sublayer of the fast Ethernet is same as that of the traditional Ethernet.

2. Access method: The access method also remains the same. It is same as CSMA/CD. However, the fast Ethernet is a full duplex protocol and does not need the CSMA/CD. But, the CSMA/CD is used for backward compatibility, with the traditional Ethernet.

> **DO YOU KNOW?**
>
> Switched Ethernet, Fast Ethernet, and Gigabit Ethernet are Ethernet implementations with improved performance and data rates.

3. Frame format: Frame format of fast Ethernet is same as that of the traditional Ethernet.

4. Minimum and maximum frame lengths: Minimum and maximum frame lengths of the fast Ethernet frame are same as those of traditional Ethernet.

5. Addressing: Addressing is also same as that for the traditional Ethernet.

5.21.1. Auto-negotiation

This is the new feature of the fast Ethernet. The auto-negotiation makes the negotiation on the mode or data rate of operation between devices possible.

Features of the auto-negotiation

The important features of auto-negotiation may be listed as under:

(i) The non-compatible devices can be connected to each other.

(ii) One device can be allowed to have multiple capabilities.

(iii) A station can check hub's capabilities.

5.21.2. Physical Layer

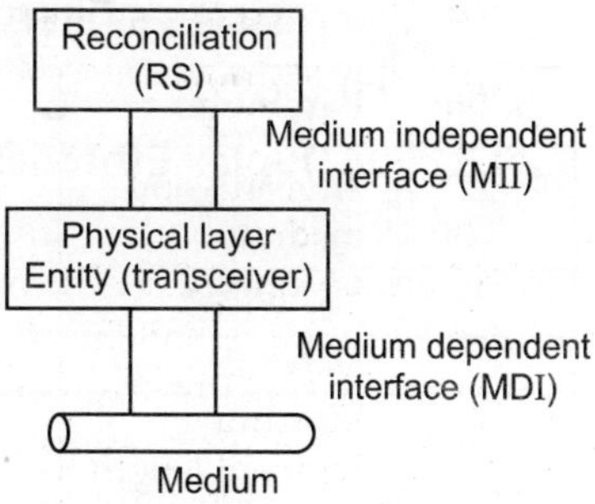

Fig. 5.49. Physical layer of fast Ethernet

The physical layer of fast Ethernet is shown in figure 5.49. As shown in figure 5.49, the physical layer consists of the following sublayers:

(i) RS i.e. reconciliation.

(ii) MII i.e. medium independent interface

(iii) Physical layer entity PHY or transceiver.

(iv) Medium dependent interface (MDI).

Let us discuss these sublayers one by one.

1. Reconciliation RS: This sublayer performs the last of passing the data in the 4-bit format to the MII.

2. MII: This is the counterpart of AUI in the traditional Ethernet. It is an improved interface and can be used for the 10 Mbps as well as the 100 Mbps data rate.

Features of MII: It can operate at 10 Mbps or 100 Mbps giving the backward compatibility with AUI. It constitutes a 4 bit parallel path between RS and PHY sublayers. It adds management functions.

3. PHY (Transceiver): PHY sublayer in the fast Ethernet is also called as the transceiver. It carries out all the basic functions mentioned for the traditional Ethernet. In addition to the basic functions here, it carries out the encoding and decoding. The transceiver can be external or internal, and it is medium dependent.

4. MDI: The medium dependent interface (MDI) is required to connect the transceiver to the medium. MDI is a piece of hardware which is specified according to implementation.

5.21.3. Physical Layer Implementation

Figure 5.50 shows the categories of fast Ethernet. As shown it can be either a two-wire or four-wire implementation. The 100 Base X is a two wire implemen-tation. It can be either a twisted pair cable (100 Base – TX) or fiber optic cable (100 Base – FX). The 100 Base T4 is a four-wire specification and it is designed only for the twisted pair cable.

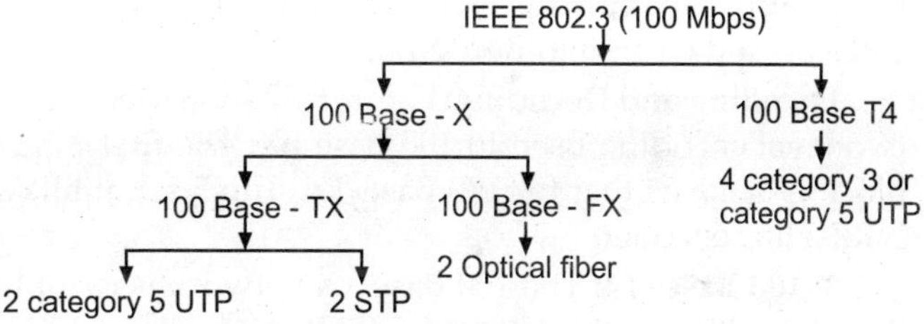

Fig. 5.50. IEEE 802.3 100 BASE–T options

All of the 100 BASE – T options shown in figure 5.50 use the IEEE 802.3 MAC protocol and frame format. 100 BASE – X indicates the options which use the medium specifications defined by FDDI. All the 100 BASE – X types use two physical links between any two nodes, one of them is used for transmission and the other one for the reception. Let us consider figure 5.50. The 100 BASE – TX uses either the shielded twisted pair (STP) or a high quality (category 5) unshielded twisted pair (UTP) whereas 100 BASE-FX uses optical fiber. There is a disadvantage of using any of the 100 BASE-FX option because a new cable needs to be installed.

Hence, 100 BASE-T4 provides a low cost option because it uses category-3 voice grade UTP or a higher quality (category 5) UTP. 100 BASE – T4 uses four twisted pair lines between any two nodes in order to achieve 100 Mbps data rate over a low quality cable. For all the 100 BASE – T options, the star topology is used.

Some important characteristics of 100 BASE – T options are summarized in Table 5.5.

Table 5.5. IEEE 802.3 100 Base –T physical layer medium alternatives

S. No.	Parameter	100 BASE–TX		100 BASE-FX	100 Base –T4
1.	Transmission medium	2 pairSTP	2 pair category 5 UTP	2 optical fibers	4 pair, category 3,4 or 5 UTP
2.	Signalling	MLT-3	MLT-3	4B5B, NRZI	8B6T, NRZ
3.	Data rate	100 Mbps	100 Mbps	100 Mbps	100 Mbps
4.	Maximum segment length	100 m	100 m	100 m	100 m

1. 100 Base –TX: We have already taken the information about the cables used. The topology is used in physical star. The transceiver can be internal or external. It is supposed to perform transmitting, receiving, detection of collision and encoding/decoding of data. The encoding and decoding is implemented in two steps. It is demonstrated in figure 5.51.

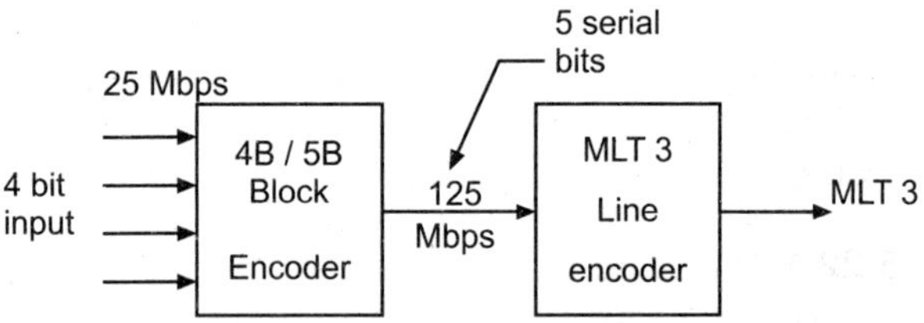

Fig. 5.51. Encoding and decoding in 100 Base TX

The block encoding is performed first. The 4 parallel input bits received from NIC are encoded into 5 serial bits. Hence, the block encoder output is at 125 Mbps. It is then applied to a MLT line encoder. The MLT 3 signal equivalent to the 4 bit input is thus obtained.

2. 100 Base–FX: The 100 Base-FX uses two pairs of fiber-optic cables. They are connected in the physical star topology. The transceiver can be either external or internal. The transceiver is supposed to carry out transmitting sending, detecting the collision and encoding/decoding.

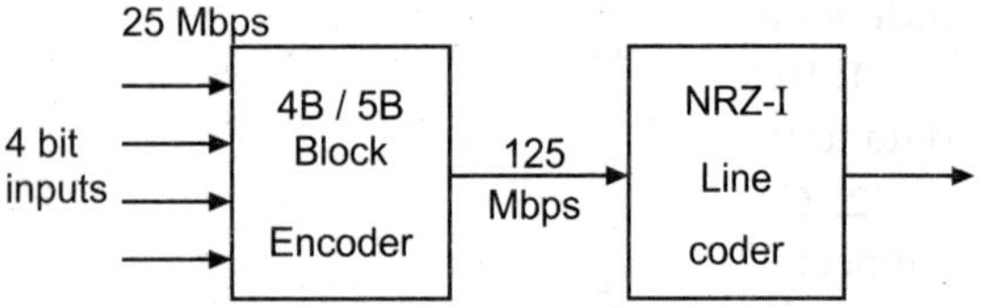

Fig. 5.52. Encoding and decoding in 100-Base-FX

Encoding and Decoding: Figure 5.52 shows the two level encoding used in 100 Base FX. The first block is same as that for 100 Base-TX. The 5 serial bits are converted into NRZ-I signal using the NRZ-I line encoder.

3. 100 Base–T4: The 100 Base TX network is capable of providing the data rates of 100 Mbps. But, it need to use the category 5 UTP or STP cable, which are expensive. 100 Base T4 can support the 100 Mbps rate using the category 3 or higher UTP cable. Four pairs of UTP are used for transmitting 100 Mbps. The transceiver is similar to the other implementations.

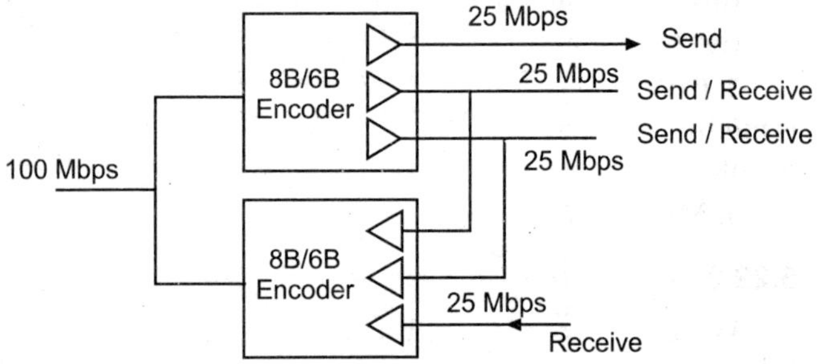

Fig. 5.53.

Encoding and decoding: The encoding and encoding of 100 Base T4 is illustrated in figure 5.53. The 100 Mbps input is applied to two 8B/6T encoders. They reduce the bandwidth from 100 to 75 Mbps due to the ratio 8/6. But, even the 75 Mbps rate cannot be supported by a voice grade UTP. So, 100 Base – T4 network is designed to operate on 25 M baud bandwidths as shown in figure 5.53. For transmission (which is unidirectional), six pairs of cable are required. To reduce the number of pairs to four, two pairs are used for unidirectional transmission and the other two are used for bi-directional transmission. The two unidirectional pairs can carry the collision signals because they are always free in one direction.

5.22 GIGABIT ETHERNET

The Gibabit Ethernet protocol has been designed in order to support the data rates upto 1000 Mbps. The MAC layer was supposed to remain unchanged throughout the evolution of the Ethernet but it does not remain so when the rate of 1 Gbps is to be supported. The Gigabit Ethernet is capable of operating in either half duplex or full duplex modes. If it operates in the half duplex mode, then the access method used is CSMA/CD. But, if the full duplex mode is used then CSMA/CD is not required. Almost all the implementations in Gigabit Ethernet use the full duplex mode.

Key Point: Just as the dust was beginning to settle on the Fast Ethernet standard, the IEEE, in March 1996, commissioned the Higher Speed Study Group (HSSG) to investigate increasing Fast Ethernet's data transmission rate tenfold to 1000 Mbps. That's one billion bits per second.

5.22.1. Physical Layer

The physical layer of Gigabit Ethernet is shown in figure 5.54. It consists of four sublayers, namely reconciliation, GMII, PHY and MDI. The sublayers are of two types. First the medium dependent sublayers and then the medium independent sublayers. The reconciliation (RS) sublayer is common for all the implementations. The PHY and MDI are medium dependent. GMII is medium independent.

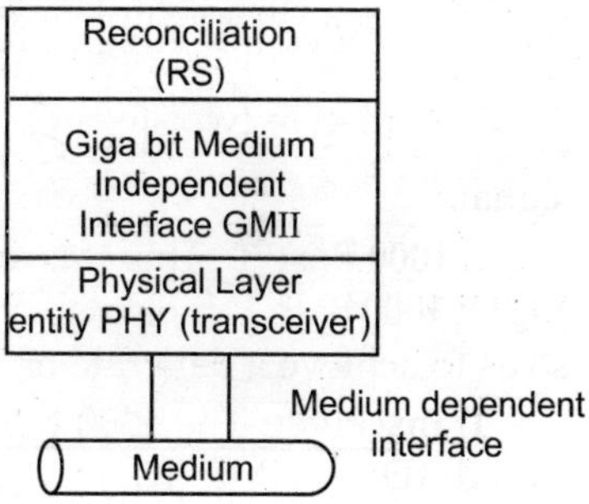

Fig. 5.54. Physical layer in Gigabit Ethernet

1. Reconciliation sublayer (RS): The RS sends an 8-bit parallel data to the PHY sublayer (transceiver) via a GMII interface.

2. GMII sublayer: This layer specifies how the RS will be connected to PHY (transceiver) sublayer. GMII is never an external physical component. It is always an internal one. So, it is not a physical interface, rather, it is a logical interface.

Features of GMII sublayer

(i) It can operate only at 1000 Mbps.

(ii) It establishes an 8-bit parallel path between RS and transceiver.

(iii) Management functions are included.

(iv) No GMII cable or connector.

3. PHY sublayer (transceiver) : Similar to the fast Ethernet, the transceiver in the Gigabit Ethernet is medium dependent and it encodes as well as decodes. The Gigabit Ethernet transceivers can only be internal since there is no external GMII for providing the connection.

4. MDI: It is used for connecting the transceiver to the medium.

5.22.2. Physical Layer Implementation

We can categorize the Gigabit Ethernet as either a two-wire or a four-wire implementation. The two-wire implementation is known as 1000 Base X and the four-wire implementation is known as 1000 Base-T. The four-wire implementation uses twisted pair cable. Figure 5.55 shows the physical layer implementations for the Gigabit Ethernet.

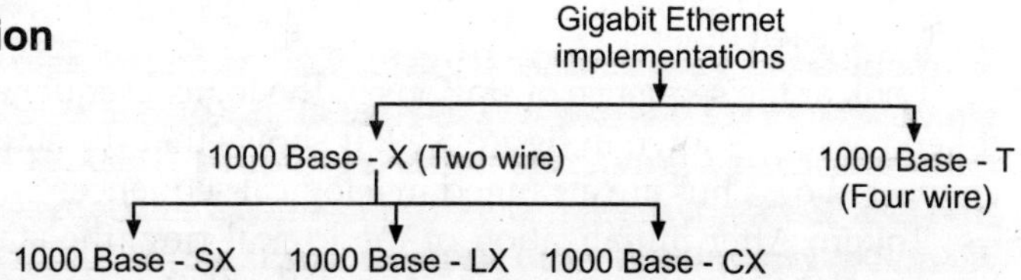

Fig. 5.55. Physical layer implementations of Gigabit Ethernet

1. 1000 Base X: The 1000 Base-X is further classified into three categories, out of which the 1000 Base–SX and 1000 Base LX use two fiber optic cables. The 1000 Base –SX uses a shortwave laser whereas 1000 Base – LX uses the longwave laser. The 1000 Base – CX implementation was supposed to use STP cable. But, it is not yet implemented in practice.

Transceiver for 1000 Base X: A transceiver in Gigabit Ethernet is internal and it is supposed to carry out the functions such as encoding and decoding, transmission and detection and collision detection.

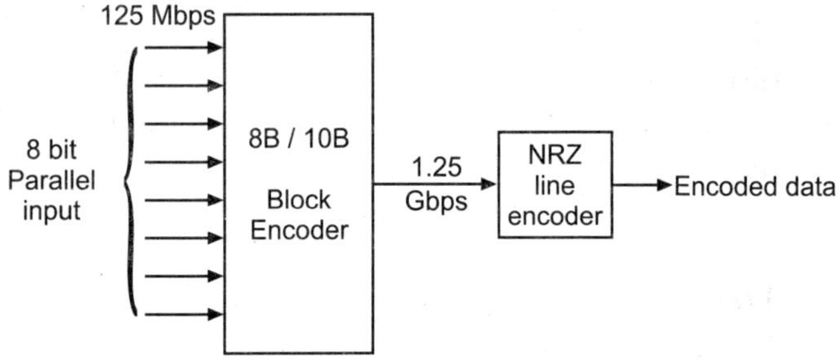

Fig. 5.56. Encoding in 1000 Base X

Encoding: Figure 5.56 shows the encoding in 1000 Mbps Ethernet. It takes place in two steps. The encoder first performs the block encoding in order to maintain synchronization. The 8 parallel bits are encoded into 10 serial bits using the 8B/10B block encoder. Due to the ratio 8/10, a bandwidth of 1.25 GHz is required. This data (at 1.25 Gbps) is then encoded by the NRZ line encoder to produce the encoded signal.

2. 1000 Base T: This was designed for using the category 5 UTP. 1000 Base T implementation uses for twisted pair cables so as to achieve a data rate of 1 Gbps.

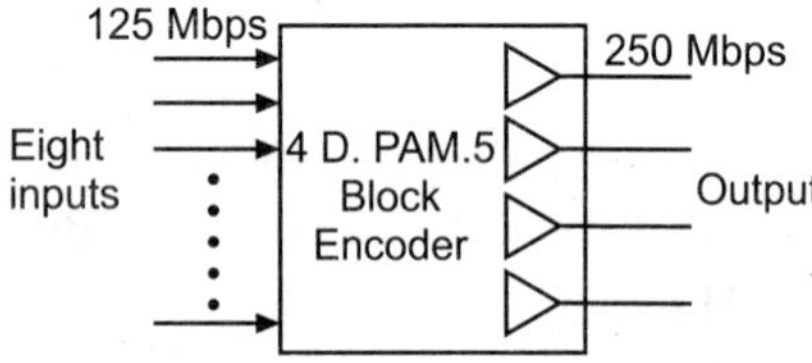

Fig. 5.57. Encoder for 1000 Base T

Transceiver: The 1000 Base T uses an encoding scheme called 4D – PAM5. It is a 4 – dimensional, 5 level pulse amplitude modulation. Five levels of pulse amplitude modulation are used. Figure 5.57 shows the encoder for 1000 Base T.

5.23 TOKEN BUS : IEEE 802.4*

The IEEE 802.4 standard for media access control is known as **Token bus**. A token bus network is characterized as a logical ring on a physical bus. Physically the network resembles a bus topology, but logically it is arranged as a ring with respect to passing the token from lobe to lobe. Token bus is a linear or tree-shaped cable through which different stations are interconnected. Logically, the interconnected stations form a ring as shown in figure 5.59. The physical topology has been shown in figure 5.58. Each station knows its own identification number and the identity of the stations preceding and following it. The sequence number and the physical location of a station on the bus are not related to each other.

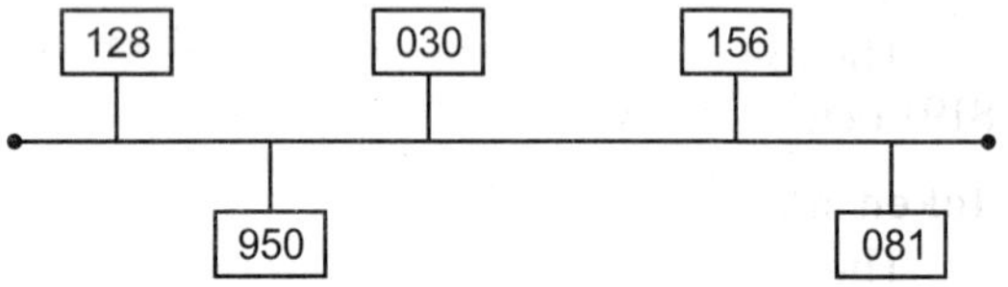

Fig. 5.58. Physical topology in token passing

Look at the sequence of stations in the logical sequence of token passing, shown in figure 5.59. It shows that the stations connected on a bus are arranged in a logical sequence.

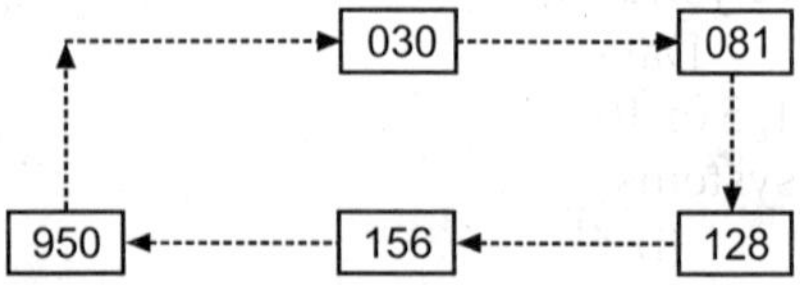

Fig. 5.59. Logical sequence of token passing

Token: After initialization of the logical ring, the station bearing the highest number sends out the first frame. After doing so, it passes a permission to its neighbouring station that

* Another token passing protocol LAN is token bus, which is defined in IEEE 802.4.

now the neighbouring station can send its frame. This permission is passed by sending a special control frame called *Token*.

Media access control: The operation of token bus takes place as follows:

(i) At any time, the station which holds the token only can transmit its data frames on the bus. Every frame contains source and destination addresses.

(ii) All other stations are ready to receive these data frames.

(iii) As soon as the transmission time of a station is over, it passes the token to the next station in the logical sequence. The transmission is then taken over by the next station.

(iv) In one cycle of operation, each situation gets an opportunity to transmit once. The same station can get more number of chances to transmit in one cycle if more than one addresses are assigned to it.

Frame format: The frame format as specified by IEEE 802.4 is as shown in figure 5.60.

Number of bytes →	1(min)	1	1	2-6	2-6		4	1
	Preamble	SD	FC	DA	SA	DATA	FCS	ED

Preamble	:	Bit synchronization	SA	:	Source address
SD	:	Frame Start Delimiter	DATA	:	Data field
FC	:	Frame Control (Type)	FCS	:	Frame check sequence
DA	:	Destination Address	ED	:	End delimiter

Fig. 5.60. Format of IEEE 802.4 frame

The frame consists of following fields :

(i) **Preamble :** Preamble is at least one octet long and used for bit synchronization.

(ii) **Start Delimiter (SD) :** It is a one byte unique pattern which marks the beginning of a frame.

(iii) **Frame Control (FC) :** This field indicates the type of frame. It is one octet long and indicates if the frame is a data frame or control frame. Token is one of the control frames.

(iv) **Destination Address (DA) :** It contains the destination address and it is 2 to 6 bytes long.

(v) **Frame Check Sequence (FCS) :** This field contains a CRC code. It is 4 byte long and used to check on DA, SA, FC and Data fields.

(vi) **End Delimiter (ED) :** This a one byte unique bit pattern which marks the end of the frame.

The total length of the frame from FC to FCS field including the Data field can be, at the most, 8191 octets.

Token management

The active stations control and manage the token. Each one of them can initiate and respond to the control frames such as claim token frame, solicit successor frame, set successor frame and who follows frame. The function of these frames is to initialize the bus and for adding or removing a station.

> **DO YOU KNOW?**
>
> Token Bus (IEEE 802.4), used in factory automation and process-control, combines features of Ethernet and Token Ring.

Physical specifications

Data rates at which a token passing LAN operates can be 1, 5 or 10 M bits/sec using analog signalling over 75 ohm co-axial cable. Two types of transmission systems are used:

(i) Carrierband (single channel)

(ii) Broadband (multiple channel)

Both these systems use some kind of modulation to reduce the effect of noise. The carrier band system FSK (frequency shift keying) modulation is used. It is a bi-directional transmission system. 1 M bits/sec bus is implemented using a flexible semi-rigid coaxial cable. The more expensive versions of the bus can operate at 5 to 10 M bits/sec. The broadband system uses a unidirectional transmission. It uses a combination of the phase and amplitude modulation. Separate carriers are used for the transmit and receive directions. As this is a multiple channel system, there will be several transmit and receive carriers. Each carrier provides a data rate of 5 or 10 M bits/sec. Broadband LANs can cover a span of several kilometres.

5.23.1. Bus LANs

We have already seen the basic structure of the bus topology of LAN. In this section, we shall provide some technical details of the bus topology.

5.23.2. Characteristics of the Bus Topology

Following are some of the important characteristics of the bus topology :

(i) This is a multipoint configuration. There are more than two devices connected to the medium and they are capable of transmitting on the medium. Hence, the medium access control is essential for bus topology.

(ii) The signal strength of the transmitted signal should be adequately high so as to meet the minimum signal strength requirements of the receiver.

(iii) Adequate signal to noise ration should be maintained.

(iv) The signal should not be too strong. This is necessary to avoid the overloading of transmitter and hence the possibility of signal distortion.

(v) This is called signal balancing which is not an easy task at all. Specially, the signal balancing becomes increasingly difficult with increase in the number of stations.

(vi) The solution to this problem is to divide the transmission medium into small segments and within each such segment, a pair-wise balancing is done using amplifiers or repeaters between the segments.

5.23.3. Transmission Media for Bus LANs

We can use the following transmission media for the bus LANs :

(i) Twisted pair
(ii) Baseband coaxial cable
(iii) Broadband coaxial cable
(iv) Optical fiber.

5.23.4. Twisted Pair

The twisted pair was used in early days of LANs. It could support the data rates of upto 1 Mbps only. Hence, now a days, it is not being used.

5.23.5. Baseband Co-axial Cable

The baseband co-axial cable is the one that makes use of digital signalling the original Ethernet scheme makes use of baseband coaxial cable.

Characteristics of baseband co-axial cable

The baseband coaxial cables are used for digital signalling. The digital signal inserted on these cables is encoded using Manchester or Differential Manchester coding. The digital signal consumes the entire frequency spectrum of the cable. So, it is not possible to transmit multiple channel using FDM. The transmission of digital signal on the cable is bi-directional. The baseband coaxial cable was originally used for the Ethernet system that operates at 10 Mbps. These cables have a characteristic impedance of 50 Ω rather than 75 Ω of the cable TV coaxial cables.

This because for 50 Ω cables the reflections taking place are less intense, they provide better noise immunity against low frequency electromagnetic noise as compared to the 75 Ω cable. The maximum length of baseband coaxial cable between two repeaters is dependent on the data rates. Lower the data rate longer is the cable. The length has to be reduced with increased data rates so as to reduce the probability of errors getting introduced. There are two systems that use the bus LANs, namely 10 BASE 5 and 10 BASE 2 which are compared based on various factors in Table 5.6.

Table 5.6. IEEE 802.3 specification for 10 Mbps baseband coaxial cable Bus LAN

S.No.	Parameter	10 BASE 5	10 BASE 2
1.	Data rate	10 Mbps	10 Mbps
2.	Maximum segment length	500 m	185 m
3.	Network span	2500 m	1000 m
4.	Nodes per segment	100	30
5.	Node spacing	2.5 m	0.5 m
6.	Cable diameter	1 cm	0.5 cm

5.23.6. Broadband Coaxial Cable

This is the co-axial cable which is used in the cable TV system. The type of signalling is analog at radio frequencies. This system has certain disadvantages such as it is more expensive, more difficult to install and maintain, as compared to the baseband co-axial cable. IEEE 802.3 standards have specified this as an option but practically the broadband co-axial cables are not popular.

5.23.7. Optical Fiber

Even this medium is not practically used because of the expense of the optical fibre taps and the availability of better alternatives. Hence, for the bus topology, the medium used is baseband coaxial cable, primarily for Ethernet and IEEE 802.3 systems.

5.23.8. Repeaters

It is possible to extend the length of the network by using repeaters. The bus repeaters are special type devices which can transmit in both the directions. A repeater joins two segments of cables and can pass digital signals in both the directions. If two stations on different segments are transmitting at the same time then their packets is likely to collide. To avoid this, only one path of segments and repeaters is allowed between any two stations. Figure 5.61 shows a multiple segment baseband bus LAN.

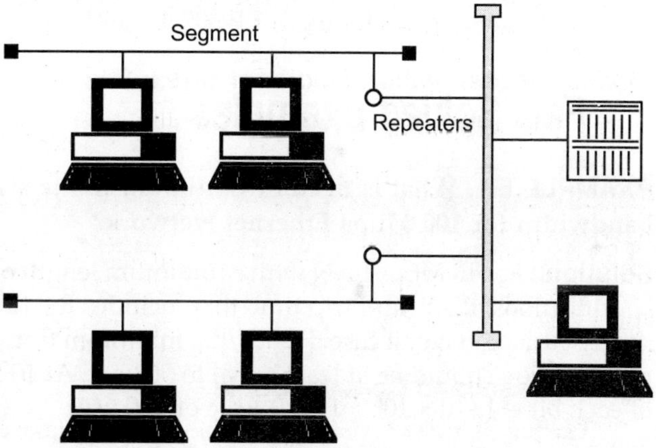

Fig. 5.61. Multiple segment baseband bus LAN

Use of BNC barrel connector

To connect the coaxial cable to device, we need to use co-axial connectors. The most common type of connector used is Bayone-Neill-Concelman or BNC connector. The three types of BNC connectors are :

(i) BNC connector

(ii) BNC T connector

(iii) BNC terminator.

The BNC connector is used to connect the end of the cable to a device such as TV set. The BNC T connector is used in Ethernet networks, for branching out a cable for connection to computer or other devices. The BNC terminator is used of the end of cable to avoid reflection of the signal.

Advantages of bus LANs

(i) The bus topology is easy to understand, install, and use for small networks.

(ii) The cabling cost is less as the bus topology requires the least amount of cable to connect the computers.

(iii) The bus topology is easy to expand by joining two cables with a BNC barrel connector.

(iv) In the expansion of a bus topology repeaters can be used to boost the signal and increase the distance.

DO YOU KNOW?

Token Ring (IEEE 802.5) employs token passing as its method of transmission initiation.

Disadvantages of bus LANs

(i) Heavy network traffic slows down the bus speed. In bus topology only one computer can transmit and other have to wait till turn comes and there is no co-ordination between computers for reservation of transmitting time slot.

(ii) The BNC connectors used for expansion of the bus attenuates the signal considerably.

(iii) A cable break or loose BNC connector will cause reflections and bring down the whole network causing all network activity to stop.

Important Point:

Ethernet 10 base 2 also known as thinnet is an inexpensive network based on the bus topology. A bus network behaves erratically if it is not terminated or improperly terminated. Token bus networks are defined by the IEEE 802.4 standard.

Few Solved Examples

EXAMPLE 5.1. What is need of minimum frame size ? If is 64 byte then calculate the loss of bandwidth for 100 Mbps Ethernet Network?

Solution : For 10 Mbps LAN with a minimum length of 2500m and four repeaters as per IEEE 802.3 specifications, the round trip time that includes the time to propagate through the four repeater is nearly 50 μsec in worst case. Hence, the minimum frame length must be such that the time taken for transmission should be at least equal to 50 μsec. At 10 Mbps (for traditional Ethernet) the duration of each bit is $1/10 \times 10^6 = 1 \times 10^{-7}$ sec or 100 nsec.

So 50 μsec will correspond to $\dfrac{50\times10^{-6}}{100\times10^{-9}} = 500$ bits.

Hence, the minimum frame size was decided to be of 500 bits. But in practice, to add some margin of safety this number was rounded to 512 bits or 64 bytes.

Hence, the minimum frame size in traditional Ethernet is 64 bytes. As the speed of the network increases, the minimum frame length should go up or the maximum cable length should proportionally reduce. Thus, for the 100 Mbps Ethernet, for the same maximum cable length, the minimum frame length should be 640 bytes. This is calculated as follows :

Minimum time for transmission of a frame = 50 μs.

Duration of each bit (at 100 Mbps) = $\dfrac{1}{100\times10^6}$ 1×1^{-8} sec.

Therefore, number of bits corresponding to 50 μs is $\frac{50\times10^{-6}}{1\times10^{-8}}$ = 5000 bits.

Round it to 5120 bits or 640 bytes.

Hence, minimum frame size for 100 Mbps Ethernet is 640 bytes. **Ans.**

Loss of Bandwidth

All the Ethernets use the Manchester encoding due to its simplicity. But disadvantage of Manchester encoding is that it requires twice as much bandwidth as the straight binary encoding.

$$\text{Bandwidth} = \frac{1}{2}\times \text{Bit Rate}$$

Hence, for a 10 Mbps Ethernet.

$$\text{Bandwidth} = \frac{1}{2}\times 2\times 10 \text{ MHz}$$

Here, it may be noted that multiplication by 2 is for Manchester encoding.

Therefore, BW = 10 MHz

If the frame size is 64 bytes for 100 Mbps Ethernet then 640 – 64 = 576 bytes are transmitted less.

576 bytes = 4608 bits less

If 640 bytes = 100 MHz bandwidth.

576 bytes = ?

$$\frac{640}{576} = \frac{10\text{MHz}}{x}$$

or $$x = \frac{576\times100}{640} = 90 \text{ MHz}$$

Hence the loss of Bandwidth = 90 MHz.

EXAMPLE 5.2. Fast Ethernet has the same 64-byte minimum frame size but can get bits out ten times faster. How is it possible to maintain the same minimum frame size ? How much bandwidth loss will occur in this case?

Solution: If the fast Ethernet has the same minimum frame size of 64 bytes i.e. 512 bits, as that of the traditional Ethernet, then during the minimum frame transmission time of 50 μsec, at 100 Mbps rate, the number of transmitted bits will be ten times more as compared to the traditional Ethernet.

The calculation follows :

Minimum time required to transmit a frame = 50 μs.

Duration of each bit at 100 Mbps is $\frac{1}{100\times10} = 1\times10^{-8}$ sec.

Therefore, number of bits transmitted during 50 μsec is

$$x = \frac{50\times10^{-6}}{1\times10^{-8}} = 5000 \text{ bits}$$

If we round them to 5120 bits or 640 bytes then it is ten times higher than that in a traditional Ethernet.

The same minimum frame size can be maintained by reducing the maximum cable length proportionally. That means maximum cable length should be 250 m for the fast Ethernet with 64 byte frame length.

EXAMPLE 5.3. Evaluate ring latency of 20 stations separated by 100 meters and operate at a speed of 4 Mbps. Assume the delay introduced by each station to be 2.5 bit.

Solution: Given that,

Number of station N = 20

Length of ring d = 100 m

Propagation speed V = 2×10^8 m/sec.

Rate of transmission R = 4 Mbps

Delay introduced by each station = b = 2.5 bits

(i) First, let us calculate the total delay.

$$\text{Delay introduced by N stations} = \frac{N \times b}{R} = \frac{20 \times 2.5}{4 \times 10^6} = 12\ \mu\text{sec.} \qquad \text{... (i)}$$

$$\text{Also, additional delay introduced by the ring is given by} = \frac{d}{V}$$

$$= \frac{100\ \text{m}}{2 \times 10^8\ \text{m/s}} = 0.5\ \mu\text{s.} \qquad \text{... (ii)}$$

Therefore, Total delay = 12.5 + 0.5 = 13 μsec.

(ii) Next, let us evaluate calculate the ring latency.

Ring latency is defined as the number of bits that can be simultaneously in transit around the ring.

Therefore Ring latency = Total delay × Rate of transmission

$= 13 \times 10^{-6} \times 4 \times 10^6 = 52$ bits. **Ans.**

EXAMPLE 5.4. ALOHA protocol is used to share 56 kbps satellite channel. If each packet is 1000 bits long, find maximum throughput in packets/sec.

Solution: Given

Rate of transmission = 56 kbps = 56000 bps

Frame length = 1000 bits

1. For pure ALOHA

$$\text{Number of frames/sec} = \frac{56000\ \text{bits}}{1000\ \text{bits/frame}} = 56\ \text{frames/sec}$$

The maximum throughput for pure ALOHA = 0.184

Hence, Throughput = 56 × 0.184 = 10.304 frames/sec **Ans.**

2. For slotted ALOHA

Maximum throughput = 0.368

Hence, throughput = 0.368 × 56 = 20.608 Frames/sec

EXAMPLE 5.5. A group of N users share 56 kbps pure ALOHA channel. Each station outputs 1000 bits frame on an average of once 100 sec. Even if the previous has not yet been sent (buffered) what is maximum value of N.

Solution: For pure ALOHA

The maximum throughput = 0.184

Hence, the maximum usable channel bandwidth is given by,

R = 0.184 × 56 kbps = 10.3 kbps

$$\text{Transmission rate of stations} = \frac{1000 \text{ bits}}{100 \text{ sec}} = 10 \text{ bits/sec.}$$

Let N be the number of stations that can use the channel.

$$\text{Therefore,} \quad N = \frac{R}{10 \text{ bits/sec}} = \frac{10.3 \text{ kbps}}{10} = 1030$$ **Ans.**

EXAMPLE 5.6. An ALOHA network user 19.2 kbps channel for sending message packets of 100 bit long size. Calculate the maximum throughput for pure ALOHA network.

Solution: Given

Rate of transmission = 19200 bits

Frame length = 100 bits

$$\text{Hence, number of frames per second} = \frac{\text{Rate of transmission}}{\text{Frame length}} = \frac{19200}{100} = 192 \text{ frames/sec}$$

The maximum throughput for a pure ALOHA system is 0.184.

Therefore, throughput = 0.184 × Number of frames/sec.

= 0.184 × 192 = 35.328 frames/sec **Ans.**

EXAMPLE 5.7. Using the binary count down protocol, find the highest priority station. The station addresses are as follows:

Station	Address
A	0010
B	0100
C	1010
D	1001
E	1011

Solution:

1. All stations broadcast their MSBs

Station	A	B	C	D	E
MSB	0	0	1	1	1

Stations A and B will give up.

2. Stations C, D, E broadcast their next bit

Station	C	D	E
Next bit	0	0	0

3. Stations C, D, E broadcast their next bit

Station	C	D	E
Next bit	1	0	1

Station D will give up.

4. Station C and E broadcast their LSBs

Station	C	E
LSB	0	1

Station C will give up.

Hence, station E has the highest priority. Its station address is 1011.

EXAMPLE 5.8. Measurement of slotted ALOHA channel with an infinite number of users show that 20% slots are idle.

1. **What is the channel load?**
2. **What is the throughput?**
3. **Is the channel underload or overload? Show with graph.**

Solution: Given $P_0 = 20\%$ i.e., 0.2, Type : slotted ALOHA

To find:

1. Channel load G
2. Throughput S.
3. Decide the status of the channel

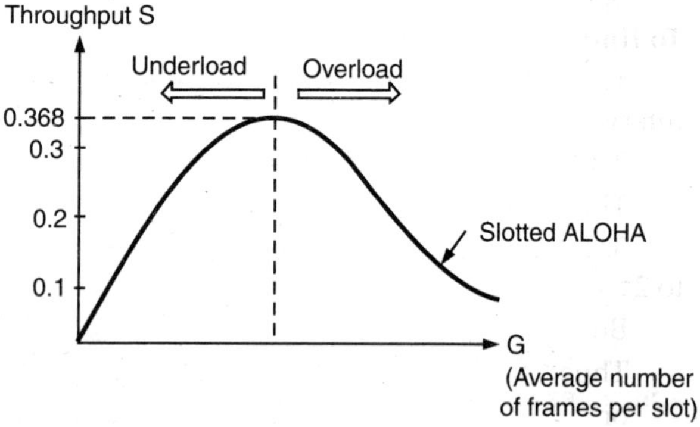

Fig. 5.62. Graph for slotted ALOHA

1. Channel load (G)

For the slotted ALOHA, we have

$$P_0 = e^{-G}$$

Thus, $0.2 = e^{-G}$

Hence, $G = 1.6094$ **Ans.**

2. Throughput (S)

We have $S = Ge^{-G}$

$= P_0G = 0.2 \times 1.6094$

Thus, $S = 0.3218$ **Ans.**

3. Status of the channel

From figure 5.62, it is evident that the maximum througput $S_{max} = 0.368$ corresponds to $G = 1$. Since the value of of $G = 1.6094$ which is greater than 1, the channel is overloaded.

EXAMPLE 5.9. A large population of ALOHA users manages to generate 50 requests/sec, including both originals and retransmissions. Time is slotted in units of 40 msec.

(a) What is the chance of success on the first attempt?

(b) What is the probability of exactly k collisions and then a success?

(c) What is the expected number of transmission attempts needed?

Solution:

1. There are 50 requests/sec and time is slotted in units of 40 msec.

1 sec = 50 requests (transmissions)

Hence, 40 msec = x transmissions

Thus, $$\frac{1}{40 \times 10^{-3}} = \frac{50}{x}$$

or $x = 50 \times 40 \times 10^{-3}$

or $x = 2$

2. But, number of transmissions (x) = e^G

Thus, $2 = e^G$ or $G = 0.693$

3. Probability of k collisions and then a success is

$$P_k = e^{-G} (1 - e^{-G})^{k-1}$$

Thus, $P_k = e^{-0.693} (1 - e^{-0.693})^{k-1}$

4. Chance of success in the first attempt is $G e^{-G}$.

i.e., $0.693 e^{-0.693} = 0.3465$ or 34.65% **Ans.**

EXAMPLE 5.10. Consider building a CSMA/CD network running at 1 Gbps over a 1 km cable with no repeaters. The signal speed in the cable is 2,00,000 km/sec, what is the minimum frame size?

Solution: Given:

Bit rate $R = 1 \times 10^9$ Bits/sec, No repeaters used.

Length $L = 1 \text{ km} = 1 \times 10^3$ m

Speed $v = 2,00,000 \text{ km/s} = 2 \times 10^8$ m/s

To find: Minimum frame size

1. Let the time for a signal to propagate between two farthest stations be τ. The contention interval is such that width of each slot is 2τ.

2. On a 1 km long cable, $\tau \cong 5\ \mu$ sec.

Thus, $2\tau = 10\ \mu$sec.

3. To make CSMA/CD work, it must be ensured that the minimum frame size should be equal to $2\tau = 10\ \mu$ sec.

But $R = 1 \times 10^9$ bits/sec

Thus, $1 \text{ sec} \equiv 1 \times 10^9$ bits

or 10×10^{-6} sec = ? bits

Therefore, $$\frac{1}{10 \times 10^{-6}} = \frac{1 \times 10^9}{x}$$

Thus, $x = 1 \times 10^9 \times 10 \times 10^{-6} = 10 \times 10^3 = 10,000$ bits.

Therefore, minimum frame size = 10,000 bits or 1250 bytes.

GLOSSARY

- **Bluetooth** An emerging wireless convergence technology which has been designed to interconnect various devices, including computers, mobile phones, mobile computers, and hand-held or portable terminals using short-range (approximately 10 m) radio links.
- **Carrier Sense Multiple Access (CSMA)** A protocol that serves as the basis for different random access protocols. CSMA-based protocols include I-persistent CSMA, non-persistent CSMA, CSMA with Collision Detection (CSMA/CD) and CSMA with Collision Avoidance (CSMA/CA).
- **Carrier Sense Multiple Access with Collision Avoidance (CSMA/CA)** A variant of CSMA/CD except that it specifies an implementation technique for collision avoidance instead of collision detection.
- **Carrier Sense Multiple Access with Collision Detection (CSMA/CD)** A variant of either I-persistent or non-persistent CSMA that specifies what a node is to do upon detecting a collision. I-persistent CSMA/CD is the MAC sublayer protocol that is used in Ethernet 802.3 LANs.
- **Carrier Sense Protocol** A network protocol that requires nodes to listen (sense) for the sound of another node's transmission prior to accessing a shared channel.
- **Collision** What happens when two or more nodes try to transmit data simultaneously on an Ethernet/802.3 network.
- **Collision Domain** A field within a single Ethernet/802.3 network where two nodes can cause a collision. In case of a single segmented Ethernet/802.3 LAN, an independent segment represents the collision domain; in a multi-segmented Ethernet/802.3 LAN, the collective segments comprise the collision domain.
- **Contention Protocol** A network protocol which specifies access to the same communication channel at the same time. It is also called Random Access Protocol.

- **Copper Distributed Data Interface (CDDI)** An interface which provides a 100 Mbps data transmission rate over copper. A CDDI network is quite similar to an FDDI network. CDDI also is restricted to connections between concentrators on the ring and single-attachment devices, not for the ring itself.
- **Distributed Queue Dual Bus (DQDB)** A data link layer protocol (IEEE 802.6) which specifies the medium access method for MANs. It is used in SMDS.
- **Ethernet** A local area network protocol developed jointly by Xerox, Intel, and Digital Equipment Corporation (DEC) at the Xerox Palo Alto Research Center (PARC) in the mid-1970s. The name Ethernet was derived from the old electromagnetic (EM) theoretical substance called luminiferous ether, which was formerly believed to be the invisible universal element that bound together the entire universe and all its associated parts. Hence, an Ether net is a network which connects all components attached to the net.
- **Extranet** An interconnection from an internal Intranet to a customer or non-company network which is not the Internet connection.
- **4B/5B** A data encoding method, that stands for four bits in five baud, or four-bits to five-bit, used in FDDI networks.
- **Fast Ethernet** 100 Mbps Ethernet (IEEE 802.3u). Three different media specifications are defined: 100BASE-TX, 100BASE-T4, and 100BASE-FX.
- **Full-Duplex Ethernet** Ethernet/802.3 LANs that are switched-based and hence collision-free. In full-duplex Ethernet, both transmitting and receiving channels can operate simultaneously since the receiving channel does not have to listen for collisions.
- **Full-Duplex Transmission** A data transmission method which involves the simultaneous sending and receiving of data in both directions.
- **Gateway** A software application which converts between different application protocols. The host on which this software resides is called a gateway machine. Basically, this term also refers to a router in the IP community.
- **IEEE 802** The primary IEEE standard for the 802 x series for LANs and MANs.
- **IEEE 802.1** IEEE standard which defines an architectural overview of LAN.
- **IEEE 802.2** IEEE standard which defines the logical link control, which describes services for the transmission of data between two nodes.
- **IEEE 802.3** IEEE standard that defines the carrier sense multiple access/collision detection (CSMA/CD) access method commonly referred to as Ethernet. In fact supplements include 802.3c (10-Mbps Ethernet), 802.3u (100-Mbps Ethernet, known as Fast' Ethernet), and 802.3z and 802.3b (100 Mbps Ethernet, known as Gigabit Ethernet).
- **IEEE 802.4** IEEE standard which defines the token bus network access method.
- **IEEE 802.5** IEEE standard which defines the logical ring LAN that uses a token passing access method; known also as token ring.
- **IEEE 802.6** IEEE standard which defines metropolitan area networks (MANs).
- **IEEE 802.7** IEEE standard which defines broadband LANs (capable of delivering video, data, and voice traffic).
- **IEEE 802.9** IEEE standard which defines integrated digital and video networking—integrated services LANs (ISLANs).
- **IEEE 802.10** IEEE standard which defines standards for interoperable LAN/MAN security services.
- **IEEE 802.11** IEEE standard which defines standards for wireless media access control and physical layer specifications.
- **IEEE 802.12** IEEE standard which defines the demand priority access method for 100-Mbps LANs; known also as 100 Base-VG or 100VG-AnyLAN.
- **IEEE 802.13** Define nothing—IEEE was concerned about the superstitious overtones associated with 1.3.
- **IEEE 802.14** IEEE standard which defines a standard for cable TV-based broadband communication.
- **Thick Ethernet** IEEE 802.3 10BASE5, that uses thick coaxial cable (outer diameter between 0.375 inch and 0.405 inch) as its physical medium.

- **Thin Ethernet** IEEE 802.3 10BASE2, which uses thin coaxial cable (outer diameter between 0.175 inch and 0.195 inch) as its physical medium.
- **Throughput** A realistic measure of the amount of data transmitted between two nodes in a given time period. It is a function of hardware/software, speed, CPU power, overhead, and many other items. Compared to bandwidth, throughput is what the channel really achieves, whereas bandwidth is what is theoretically possible.

SUMMARY

- LAN protocols with the OSI architecture. This architecture was developed by IEEE 802 committee and it has been accepted as LAN standard. It is called as IEEE 802 reference model.
- In a broadcast network, the single broadcast channel is to be allocated to one transmitting user at a time. This is called channel allocation.
- The traditional way of allocating a single channel, such as a telephone channel among many users is FDM. The Frequency Division Multiplexing (FDM) and Time Division Multiplexing (TDM) are the examples of static channel allocation.
- When a number of stations (users) use a common link of communication system we have to use a multiple access protocol in order to coordinate the access to the common link. The three techniques used to deal with the multiple access problem are as follows :
  (i) Random Access
  (ii) Controlled Access
  (iii) Chanelization.
- In the random access, there is no control station. Each station will have the right to use the common medium without any control over it. With increase in number of stations, there is an increased probability of collision or access conflict.
- The first method, known as ALOHA used a simple procedure called multiple access (MA). It was improved to develop the carrier sense multiple access (CSMA).
  The CSMA further evolved into two methods namely CSMA/CD (CSMA with collision detection) and CSMA/CA(CSMA with collision avoidance). which avoid collisions.
- Systems in which multiple users share a common channel in a way that can lead to conflicts are widely known as contention systems. The ALOHA system is a contention protocol which was developed at the University of Hawaii in the early 1970's by Norman Abramson and his colleagues.
- The ALOHA system has two versions :
  (i) Pure ALOHA – does not require global time synchronisation.
  (ii) Slotted ALOHA – requires time synchronisation.
- It works on a very simple principle. Essentially, it allows for any station to broadcast at any time. If two signals collide, each station simply waits a random time and tries again. Collisions are easily detected.
- If a user station receives an acknowledgement, it assumes that the transmitted frame was successfully received and if it does get an acknowledgement it assumes that collision had occurred and is ready to retransmit.
- The advantage of pure ALOHA is its simplicity in implementation but its performance becomes worse as the data traffic on the channel increases.
- To overcome the disadvantage of the pure ALOHA system, Robert published a method for doubling the capacity of traffic on the channel. In this method, time is divided up into discrete intervals, each interval corresponding to one frame.
- The CSMA protocol operates on the principle of carrier sensing. In this protocol, a station listens to see the presence of transmission (carrier) on the cable and decides to act accordingly.
- The CSMA/CD specifications have been standardized by IEEE 802.3 standard. It is a very widely used MAC protocol.
- The systems in which multiple users share a common channel in a way that can lead to conflicts are known as contention systems.

- In LANs, all the stations share the common cable (i.e., medium). Therefore, IEEE adopted three mechanisms of media access control, namely :
  (i) Carrier sense multiple access/collision detection (CSMA/CD)
  (ii) Token bus and
  (iii) Token ring
- The name Ethernet refers to the cable. Four types of cabling are commonly used for Ethernet as follows :
  1. 10 Base 5
  2. 10 Base 2
  3. 10 Base – T
  4. 10 Base – F.
- The MAC is a protocol which controls the access to the transmission medium for an orderly and efficient use of the transmission capacity of the network.
  Such a control can be exercised in two different ways.
  (i) Centralized control
  (ii) Decentralised control.
- The LLC sublayer provides the following three types of services to the network layer entity :
  Type 1 : Unacknowledged connectionless-mode service
  Type 2 : Connection-mode service
  Type 3 : Acknowledged connectionless-mode service.
- There are various technologies available in the LAN market but the most popular of them is Ethernet following are the three generations of Ethernet :
  (i) Traditional Ethernet (10 Mbps)
  (ii) Fast Ethernet (100 Mbps)
  (iii) Gigabit Ethernet (1000 Mbps).
- Traditional Ethernet was created in 1976 and has a data rate of 10 Mbps. The fast Ethernet is its next version and has a data rate of 100 Mbps. The Gigabit Ethernet operates at the data rate of 1000 Mbps.
- The traditional Ethernet is the oldest version of Ethernet created in 1976 which is designed to operate at 10 Mbps. The access to the network by a device is through the CSMA/CD and the media are shared between all the stations.
- The MAC layer governs the operation of the access method (CSMA/CD). It receives the data from the upper layer, frames it and passes it to the PLS sublayer for encoding. The access method used is 1-persistent CSMA/CD.
- Fast Ethernet is the protocol designed to work upto 100 Mbps. The traditional Ethernet can operate only upto 10 Mbps. Hence for higher data rates, fast Ethernet has been developed.
- The operation of token bus takes place as follows :
  (i) At any time, the station which holds the token only can transmit its data frames on the bus. Every frame contains source and destination addresses.
  (ii) All the other stations are ready to receive these data frames.
  (iii) As soon as the transmission time of a station is over, it passes the token to the next station in the logical sequence. The transmission is then taken over by the next station.
  (iv) In one cycle of operation, each situation gets an opportunity to transmit once. The same station can get more number of chances to transmit in one cycle if more than one addresses are assigned to it.
- We can use the following transmission media for the bus LANs :
  (i) Twisted pair
  (ii) Baseband coaxial cable
  (iii) Broadband coaxial cable
  (iv) Optical fiber.

SHORT QUESTIONS WITH ANSWERS

Q.1. Write the Functions of Media Access Control Sublayer.

Ans. (i) To perform the control of access to media.
(ii) It performs the unique addressing to stations directly connected to LAN.
(iii) Detection of errors.

Q.2. Write the Functions of Logical Link Control (LLC) sublayer.

(i) Error Recovery
(ii) It performs the flow control operation
(iii) User addressing.

Q.3. What do you mean by multiple access communications ?

Ans. When a number of stations (users) use a common link of communication system we have to use a multiple access protocol in order to coordinate the access to the common link. The three techniques used to deal with the multiple access problem are as follows :

(i) Random Access (ii) Controlled Access
(iii) Channelization.

Q.4. What is Random Access ? Explain.

Ans. In the random access, there is no control station. Each station has the right to use the common medium without any control over it. With increase in number of stations, there is an increased probability of **collision or access conflict.** The collisions occur when more than one user try to access the common medium simultaneously. As a result of such collisions, some frames can be either modified or destroyed.

Q.5. How can you compare Ring and Star topology ?

Ans. Following table presents the performance comparison of Ring and Star topologies.

S.No.	Ring topology	Star topology
1.	Media failure on uni-directional or single loops causes complete network failure.	Media faults are automatically isolated to the failed segment.
2.	Relatively difficult to reconfigure.	Relatively easy to configure.
3.	It is difficult to troubleshoot.	Easy to troubleshoot.
4.	The failure of one computer can affect the whole network.	The failure of single computer or cable doesn't bring the network down.
5.	One computer cannot monopolize the network.	Failure of the central hub causes the whole network failure.
6.	Adding and removing computers distrpts the network.	Adding and removing the computers is relatively easier.
7.	Fault diagnosis is difficult.	Fault diagnosis is relatively easy.

Q.6. What is IEEE 802.3?

Ans. A standard for the MAC and Physical layers in Ethernet networks.

Q.7. Show the frame format of IEEE 802.2.

Ans.

1 byte	1 byte	1 or 2 bytes	
DSAP	SSAP	Control	Information

Q.8. What is IEEE 802.2?

Ans. A standard for Logical Link Control (LLC).

Q.9. What is the MAC address?

Ans. Also called the physical address, the MAC address is the 48 bits assigned to the NIC by its manufacturer.

Q.10. Explain collision in Ethernet.

Ans. When two stations transmit information on the Ethernet network at the same time, a collision will occur.

Q.11. Determine the maximum size of a network using three repeaters for following networks.

(a) 10 Base-T **(b) 10 Base-5** **(c) 10 Base-2**

Ans. (a) 400 meters, (b) 2000 meters, (c) 740 meters

Q.12. What is the maximum size of a frame determined by IEEE 802.3?

Ans. 1512 bytes.

Q.13. How do computers distinguish one another on an Ethernet network?

Ans. From their hardware, or MAC address.

Q.14. What is the function of the length field in an Ethernet frame?

Ans. Identifying the number of bytes in an information field.

Q.15. What is the IEEE number for Gigabit Ethernet?

Ans. IEEE 802.3z.

Q.16. What type of frame is used by Gigabit Ethernet?

Ans. Ethernet frame format.

Q.17. List the transmission media used in Gigabit Ethernet.

Ans. Multi-mode fiber, single mode fiber, UTP, snd shielded balanced twisted pair.

Q.18. What are the hardware components of Gigabit Ethernet?

Ans.
1. Gigabit Ethernet Interface card
2. Switches that can handle 100-Mbps Fast Ethernet and 1000-Mbps Ethernet.

Q.19. What is the function of an access point device in WLAN?

Ans. An access point acts like a bridge, receiving information and retransmitting through the air and its wired port.

REVIEW QUESTIONS

1. How can you compare 1-persistent CSMA and P-persistent CSMA ?
2. What part of the 802 project makes use of CSMA/CD ?
3. What are the types of controlled access methods? Explain.
4. Explain about polling.
5. What are the channelization techniques? Explain.
6. Listy the three techniques used to deal the multiple access problem.
7. Write various advantages and disadvantages of FDMA.
8. What are the various functions of logical link control? Explain.
9. What is the maximum and minimum length of frame of traditional ethernet? Explain.
10. State the types of addresses.
11. Explain the physical properties of ethernet.
12. How can you compare thick and thin ethernet?
13. What are the properties of autonegotiation? Explain.
14. Why can not ethernet and token ring be used in a WAN? Explain.
15. When a remote bridge is used? Explain.
16. What is transparent bridge? Explain.
17. Which are three generations of ethernet? Explain.
18. How yoyu can compare pure ALOHA and slotted ALOHA?
19. Write a short technical note on CSMA.
20. Explain about CSMA/CD and its uses.
21. Write a note on CSMA/CD.
22. Write a short technical note on Ethernet.
23. Explain about the MAC frame format.
24. Write a short technical note on Fast ethernet.
25. Explain encoding and decoding used in 100-base-Fx.
26. Write a short technical note on Gigabit ethernet
27. Differentiate between 802.3, 802.4 and 802.5 IEEE standard.
28. Explain about frame format of IEEE 802.5.

UNIT – III

SYLLABUS:

Wireless Protocols: The 802.11, the 802.16, Bluetooth, RFID, Data link layer switching- uses of repeaters, hubs, bridges, switches, routers and gateways.

The Network Layer: Design Issues, Virtual Circuit and datagram networks, routing algorithms-adaptive and non-adaptive algorithms, congestion control algorithms, quality of service, internetworking, Network layer in the Internet- IPv4 protocol, IP addresses, IPv6 protocol, Internet control protocols, Mobile IP.

The above syllabus has been covered in following chapters:

Chapter 6 : Wireless Protocols

Chapter 7 : The Network Layer

Chapter 6

Wireless Protocols

Syllabus

Wireless Protocols: The 802.11, the 802.16, Bluetooth, RFID, Data link layer switching- uses of repeaters, hubs, bridges, switches, routers and gateways.

Chapter Outline

6.1. WIRELESS LAN

Wireless Local-Area Network (WLAN) is a technology which uses radio frequency to transmit and receive data over the air, providing all the features and benefits traditional LANs but without the limitations of a cable. It is a data communication system providing wireless peer-to-peer (PC-to-PC, PC-to- hub, or printer-to-hub) and point-to-point (LAN-to-LAN) connectivity within a building or campus. In place of TP or coaxial wires or optical fiber as used in a conventional LAN, WLANs transmit and receive data over electromagnetic waves. WLANs perform traditional network communication functions such as file transfer, peripheral sharing, e-mail, and database access as well as augmenting wired LANs. WLANs have gained strong popularity in a number of vertical markets, including the health-care, retail, manufacturing, warehousing, and academia. These industries have profited from the WLANs productivity gains of using hand-held terminals and notebook computers to transmit real-time information. Today, WLANs are becoming more widely recognized as the connectivity alternative for a broad range of business customers.

Wireless LAN technology standard 802.11b has the strongest momentum to becoming the main standard for corporate internal wireless LAN networks. The bandwidth of 802.11b is 11 Mbps and operates at 2.4 GHz frequency. The successor of this current 802.11b standard is 802.11a and it is designed to be faster speed and operate at a different frequency. Its bandwidth is 54 Mbps. While 802.1 la standard and the technology behind it has become available, 802.11b is still widely used today and many companies and individuals are deploying it or deploying dual 802.11b and 802.1 la devices. As more wireless technology is developed and implemented, the complexity of the types of attacks will increase. These attacks may be very similar against other wireless type technologies and is not unique to 802 11b. By understanding these risks and how to develop security solution for 802.11b, this will be a good stepping-stone for providing a good secure solution to any wireless solution.

6.1.1. How Wireless LANs Work

Wireless LANs use electromagnetic air waves (radio or infrared) to communicate information from one point to another without relying on any physical connection. Radio waves are often referred to as radio carriers because they simply perform the function of delivering energy to a remote receiver. The data being transmitted is superimposed on the radio carrier so that it can be accurately extracted at the receiving end. This is generally referred to as modulation of the carrier by the information being transmitted. Once data is superimposed onto the radio carrier, the radio signal occupies more than a single frequency, since the frequency or bit rate of the modulating information adds to the carrier. Multiple radio carriers can exist in the same space at the same time without interfering with each other if the radio waves are transmitted on different radio frequencies. To extract data, a radio receiver tunes in one radio frequency while rejecting all other frequencies.

In a typical wireless LAN configuration, a transmitter/receiver (transceiver) device, called an access point, connects to the wired network from a fixed location using standard cabling. At a minimum, the access point receives, buffers, and transmits data between the wireless LAN and the wired network infrastructure.

A single access point can support a small group of users and can function within a range of less than-one hundred to several hundred feet. The access point (or the antenna attached to the access point) is usually mounted high but may be mounted essentially anywhere that is practical as long as the desired radio coverage is obtained. End users access the wireless LAN through wireless-LAN adapters, which are implemented as PC cards in notebook or palmtop computers, as cards in desktop computers, or integrated within hand-held computers. Wireless LAN adapters provide an interface between the client network operating system (NOS) and the airwaves via an antenna. The nature of the wireless connection is transparent to the NOS.

6.1.2. Wireless LAN Technology

A wide range of wireless technologies are available. Some may be listed as under :

1. **Spread spectrum technology:** Most wireless LAN systems use spread-spectrum technology, a wideband radio frequency technique developed by the military for use in reliable, secure, mission-critical communication systems. Spread-spectrum is designed to trade off bandwidth efficiency for reliability, integrity, and security. If a receiver is not tuned to the right frequency, a spread-spectrum signal looks like background noise. There are two types of spread spectrum radio: frequency hopping and direct sequence.

2. **Narrowband Technology:** A narrowband radio system transmits and-receives user information on a specific radio frequency. Narrowband radio keeps the radio signal frequency as narrow as possible just to pass the information. Undesirable crosstalk between communication channels is avoided by carefully coordinating different users on different channel frequencies.

3. **Infrared Technology:** The third technology, little used in commercial wireless LANs, is infrared. Infrared (IR) systems use very high frequencies, just below visible light in the electromagnetic spectrum, to carry data. Like light; IR cannot penetrate opaque objects; it is either directed (line-of-sight) or diffuse technology. Inexpensive directed systems provide very limited range (3 feet) and typically are used for personal area networks but occasionally are used in specific wireless LAN applications.

6.1.3. Types of Wireless LAN

The wireless LANs are becoming more and more popular since they can satisfy the requirements like mobility, relocation of user, ad-hoc networking and coverage of locations which are difficult to wire. Earlier the wireless LANs were costly, could support only low data rates, a license was required. Hence, there were limitations on the practical utility of wireless LANs. However, all these problems are being addressed now which is increasing the popularity of wireless LANs day by day.

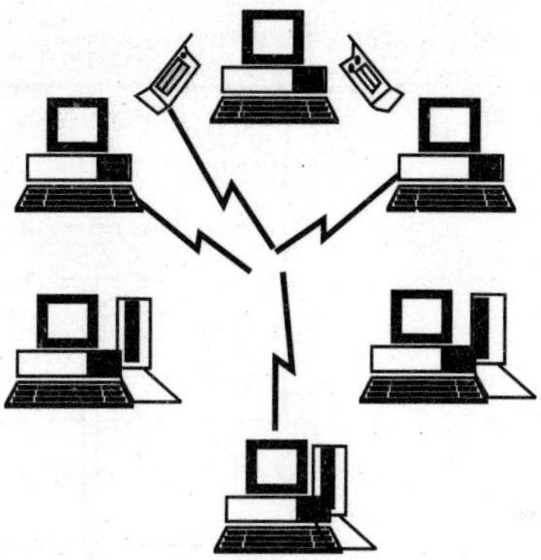

Fig. 6.1. Wireless LAN

6.1.4. Wireless LAN Configuration

Wireless LAN is a more latest example that uses wireless communication. It has been shown in figure 6.1. Devices such as workstations, Laptop, computers, cordless telephones and other communication appliances share the wireless medium such as 5 GHz radio or infrared light.

6.1.5. Applications of Wireless LAN

There are four important applications of wireless LAN which may be listed as under:

(i) LAN extension (ii) Nomadic access

(iii) Cross building interconnection (iv) Ad-hoc networks

6.1.6. LAN Extension

1. Basic Concepts: Wireless LANs save the cost of installation of LAN cabling, they can be easily relocated and the network structures can be easily modified. The wireless LANs are preferred for buildings having large open areas, historical buildings and small offices. Many a times, a wireless LAN is linked into a wired LAN on the same premises. Thus, this application area is referred to as LAN extension. Figure 6.2 indicates a simple wireless LAN that can be used in many applications. There is a backbone wired LAN such as Ethernet which is used for supporting the servers, workstations and bridges or routers. CM is a control module which acts as an interface to wireless LAN. It consists of bridges or routers to link the wireless LAN to the wired backbone LAN. The control module (CM) also consists of some kind of access control logic like polling or token passing scheme for regulating the access. The user module (UM) in figure 6.2 controls a number of stations which are not connected by the wired LAN. The UMs can be a part of wireless LAN configuration. Figure 6.2 is known as a single cell wireless LAN. All the wireless end systems are within the range of single control module.

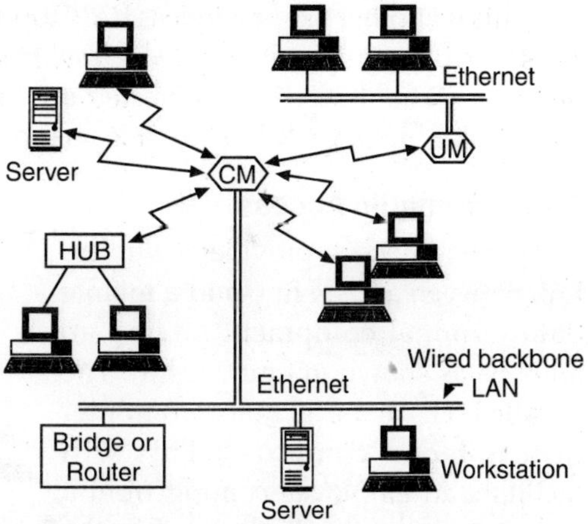

Fig. 6.2. A Single cell wireless LAN

2. Multiple Cell Wireless LAN: Figure 6.3 illustrates a multiple cell wireless LAN. This is because there are more than one control modules (CMs) that are interconnected by a wireless LAN. Every CM controls many wireless systems within its range. Every infrared LAN has a limited range which typically corresponds to a single room. Therefore, one cell is required for each room in an office building. The frequency used in every room for the wireless LAN is different.

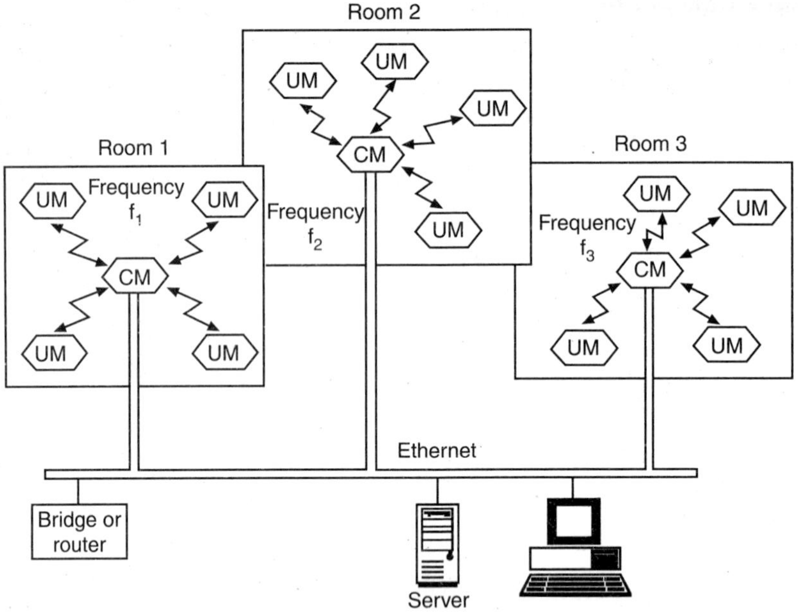

Fig. 6.3. Illustration of Multiple-cell wireless LAN configuration

6.1.7. Cross Building Interconnect

This is another use of wireless LAN topology. It is used for connecting LANs in nearby buildings. These LANs can be wired or wireless. Here, a point-to-point wireless link is used between two buildings. The devices so connected are bridges or routers. Strictly speaking, the single point to point link is not a LAN but still it is included under the heading of wireless LAN.

6.1.8. Nomadic Access

Nomadic access provides a wireless link between a LAN hub and a mobile data terminal equipment having an antenna as shown in figure 6.4(a). The mobile DTE can be a Laptop computer or a notepad computer. This will facilitate an employee coming from a trip to transfer his collected data to the server. Hence, in general, Normadic access is quite useful whenever a user moving around with his portable computer want to access a server that is connected in a wired LAN.

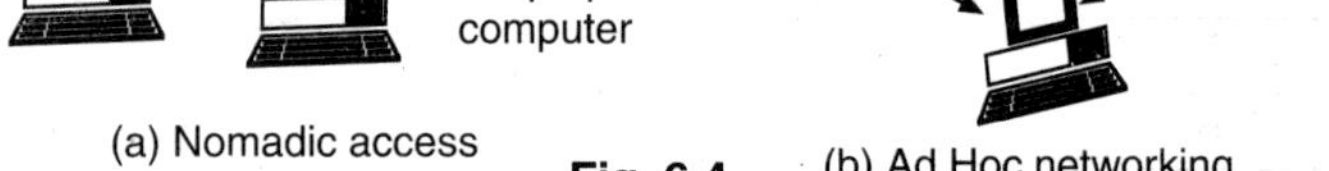

(a) Nomadic access **Fig. 6.4.** (b) Ad Hoc networking

6.1.9. Ad-Hoc Networking

It is a peer-to-peer network without any centralized server. Figure 6.4(b) illustrates the ad-hoc networking scheme.

The Ad Hoc LAN is set up temporarily to meet some immediate requirements. Such as a group of people with laptops conferencing with each other in a room. The difference between the nomadic access and ad-hoc networking is evident from figure 6.4(a). As observed from figure 6.4(b), the ad-hoc LAN does not have any infrastructure. It is just a collection of few stations within range of each other dynamically configuring themselves into a temporary network.

6.1.10. Wireless LAN Requirements/Properties

Some of the important requirements of a wireless LAN may be listed as under:

1. Throughput: The throughput capacity of the wireless LAN must be as high as possible. This is possible if the medium access control protocol is made as efficient as possible.

2. Number of Nodes: Usually, a few hundred nodes are required to be supported using the wireless LAN.

3. Connection to Backbone LAN: A wireless LAN in most of the cases is interconnected with a wired backbone LAN. The ad hoc LANs and mobile users must also be connected to the backbone LAN.

4. Service Area: A wireless LAN has a typical coverage area which range from 100 m to 300 m.

5. Longer Battery Life: The mobile users use battery power. In order to have a longer battery life, it is essential to minimize the battery power consumption.

6. Robustness and Security: Wireless LAN can be prone to interference. Hence, it must be properly designed to ensure reliable transmission even in a noisy atmosphere. Also, a minimum level of security must be provided from eaves dropping.

7. License-Free Operation: For the wide spread use of wireless LAN, it is essential to eliminate the requirement of license.

8. Allocated Network Operation: If two or more wireless LANs operate in the same area, then interference between them is possible. This may result in malfunctioning of a MAC algorithm and may allow unauthorized access to a LAN.

9. Roaming Facility: A roaming facility must be provided to the mobile stations. When a mobile station moves from one cell to the other, they must remain in the network.

10. Dynamic configuration: The MAC addressing and the network management aspects of the LAN must permit automatic addition, deletion and relocation without any problem to the other users.

6.1.11. Performance Comparison of Ethernet and Wireless Network

S.No.	Ethernet	Wireless Network
1.	IEEE standard 802.3	IEEE standard 802.11
2.	Communication medium is coaxial cable	Infrared on radio frequencies act as medium
3.	Spread spectrum is not used	Spread spectrum is used
4.	Uses MAC	Uses two MAC sublayers
5.	Uses CDMA/CD	Uses CSMA/CA
6.	Efficiency is high	Efficiency is low
7.	Addressing is simpler	Addressing is complicated
8.	Large range	Short range

6.1.12. Few Problems Associated with Wireless LAN

If we try to use CSMA for the wireless LAN, then it uses the principle of simply listening to other transmission and only transmit if no one else is transmitting. However, there are two problems in using CSMA. They are hidden station problem and exposed station problem.

1. Hidden Station Problem: Let us consider figure 6.5 where station-1 is transmitting to station-2. Now, if station-3 senses the medium to see if anyone is transmitting, it will not hear station 1

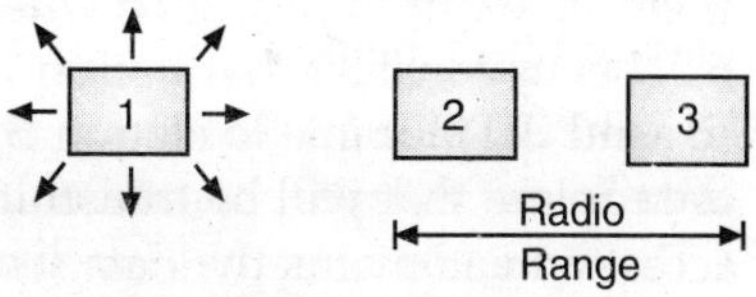

Fig. 6.5. Hidden station problem

since it is out of range. Therefore station-3 will come to a wrong conclusion that no one is transmitting and so it can transmit to station-2. If station-3 starts transmitting, it will create an interference at station-2 and will wipe out the frames from station-1.

This problem in which a station is not able to detect a potential competitor which is too far away is known as the *hidden station problem.*

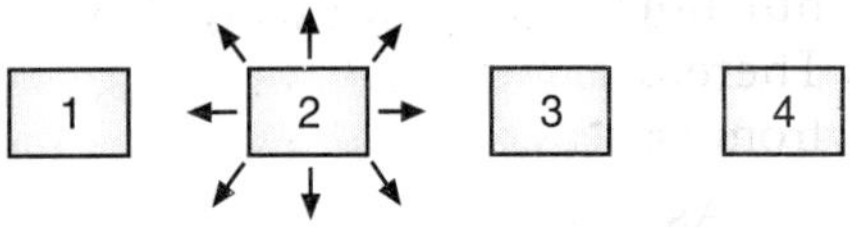

Fig. 6.6 Exposed station problem

2. Exposed Station Problem: Now, let us consider figure 6.6 in which station-2 is transmitting to station-1. If station-3 senses the medium, then since station-2 comes in range with station-3, so station-3 will sense the transmission going on. Therefore, it will falsely decide that it must not transmit to station-4. Here, it may be noted that the transmission from 3 to 4 is ok, even when 2 is transmitting. This problem is known as the *exposed station problem.*

6.1.13. Design Issues for Wireless Networks

Wireless networks make use of electromagnetic waves or infrared red light for communicating between two or more wireless devices. The data to be transmitted is superimposed on the radio carrier (modulation).

Some of the important issues that must be addressed while designing a wireless network may be listed as under:

(i) Bandwidth
(ii) Security
(iii) Mobility
(iv) Reliability
(v) Scalability
(vi) Ease of installation.

Out of these, the bandwidth and security issues have already been discussed.

(i) Mobility: Wireless networks provide mobility to its users. They provide access to Internet and contact with the other users without using any wires.

(ii) Installation: Installing a wireless network is quite simple. It eliminates the requirement of pulling connecting wires.

(iii) Reliability: In wireless systems, the EM waves are used to carry the information. However, these waves undergo fading (loss of signals) because of various reasons. This shall lead to reduced reliability. Hence, some minimum level of reliability must be ensured.

(iv) Scalability: Wireless systems can be configured in different types of topologies depending upon the requirement of applications. It is possible to change the configuration easily right from peer to peer which is suitable for small networks to large infrastructural network used for large areas.

6.2. WIRELESS LAN PROTOCOLS

A system of portable computers that communicate by using radio frequency (RF) or infrared waves are regarded as a wireless LAN. The wireless LAN requires special MAC sublayer protocols. Multiple Access with Collision Avoidance (MACA) is the protocol designed for wireless LANs. It is used as the basis for the IEEE 802.11 wireless LAN standard. In this protocol, the sender stimulates the receiver to send a short frame, so stations nearby can detect this transmission and avoid transmitting themselves for the duration when data is being transmitted. The MACA protocol is shown in figure 6.7. Let us consider that station 1 wants to send data to station-2. Station 1 starts by sending request to send (RTS) frame to station B. This short frame consisting of 30 bytes contains the length of the data frame that will be transmitted. Then, station-2 replies with a clear to send (CTS) frame. The CTS frame contains the data length copied from the RTS frame. When station 1 receives the CTS frame, it starts transmission.

As illustrated in figure 6.7, station 3 is within range of 1 but not within range of 2. Therefore, it hears the RTS from 1 but not the CTS from 2.

As long, as it does not interfere with the CTS, it is free to transmit while the data frame is being sent. In contrast, station 4 is within the range of 2 but not 1. It does not hear the RTS but does hear the CTS. On hearing CTS, it does not send anything until that frame is expected to be finished. The station 5 hears both the RTS and CTS control messages and remains silent until the data frame is completed. Despite the above precaution, it is possible that both stations 2 and 3 may send a RTS of station 1. These will collide and will be lost. In this case, if station-1 does not receive a CTS within a specified time, it waits for a random amount of time and then again transmits the RTS command. The MACA was latter fine tuned for improvement in performance and renamed as MACAW (MACA for wireless). In MACAW, an ACK frame is introduced after each successful data frame. Carrier sensing (CSMA) is also added and the back off algorithm is run separately for each data stream rather than for each station.

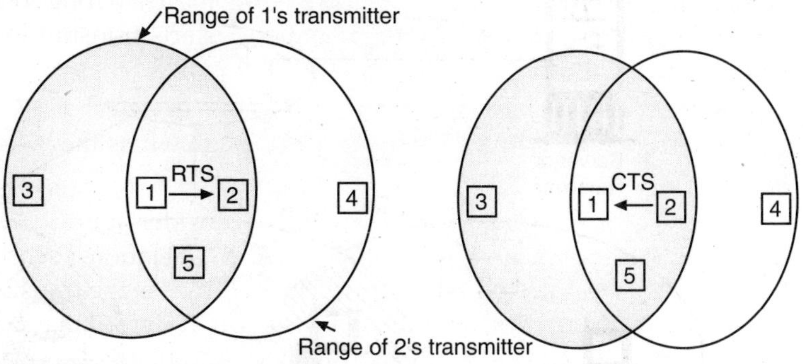

(a) I sending an RTS to 2 (b) 2 responds with a CTS to 1

Fig. 6.7. The multiple access with collision avoidance

6.3. THE IEEE 802.11 *(RGTU, Bhopal, Sem. Exam., June, 2014)*

As a matter of fact, 802.11 are the IEEE specifications for a wireless LAN. It covers the physical and data layers.

Architecture : IEEE 802.11 defines following two types of services :

1. Basic service set (BSS)
2. Extended service set (ESS)

1. Basic Service Set (BSS): IEEE 802.11 has defined the BSS as the basic building block of wireless LAN. A BSS is made of stationary or moving wireless stations and a central base station called as the access point (AP). A BSS can be either without AP or with AP as shown in figures 6.8 (a) and (b). The BSS without AP cannot send data to another BSS. Hence, it is known as a stand alone network or ad hoc architecture. In such an architecture, the stations can form a network without using AP.

2. Extended Service Set (ESS): An extended service set (ESS) consists of two or more BSSs with APs. The BSSs in this system are connected to each other via a **distribution system** which is generally a wired LAN as shown in figure 6.9. The distribution system connects the APs to each other. The distribution system can be any type of LAN such as Ethernet.

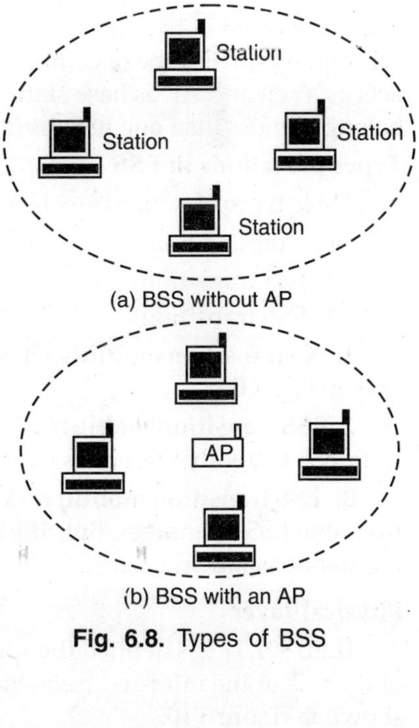

(a) BSS without AP

(b) BSS with an AP

Fig. 6.8. Types of BSS

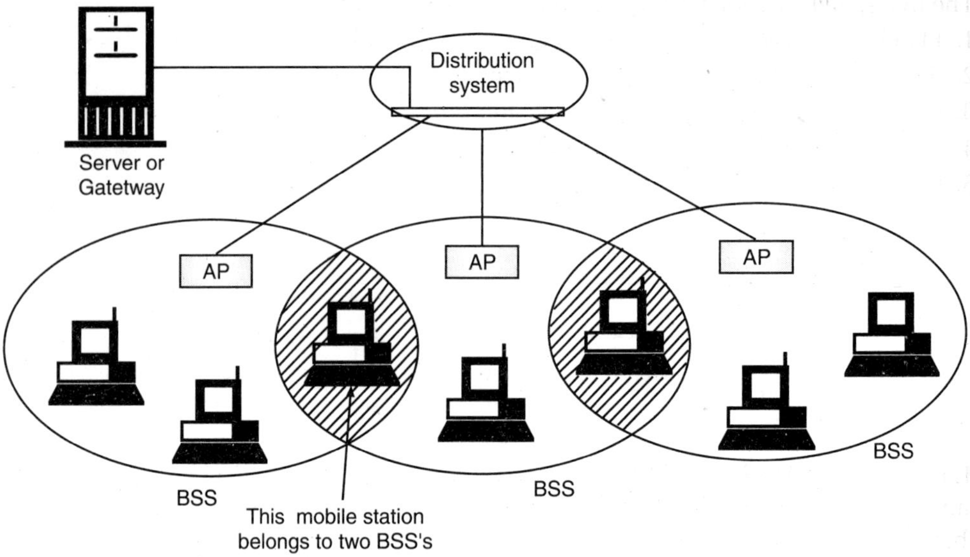

Fig. 6.9. ESS

Thus, ESS contains following two types of stations :

(i) Mobile stations

(ii) Stationary of non-moving stations.

Out of these, the non-moving stations are the APs which are a part of the wired LAN. Whereas the mobile stations are those contained in the BSS. Figure 6.9 shows an ESS.

The BSSs are connected to each other to form a network called infrastructure **network.** In such networks, the stations close to each other can communicate without taking help of AP. But if two stations located in two different BSS wish to communicate with each other than they have to do so through APs. This type of communication is very similar to that in the cellular communication. BSS acts as a cell and AP as base station. As shown in figure 6.9, it is possible that a mobile station can belong to more than one BSSs simultaneously.

Types of stations in ESS

Three types of stations are defined by IEEE 802.11 depending on their mobility in wireless LAN as:

1. No transition
2. BSS transition
3. ESS transition

1. No transition mobility : It is defined as a station which is non-moving (stationary) or moving only inside a BSS.

2. BSS transition mobility : A station having BSS transition mobility is the one which can move from BSS to another but does not move outside one ESS.

3. ESS transition mobility : A station having ESS transition mobility is the one which can move from one ESS to another. But, IEEE 802.11 does not guarantee a continuous communication when the station is moving.

Physical layer

IEEE 802.11 has defined the specification for converting bits to a signal in the physical layer one of them is in the infra-red frequency spectrum and the other five specifications are in RF range as shown in figure 6.10.

The five specifications in RF range may be listed as under :

1. FHSS - Frequency Hopping Spread Spectrum (802.11).
2. DSSS - Direct Sequence Spread Spectrum (802.11)
3. OFDM - Orthogonal Frequency Division Multiplexing (802.11 a)
4. HR-DSSS-High Rate-DSSS (802.11 b).
5. OFDM (802.11g)

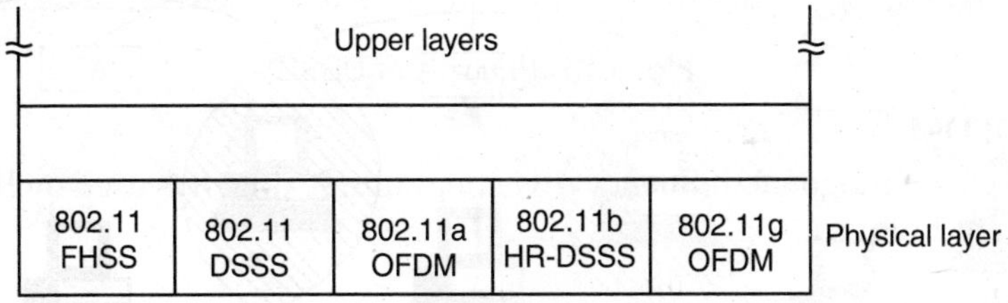

Fig. 6.10.

1. IEEE 802.11 FHSS: The IEEE 802.11 describes a method called FHSS i.e. Frequency Hopping Spread Spectrum for conversion of bits into a signal. The frequency band used for this is 2.4 GHz ISM band.

2. Principle of FHSS: In FHSS, the sender sends one frequency for a short period of time. Then hope to another carrier frequency and transmits it for the same amount of time. Then, it hops again to a new carrier frequency and transmits it for the same duration and so on. In all, there are N-such hoppings in one cycle as shown in figure 6.11. The cycle repeats itself after N-hoppings. The bandwidth of FHSS signal is equal to NB Hz where B is the bandwidth of the original signal that is being converted to FHSS signal. The techniques of spreading used in FHSS makes it difficult for an unauthorised person to understand the transmitted data.

The frequency of allocated bands in FHSS is decided with mutual agreement between the sender and the receiver. The frequency band by FHSS is 2.4 GHz ISM (industrial, scientific and medical). This band is divided in 79 equal subbands of 1 MHz each. The hopping frequency is selected by a pseudorandom number generator.

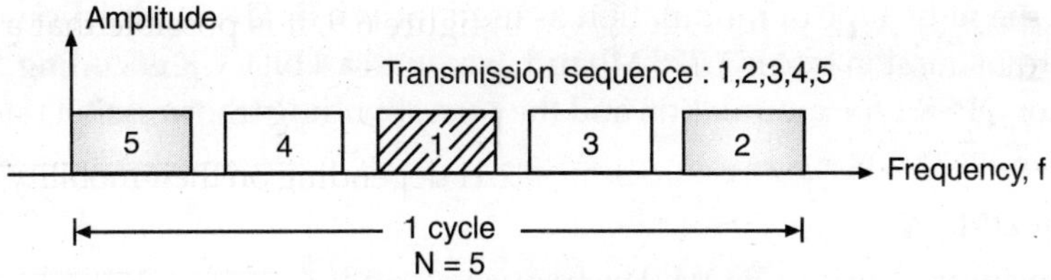

Fig. 6.11. Principle of FHSS

The modulation techniques used is FSK (frequency shift keying) at 1 M baud / sec. And the data rate is 1 or 2 Mbps.

IEEE 802.11 DSSS

IEEE 802.11 has defined the DSSS i.e. direct sequence spread spectrum technique in order to convert the bits into signal. The DSSS also uses the same frequency band as that of the FHSS i.e., 2.4 GHz ISM band.

Principle of DSSS

In DHSS, each bit being sent by the sender is first converted .into a group of bits called as the chip code. The time required to send each chip code should be equal to the time period of the original bit in order to avoid the buffering. Let N represent the number of bits in each chip code. Then, the data rate of DSSS would be equal to N times the data rate of the original signal. Figure 6.12 demonstrates the principle of DSSS. DSSS even though similar to CDMA, is not a multiple access method. The bit sequence used in DSSS uses the entire frequency band of 2.4 GHz (ISM band). DSSS uses the PSK (phase shift keying) as its modulation technique at 1 M baud/s and the data rate is 1 or 2 Mbps.

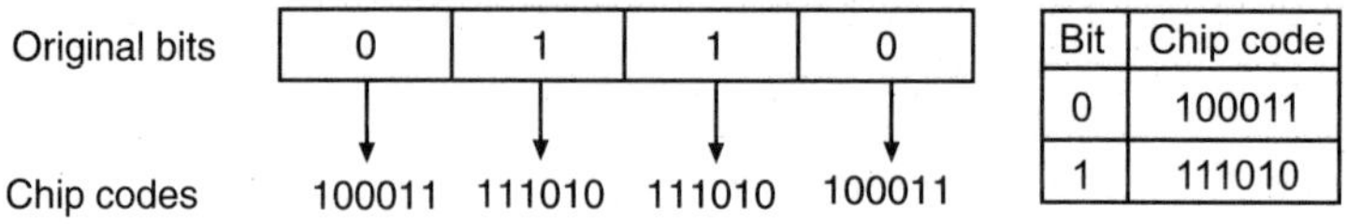

Fig. 6.12. Principle of DSSS

IEEE 802.111 a OFDM

OFDM stands for orthogonal frequency division multiplexing. It is used by IEEE 802.11 a as the signal conversion technique.

Principle of OFDM

The basic principle of OFDM is same as that of FDM. But, the major difference is that all the frequency subbands are used by one source at a given time. OFDM uses the 5 GHz ISM band for its operation. This band is subdivided into 52 subbands. Out of these 52 subbands, 48 subbands are used for sending 48 groups of bits at a time and the remaining 4 subbands are used for sending the control information. These subbands can be used randomly in order to increase the security of transmitted data. The type of modulation used by OFDM is PSK (Phase shift keying) and QAM (Quadrature Amplitude Modulation). The data rates are 18 Mbps (PSK) and 54 Mbps (QAM).

IEEE 802.11 b HR- DSSS

HR-DSSS is the short form for high speed direct sequence spread spectrum. IEEE 802.11 b uses HR-DSSS method for signal generation. It uses the frequency band of 2.4 GHz (ISM band). HR-DSSS is conceptually very similar to that of a DSSS technique. The only difference is the encoding method used. HR-DSSS uses a method called complementary code keying (CCK) which encodes 4 or 8 bits of original data into one CCK symbol. The HR-DSSS needs to backward compatible with DSSS. Hence, HR-DSSS defines four data rates : 1,2,5.5 and 11 Mbps. The 1 and 2 Mbps data rates are allocated for the same type of modulation as used for DSSS. The modulation used for 5.5 Mbps is BPSK and the transmission rate is 1.375 Mbaud /sec with a 4 bit CCK encoding. The 11 Mbps data rate is assigned for QPSK type modulation and the corresponding transmission rate is 1.375 Mbaud/s. It uses 8 bit CCK encoding.

IEEE 802.11 g OFDM

It is a new specification using OFDM. The frequency band used is 2.4 GHz ISM with a complex modulation technique used, it is possible to achieve data rate of upto 54 Mbps.

MAC layer

The IEEE 802.11 has defined two MAC sublayers as under:

(i) The distributed coordination function (DCF)

(ii) The point coordination function (PCF).

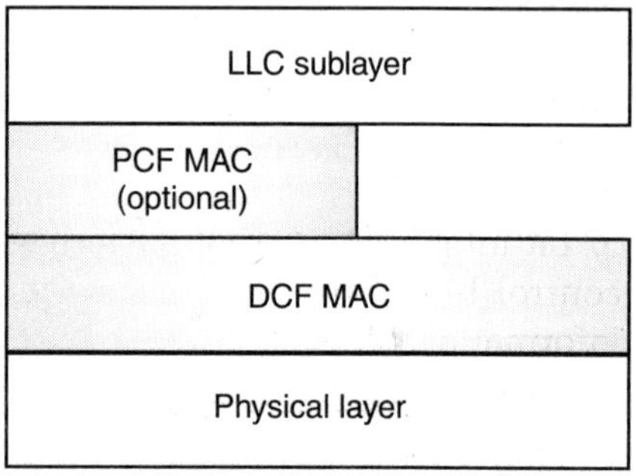

Fig. 6.13. MAC Layers in 802.11 standard

Both these have been shown in figure 6.13. Out of the two, the PCF is an optional and complex access method. It can be implemented in an infrastructure network (not in an ad hoc network). The other access method i.e., DCF is very similar to the CSMA/CA though DCF has some additional features.

6.4. DIGITAL CELLULAR RADIO

Digital cellular radio is another form of wireless networking. It is somewhat differ from the wireless LAN and uses different protocols. It is more oriented towards telephony and requires connections lasting for minutes rather than few milliseconds. It is more efficient to do allocation per call rather than per frame. This technique also supports data traffic along with voice.

Digital transmission has the following advantages over analog for mobile transmission :

(i) Voice, data and fax can be integrated into a single channel.

(ii) Less bandwidth per channel is required because of speech compression techniques.

(iii) Error detecting and correcting codes can be used to improve transmission quality.

(iv) Digital signals can be encrypted for security.

In digital cellular radio systems, three different approaches are used for channel allocation

(i) Global System for Mobile Communication (GSM)

(ii) Cellular Digital Packet Data (CDPD)

(iii) Code Division Multiple Access (CDMA)

6.4.1. GSM (Global System for Mobile Communications)

It is a full digital system used in Europe. It handles channel access using a combination of slotted ALOHA, FDM and TDM. It was initially designed to be used in the 900 MHz band, later the 1800 MHz band was allotted and it is also called as DCS 1800. A GSM system has upto a maximum of 200 full-duplex channels per cell. Each channel consists of a downlink frequency (935.2 MHz to 959.8 MHz) i.e., from base station to mobile station and an uplink frequency (890.2 MHz to 914.8 MHz) i.e., from mobile station to base station.

Each frequency band is 200 kHz (0.2 MHz) wide. Both the uplink and downlink frequency bands consists of 124 channels as shown in figure 6.14. As shown in the figure 6.14, each frequency channel supports eight separate connections using TDM. Each TDM slot has a specific structure as shown in figure 6.15. Each TDM slot consists of 148 bit data frame. Each data frame starts and ends with 3 zero bits, for frame delineation purpose. It also contains two 57 bit information fields, each having a control bit that indicates whether the information field is for voice or data.

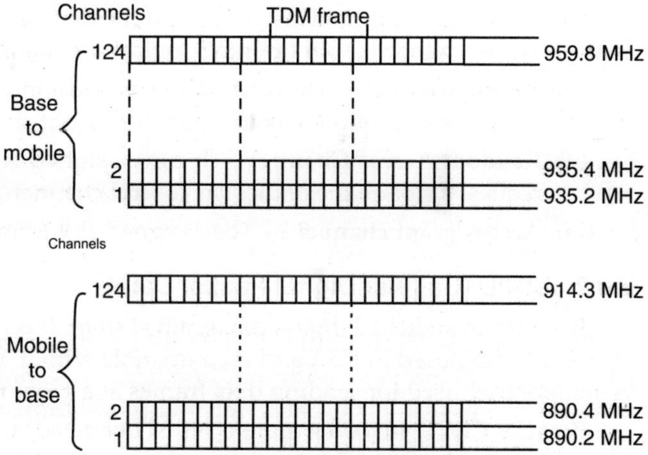

Fig. 6.14. GSM frequency band allocation

Between the information fields is a 26-bit sync field which is used by the receiver to synchronize to the sender's frame boundaries. As shown in the figure 6.15, eight data frames make up a TDM frame and 26 TDM frames make up a 120 m sec multiframe, of the 26 TDM frames in a multiframe, slot 12 is used for control and slot 25 is reserved for future use.

The broadcast control channel is a continuous stream of output from the base station containing its identity and the channel status.

All mobile stations monitor their signal strength to see when they have moved into a new cell. The dedicated control channel is used for location updating, registration and call setup. Each base

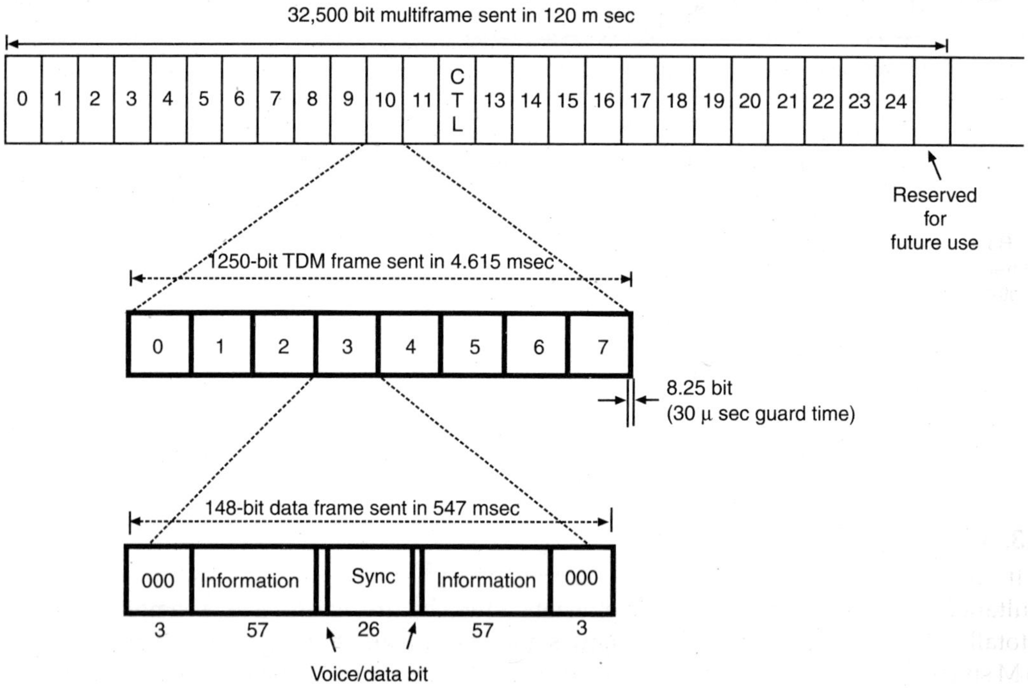

Fig. 6.15. GSM TDM structure

station maintains a database of mobile stations currently under its jurisdiction and the information needed to maintain this database is sent on the dedicated control channel. There is also a common control channel which is split up into 3 logical subchannels as under :

(i) **Paging channel** — The base station uses to announce incoming calls and each mobile station monitors it continuously to watch for calls it should reply to.

(ii) **Random Access Channel** — It runs a slotted ALOHA system to allow a mobile station to request a slot on the dedicated control channel. Using this slot the station can set up a call.

(iii) **Access grant channel** — The assigned slot is announced on this channel.

6.4.2. CDPD (Cellular Digital Packet Data)

It is packet switched digital datagram system. It is built on top of the Advanced Mobile Phone System (AMPS) used in USA and is compatible with it. In this system, any idle 30 kHz channel can be temporarily used for sending data frames at a gross rate of 19.2 kbps.

Because CDPD involves quite a bit of overhead, the net data rate is closer to 9600 bps. This system follows the OSI model for its working. A CDPD system consists of 3 types of stations, a base station, a mobile station and a base interface station as shown in figure 6.16. As shown in figure 6.16, kinds of interfaces are defined in this system.

(i) E - interface → external to the CDPD provider and connects a CDPD area to a fixed network.

(ii) I – interface → internal to the CDPD provider and connects two CDPD areas together.

(iii) A – interface → air interface between the base station and mobile hosts.

Data over the air interface are sent using compression, encryption and error correction. It uses Reed – Solomon error correcting code. Each CDPD cell has only one downlink/uplink pair available for data.

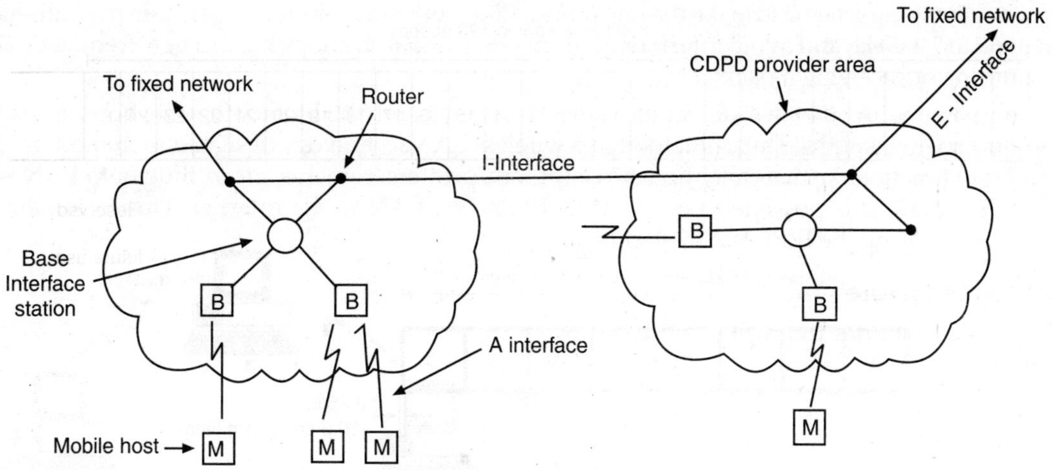

Fig. 6.16. CDPD system

6.4.3. CDMA

It allows each station to transmit over the entire frequency spectrum all the time. Multiple simultaneous transmission are separated using coding theory. It also assumes that colliding frames are totally distorted but that multiple signals add linearly. In CDMA, each bit time is subdivided into M short intervals called chips. There are typically 64 or 128 chips per bit. Each station is assigned a unique m-bit code or chip sequence.

To transmit a 1 bit, a station sends its chip sequence (e.g. 10111001). To transmit a 0 bit it sends the one's complement of its chip sequence (01000110). If we have a 1 - MHz band available for 100 stations, with FDM, each one would have 10 kHz and could send at 10 kbps (assuming 1 bit/Hz). With CDMA, each station uses the full 1 MHz, hence, the chip rate is 1 megachip per second. The effective bandwidth per station is higher for CDMA than FDM and the channel allocation problem is solved. The sender and receiver synchronize by having the sender transmit a long enough known chip sequence that the receiver can lock onto. All other unsynchronised transmissions are then seen as random noise. The receiver can listen to all the senders at once by running the decoding algorithm for each of them in parallel.

6.4.4. Common Properties of WDMA and GSM Channel Access Protocol

There are certain properties common to WDMA and GSM channel access protocol. Some of them may be listed as under :

(i) Both divide the channel into subchannels using FDM or TDM.

(ii) Both protocols support constant data rate traffic, variable data rate and datagram traffic.

(iii) Both handle channel access using a combination of FDM, TDM and Slotted ALOHA.

(iv) Both allow the transmission to take place at the same time by dividing the spectrum into channels.

6.5. BLUETOOTH

Bluetooth is the name given to a new technology using short-range radio links, intended to replace the cable(s) connecting portable and/or fixed electronic devices. It is envisaged that it will allow for the replacement of many propriety cables that connect one device to another with one universal radio link. Its key features are robustness, low complexity, low power and low cost. Designed to operate in noisy frequency environments, the Bluetooth radio uses a fast acknowledgement and frequency-hopping scheme to make the link robust. Bluetooth radio modules operate in the unlicensed ISM band at 2.4 GHz, and avoid interference from other signals by hopping to a new frequency after transmitting or receiving a packet.

Compared with other systems in the same frequency band, the Bluetooth radio hops faster and uses shorter wavelengths. Thus, Bluetooth is a wireless LAN technology designed to connect devices of different functions such as telephones, computers, printers, cameras, etc. A Bluetooth LAN is an Ad hoc network. It is possible to connect the Bluetooth LAN to the Internet. This technology is implemented using the IEEE 802.15 standard.

6.5.1. Architecture

Bluetooth defines following two types of networks :

(i) Piconets, and

(ii) Scatternets

6.5.2. Piconets

A Bluetooth network is called as a piconet or a small net. It can have at the most eight stations. One of them is called as a master and all others are called as slaves. All the slave stations are synchronised with the master. A piconet can have only one master station. Figure 6.17 shows a piconet. The communication between a master and slaves can be one-to-one or one-to-many.

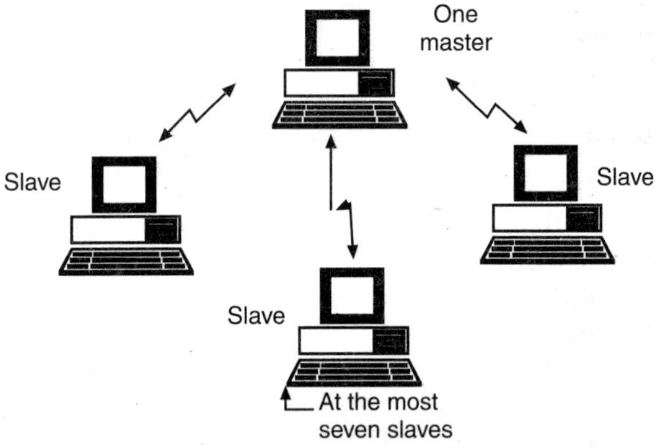

Fig. 6.17. A piconet

6.5.3. Scatternet

Piconets are combined to form a scatternet. It has been shown in figure 6.18.

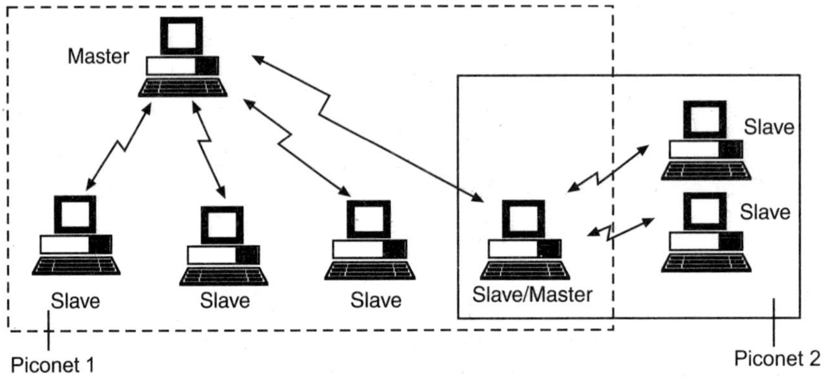

Fig. 6.18. Scatternet

Figure 6.18 shows a scatternet consisting of two piconets. A slave in the first piconet can act as a master in the second piconet. It will receive the messages from the master in the first piconet by acting as a slave and then delivers the message to the slaves in the second piconet as shown in figure 6.94.

6.5.4. Bluetooth Devices

Every Bluetooth device consists of a built in short range **radio transmitter**. The cure data rate is 1 Mbps and the bandwidth is 2.4 GHz. Hence, an interface between the IEEE 802.11 wireless LAN and Bluetooth LAN is possible. The Bluetooth specification standard defines a short-range (10-meter) radio link. The devices carrying Bluetooth-enabled chips can easily transfer data at a rate of about

1 Mbps (Megabits per second) within 10 meters (33 feet) of range through walls, clothing and luggage bags. The interaction between devices occurs by itself without direct human intervention whenever they are within each other's range. In this process, the software technology embedded in the Bluetooth transceiver chip triggers an automatic connection to deliver and accept the data flow.

Since Bluetooth is of short range with limited speed and low-power technology. It is less; attractive to corporate wireless local area networks that are generally powered with 802.11 wireless LAN technologies. Each Bluetooth-enabled device contains a 1.5-inch square transceiver chip operating in the ISM (industrial, scientific, and medical) radio frequency band of 2.40 GHz to 2.48 GHz. This frequency is generally available worldwide for free without any licensing restrictions. The ISM band is divided into 79 channels with each carrying a bandwidth of 1 MHz.

6.5.5. Bluetooth Layers (Bluetooth Protocol Architecture)

The layers of Bluetooth as shown in figure 6.19 do not exactly match with the Internet layers.

1. **Radio :** The Bluetooth Radio (layer) is the lowest defined layer of the Bluetooth specification. It defines the requirements of the Bluetooth transceiver device operating in the 2.4 GHz ISM band. This band is divided into 79 channels of 1 MHz each.

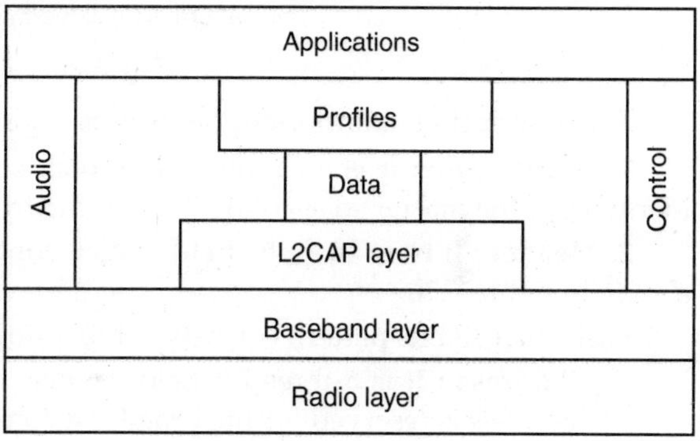

Fig. 6.19. Bluetooth layers

2. **Baseband** : The baseband is the physical layer of the Bluetooth. It manages physical channels and links apart from other services like error correction, data widening, hop selection and Bluetooth security. The baseband layer lies on top of the Bluetooth radio layer in the bluetooth stack. The baseband protocol is implemented as a link controller, which works with the link manager for carrying out link level routines like link connection and power control. The baseband also manages asynchronous and synchronous links, handles packets and does paging and inquiry to access and inquire Bluetooth devices in the area. The baseband transceiver applies a Time-Division Duplex (TDD) scheme (alternate transmits and receives). Therefore, apart from different hopping frequency (frequency division), the time is also slotted.

6.5.6 TDMA

Bluetooth uses a form of TDMA that is called TDD-TDMA (Time Division Multiple Access).

6.5.7 Logical Link Control and Adaptation Protocol (L2CAP)

The Logical Link Control and Adaptation Layer Protocol (L2CAP) is layered over the baseband protocol and resides in the data link layer. L2CAP provides connection-oriented and connectionless data services to upper layer protocols with protocol multiplexing capability, segmentation and reassembly operation, and group abstractions. L2CAP permits higher-level protocols and applications to transmit and receive L2CAP data packets up to 64 kilobytes in length. Two link types are supported for the baseband layer : Synchronous Connection-Oriented (SCO) links and Asynchronous Connection-Less (ACL) links. SCO links support real-time voice traffic using reserved bandwidth. ACL links support best effort traffic. The L2CAP specification is defined for only ACL links and no support for SCO links is planned.

6.5.8 Frame Format

A frame in the baseband layer can be one of following three types :

1. One slot
2. Three slot
3. Five slot

The frame format of the three frame types is shown in figure 6.20.

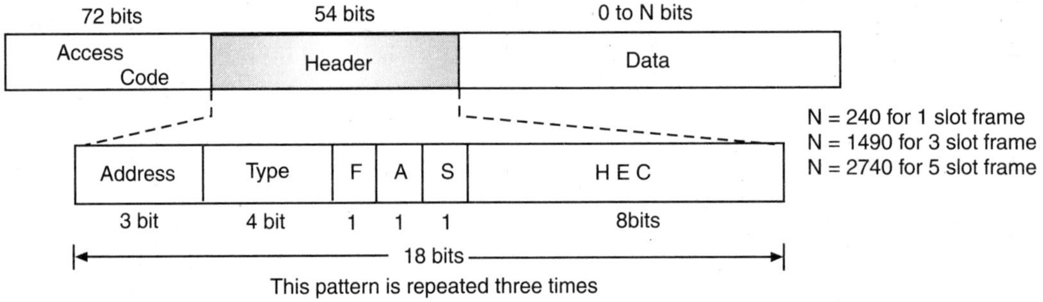

Fig. 6.20. Frame format

The description of important fields is as under :

1. Access code: It is a 72 bits field which contains synchronization bits. It also contains the identifier of the master so as to distinguish the frame of one piconet from another.

2. Header : It is a 54 bits field which contains an 18 bits pattern repeated three times as shown in figure 6.20.

Each such 18 bits pattern consists of the following fields :

(i) **Address :** It is a three bit field. Hence, it can define upto seven slaves (1 to 7). The 000 address is reserved for the broadcast communication between a master and the slaves.

(ii) **Type :** This is a four bit subfield used for defining the type of data, coming from the upper layers.

(iii) **F :** This bit is used for the flow control. F = 1 is an indication of buffer full.

(iv) **A :** This bit is for acknowledgement, since Bluetooth uses the stop-N-wait ARQ.

(v) **S :** This bit is used to hold the sequence number.

(vi) **HEC :** This is an eight bit header error correction subfield. It contains the checksum to detect errors in the 18 bit header section.

6.5.9. Data (Payload) Field

This field contains the data or control bits. It can be 0 to 2740 bits long. It contains data or control bits coming from the upper layers.

6.5.10. Security Limitations in Bluetooth

Due to the aspect of radio waves, experts fear a security concern with Bluetooth. The issue can be addressed with three aspects: specific sequence of channel hopping known only to the sending and receiving devices, challenge-response authentication routine to verify the validity of the receiving unit, and the 128-bit key encryption standard for securing transmission between devices.

6.5.11. Bluetooth Advantages

(i) One can create a personal area network at home or on the road with Bluetooth-enabled devices such as keyboard, mouse, scanner, PDA, laptop, cell phone, etc.

(ii) This network can automatically help synchronize notes, calendar, address book and also print pictures, receive emails, access cell phones messages, etc. It can even help consumers pay bills with credit card through Bluetooth cash register if a Bluetooth PDA stores the card information.

6.6. DIFFERENCE BETWEEN BLUETOOTH AND WLAN IEEE 802.11x

Sr. No.	Bluetooth	IEEE 802.11 x
1.	Bluetooth hop frequency is 1600 hops/second	IEEE 802.11x hop frequency is 2.5 hops/second
2.	Data transfer rate is 1 Mbps	Data transfer rate is 11 Mbps
3.	Transmission range is 10 m	Transmission range is 15-150 m indoor and 300 m outdoor
4.	Bluetooth uses lower transmission power	IEEE 802.11 uses more transmission power than Bluetooth
5.	It is used to connect devices that are in close proximity such as palm computing devices attached to smart phones, notebooks to printers.	It is a full LAN connectivity solution designed to provide full network service at Ethernet data rate
6.	Bluetooth is being a standard for short time network	IEEE 802.11 is a standard for LAN and is for longer time network.
7.	Bluetooth uses GFSK (Gaussian Frequency Shift Keying) modulation technique	IEEE 802.11 uses CCK (Complementary Code Keying) modulation technique

6.7. THE IEEE 802.16: THE BROADBAND WIRELESS

Figure 6.21 shows the structure of the wireless MAN. The IEEE developed the 802.16 standard as a replacement to the local network operators.

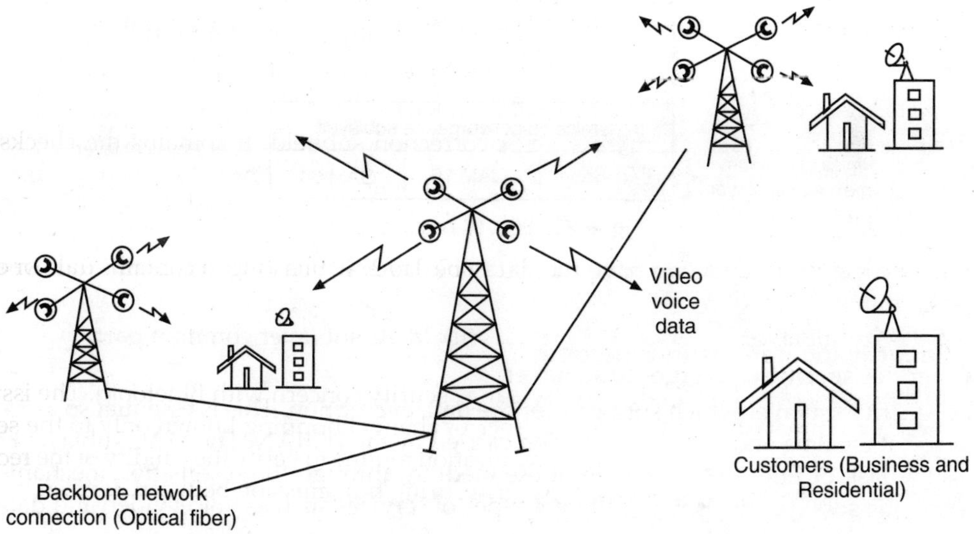

Fig. 6.21. Broadband wireless MAN IEEE 802.16

The 802.16 WMAN provides a standard solution for a cable free telecommunication service market. The IEEE 802.16 standard was published in 2002. Before this time, there were two other wireless MAN solutions already existing.

They were,

(i) Multichannel Multipoint Distribution System (MMDS)

(ii) Local Multipoint Distribution System (LMDS)

These two operated at different frequencies in the millimeter wavelength range. Both these WMAN solutions suffered due to lack of standardization. Hence, their use remained restricted.

The MMDS solution works at 2.4 GHz or 5 GHz band and has a range of upto 50 km. MMDS was designed originally for the wireless GATV solution, without a backward channel from the customer. But in order to make as much revenue as possible the MMDS solutions are used for voice and Internet services. The data rates for such applications are 128 kbps upto 3 Mbps downstream from base station to subscriber and 128 kbps from subscriber to the base station. The subscribers can access telephony and ISP through a modem. 802.16 MMDS and LMDS use the frequencies in the band 2 GHz to 60 GHz. With increase in frequency these waves become more and more directional. They travel in a straight line and can be easily absorbed by water. Hence, rain, snow, trees absorb these electromagnetic waves and create errors in the received signal. In order to overcome this problem, the signal produced by the base station and the customer stations are encoded using Hamming codes.

6.7.1. The 802.16 Protocol Stack

Figure 6.22 shows the 802.16 protocol stack. The physical layer is the lowermost layer with three modulation options i.e. QPSK (Quadrature Phase Shift Keying), 16-QAM (Quadrature Amplitude Modulation) and 64-QAM. These modulation schemes are used to counterbalance the range problem and the signal degradation due to absorptions.

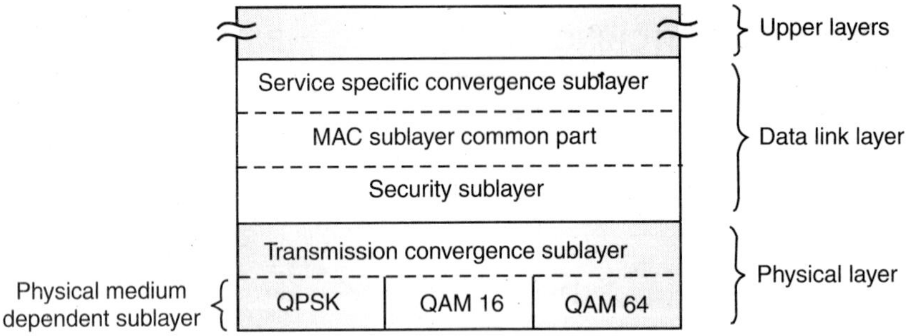

Fig. 6.22. 802.16 protocol stack

Above the physical layer, we have the data link layer which is divided into following three sublayers:

(i) Security sublayer (ii) MAC sublayer common part

(iii) Service specific convergence sublayer.

The security sublayer which supports for the data encryption. This is essential so as to protect the privacy of the data and video information carried by the WMAN. The MAC sublayer common part has the task of time slot allocation of the medium through the reservation mechanism. The service specific sublayer deals with different types of services such as voice, video and data.

6.7.2. 802.16 Frame Format

Figure 6.23 shows the frame formats of 802.16 WMAN system. There are two frames, namely the data frame and the control frame. In figures 6.23 (a) and (b), we have

EO	=	Encryption payload
Type	=	Type of frame
CI	=	Check sum indication

EK	=	Which of the 4 encryption keys used
Length	=	Including Header
Header CRC	=	$X^8 + X^2 + X + 1$

The control frame has been shown in figure 6.23(b). It is used to request time slots for different types of services. The control frame identifies the terminal through the connection ID. The type field which is 6 bit long identifies the type of control packet. Bytes needed field identifies how much data (in terms of number of bytes), the terminal wants to transmit. Figure 6.23(a) shows the data frame format. The first bit in the frame decides whether this frame is a data frame or control frame.

0 - Data frame

1 - Control frame

The 6 bit type field identifies the type of frame connection ID and Header CRC fields are same as those in the control frame.

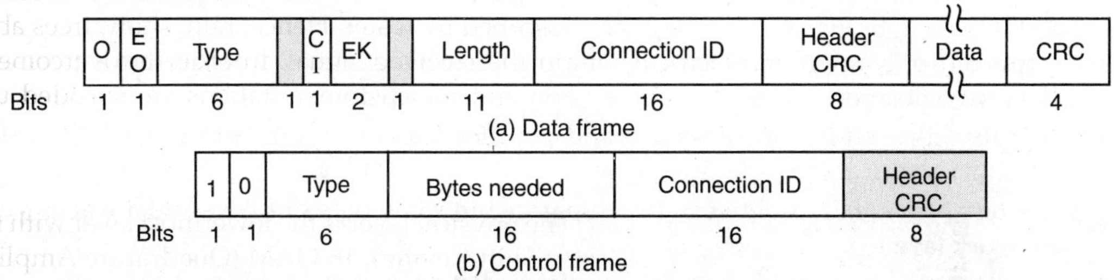

Fig. 6.23. 802.16 Frame formats

6.8. DATA LINK LAYER SWITCHING: NETWORK CONNECTING DEVICES*

(RGTU, Bhopal, Sem. Exam., June, 2014)

1. **Classification:** Figure 6.24 illustrates the different types of network connecting devices.

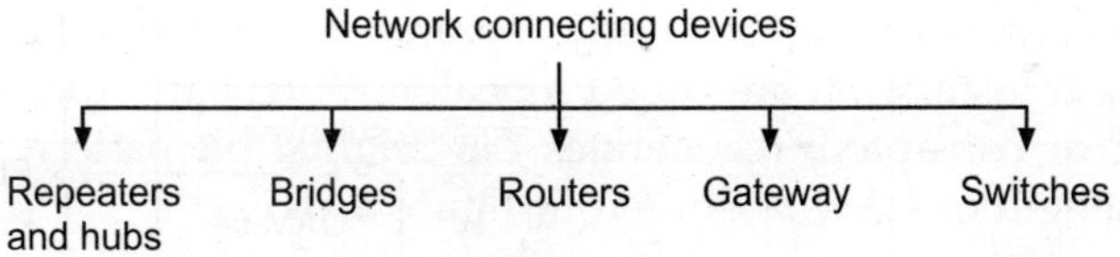

Fig. 6.24.

2. **Relation between OSI model and Connecting Devices:** Figure 6.25 illustrates the relation between OSI reference model and various connecting devices.

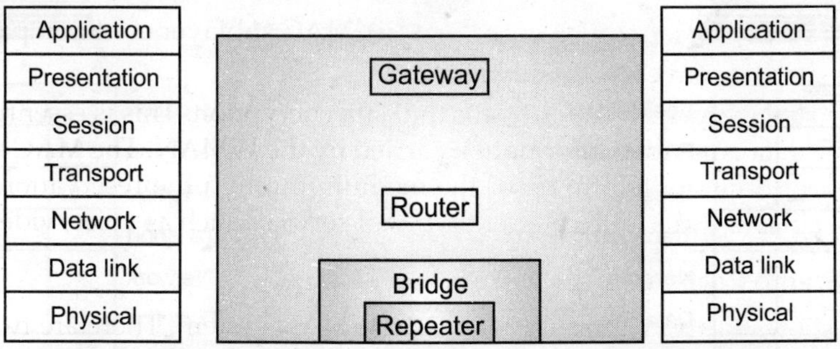

Fig. 6.25. Illustration of connecting devices and OSI model

* Explain uses of different network devices. *(RGTU, Bhopal, Sem. Exam., June, 2014)*

3. Network Connecting Devices : Brief Introduction

(i) We know that the two or more devices connected for the purpose of sharing data or resources form a network. In fact, putting together a network is more complicated than just simply plugging cable into hub. A LAN may require to cover more distance than its media can handle effectively.

(ii) The number of stations may be many for efficient frame delivery or management of the network and the network requires to be subdivided.

(iii) A device called a repeater is inserted into the network to increase the coverable distance or a device called a bridge can be inserted for traffic management.

(iv) When two or more separate networks are connected for exchanging data or resources, they become an Internetwork. Routers and gateways are used for Internetworking. Each of these device type interacts with protocols at different layers of the OSI model.

(v) Repeaters only upon the electrical components of a signal and are therefore active only at the physical layer.

(vi) Bridges utilize addressing protocols and can affect the flow control of a single LAN. Bridges are most active at the data link layer.

(vii) Routers provide links between two separate but same type of LANs and are active at the network layer.

(viii) Finally, gateways provide translation services between incompatible LANs or applications and are active in all of the layers.

Connecting devices and the OSI model has been shown in figure 6.26.

6.9. REPEATERS

Basically, a repeater may be defined as a device which operates only in the physical layer. We know that all transmission media weaken the electromagnetic waves that travel through them. Attenuation of signals limits the distance any medium can carry data. Devices that amplifies signals to ensure data transmission are called repeaters. A repeater receives a signal and before it becomes too weak or corrupted, regenerates the original bit pattern. Hence, a repeater can extend the physical length of LAN as shown in figure 6.26.

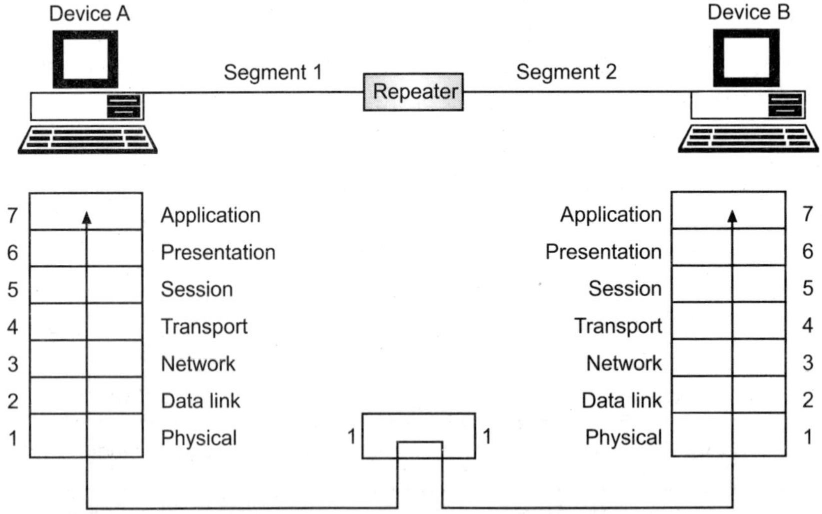

Fig. 6.26. Repeater in OSI model

Repeater is not an amplifier because amplifiers simply amplify the entire incoming signal along with noise. Signal-regenerating repeaters create an exact duplicate of incoming data by identifying it amidst the noise, reconstructing it and retrans-mitting only the desired information. The original signal is duplicated, boosted to its original strength and sent as shown in figures 6.26 and 6.27. A repeater does not actually connect two LANs. It connects only two segments of the same LAN. It cannot connect two LANs of a different protocol. A repeater forwards every frame, it has no filtering capability. A repeater must be placed at a precise point on the link. Such that the signal reaches it before the noise has induced an error in any of the transmitted bits. Figure 6.27 illustrates the function of a repeater. Repeaters operate at the physical layer of the OSI model and they deal with the actual physical signals.

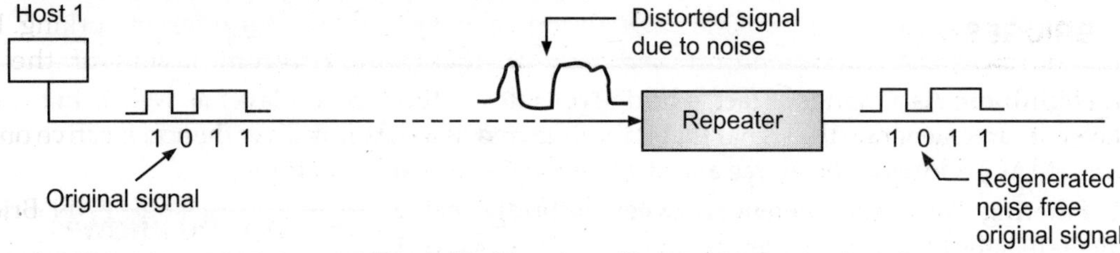

Fig. 6.27. Function of a repeater.

6.10. HUBS

1. Definition: Usually, the word *hub* can refer to any connecting device. However, its specific meaning is multiport repeater. It is basically used for connecting stations in a physical star topology. All networks require a central location to bring media segments together. These central locations are called *Hubs*. A hub organizes the cables and relays signals to the other media segments as shown in figure 6.28.

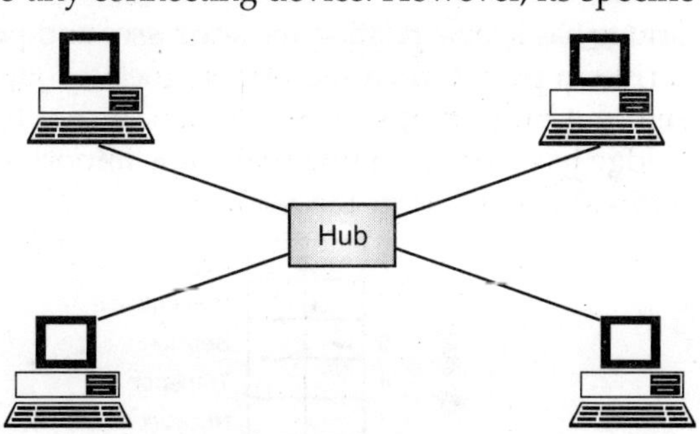

Fig. 6.28. Illustration of Hubs

2. Types: There are following three main types of hubs:

(i) Passive
(ii) Active
(iii) Intelligent

(i) Passive Hubs: A passive hub simply combines the signals of a network segments. There is no signal processing or regeneration. A passive hub reduces the cabling distance by half because it does not boost the signals and in fact absorbs some of the signal. With a passive hub, each computer receives the signals sent from all the other computers connected to the hub.

(ii) Active Hubs: They are like passive hubs but have electronic components for regeneration and amplification of signals. By using active hubs, the distance between devices can be increased. The main drawback of active hubs is that they amplify noise along with the signals. They are also much expensive than passive *hubs*.

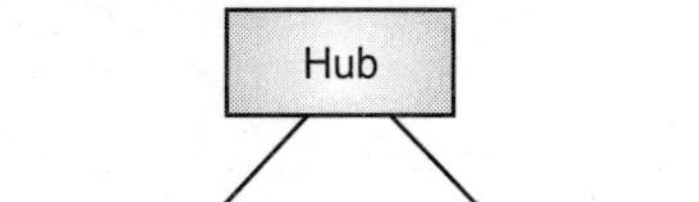

Fig. 6.29. Hubs to create multiple levels of hierarchy

(iii) Intelligent Hubs: In addition to signal regeneration, intelligent hubs perform some network management and intelligent path selection. A switching hub chooses only the port of the device where the signal needs to go, rather than sending the signal along all paths. One advantage to this is that all transmission media segment can be connected permanently because each segment is used only when a signal is sent to a device using that particular segment. Hubs can also be used to create multiple levels of hierarchy as shown in figure 6.29.

Important Point:
The number of hubs that can be connected to each other is typically four, but the maximum number of hubs depends on the type of network topology used. Whenever possible each hub should be connected directly to a server network card rather than to another hub. The more data hub passes through, the slower the connection.

6.11. BRIDGES

1. Definition: As a matter of fact, a bridge operates in the physical layer as well as in the data link layer. It can regenerate the signal that it receives and as a data link layer device, it can check the physical (MAC) addresses of source and destination contained in the frame.

2. Filtering: The major difference between the bridge and repeater is that the bridge has a filtering capability. This means that it can check the destination address of a frame and decide if the frame must be forwarded or dropped.

DO YOU KNOW?
Bridges interconnect LANs by passing frame from one network to another.

If the frame is to be forwarded, then the bridge must specify the port over which it must be forwarded. Hence, a bridge has a table relating the addresses and ports as shown in figure 6.30. If a frame for 723B134561 arrives a port 2, then the bridge goes through its table and understands that the frame is to be sent out on port 1, so it will do so. In figure 6.30, a two port bridge is shown, but, in reality a bridge has more than two ports. It is important to note that the bridges do not change the physical address contained in the frame.

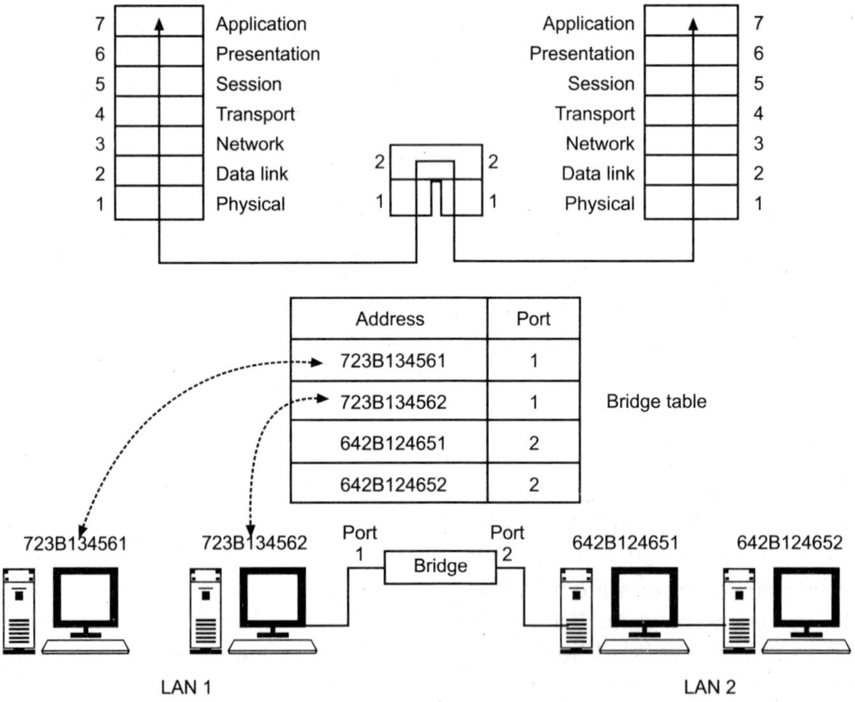

Address	Port
723B134561	1
723B134562	1
642B124651	2
642B124652	2

Fig. 6.30. Illustration of bridge and bridge table

3. Classification of bridges: The bridges may be of following two types:
(i) Transparent bridges, and
(ii) Routing bridges

Transparent bridge is a bridge in which the stations are not at all aware of the existence of the bridge. Transparent bridges keep a table of addresses in memory to determine where to send data. The duties of a transparent bridge may be listed as under :
(i) Filtering frames
(ii) Forwarding, and
(iii) Blocking

In source routing, a sending station defines the bridges that must be visited by the frames. The addresses of these bridges are included in the frame. Hence, a frame contains not only the source and destination address but also the bridge addresses. Source routing bridges are used to avoid a problem called looping. These bridges were designed for the token ring LANs. However, these LANs are not very common nowadays.

6.12. ROUTERS

Basically, Routers are devices that connect two or more networks as shown in figures 6.31 and 6.32. They consist of a combination of hardware and software.

The hardware can be a network server, a separate computer or a special device. The hardware includes the physical interfaces to the various networks in the Internetwork. These interfaces can be Token Ring, Ethernet, T1, Frame Relay, ATM or any other technology.

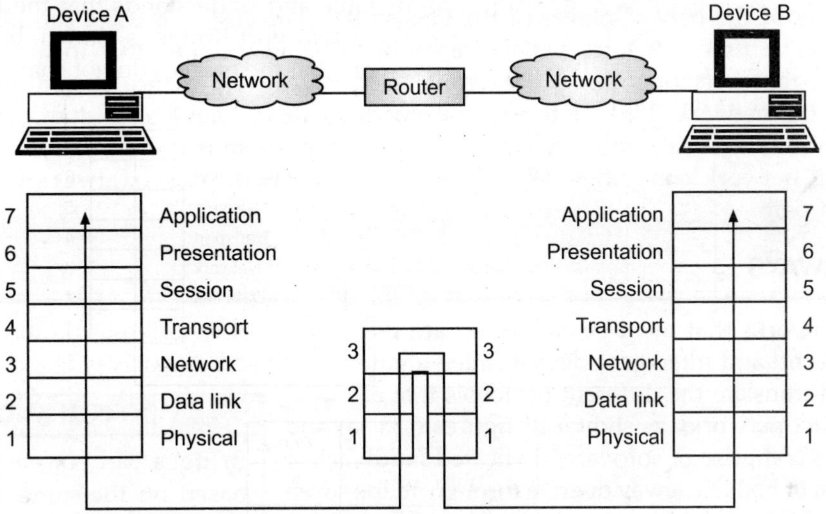

Fig. 6.31. Router in the OSI model

The software in a router are the operating system and the routing protocol. Management software can also be used. Routers use logical and physical addressing to connect two or more logically separate networks. They accomplish this connection by organising the large network into logical network segments called as **subnets**. Each of the subnet is given a logical address. This allows the networks to be separate but still access each other and exchange data. Data is grouped into packets, or blocks of data. Each packet has a physical device address as well as logical network address. The network address allows routers to calculate the optimal path to a workstation or computer.

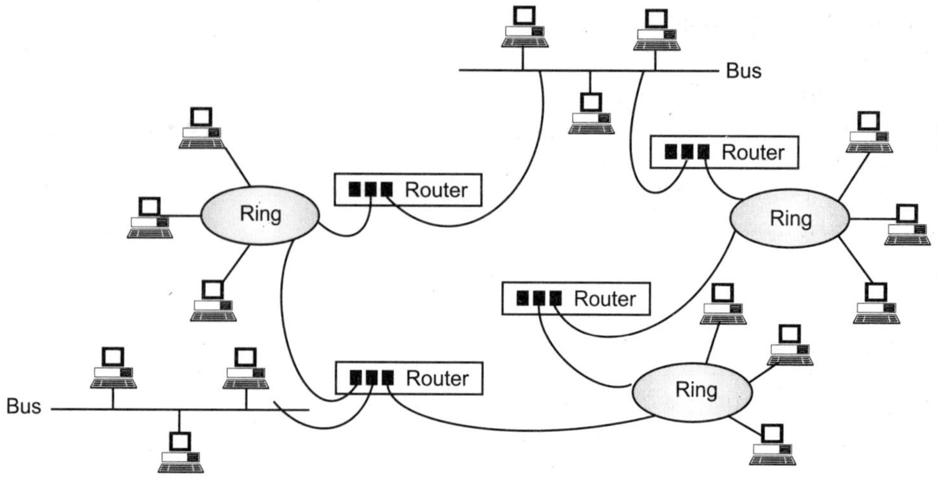

Fig. 6.32. Routers in an Internet

Route discovery is the process of finding the possible routes through the Internetwork and then building routing tables to store that particular information. The two methods of route discovery are as under:

(i) Distance vector routing.

(ii) Link state routing.

Important Note:

Routers work at the network layer of the OSI model. The cost of a route can be defined in several terms: a time estimate, a distance estimate or an estimate that includes monetary terms. In distance vector routing, each router advertises its presence periodically to other routers on the network. Link state routers broadcast their complete routing tables only at startup and at certain intervals. Dynamic route selection permits routers to constantly adjust to changing network conditions. With static route selection, packets always follow a pre-determined path.

6.13. GATEWAYS

When the networks that must be connected are using completely different protocols from each other, a powerful and intelligent device called a gateway is used. A gateway is a device that can interpret and translate the different protocols that are used on two distinct networks as shown in figures 6.33 (a) and (b). Gateways comprise of software, dedicated hardware or a combination of both. Gateway operate through all the seven layers of the OSI model. A gateway can actually convert data so that it works with an application on a computer on the otherside of the gateway. For example a gateway can receive e-mail message in one format and convert them into another format.

DO YOU KNOW?

Bridges can connect networks based on the same architecture (e.g., Ethernet to Ethernet) or interconnect networks of different architectures (e.g., Ethernet to token ring).

Gateways can connect systems with different communication protocols, languages and architecture. For example IBM networks using **Systems Network Architecture (SNA)** can be connected to LANs using a gateway.

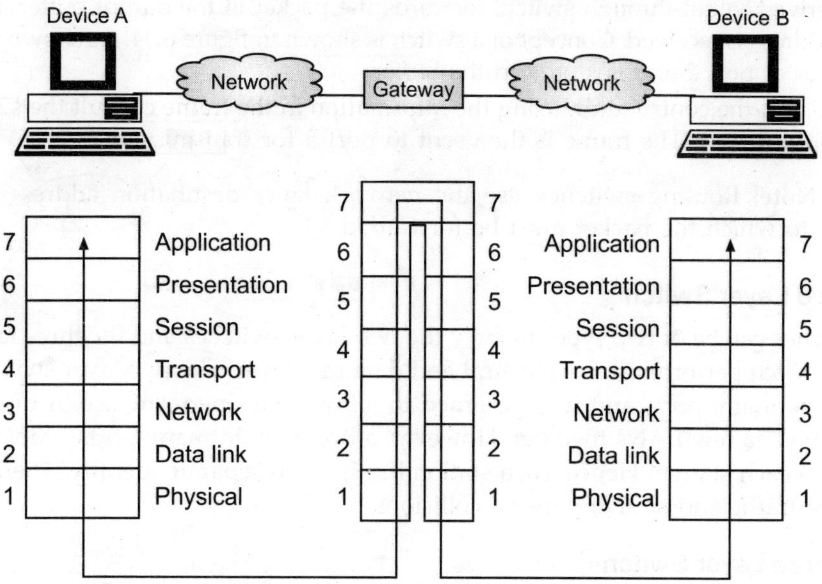

(a) A gateway in the OSI model

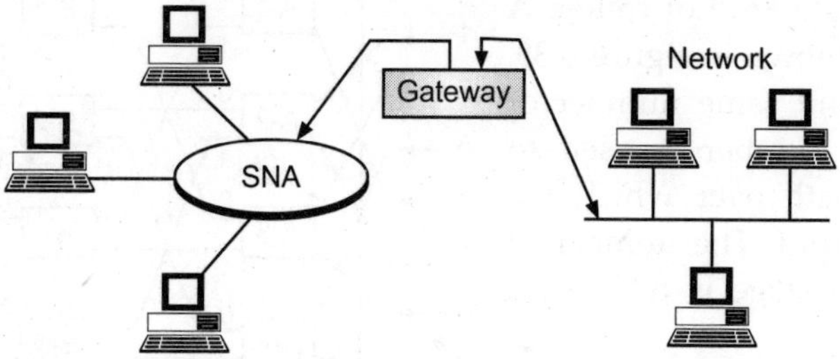

Fig. 6.33. (b)

Important Point:

Gateways are slow since they require to perform intensive conversions.

6.14. SWITCHES

1. Definition: A switch is a device which provides bridging functionality with greater efficiency. A switch acts as a multiport bridge to connect devices or segments in a LAN. The switch has a buffer for each link to which it is connected. When it receives a packet, it stores the packet in the buffer of the receiving link and checks the address to find the outgoing link. If the outgoing link is free, the switch sends the frame to that particular link.

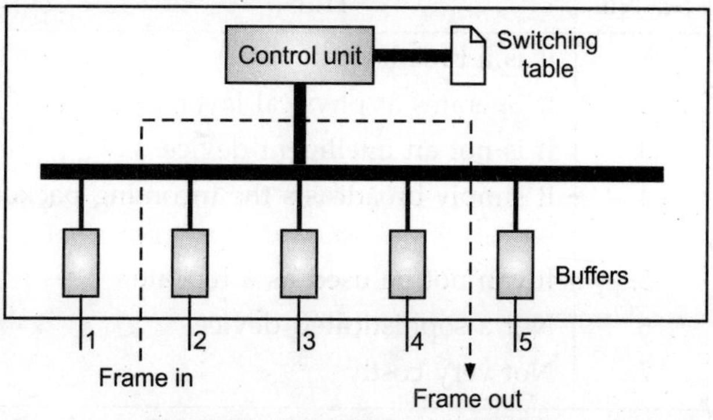

Fig. 6.34. Illustration of a Switch

2. Type: Switches are of following two types:

(i) Store-and-forward switch

(ii) Cut-through switch

A store-and-forward switch stores the frame in the input buffer until the whole packet has arrived. A cut-through switch, forwards the packet to the output buffer as soon as the destination address is received. Concept of a switch is shown in figure 6.34. As shown in figure 6.34, a frame arrives at port 2 and is stored in the buffer.

The CPU and the control unit, using the information in the frame consult the switching table to find the output port. The frame is then sent to port 5 for transmission.

Important Note: Routing switches use the network layer destination address to find the output link to which the packet must be forwarded.

6.14.1. A Two Layer Switch

The switches can be of two types namely the two layer switches and the three layer switches. A two layer switch performs at the physical and data link layer. The two layer switch is basically a bridge. It has many ports and it is designed to allow better performance. A bridge with few ports can connect a few LANs together. However, a bridge with many ports is able to allocate a unique port to each station. Hence, each station has its own separate identity. Therefore, there is no competing traffic and so there are no collisions.

6.14.2. A Three Layer Switch

A three layer switch is used at the network layer and it is a kind of router. A three layer switch is shown in figure 6.35. It has n = 8 inputs and same number of outputs. A three bit number is used to decide the internal path over which the input is passed to output. The number of microswitches at each stage is n/2 *i.e.*, 4 switches.

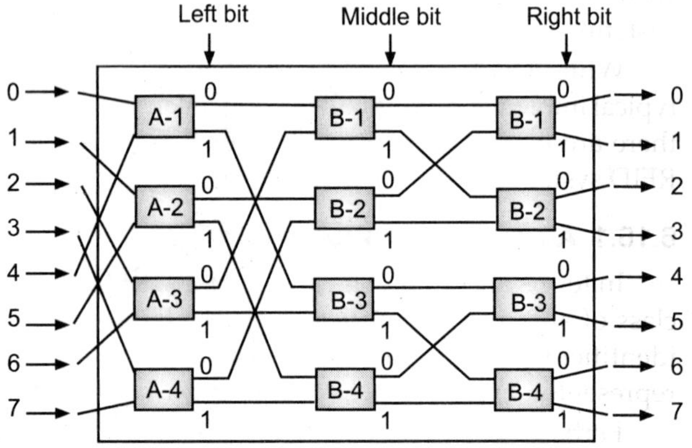

Fig. 6.35. A three layer switch

The first stage routes the cell based upon the high order bit in the binary bit string. The second stage routes the cell based upon the middle bit and last stage routes it based upon the low order bit. Here, it may be noted that the number of stages = $\log_2 (n) = \log_2 8 = 3$.

6.15. PERFORMANCE COMPARISON OF HUB AND SWITCH

Sr. No.	Hub	Switch
1.	It is a broadcast device	It is a point to point device
2.	It operates at physical layer	It operates at datalink layer
3.	It is not an intelligent device	It is an intelligent device
4.	It simply broadcasts the incoming packet	It uses switching table to find the correct destination
5.	It can not be used as a repeater	It can be used as a repeater
6.	Not a sophisticated device	It is a sophisticated device
7.	Not very costly	Costly

6.16. RFID

Radio frequency identification (RFID) is a rapidly growing technology that has the potential to make great economic impacts on many industries. While RFID is a relatively old technology, more recent advancements in chip manufacturing technology are making RFID practical for new applications and settings, particularly consumer item level tagging. These advancements have the potential to revolutionize supply-chain management, inventory control, and logistics.

As its basic, RFID systems consist of small transponders, or *tags,* attached to physical objects. RFID tags may soon become the most pervasive microchip in history. When wirelessly interrogated by RFID transceivers, or *readers,* tags respond with some identifying information that may be associated with arbitrary data records. Thus, RFID systems are one type of automatic identification system, similar to optical bar codes.

There are many kinds of RFID systems used in different applications and settings. These systems have different power sources, operating frequencies, and functionalities. The properties and regulatory restrictions of a particular RFID system will determine its manufacturing costs, physical specifications, and performance. Some of the most familiar RFID applications are item-level tagging with *electronic product codes,* proximity cards for physical access control, and contact-less payment systems. Many more applications will become economical in the coming years.

While RFID adoption yields many efficiency benefits, it still faces several hurdles. Besides the typical implementation challenges faced in any information technology system and economic barriers, there are major concerns over security and privacy in RFID systems. Without proper protection, RFID systems could create new threats to both corporate security and personal privacy.

6.16.1. Auto-Identification and RFID

In terms of commercial applications, RFID systems may be considered an instance of a broader class of automatic identification (auto-ID) systems. Auto-ID systems essentially attach a name or identifier to a physical object by some means that may be automatically read. This identifier may be represented optically, electromagnetically, or even chemically.

Perhaps the most successful and well-known auto-ID system is the Universal Product Code (UPC). The UPC is a one-dimensional, optical barcode encoding product and brand information. UPC labels can be found on most consumer products in the United States. Similar systems are deployed worldwide.

Auto-ID systems that transmit data via RF signals, i.e. RFID, do not have the same performance limitations as optical systems. Data may be read without line-of-sight and without human or mechanical intervention. A key advantage in RF-based auto-ID systems is parallelism. Modern RFID systems may offer read rates of hundreds of items per second.

6.16.2. Applications

Early commercial examples of RFID applications include automatic tracking of train cars, shipping containers, and automobiles. Railroad cars were originally labeled with optical bar code labels for tracking. These labels began to deteriorate and be obscured by dirt, causing reads to fail. As a solution, railroad companies began to tag railcars with RFID devices. By 1994, these devices were mandatory and nearly every railcar in the United States was tagged.

RFID devices began to be used for automated toll collection in the late 1980s and early 1990s. Electronic toll systems have since been adopted around the world. Like railway and shipping applications, electronic toll systems may use sturdy, self-powered RFID devices. Automobiles, railcars, and shipping containers are all high-value items, with ample physical space that can accommodate more expensive and bulky RFID devices. These types of tags could offer much more functionality than simple identification. For example, shipping containers might have accelerometer sensors, tamper alarms, or satellite tracking integrated into an identification device.

As manufacturing costs dropped, RFID systems began to be used for lower-value items in industries besides transport. An example is in animal identification of both pets and livestock. Glass-encapsulated RFID devices have been implanted in millions of pets throughout the United States. These tags allow lost animals to be identified and returned to their rightful owners. These tags have a very short read range.

Livestock, particularly cattle, are often labeled with a RFID device that is clamped or pierced through their ear, attached to a collar, or swallowed. Unlike implanted pet tags, these RFID devices are rugged and able to be read from greater distances.

6.16.3. Principles

This sub section discusses basics of RFID systems and offers taxonomy of the many various types of RFID systems. We briefly discuss two major RFID standards and how they relate to practice.

1. System Essentials: Discussion of RFID technology tends to focus only on tag devices. It is more accurate to view RFID as a complete system that includes not only tags, but also other important components. RFID systems are composed of at least three core components:

- RFID tags, or transponders, carry object-identifying data.
- RFID readers, or transceivers, read and write tag data.
- Databases associate arbitrary records with tag identifying data.

We illustrate the interaction of these components in figure 6.36. In this figure, three tags are readable by one or both of two readers, A and B. For instance, tag 1 is only readable by A, while 2 is readable by both A and B, perhaps due to access control restrictions. The readers then may connect to databases with records associated with particular tag identifiers. In this case, two databases each have their own record for tag 1 .

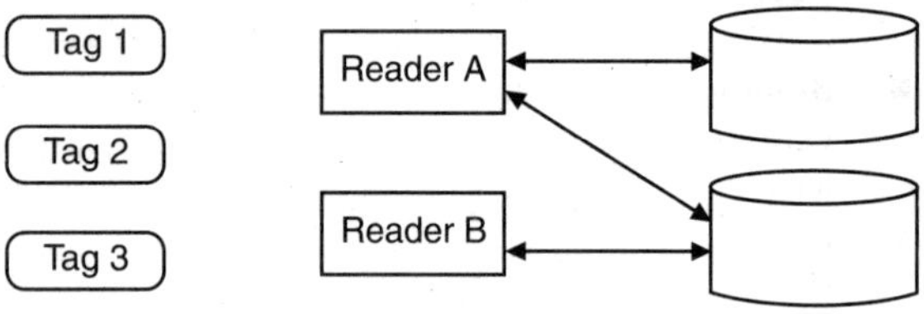

Fig. 6.36. Illustration of RFID System Interaction

2. Tags: Tags are attached to all objects to be identified in an RFID system. A tag is typically composed of an antenna or coupling element, and integrated circuitry. Often tags carry no on-board power source and must passively harvest all energy from an RF signal.

3. Readers: RFID readers communicate with tags through an RF channel to obtain identifying information. Depending on the type of tag, this communication may be a simple *ping* or may be a more complex multi-round protocol. In environments with many tags, a reader may have to perform an *anti-collision* protocol to ensure that communication conflicts to not occur. Anti-collision protocols permit readers to rapidly communicate with many tags in serial order.

4. Databases: RFID databases associate tag-identifying data with arbitrary records. These records may contain product information, tracking logs, sales data, or expiration dates. Independent databases may be built throughout a supply chain by unrelated users, or may be integrated in a centralized or federated database system.

5. Power Sources: As briefly mentioned before, tags may obtain their power in several different ways. The power source is an essential property of a tag, since it will determine a tag's potential read range, lifetime, cost, and what kind of functionalities it may offer. The power source will also be important in determining how a tag may be oriented and what physical forms it may take.

6.16.4. Operating Frequencies

Different RF1D systems operate at a variety of radio frequencies. Each range of frequencies offers its own operating range, power requirements, and performance. Different ranges may be subject to different regulations or restrictions that limit what applications they can be used for.

The operating frequency determines which physical materials propagate RF signals. Metals and liquids typically present the biggest problem in practice. In particular, tags operating in the ultra-high frequency (UHF) range do not function properly in close proximity to liquids or metal.

Operating frequency is also important in determining the physical dimensions of an RFID tag. Different sizes and shapes of antennas will operate at different frequencies. The operating frequency also determines how tags physically interact with each other. For instance, stacking flat foil inlay tags on top of each other may interfere or prevent tags from reading properly. Table 6.2. lists standard frequencies and their respective passive read distances.

Table 6.2. Common RFID operating frequencies

Frequency Range	Frequencies	Passive Read Distance
Low Frequency (LF)	120- 140 KHz	10-20 cm
High Frequency (HF)	13.56MHz	10-20 cm
Ultra- High Frequency (UHF)	868-928 MHz	3 meters
Microwave	2.45 & 5.8 GHz	3 meters
Ultra-Wide Band (UWB)	3.1-10.6 GHz	10 meters

GLOSSARY

- **Bluetooth** An emerging wireless convergence technology which has been designed to interconnect various devices, including computers, mobile phones, mobile computers, and hand-held or portable terminals using short-range (approximately 10 m) radio links.
- **Gateway** A software application which converts between different application protocols. The host on which this software resides is called a gateway machine. Basically, this term also refers to a router in the IP community.
- **Global Area Network (GAN)** A collection of WANs which spans the globe.
- **Hertz** A measure of frequency in cycles per second. A frequency rate of one cycle per second is defined as 1 hertz (Hz). It has been named in honour of Heinrich Rudolph Hertz, a German physical who in the late 1880s was the first to discover radio waves artificially.
- **IEEE 802** The primary IEEE standard for the 802 x series for LANs and MANs.
- **IEEE 802.1** IEEE standard which defines an architectural overview of LAN.
- **IEEE 802.2** IEEE standard which defines the logical link control, which describes services for the transmission of data between two nodes.
- **IEEE 802.3** IEEE standard that defines the carrier sense multiple access/collision detection (CSMA/CD) access method commonly referred to as Ethernet. In fact supplements include 802.3c (10-Mbps Ethernet), 802.3u (100-Mbps Ethernet, known as Fast' Ethernet), and 802.3z and 802.3b (100 Mbps Ethernet, known as Gigabit Ethernet).
- **IEEE 802.4** IEEE standard which defines the token bus network access method.
- **IEEE 802.5** IEEE standard which defines the logical ring LAN that uses a token passing access method; known also as token ring.
- **IEEE 802.6** IEEE standard which defines metropolitan area networks (MANs).
- **IEEE 802.7** IEEE standard which defines broadband LANs (capable of delivering video, data, and voice traffic).

- **IEEE 802.9** IEEE standard which defines integrated digital and video networking—integrated services LANs (ISLANs).
- **IEEE 802.10** IEEE standard which defines standards for interoperable LAN/MAN security services.
- **IEEE 802.11** IEEE standard which defines standards for wireless media access control and physical layer specifications.
- **IEEE 802.12** IEEE standard which defines the demand priority access method for 100-Mbps LANs; known also as 100 Base-VG or 100VG-AnyLAN.
- **IEEE 802.13** Define nothing—IEEE was concerned about the superstitious overtones associated with 1.3.
- **IEEE 802.14** IEEE standard which defines a standard for cable TV-based broadband communication.
- **Router** A layer-3 device which is responsible for determining the appropriate path a packet takes to reach its destination. It is commonly referred to as gateway.

SUMMARY

- LAN protocols with the OSI architecture. This architecture was developed by IEEE 802 committee and it has been accepted as LAN standard. It is called as IEEE 802 reference model.
- In a broadcast network, the single broadcast channel is to be allocated to one transmitting user at a time. This is called channel allocation.
- Wireless Local-Area Network (WLAN) is a technology which uses radio frequency to transmit and receive data over the air, providing all the features and benefits traditional LANs but without the limitations of a cable.
- Wireless LAN technology standard 802.11b has the strongest momentum to becoming the main standard for corporate internal wireless LAN networks.
- Wireless LANs use electromagnetic air waves (radio or infrared) to communicate information from one point to another without relying on any physical connection.
- In a typical wireless LAN configuration, a transmitter/receiver (transceiver) device, called an access point, connects to the wired network from a fixed location using standard cabling. At a minimum, the access point receives, buffers, and transmits data between the wireless LAN and the wired network infrastructure.
- There are four important applications of wireless LAN which may be listed as under:
  (i) LAN extension (ii) Nomadic access
  (iii) Cross building interconnection (iv) Ad-hoc networks
- Bluetooth is the name given to a new technology using short-range radio links, intended to replace the cable(s) connecting portable and/or fixed electronic devices. It is envisaged that it will allow for the replacement of many propriety cables that connect one device to another with one universal radio link. Its key features are robustness, low complexity, low power and low cost.
- Bluetooth defines following two types of networks :
  (i) Piconets, and (ii) Scatternets
- Bluetooth uses a form of TDMA that is called TDD-TDMA (Time Division Multiple Access).
- The 802.16 WMAN provides a standard solution for a cable free telecommunication service market. The IEEE 802.16 standard was published in 2002. Before this time, there were two other wireless MAN solutions already existing.

SHORT QUESTIONS WITH ANSWERS

Q.1. What does WLAN stand for?

Ans. Wireless LAN.

Q.2. What are the WLAN topologies?

Ans. Managed wireless network (BSS and ESS) and wireless unmanaged network (Ad-Hoc).

Q.3. What is an Ad-Hoc topology?

Ans. Ad-Hoc topology is made of wireless clients without access points.

Q.4. What is the function of an access point device in WLAN?

Ans. An access point acts like a bridge, receiving information and retransmitting through the air and its wired port.

Q.5. What are the IEEE standards for WLAN and their data rates?

Ans. 802.11 with data rate of 1 and 2 Mbps
802.11b with data rate 1 , 2, 5.5, and 11 Mbps
802.11a with data rate 6, 12,18, 24, 36, 45, and 54 Mbps
802.11g with data rate 6, 12, 18, 24, 36, 45, and 54 Mbps.

Q.6. What are the types of frame formats for WLAN?

Ans. Management frame, control frame, and data frame.

Q.7. How many devices can be in one Piconet?

Ans. 7 devices.

Q.8. How many slaves are in a Piconet?

Ans. 6 slaves.

Q.9. What are the types of physical connections supported by Bluetooth?

Ans. The physical connections supported by Bluetooth are synchronous connection oriented and asynchronous connectionless links.

Q.10. What is the IEEE number for Wireless MAN?

Ans. IEEE 802.16.

Q.11. What is the topology for IEEE 802.16?

Ans. IEEE 802.16 uses PMP and optional mesh topology.

Q.12. What is the frequency spectrum of IEEE 802.16?

Ans. 10–66 GHz.

Q.13. What is difference between IEEE 802.16a and the 802.16 Physical layer?

Ans. 802.16 operates in line-of-sight but 802.16a operates in non-line-of-sight.

Q.14. What are the types of connections that are used by IEEE 802.16?

Ans. Basic connection, primary management connection, and secondary management connection.

REVIEW QUESTIONS

1. What are the properties of autonegotiation? Explain.
2. Why can not ethernet and token ring be used in a WAN? Explain.
3. When a remote bridge is used? Explain.
4. What is transparent bridge? Explain.

Chapter 7

The Network Layer

Syllabus

The Network Layer: Design Issues, Virtual Circuit and datagram networks, routing algorithms- adaptive and non-adaptive algorithms, congestion control algorithms, quality of service, internetworking, Network layer in the Internet- IPv4 protocol, IP addresses, IPv6 protocol, Internet control protocols, Mobile IP.

Chapter Outline

- Network Layer
- Network Layer issues
- Introduction to Routers
- Routing Algorithms
- Static Algorithms
- Dynamic Routing Algorithms
- Hierarchical Routing
- Routing for Mobile Hosts
- Broadcast Routing
- Multicast Routing
- Congestion
- Congestion Control in Datagram Subnets
- Quality of Service (QoS)
- Queueing Disciplines
- QoS in ATM
- Internetworking
- Tunneling
- Internetwork Routing
- Fragmentation
- Firewalls
- Introduction to Various Network Layer Protocols
- ARP (Address Resolution Protocol)
- Reverse Address Resolution Protocol (RARP)
- Network layer in the internet: Internet Protocol (IP)* : IPv4 Protocol

7.1. NETWORK LAYER

As a matter of fact, the network layer is responsible for carrying the packet from the source all the way to destination. In short, it is responsible for host-to-host delivery. We can also say that the network layer is responsible for an important job called routing and the IP protocol in the TCP/IP protocol suite handles this important assignment. Once the router knows the location of other routers, it can send the packets to a given destination. It is known as **forwarding**. In other words, routing is about knowing where all other routers are located and to which networks they are connected to forwarding relates to finding one's own neighbouring routers and sending specified packet to its nearest neighbour router. The network layer has a higher responsibility than the data link layer, because the data link layer is only supposed to move the frames from one end of the wire to the other end. Thus, network layer is the lowest layer that deals with the end to end transmission.

7.1.1. Position of Network Layer

Figure 7.1 shows the position of network layer in the five layer internet model. It is the third layer as shown in figure 7.1.

A network layer receives services from the data link layer and provides services to the transport layer.

- IPv6 Protocol*
- Extension Headers
- Internet Control Protocols
- Mobile IP
- Terms used in Mobile IP
- Mobile IP Routing Operation
- Functionality in the Home Network
- Moving to Foreign Network
- Functionality in the Foreign Network
- Mobile IP Route Optimization
- Limitations of Mobile IP

Pioneer in Communications: Vinton G. Cerf

Vinton G. Cerf is widely popular as a Father of Internet. Cerf initially got involved with the development of the Internet when, as a graduate student at UCLA from 1967-72, he had been a part of a team that worked upon one component of the U.S. Department of Defence Advanced Research Projects Agency's (DARPA) packet-switching network project, ARPANET.

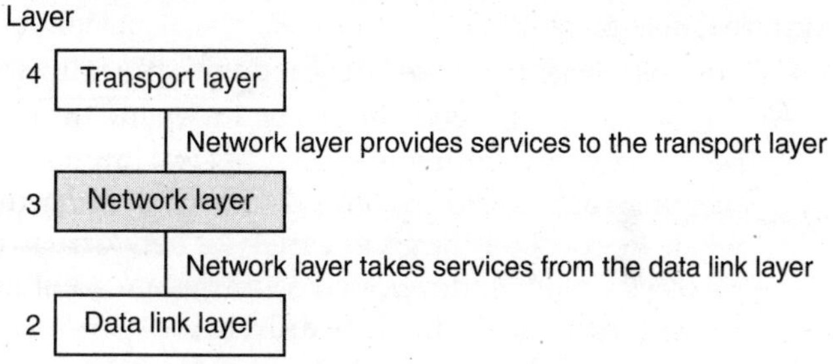

Fig. 7.1. Position of network layer

7.1.2. Duties of Network Layer

Figure 7.2 illustrates the set of duties of the network layer.

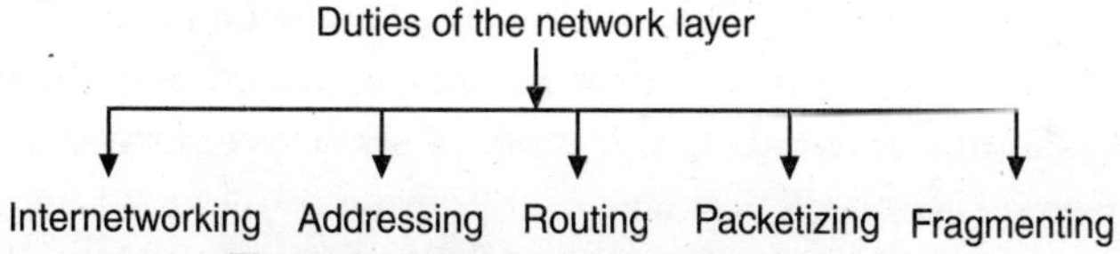

Fig. 7.2. Duties of the network layer

1. Internetworking: This is the main duty of network layer. It provides the logical connection between different types of network.

2. Addressing: Addressing is necessary to identify each device on the Internet uniquely. This is similar to a telephone system. The addresses used in the network layer should uniquely and universally define the connection of a computer.

3. Routing: In a network, there are multiple roots available from a source to a destination and one of them is to be chosen. The network layer decides the root to be taken. This is called as routing and it depends on various criterions.

4. Packetizing*: As discussed earlier, the network layer encapsulates the packets received from upper layer protocol and makes new packets. This is called as packetizing. It is done by a network layer protocol called IP (Internetworking protocol).

5. Fragmenting: The datagram can travel through different networks. Each router decapsulates the IP datagram from the received frame. Then the datagram is processed and encapsulated in another frame.

6. Other issues: The other issues which are not directly related to the duties of network layer but need to be discussed are as under:

*At the network layer, the content is known as a packet.

(i) Address resolution.
(ii) Multicasting.
(iii) Routing protocols.

7. Other Supporting Protocols: The internetworking protocol (IP) needs the support of another protocol ICMP (Internet Control Message Protocol) to achieve the host-to-host delivery.

8. How to Achieve the Goals? : In order to achieve the goals, the network layer must know about the topology of the communication subnet *i.e.* the set of all routers. It also should choose appropriate paths for communication. The routes should be chosen in such a way that overloading of some routers and idle operation of others should be avoided.

7.2. NETWORK LAYER DESIGN ISSUES

The important network layer design issues include the service provided to the transport layer and the internal design of subnet.

The network layer has been designed with the following goals:

(i) The services provided should be independent of the underlying technology. Users of the service need not be aware of the physical implementation of the network - for all they know, their messages could be transported via carrier pigeon. This design goal has great importance when we consider the great variety of networks in operation. In the area of Public networks, networks in under developed countries are nowhere near the technological prowess of those in the countries like the U.S. or Ireland.
(ii) The design of the layer must not disable us from connecting to networks of different technologies.
(iii) The transport layer (that is the host computer) should be shielded from the number, type and different topologies of the subnets user uses. That is, all that transport layer wants is a communication link, it need not know how that link is made.
(iv) Finally, there is a need for some uniform addressing scheme for network addresses.
(v) With these goals in mind, two different types of services emerged as follows:
(a) Connection oriented Network Services
(b) Connectionless Network Services.

A connection-oriented service is one in which the user is given a reliable end to end connection. To communicate, the user requests a connection, then uses the connection to his hearts content, and then closes the connection. A telephone call is the classic example of a connection oriented service. In a connectionless service, the user simply bundles his information together, puts an address on it, and then sends it off, in the hope that it will reach its destination. There is no guarantee that the bundle will arrive. Therefore, a connectionless service is one similar to that of the postal system. A letter is sent, that is, put in the post box. It is then in the post network where it gets bounced and hopefully will leave the network in the correct place, that is, in the addressee's letter box. We can never be totally sure that the letter will arrive, but we know that there is a high probability that it will, and so we place our trust in the postal network.

7.2.1. Services Provided to the Transport Layer

The network layer services are designed to achieve the following goals :

(i) The services should be independent of the subnet technology.
(ii) Transport layer should be shielded from the number, type and topology of the subnet.
(iii) The network address which is made available to the transport layer must use a uniform numbering plan.

The network service can be connectionless or connection oriented. The internet has a connectionless network layer whereas the ATM networks have a connection oriented network layer. The connection oriented and connectionless services both have their own sets of advantages and disadvantages. Finally, we can say that the network layer should provide a raw means to send packets from *a* to *b* and that is all.

7.2.2. Virtual Circuit and Datagram Networks: Internal Organization of the Network Layer

1. Basic Concepts: Basically, there are two methodologies for organizing the subnet as under:

(i) To use connection oriented service.

(ii) To use connectionless service.

In the connection oriented service, a connection is called as virtual circuit. It is similar to a physical connection.

In the connectionless organization, the independent packets are called as **datagrams,** in analogy with telegrams.

2. Virtual circuits: The principle behind the virtual circuits is to choose only one route from source to destination. When a connection is established, it is used for all the traffic flowing over the connection. When the connection is released, the virtual circuit is terminated.

3. Datagram: With a datagram, the routes from source to destination are not worked out in advance. Each packet sent is routed independently. Successive packets can follow different routes. The datagram subnets have to do more work but they are more robust and deal with failures and congestion more easily as compared to virtual circuit subnets.

4. Features of Virtual Circuits: In virtual circuits, every router has to maintain a table. Each packet must have a virtual circuit number field in its header in addition to sequence number checksum etc. Circuit set up is required in virtual circuits. The virtual circuits can be initiated from either end. But problem can arise if call set ups are propagating in both directions simultaneously. The users are charged for connect time as well as for data transported.

5. Features of a Datagram: The routers do not have to maintain any table. Each datagram must contain full destination address. These addresses can be very long. When a packet comes in, the router finds an available outgoing line and sends the packet out on that line.

7.2.3. Performance Comparison of Virtual Circuit and Datagram Subnets

Table 7.1 shows the performance comparison of VC subnet and datagram subnets.

Table 7.1.

S. No.	Parameter	VC subnet	Datagram subnet
1.	Circuit set up	Required	Not required.
2.	Addressing	Each packet contains a short VC number	Each packet contains the source as well as destination address.
3.	Repairs	Harder to repair	Easy to repair.
4.	State information	A table is required to hold the state information.	Subnet does not hold state information.
5.	Routing	Route chosen is fixed. All packets follow this route. This is static routing.	Each packet is routed independently. This is dynamic routing.
6.	Congestion control	Easy	Difficult.
7.	Effect of router failure.	All VCs which passed through failed router are terminated.	No other effect except for the packets lost at the time of crash.

7.3. INTRODUCTION TO ROUTERS

Routers are devices that connect two or more networks as shown in figures 7.3 (a) and (b). They consist of a combination of hardware and software. Whenever a router encounters a packet, it must decide where to pass that packet. A router might be connected to multiple routers. Routers need to find out where other routers are and how they are connected to each other. They are supposed to know the map of the network in terms of routers connected to each other. Routers learn about each other by applying special algorithms which are known as routing algorithms. Each router has a routing table that decides the route to be followed for each packet. Routers look into the routing table when a new packet arrives.

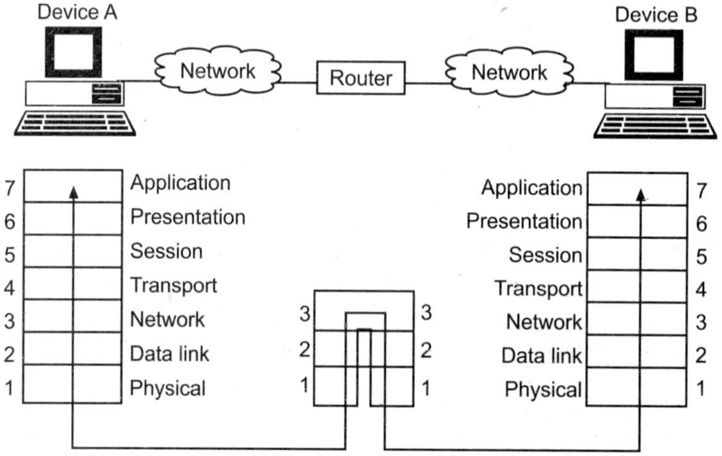

Fig. 7.3. (a) A router in the OSI model

The hardware can be a network server, a separate computer or a special device. The hardware includes the physical interfaces to the various networks in the internetwork. These interfaces can be Token Ring, Ethernet, TI, Frame Relay, ATM or any other technology.

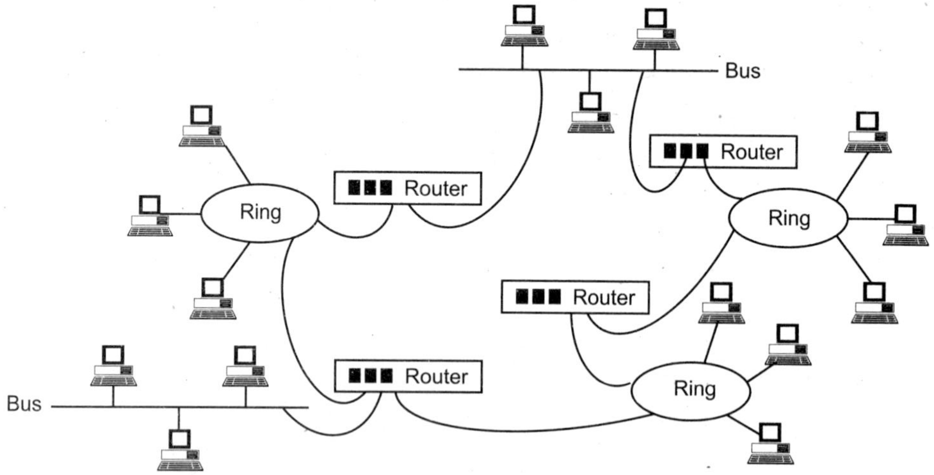

Fig. 7.3. (b) Routers in an Internet

The software in a router are the operating system and the routing protocol. Management software can also be used. Routers are logical and physical addressing to connect two or more logically separate networks. They accomplish this connection by organising the large network into logical segments called as subnets. Each of the subnet is given a logical address. This allows the networks to be separate but still access each other and exchange data. Data is grouped into packets, or blocks of data. Each packet has a physical device address as well as logical network address. The network address allows routers to calculate the optimal push to a workstation or computer. Route discovery is the process of finding the possible routes through the internetwork and then building routing tables to store that information. The two methods of route discovery are as under:

> **DO YOU KNOW?**
>
> With a datagram subnet, the congestion avoidance is more difficult.

(i) Distance vector routing.

(ii) Link state routing.

Important Point :

Routers work at the network layer of the OSI model. The cost of a route can be defined in several terms : a time estimate, a distance estimate or an estimate that includes monetary terms. In distance vector routing, each router advertises the presence periodically to other routers on the network. Link state routers broadcast their complete routing tables only at startup and at certain intervals. Dynamic route selection permits routers to constantly adjust to changing network conditions. With static route selection, packets always follow a pre-determined path.

7.4. ROUTING ALGORITHMS

Routing is a major component of the network layer and is concerned with the problem of determining feasible paths (or routers) or packets to follow from each source to each destination. One of the important functions of the network layer is to route the packets from the source machine to the destination machine. The major area of network layer design includes the algorithms which choose the routes and the data structures which are used. Routing algorithm is a part of network layer software. The network layer must determine the route or path taken by packets as they flow from a sender to a receiver. The algorithms that calculate these paths are referred to as **routing algorithms**. It is responsible for deciding the output line over which a packet is to be sent. Such a decision is dependent on whether the subnet is a virtual circuit or it is datagram switching.

7.4.1. Desired Properties of a Routing Algorithm

The main ingredients of a good routing algorithm depend on the objective function that one is trying to optimize.

There are certain desirable properties of a routing algorithm as under :

(i) Correctness. (ii) Robustness.

(iii) Stability. (iv) Fairness and

(v) Optimality.

> **DO YOU KNOW?**
>
> Regardless of whether routes are chosen independently for each packet or only when new connections are established, certain properties are desirable in a routing algorithm: correctness, simplicity, robustness, stability, fairness, and optimality.

7.4.2. Adaptive and Non-Adaptive Algorithms: Types of Routing Algorithms

Routing algorithms can be divided into two groups as under:

1. Non-adaptive algorithms, 2. Adaptive algorithms.

1. Nonadaptive algorithms: For this type of algorithms, the routing decision is not based on the measurement or estimation of current traffic and topology. However, the choice of the route is done in advance, off line and it is downloaded to the routers. This is called as static routing.

2. Adaptive algorithms: For these algorithms, the routing decision can be changed if there are any changes in topology or traffic etc. This is called as dynamic routing. In the following sections, let us discuss various static and dynamic algorithms.

3. Optimality principle: A general statement about optimality is called as optimality principle. It states that if router *J* in on the optimal path from router I to router K, then the optimal path from J to K also falls along the same route. This can be elaborated as, call the part of the route from *I* to *J* as *r*1 and rest of route as *r*2. If a route better than *r*2 existed from *J* to *K*, it could be concatenated with *r*1 to improve the route from *I* to *K*, contradicting our statement that *r*1 *r*2 is optimal.

> **DO YOU KNOW?**
>
> The routing algorithm is that part of the network layer software responsible for deciding which output line an incoming packet should be transmitted on.

4. Sink tree: A set of optimal routes from all the sources to a given destination form a tree called sink tree and it is shown in figure 7.4. Note that a sink tree need not be unique. Other trees with the same path lengths may also exist. All the routing algorithms are supposed to discover and use the sink trees for all routers. In the sink tree of figure 7.4, the distance metric is the numbers of hops. In figure 7.4 (*b*), a sink tree for router *B* has been shown. The paths from *B* to every router with minimum member of hops.

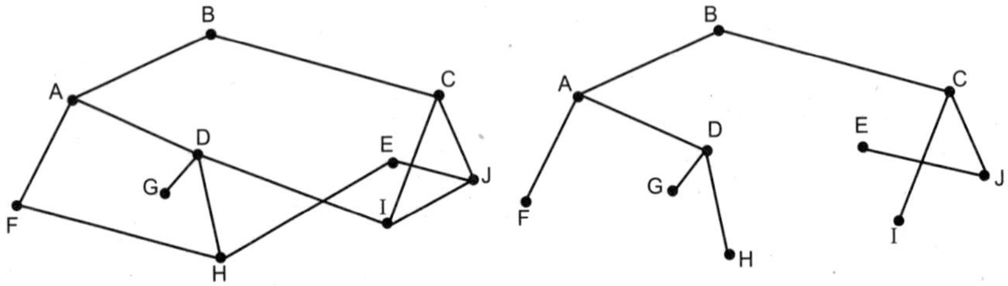

(a) A subnet (b) A sink tree for router B

Fig. 7.4.

7.5. STATIC ALGORITHMS

A broad way to classify routing algorithms is according to whether they are static or dynamic. In static routing algorithms, routes change very slowly over time, often as a result of human intervention (for example, a human manually editing a router's forwarding table). The examples of static algorithms may be listed as under :

(i) Shortest path routing.

(ii) Flooding.

(iii) Flow based routing.

> **DO YOU KNOW?**
>
> The routing algorithm should be able to cope with changes in the topology and traffic without requiring all jobs in all hosts to be aborted and the network to be rebooted everytime some router crashes.

7.5.1. Shortest Path Routing

1. Definition: This algorithm is based on the simplest and most widely used principle. How, a graph of subnet is built in which each node representing the router and each arc representing a link or a communication line. Hence, as to choose a path between a pair of routers, this algorithm simply finds the shortest path between them.

2. How to Decide the Shortest Path: One way of measuring the path length is the number of hops. Another way (metric) is the geographical distance in kilometres. Some other metrics are also possible. For example, we can label each arc (link) with the mean queueing and transmission delay and obtain the shortest path as the fastest path.

3. Labels on the Arcs: The labels on the arcs can be compared as a function of distance bandwidth, average traffic, mean queue length, cost of communication, measured delays etc. The algorithm weighs various parameters and computes the shortest path, based on any one or combination of criterions stated above.

4. Various Shortest Path Algorithms: There are many algorithms for computing the shortest path between two nodes. One of them is Dijkstra algorithm. The other one is Bellman-Ford algorithm.

7.5.2. Dijkstra's Shortest Path Algorithm

The algorithmic translation of physical process is called Dijkstra's shortest path algorithm. The algorithm is based on the following observations :

(a) The distance from ball 1 to ball n cannot decrease after ball n rises from the floor.

(b) Consequently, if we keep track of an estimated distance from ball 1 to ball n, we know that this estimated distance eventually settles to the shortest distance and that this happens when the ball rises from the floor.

1. Now, here how do we know which ball rises next ? Say that the ball k rises and let d to the distance from ball 1 to ball k. We update the estimates of the distance from ball 1 to the balls attached to ball k.
2. If some ball j is not jump yet and is attached to ball k with a string of length s, we replace the current estimated distance x from 1 to j by the minimum of x and d + s.
3. When we have updated all the neighbours of ball k, we must find the ball that will rise next. That ball is ball on the floor, say ball p, with the current smallest estimated distance away from ball 1. We can then continue the process with ball p.
4. We find explain the algorithm on the simple network shown in figure 7.5. The objective is to find the shortest path from A to all the other nodes. The lengths of the individual links are marked next to them.
5. On this small network, a simple inspection shows that the shortest path from A to bottom node has length 5 and goes through the right middle node. Dijkstra's algorithm is systematic procedure for discovering such a shortest path even in a large network. Next to each node, we mark the current estimate of the length of the shortest path from A to that node. The symbol ∞ means that the no path to that node has been found yet.
6. The algorithm starts by considering the node with the smallest label, in this case A. The algorithm explores the links going out of that node. The left link leads to the left middle node which can then be reached from A with a path of length 3. Accordingly, we reduce the label of that node from ∞ to 3. Since, the label is reduced by using that left link out of A, we mark the link as being the current candidate for the shortest path to the left middle node.
7. A similar step is performed for the right link out of A. We then shade node A to indicate that we have examined all its outgoing links. At the next step, we examine the unshaded node with the current smallest label. That node is the left middle node. We examine the outgoing link of that node and find a candidate shortest path to the bottom node with length 3 + 3 = 6. We shade that node and then examine the unshaded node with the smallest label ; the right middle node.
8. With the link out of that node, we find a shortest path to the bottom node. Accordingly, we unmark the link from the left middle node to the bottom node because we know that the link is not on the shortest path to the bottom node.
9. In addition, we mark the link from the right middle node to the bottom node. Since the bottom node has no outgoing link, the algorithm terminates. The marked links define the shortest paths from A to the nodes.
10. Next, we illustrate Dijkstra's shortest path algorithm with the example shown in figure 7.5. We start with node A as the source and we mark each node with an estimated distance

from A to the node. The initial estimates are infinite, except that of node A which is 0. The first node to rise from the floor is node A.

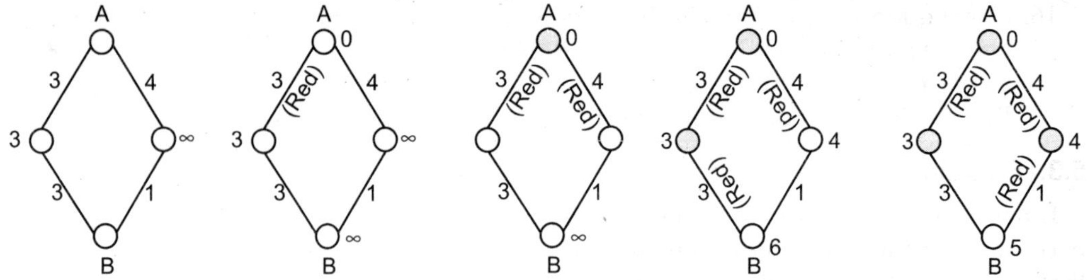

Fig. 7.5. Steps to Dijkstra's algorithm

11. We shade that node A to remember that it is off the floor and we update the estimates of all the neighbours B, C, D of A. For instance, we replace the current infinite estimate of B by the sum of the estimate of node A (equal to 0) plus the length 4 of the link from A to B.
12. The new estimate (4, 3, 2) are underlined in the second part of the figure 7.6. At that point, we determine the unshaded node with the smallest estimate. That node is node D and is therefore the next node to rise from the floor. We shade that node D.

    Here, it may be noted that all the links from node A to its neighbours are coloured red to remember that these are candidates for shortest paths from A to other nodes since these links have provided us with the smallest estimates of distances to the nodes B, C, D so far.
13. In the third part of the figure 7.6, we explore the neighbours of node *D* and we update their estimates and color the links from *D* to these neighbours red.

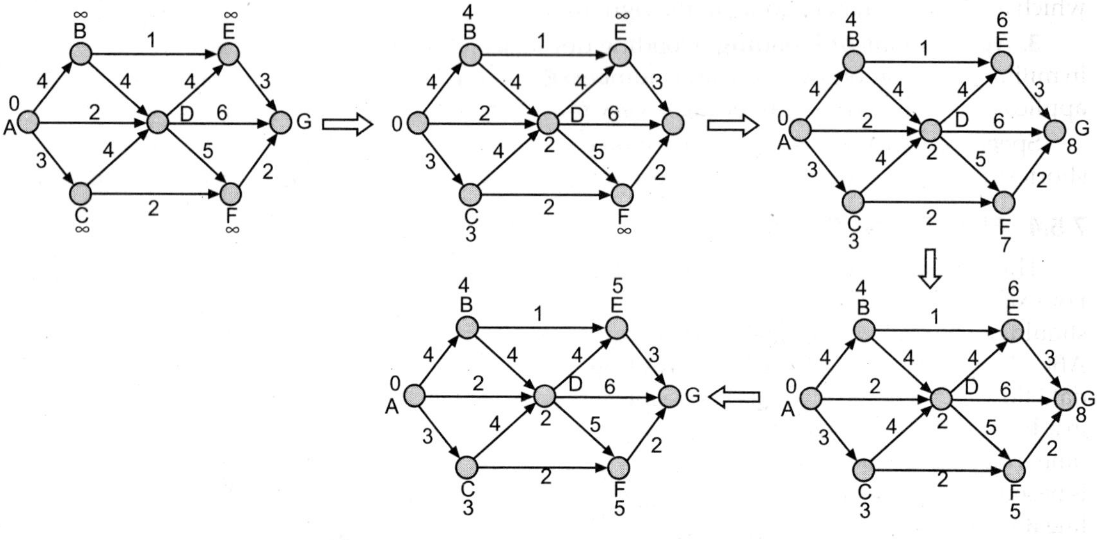

Fig. 7.6. Illustration of Dijkstra's algorithm

14. In the fourth part of the figure 7.6 we have located node C as the unshaded node with the smallest estimate; we shade C and we explore its neighbours and update their estimates.

    Here, it may be noted that the link CF produces a smaller estimate (5) for node F. Accordingly, we change the link DF from red to black we colour the link CF as red.
15. In the last part of the figure 7.6, we locate the unshaded node B with the smallest estimate; we shade node B and update the estimates of its neighbours. Note that the link from B to

D does not reduce the estimate of node D because 4 + 4 > 2. However the link from B to E reduces the estimate of E because 4 + 1 < 6.

16. Consequently, reflect this reduced estimate, we colour the link from B to E green and we change the link from D to E from red to black. Indeed, the link from D to E cannot be on a shortest path from A to E since it yields a longer path than a path that reaches E via link BE. The left most part of the figure 7.6 shows the shortest paths.

7.5.3. Flooding

1. Basic Concepts: This is another static algorithm. In this algorithm, every incoming packet is sent out an every outgoing line except the line on which it has arrived. One disadvantage of flooding is that it generate a large number of duplicate packets. In fact, it produces infinite number of duplicate packets unless we somehow damp the process. There are various damping techniques as under:

(i) Using a hop counter

(ii) To keep a track of which packets have been flooded.

(iii) Selective flooding.

To prevent endless copies of packets circulating indefinitely through the network a hop count may be used to suppress onwards transmission of packets after a number of hops exceeding the network diameter. The destination must be prepared to receive multiple copies of an incoming packet. Flooding has two interesting characteristics that arise from the fact that all possible routes are tried :

(i) As long as there is a route from source to destination, the delivery of the packet is guaranteed.

(ii) One copy of the packet will arrive by the quickest possible route.

2. Selective Flooding: This is slightly more practical variation of flooding. In this algorithm, every incoming packet is not sent out on every output line. Infact, packet is sent only on those lines which are approximately going in the right direction.

3. Applications of Flooding: Flooding does have much practical applications. But, it is useful in military applications where a large number of routers are blown into pieces at any instant. In such applications, robustness of flooding is very much desirable.

Special application is in the distributed database applications. Flooding always chooses the shortest path so it produces the shortest possible delay.

7.5.4. Flow Based Routing

This is a static algorithm which uses topology and load condition (traffic) for deciding a route. For example in figure 7.7, there is always a huge traffic from A to B. Then, the traffic from A to C should not be routed through B. Instead route it through AGEFC even though it is a longer path than ABC. This is called as a **flow based routing**. It is possible to optimise the routing by analysing the data flow mathematically. This is possible if the average traffic from one node to the other is known in advance and it is constant in time. The mathematical analysis is based on idea that for a given line if the capacity and average data flow are known, then, it is possible to calculate the mean packet delay using the Queueing theory.

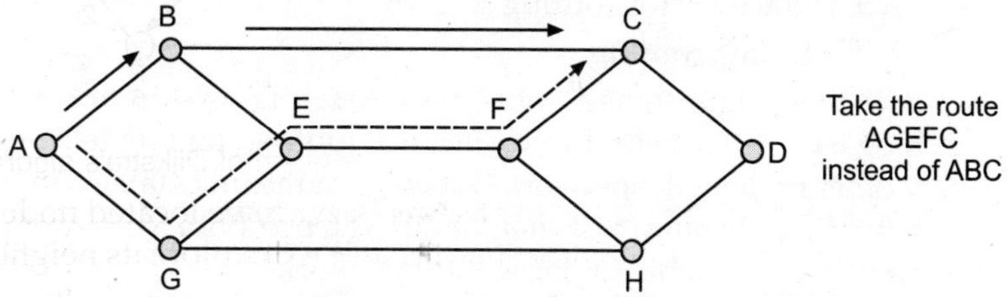

Fig. 7.7 Flow based routing

From the mean delays on all the lines, it is possible to calculate the mean packet delay for the whole subnet. To use the technique of flow based routing, the following information should be known in advance:

(i) Subnet topology.

(ii) Traffic matrix.

(iii) Line capacity matrix which specifies capacity of each line.

7.6. DYNAMIC ROUTING ALGORITHMS

We have discussed the static routing algorithms. But modern computer networks normally use the dynamic routing algorithms. Dynamic routing algorithms change the routing paths as the network traffic loads or topology change. A dynamic algorithm can be run either periodically or in direct response to topology or link cost changes. While dynamic algorithms are more responsive to network changes, they are also more susceptible to problems such as routing loops and oscillation in routes. Two dynamic routing algorithms namely distance vector routing and link state routing are popular. Both these algorithms are suitable for the packet switched networks.

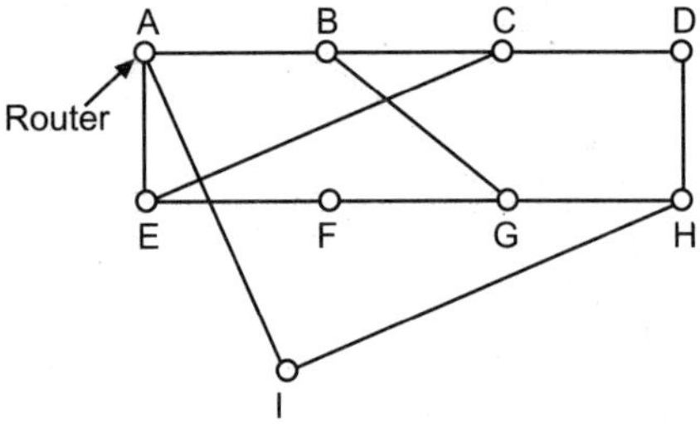

Fig. 7.8. A subnet

Both these algorithms assume that a router knows the address of each neighbour and the cost of reaching each neighbour.

In the distance routing, a node tells its neighbours about its distance to every other node in the network. Whereas in the link state routing, a node tells every other node in the network its distance to its neighbours. Therefore, both these routing algorithms are distributed type and so they are sutiable for large internetworks.

7.6.1. Distance Vector Routing Algorithm

1. Basic Concepts: In this algorithm, each router maintains a table called vector, such a table gives the best known distance to each destination and the information about which line to be used to reach there.

The algorithm is sometimes called by other names such as

1. Distributed Bellman-Ford routing algorithm
2. Ford-Fulkerson algorithm.

In distance vector routing, each router maintains a routing table. It contains one entry for each router is the subnet.

This entry has two parts :

(i) The first part shows the preferred outgoing line to be used to reach the destination and

(ii) Second part gives an estimate of the time or distance to the destination.

The two fundamental routing algorithms in packet switched networks are:

1. Distance vector routing
2. Link state routing.

Both these algorithms assume that a router knows the address of each neighbour and the cost of reaching each neighbour. In the distance routing, a node tells its neighbours about its distance to every other node in the network. Whereas in the link state routing, a node tells every other node in the network its distance to its neighbours. Hence, both these routing algorithms are distributed type and so they are suitable for large internet works.

Distance Vector

In distance vector routing, we assume that each router knows the identity of every other router in the network, but the shortest part to each router is not known. A *distance vector* is defined as the list of <destination, cost> tuples, one tuple per destination. Each router maintains a distance vector. The cost in each tuple is equal the sum of costs on the shortest path to the destination.

Updation of Router Tables

A router periodically sends a copy of its distance vector to all its neighbours. When a router receives a distance vector from its neighbour, it determines whether its cost to reach any destination would decrease if it routed packets to that destination through that neighbour. This is illustrated in figure 7.9.

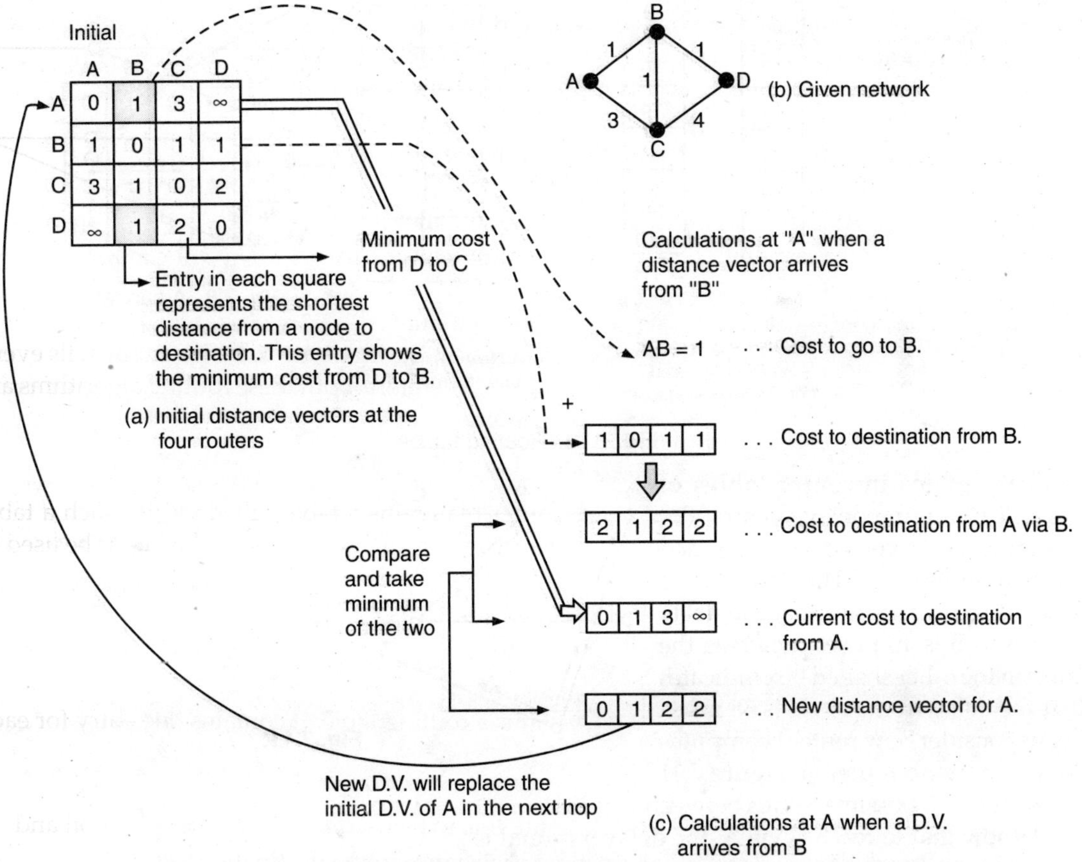

Fig. 7.9. Distance vector algorithm at router A

Figure 7.9 shows how the D.V. at A is automatically modified when a D.V. arrives from B. A similar calculation takes place at the other routers as well. In figure 7.9 (a), the initial distance vector is shown. The entries correspond to the costs corresponding to the shortest distance between the routers corresponding to that square. For example, $AC = 3$ indicates the cost corresponding to the shortest path in terms of number of hops from A to C. Even if nodes asynchronously update their distance vectors, the routing tables eventually converge. The well known example of distance vector routing is the **Bellman-Ford algorithm.**

The metric used can be one of the following:

(i) Number of hops

(ii) Time delay.

(iii) Number of packets in a queue etc.

2. Routing Procedure in distance vector routing: Example of a Subnet

The example of a subnet is shown in figure 7.8 and the routing tables are shown in figure 7.10.

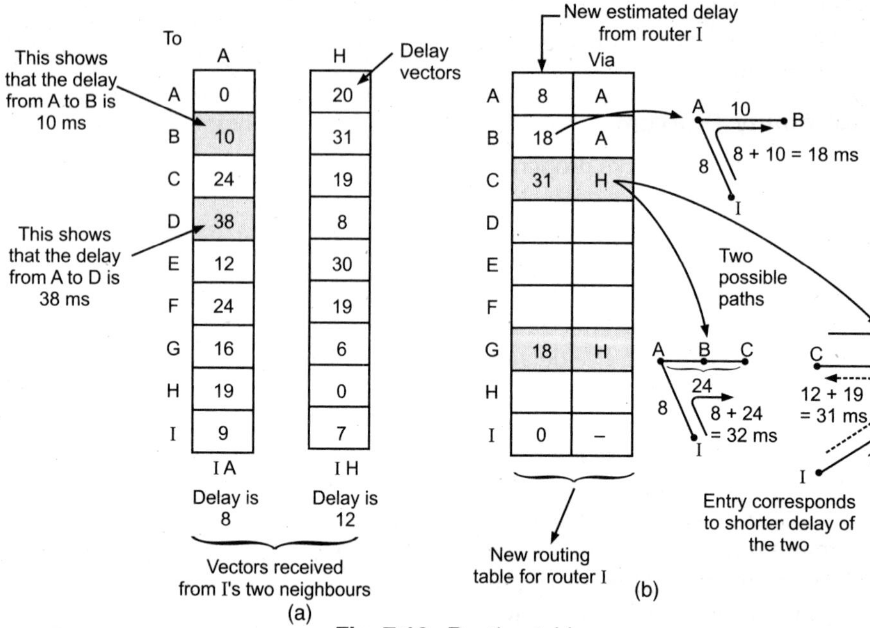

Fig. 7.10. Routing tables

The entries in router tables of figure 7.10 are the delay vectors. For example, let us consider the shaded boxes of figure 7.10. The entry in the first shaded box shows that the delay from A to B is 10 m sec, whereas the entry in the other shaded box indicates that the delay from A to D is 38 msec. Let us consider how router I computes its new route to router G. Figure 7.11 shows the two possible routes between I and G.

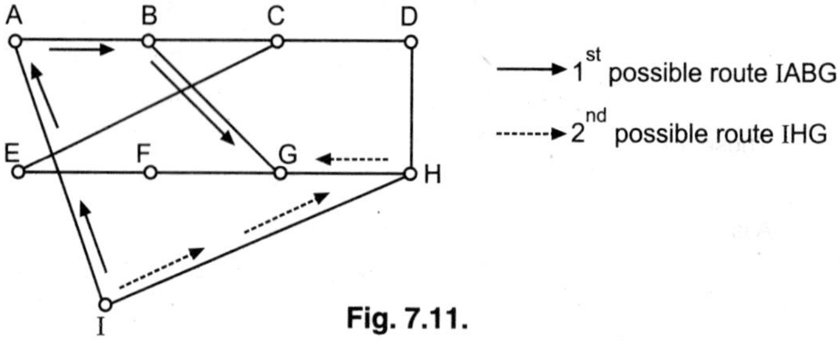

Fig. 7.11.

I know that to reach G via A, the delay required is :

I to *A* Delay = 8 ms
A to G Delay = 16 ms } therefore, I to G Delay = 8 + 16 = 24 msec

Whereas the delay between I and G via H (route IHG) is :

I to H Delay = 12 ms
H to G Delay = 6 ms } therefore, I to G Delay = 12 + 6 = 18 msec

The best of these values is 18 msec corresponding to the path IHG. Hence, it makes an entry in its routing table (I's table) that the delay to G is 18 msec and that the route to use it is via H. The new routing table for router I is shown in figure 7.10 (b). Similarly, we can calculate the delays, from I to different destinations from A to I and enter the minimum possible delay into the *I*'s router table.

3. Drawbacks

(i) The problem with distance vector routing is its slowness in converging to the correct answer. This is due to a problem called count-to-infinity problem.

(ii) This problem can be solved by using the split horizon algorithm.

(iii) Another problem is that this algorithm does not take the line bandwidth into consideration when choosing root.

(iv) This is a serious problem which led to the replacement of this algorithm by the Link State Routing algorithm.

7.6.2. Count to Infinity Problem

The distance vector routing works properly theoretically but practically it has a serious problem. Although we get a correct answer, we get it slowly. In other words it reacts quickly to good news but does so leisurely to bad news. Let us consider a router whose best route to destination X is large. If on the next exchange neighbour A suddenly reports a short delay to X, the router will switch over and start using the line to A for sending the traffic to destination X. Thus, in one vector exchange, the good news is processed. Let us see how fast does a good news propagate. Let us consider a liner subnet of figure 7.12 (a) which has five nodes. The delay metric used is the number of hops. Assume that A is initially down and that all the other routes know this. Hence, all the routers have recorded that the delay to A is infinity. When A becomes ok, the other routers come to know about it via the vector exchanges. Then suddenly a vector exchange at all the routers will take place simultaneously. At the time of first vector exchange, B comes to know that its left neighbour has a zero delay to A. So as shown in figure 7.12 (a), B makes an entry in its routing table that A is one hop away to the left. All the other routers still think that A is down. So in the second row of figure 7.12 (a), the entries below C D E are ∞. On the second vector exchange, C comes to know that B has a path of 1 hop length to A, so C updates its routing table and indicates a path of 2 hop length. But, D and E do not change their table entries.

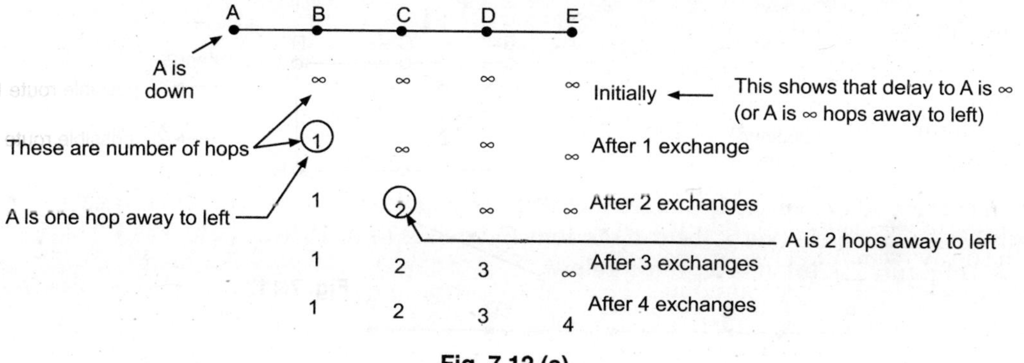

Fig. 7.12 (a).

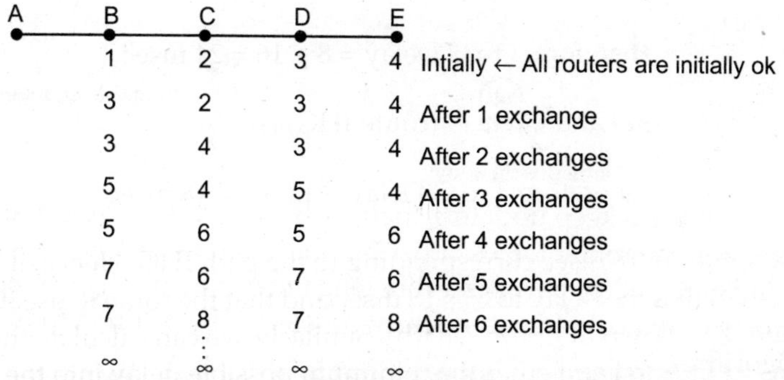

Fig. 7.12 (b).

Hence, after the second vector exchange the entries in the third row of figure 7.12 (a) are:

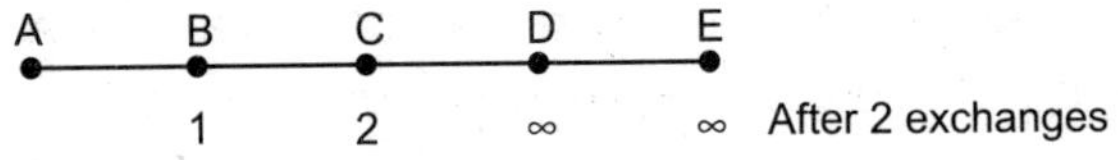

Similarly, D and E will update their routing tables after 3 and 4 exchanges respectively. Hence, we conclude that the good news of A has recovered has spread at a rate of one hop per exchange.

Explanation of figure 7.12 (b)

Now, let us consider figure 7.12 (b). Here, initially all routers are ok. The routers B, C, D and E have distances of 1, 2, 3 and 4 respectively to A. Hence, the first row of figure 7.12 (b) is as follows :

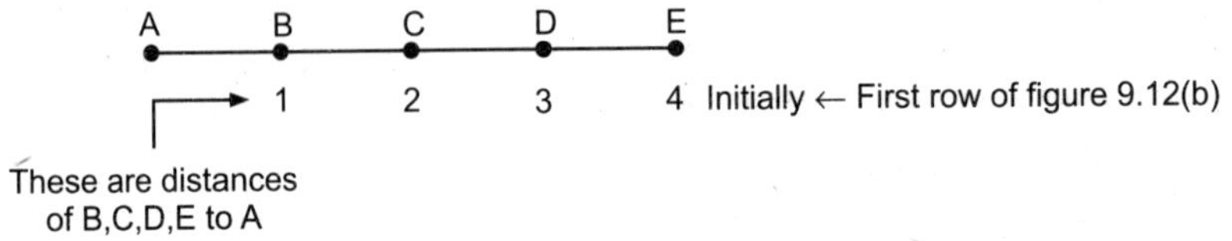

Now, let us imagine that suddenly A goes down or line between A and B is cut. At the first packet exchange B does not hear anything from A (because A is down). But C says "I have a path of length 2 to A. But poor B does not understand that this path is through B itself. So, B thinks that it can reach A via C with a path length 3. (B to C 1 hop and C to A 2 hops) so it accordingly updates its routing table. But, D and E do not update their entries. So, the second row of figure 7.12 (b) looks as follows :

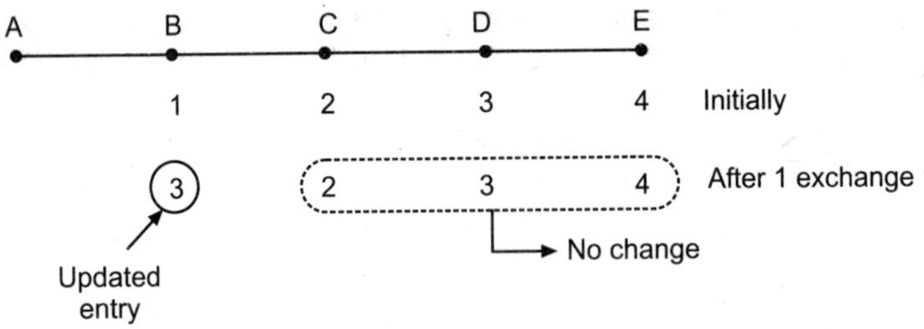

On the second exchange, C realizes that both its neighbours (B and D) claim to have a path of length 3 to A. So, it picks one of them at random and makes its new distance to A as 4. This is shown in row 3 of figure 7.12(b). It is repeated below:

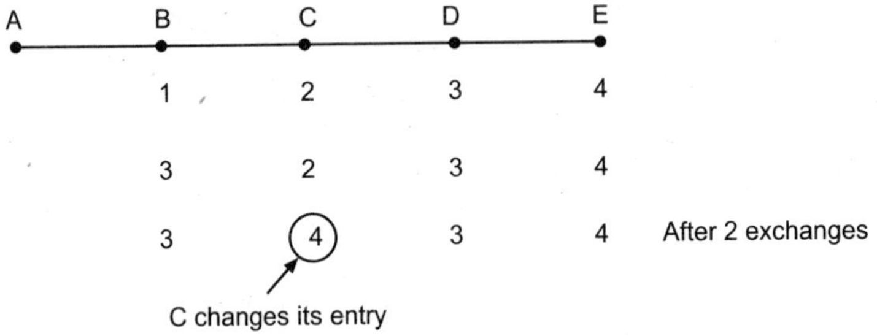

Similarly, the other routers keep updating their tables after every exchange. It is expected that finally we should get ∞ in the router tables of B, C, D and E indicating that A is down. We do reach this state at the end in figure 7.12(b) but after a very long time. The conclusion is bad news propagates slowly. This problem is called as **count-to-infinity** problem. The solution to this problem is to use the split horizon algorithm.

The Split Horizon Algorithm

This algorithm works the same way as distance vector routing, except that the distance to X is not reported on the line that packets for X are sent on.

7.6.3. Link State Routing

1. Definition: Link state routing is used in OSPF (open shortest path first) which is the current IETF standard for intra-autonomous system routing. Link state routing is better in combating the problem of slow convergence. Distance vector routing was used in ARPANET upto 1979. After that it was replaced by the link state routing. Variants of this algorithm are now widely used. The link state routing is simple and each router has to perform the following five operations :

2. Router operations:

(i) Each router should discover its neighbours and obtain their network addresses.

(ii) Then it should measure the delay or cost to each of these neighbours.

(iii) It should construct a packet containing the network address and the delays of all the neighbours.

(iv) Send this packet to all other routers.

(v) Compute the shortest path to every other router.

The complete topology and all the delays are experimentally measured and this information in conveyed to each and every router. Then a shortest path algorithm such as Dijkstra's algorithm can be used to find the shortest path to every other router.

3. Protocols: Link state routing is popularly used in practice. The OSPF protocol which is used in the Internet uses the link state algorithm. IS-IS i.e. Intermediate system – Intermediate system is the other protocol which uses the link state algorithm. IS-IS is used in Internet backbones and in some digital cellular systems such as CDPD.

7.6.4. Comparison of Link State Routing and Distance Vector Routing

S.No.	Distance vector routing	Link state routing
1.	Each router maintains routing table indexed by and containing one entry for each router in the subnet.	It is the advanced version of distance vector routing.
2.	Algorithm took too long to converge.	Algorithm is Faster.
3.	Bandwidth is less.	Wide bandwidth is available.
4.	Router measure delay directly with special ECHO packets.	All delays measured and distributed to every router.
5.	It does not line bandwidth into account when choosing the routes.	It considers the line bandwidth into account when choosing the routes.

7.7. HIERARCHICAL ROUTING

1. Basic Concepts: As the size of the network increases and the way the Internet is growing, the size of the routing tables of the routers also increases. As a result of large routing tables, a large router memory is consumed, more CPU time is required to scan the tables and more bandwidth is required to send status report about the tables. Sometimes the network becomes so large that the size of the router table becomes excessively large and practically it becomes impossible for every router to have an entry for every other router. In other words, it is getting impossible for routers to keep information about every other router in the Internet. Then the hierarchical routing, such as the one used in telephone network should be adopted. In this type of routing the total number of routers are divided into different regions. A router will know everything about the other routers in its own region only. It does not know anything about the internal structure of other regions. This reduces the size of the router table. When various networks are connected together, each network is treated as a separate region. For very large networks, the hierarchy is prepared as follows :

Level 1 : Regions

Level 2 : Clusters : it is a group of regions.

Level 3 : Zones : zone is a group of clusters.

Level 4 : Groups : group contains many zones.

2. Two level hierarchical routing: For networks of smaller size, a two level hierarchical routing is sufficient. Figure 7.13 (a) shows network containing 3 regions. Figure 7.13 (b) shows the full routing table of router 1A which has 9 entries. Now with a two level hierarchical routing, the routing table of the same router reduces to a much smaller size as shown in figure 7.13 (c). This table has only 5 entries.

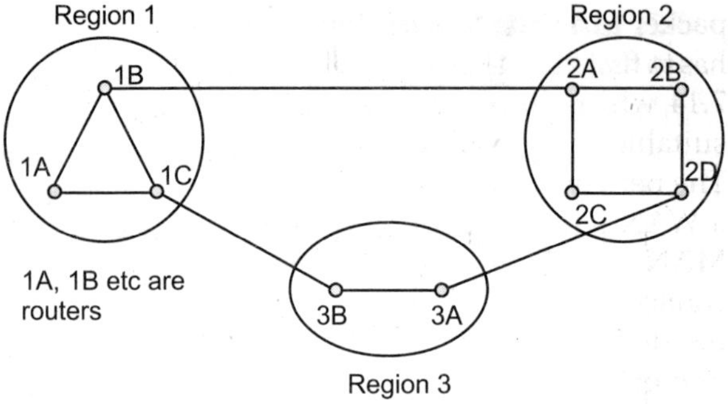

(a) A network

Full routing table for 1 A

Destination	Line	Hops
1 A	–	–
1 B	1 B	1
1 C	1 C	1
2 A	1 B	2
2 B	1 B	3
2 C	1 B	3
2 D	1 B	4
3 A	1 C	3
3 B	1 C	2

(b) Full routing table for router 1A

Hierarchical routing table For 1 A

	Destination	Line	Hops
Region 1	1 A	–	–
	1 B	1B	1
	1 C	1C	1
Region 2 →	2	1B	2
Region 3 →	3	1C	2

(c) Hierarchical routing table for router 1A

Fig. 7.13.

In the hierarchical table of figure 7.13 (c), there are entries for all local routers (1A, 1B and 1C) as before. But there are no detailed entries for the other regions. Instead all other regions have been condensed into a single router per region. For example, traffic from 1A to any router in region-2 is via 1B-2A line as shown by the shaded entry in figure 7.13(c). Similarly, all the traffic from 1 A to region 3 is routed through the line 1C - 3B. Comparison of figures 7.13 (b) and (c) shows how hierarchical routing reduces the size of routing tables.

DO YOU KNOW?

Flooding is not practical in most applications, but it does have some uses. For example, in military applications, where large numbers of routers may be blown to bits at any instant, the tremendous robustness of flooding is highly desirable.

3. Drawback: The reduced table size has a price tag attached to it. It comes at the expense of increased path length. But, it is practically acceptable.

4. How many levels a hierarchy should have ?: Kamoun and Kleinrock have discovered that for an N router subnet, the optimum number of hierarchy levels is $\log_e N$ and it requires a total of $e \log_e N$ entries per router table.

7.8. ROUTING FOR MOBILE HOSTS

1. Basic Concepts: A large number of people have portable computers and they want to work on them wherever they are in the world. Such computers are called as mobile hosts. To route a packet to a mobile host, the network has to first find it. Let us consider figure 7.14 which shows a network model suitable for mobile communication. The network of figure 7.14 consists of a WAN which has routers and hosts. MANs, LANs and wireless cells are connected to this WAN. The world is divided into areas. Each area has one or more foreign agent. A foreign agent is supposed to keep track of all the mobile users visiting his area. In addition to the foreign agents, each area has a home agent which keeps track of users whose home is in the area but who are currently visiting some other area. When a new user enters an area his computer has to register itself to the foreign agent that area and when the user leaves that area the deregistration should be carried out. The routing of packets to a mobile host takes place by following the routing procedure given below :

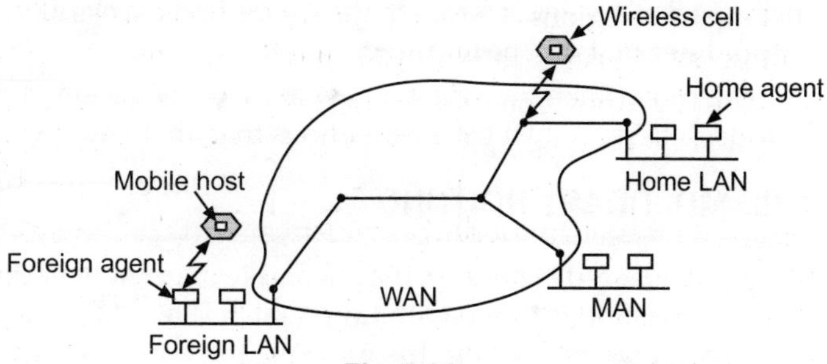

Fig. 7.14.

2. Routing procedure

(i) When a packet is to be sent to a mobile user it is first routed to the user's home LAN.

(ii) This packet is intercepted by the home agent.

(iii) This home agent then locks for the mobile user's current location and finds the address of the corresponding foreign agent.

(iv) The home agent then encapsulates the packet in the payload field of an outer packet and sends it to the foreign agent. This is called **tunneling**.

(v) This packet is received by the foreign agent, who removes the original packet from the payload field and sends it to the mobile user as a data link frame.

(vi) The home agent then tells the sender to send the packets directly to the mobile user. The packets are then routed directly to the user via the foreign agent.

DO YOU KNOW?

When a packet is sent to a mobile host, it is routed to the host's home LAN because that is what the address says should be done.

7.9 BROADCAST ROUTING

In certain applications, the host has to send packets to many or all other hosts. Sending a packet to all destinations simultaneously is called as broadcasting. In broadcast routing, the network layer provides a service of delivering a packet sent from a source node to all other nodes in the network. Various methods of broadcasting are as follows:

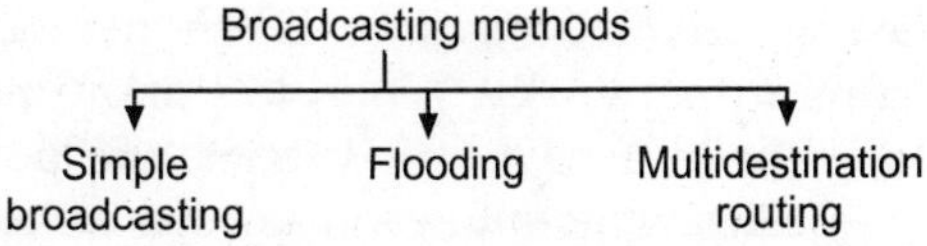

Fig. 7.15. Various methods of broadcasting

1. Simple broadcasting: In this method, the source will simply send a distinct packet to each destination. This method has two drawbacks:

(i) It wastes the bandwidth.

(ii) The source has to have a complete list of all destinations.

2. Flooding: Flooding is another method used for broadcasting. The problem with flooding is that it has a point to point routing algorithm. Because of this, it consumes a lot of bandwidth and generates too many packets.

3. Multidestination routing: This is the third algorithm used for broadcasting. In this algorithm, each packet contain a list of destinations or a bit map which indicates the desired destination. When such a packet arrives at a router, the router first checks all the destinations. Then it decides the set of output lines that will be required.

The router then generates a new copy of the received packet for each output line to be used. It includes a list of only those destinations that are to use the line in each packet going out on that line.

7.10. MULTICAST ROUTING

> **DO YOU KNOW?**
>
> For huge networks, a two-level hierarchy may be insufficient; it may be necessary to group the regions into clusters, the clusters into zones, the zones into groups, and so on, until we run out of names for aggregations.

Multicast means distributing the packets to a group of nodes. This is very similar to broadcasting, which is actually a way of multicasting only where the group is the entire network. There is not much difference in the actual forwarding mechanism of broadcasting and multicasting. For certain applications, widely separated processes need to work together in groups. A process has to send a message to all other processes in the group. For a small group, it is possible to send a point-to-point message. But, this is expensive if the group is large. So, we have to send messages to a well defined groups which are small compared to the network size. Sending message to such a group is called **multicasting** and the routing algorithm used for multicasting is **multicast routing.** Thus, multicast routing enables a single source node to send a copy of a packet to a subset of the other network nodes. To carry out multicast routing, group management is required. Multicasting provides some ways to create and destroy groups and also for a process to join and leave groups. How is it done depends on the type of routing algorithm used. When a process joins a group, it informs this to its host. The routers must know which of their hosts belong to which groups. This information is then given to the neighbouring routers. So that the information is propagated through the subnet. Each router computes a spanning tree which covers al other routers in the subnet. When a process sends a multicast packet to a group, the first router examines its spanning tree and prunes it to removes all the lines which do not lead to hosts that are members of groups. The pruning of the spanning tree can be performed using various ways. One of the ways can be to use the link state routing. Another way of pruning is by using the distance vector routing. The method will prune the spanning tree in a recursive manner.

7.11. CONGESTION

An important issue in a packet switching network is congestion. Congestion is synonymous to traffic jams in networks. When too many packets are present in a part of a subnet, the performance degrades. This situation known as **congestion**. Congestion in a network may occur when the load on the network i.e. the number of packets sent to the network is greater than the capacity of the network (i.e. the number of packets a network can handle). Figure 7.16 explain the concept of congestion graphically. Upto point A in figure 7.16, the number of packets dumped into the subnet by the host is within its carrying capacity they are all delivered. In short, the number of packets delivered is proportional to number of packets sent and no congestion takes place. But after point A, the traffic

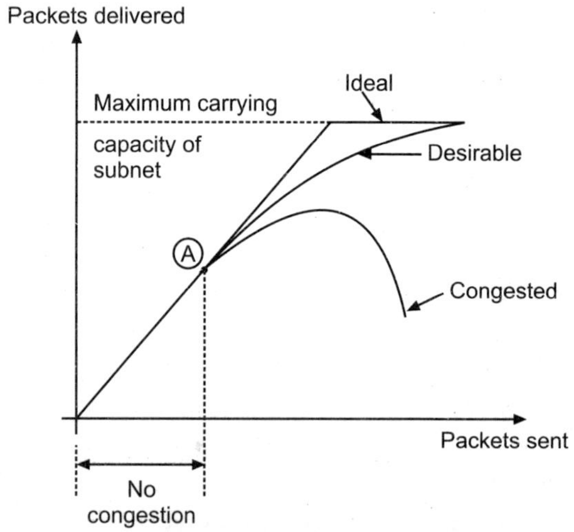

Fig. 7.16. Concept of congestion

increases too far. The routers cannot cope with the increased traffic and they begin to lose packets. The congestion begins here. As the traffic increases further, the performance degrades more and more packets are lost and congestion worsens. At very high traffic, the performance collapses completely and almost all packets are lost. This is the worst possible congestion.

7.11.1. Need of Congestion Control

It is not possible to completely avoid the congestion but it is necessary to control it. Congestion will lead to a large queue length, which results in buffer overflow and loss of packets. Therefore, congestion control is necessary to ensure that the user gets the negotiated QoS (quality of service).

7.11.2. Causes of Congestion

Some of the causes of congestion may be described as under :

(i) If suddenly a stream of packets start coming on three or four input lines which all need the same output line. Then a queue will build up. If the memory capacity is not sufficient to hold all these packets, some of them will be lost. This has been shown in figure 7.17(a).

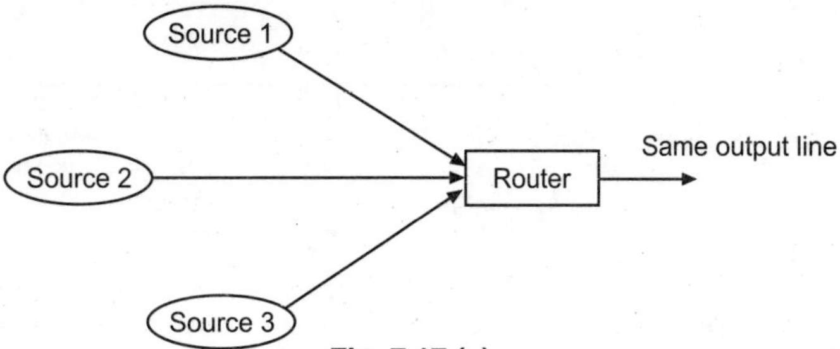

Fig. 7.17 (a)

It may be noted that increasing the memory to infinity also does not solve the problem, in fact it worsens.

(ii) Congestion is caused by slow links. The problem will be solved when high speed links become available. It is not always the case, sometimes increases in link bandwidth can aggravate the congestion problem because higher speed links may make the network more unbalanced. For the configuration showed in figure 7.17(b), if both of the two sources begin to send to destination 1 at their peak rate, congestion will occur at the switch. Higher speed links can make the congestion condition in the switch worse.

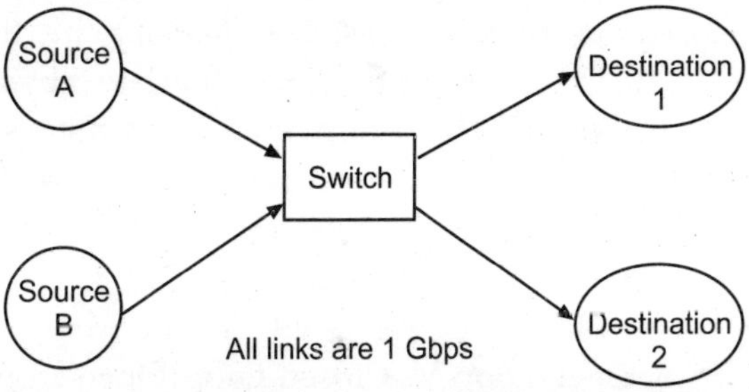

Fig. 7.17. (b) Network with high speed links

(iii) Congestion is caused by slow processors. The problem will be solved when processor speed is improved. Faster processors will transmit more data in unit time. If several nodes begin to transmit to one destination simultaneously at their peak rate, the target will be overwhelmed soon.

(iv) Congestion can make itself worse. If a router does not have any free buffers it should ignore (discard) new packets arriving at it. But, when a packet is discarded, the sender retransmit it many times because it is not receiving the acknowledgement of the packet. This multiple transmission of packets will force the congestion to take place at the sending end.

7.11.3. Difference Between Congestion Control and Flow Control

Congestion control makes it sure that the subnet is able to carry the offered traffic. Congestion control is dependent on the behaviour of all the hosts, all the routers and other factors which reduce the carrying capacity of a subnet. On the contrary, the flow control is related to point to point traffic. Flow control ensures that a fast sender does not send data at a rate faster than the rate at which the receiver can receive it. Flow control involves some kind of feedback.

7.11.4. Principle of Congestion Control

1. Definition: The solutions to congestion problems can be divided into two categories or groups as open loop solutions and closed loop solutions. Congestion control refers to the techniques and mechanisms which can either prevent congestion from happening or remove congestion after it has taken place. The open loop congestion control is based on the prevention of congestion whereas the closed loop solutions are for removing the congestion.

2. Classification: Figure 7.18 shows the classification of congestion control schemes and various policies used in open loop and closed loop groups.

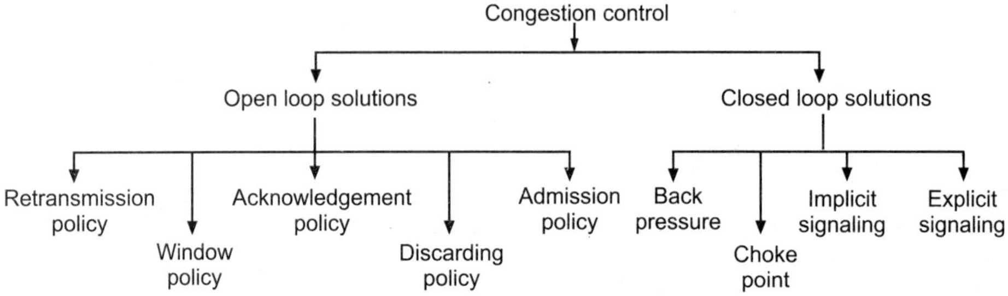

Fig. 7.18. Classification of congestion control schemes

3. Open Loop Control: Open loop solutions try to solve the problems by excellent design to prevent the congestion from happening. Open loop control is exercised by using the tools such as deciding when to accept the new packets, when to discard the packets, which packets are to be discarded and making the scheduling decisions at various points.

4. Closed Loop Control: The closed loop congestion control uses some kind of feedback. A closed loop control is based on the following three steps :

(i) Detect the congestion and locate it by monitoring the system.

(ii) Transfer the information about congestion to places where action can be taken.

(iii) Adjust the system operations to correct the congestion.

Two examples of closed loop control are :

(i) TCP flow control.

(ii) BR rate control for an ATM network.

5. Open Loop Vs Closed Loop: Open loop approach do not need-end-to-end feedback, one of the examples of this type are prior-reservation and loop-to-hop flow control. In closed-loop approaches, the source adjust its cell rate in responding to the feedback information received from the network. It has been argued that closed loop congestion control schemes are too slow in todays high-speed, large range network, by the time a source gets the feedback and reacts to it, several thousand cells may have been last. But on other hand, if the congestion has already happened and the overload is of long duration, the condition cannot be released unless the source causing the congestion is asked to reduce its rate. Furthermore, ABR service is designed to use any bandwidth that is left over the source must have some knowledge of what is available when it is sending cells.

7.11.5. Congestion Prevention Policies

In this article, let us discuss the open loop congestion control systems. These systems try to avoid congestion by using the appropriate policies at different levels. Figure 7.19 depicts various policies corresponding to different layers for avoiding congestion.

1. Policies Related to Data Link Layer

(i) Retransmission Policy: The retransmission policy and the retransmission timers must be designed to optimise efficiency and at the same time prevent congestion. The retransmission policy deals with how fast a sender times out. If a sender times out early then it will retransmit all the packets which can lead to congestion. Using retransmission policy, we can avoid this and prevent congestion.

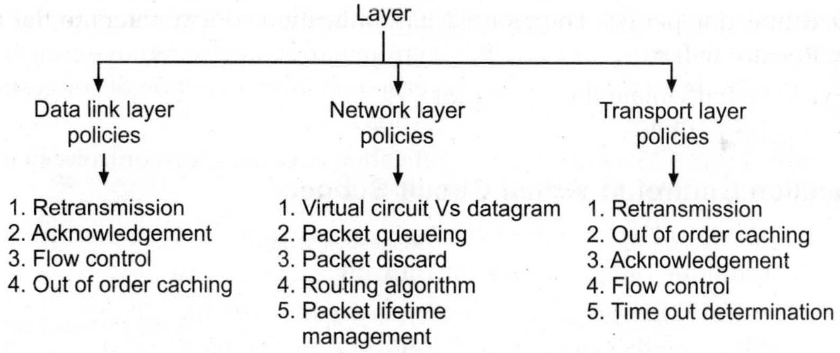

Fig. 7.19. Policies affecting the congestion

(ii) Out of Order Caching Policy: If the receiver routinely discard all the packets which are out of order, then retransmission of these packets will take place. This will increase the load and result in congestion.

(iii) Acknowledgement Policy: If each received packet is acknowledged immediately then the acknowledgement packets will increase the traffic. If the acknowledgement is delayed (piggybacking) then there is a possibility of time out and retransmission. Therefore, a tight flow control has to be exercised to avoid congestion.

(iv) Window Policy: The type of window at the sender may also affect congestion. The selective repeat window is better than the Go Back n window.

2. Policies Related to Network Layer

(i) Choice between Virtual Circuit and Datagrams: This choice at the network layer will affect the congestion because several congestion control algorithms work only with virtual circuit subnets.

(ii) Packet Queueing and Service: This policy is related to whether the routers have one queue per input line and one queue per output line or both. The policy is also related to the order in which the packets are processed e.g. round robin or priority based etc.

(iii) Discard Policy: This policy lays a rule which tells the routers about which packet is to be discarded. A good discard policy can prevent congestion and a bad one will worsen the situation.

(iv) Routing Algorithms: The routing algorithms can spread the traffic over all the lines to avoid congestion.

(v) Package lifetime management: The policy decides the time for which a packet may live before being discarded. This time should be of adequate value so that congestion can be avoided.

3. Policies Related to Transport Layer

The policies at the transport layer are same as those at the data link layer. But at transport layer determining the time out interval is more difficult. If it is too short then extra packets are sent unnecessarily whereas if it is too long, congestion will get reduced at the cost of increased response time.

Traffic Shaping

One of the important reason behind congestion is the burstly nature of the traffic. If the traffic has a uniform data rate than congestion could be less common. Traffic shaping is an open loop control. It manages the congestion by forcing the packet transmission rate to be more predictable. Thus, traffic shaping will regulate the average rate or the burstness of data transmission. Monitoring a traffic flow is called as **traffic policing.** Cleck if a packet stream (connection) obeys its descriptor, and if it violates its descriptor, give penalty. For this, the network may want to monitor the traffic flow during the connection period. The process of monitoring and enforcing the traffic flow is called traffic policing. Penalty will be :

(i) Drop packets that violate the descriptor

(ii) Give low priority to them.

7.11.6. Congestion Control in Virtual Circuit Subnets

All the congestion control techniques discussed till now were open loop techniques. Now, let us discuss a dynamic technique called **admission control.**

1. Admission Control Principle: This technique is used to keep the congestion which has already begun to a manageable level. Its principle is as follows : once congestion has been detected, do not set up any more virtual circuits until the congestion is cleared. The advantage is that it is a simple and easy to carry out control.

2. Alternative Approach: An alternative approach to admission control is to allow the virtual circuits to set up even when a congestion has taken place. However, carefully route all the new virtual circuits around the problem area.

7.12. CONGESTION CONTROL IN DATAGRAM SUBNETS

Now, let us discuss some congestion control approaches which can be used in the datagram subnets (and also in virtual circuit subnets).

(i) Choke packets

(ii) Load shedding

(iii) Jitter control.

7.12.1. Choke Packets

This approach can be used in virtual circuits as well as in the datagram subnets. In this technique, each router associates a real variable with each of its output lines. This real variable say u has a value between 0 and 1 and it indicates the percentage utilization of that line. If the value of u goes above the threshold then the output line will enter into a warning state. The router will check each newly arriving packet to see if its output line is in the warning state. If it is in the warning state then the router will send back a **choke packet** signal to the sending host. The sender host will not generate any more choke packets. Several variations on the congestion control algorithm have been proposed, depending on the value of thresholds.

Depending upon the threshold value, the choke packets can contain a mild warning, a sterm warning or an ultimatum. Another variation can be in terms of queue lengths or buffer utilization instead of using the line utilization as a deciding factor.

Drawback

The problem with choke packet technique is that the action to be taken by the source host on receiving a choke packet is voluntary and not compulsory.

7.12.2. Load Shedding

Admission control, choke packets, fair queueing are the techniques suitable for light congestion. But if these techniques cannot make the congestion to disappear, then the load shedding technique is to be used. The principle of load shedding states that when the routers are being inaundated by the packets that they cannot handle, they should simply throw the packets away. A router which is flooding with packets due to congestion can drop any packet at random. But there are better ways of doing this. The policy for dropping a packet depends on the type of packet. For file transfer, an old packet is more important than a new packet. In contrast, for multimedia a new packet is more important than an old one. Hence, the policy for file transfer called wine (old is better than new) and that for the multimedia is called milk (new is better than old). An intelligent discard policy can be decided depending on the applications. To implement such an intelligent discard policy, co-operation from the sender is essential. The application should mark their packets in priority classes to indicate how important they are.

If this is done then when the packets are to be discarded the routers can first drop packets from lowest class (i.e. the packets which are least important). Then the routers will discard the packets from next lower class and so on. One or more header bits are required to put the priority for making the class of a packet. In every ATM cell, 1 bit is reserved in the header for marking the priority. Every ATM cell is labeled either as a low priority or high priority.

7.12.3. Jitter Control

1. Definition of Jitter: Jitter may be defined as the variation in delay for the packets belonging to the same flow. The real time audio and video cannot tolerate jitter on the other hand the jitter does not matter if the packets are carrying an information contained in a file. For the audio and video transmission if the packets take 20 msec to 30 msec (delay) to reach the destination, it does not matter, provided that the delay remains constant. The quality of sound or video will be hampered if the delays associated with different packets have different values. Therefore, practically we can say that 99% packets should be delivered with a delay ranging from 24.5 msec to 25.5 msec.

2. Jitter Control: When a packet arrives at a router, the router will check to see whether the packet is behind or ahead and by what time. This information is stored in the packet and updated at every hop. If the packet is ahead of the schedule (early) then the router will hold it for a slightly longer time and if the packet is behind the schedule (late), then the router will try to send it out as quickly as possible. This will help in keeping the average delay per packet constant and will avoid time jitter.

7.12.4. Difference between End to End Delay and Jitter

1. End to End Delay: As a matter of fact, end to end delay is the time required for the signal to travel from transmitter to receiver. This delay is due to the time required for buffering, queueing, switch and routing. This time delay remains the same for all the types of packets.

2. Jitter: Jitter is defined as the variation in delay for the packets belonging to the same flow. This is the difference between end to end delay and jitter.

7.13. QUALITY OF SERVICE (QoS)

QoS is an internetworking issue that has been discussed more than defined. We can informally define quality of service as something flow seeks to attain.

Flow Characteristics

Traditionality, four types of characteristics are attributed to a flow: reliability, delay, jitter and bandwidth. These characteristics are shown in figure 7.20.

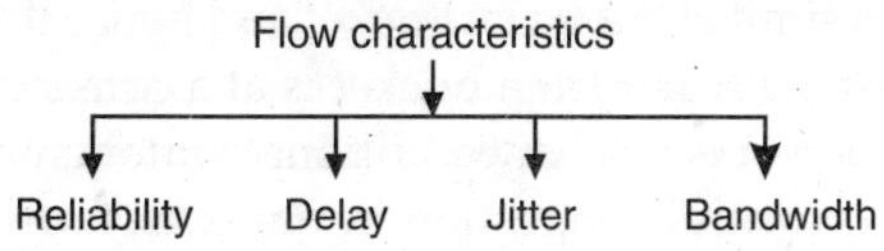

Fig. 7.20. Illustration of flow characteristics

(i) Reliability: Reliability is a characteristic that a flow needs. Lack of reliability means losing a packet or acknowledgement, which entails retransmission. However, the sensitivity of application programs to reliability is not the same. For example, it is more important that electronic mail, file transfer, and Internet access have reliable transmissions than telephony or audio conferencing.

(ii) Delay: Source-to-destination delay is another flow characteristic. Again applications can tolerate delay of different degrees. In this case, telephony, audio conferencing, video conferencing, and remote log-in and minimum delay, while delay in file transfer or email is less important.

(iii) Jitter: Jitter is the variation in delay for packets belonging to the same flow. Real-time audio and video cannot tolerate high jitter. For example, a real-time video broadcast is useless if there is a 2-ms delay for first and second packets and a 60-ms delay for the third and fourth. On the other hand, it does not matter if packet carrying information in a file have different delays. The transport layer at the destination waits until all packets arrive before delivery to the application layer.

(iv) Bandwidth: Different applications require different bandwidths. In video conferencing, we need to send millions bits per second to refresh a color screen while the total number of bits in an email may not reach even million.

7.13.1. Congestion Control Algorithms: Techniques for Achieving Good QoS

Some of the techniques useful in achieving good QoS may be listed as under :

1. Buffering.
2. Traffic shaping.
3. Leaky bucket algorithm.
4. Token bucket algorithm.
5. Resource reservation.
6. Admission control.
7. Proportional routing.
8. Packet scheduling.

7.13.2. Traffic Shaping

As discussed earlier that one of the important reason behind congestion is the bursty nature of the traffic. If the traffic has a uniform data rate then congestion could be less common. Traffic shaping is an open loop control. It manages the congestion by forcing the packet transmission rate to be more predictable. Thus, traffic shaping will regulate the average rate or the burstiness of data transmission. Monitoring a traffic flow is called as **traffic policing.** Check if a **packet** stream (connection) obeys its descriptor, and if it violates its descriptor, give penalty. For this the network may want to monitor the traffic flow during the connection period. The process of monitoring and enforcing the traffic flow is called **traffic policing**. Penalty will be :

(i) Drop **packets** that violate the descriptor.

(ii) Give **low priority** to them.

Traffic shaping is a mechanism to control the amount and rate of the traffic sent to the network.

The two traffic shaping techniques are as under:

(i) Leaky bucket.

(ii) Token bucket.

7.13.3. Leaky Bucket Algorithm*

1. Basic Concepts: It is the algorithm used to control congestion in network traffic. It's working is similar to a leaky bucket and hence the name. Leaky bucket is a bucket with a hole at bottom. Flow of the water from bucket is at a constant rate which is independent of water entering the bucket. If bucket is full, any additional water entering in the bucket is thrown out. Same technique is applied to control congestion in network traffic. Every host in the network is having a buffer with finite queue length. Packets which are put in the buffer when buffer is full are thrown away. The buffer may drain onto the subnet either by some number of packets per unit time, or by some total number of bytes per unit time (helpful if packets vary greatly in size).

A variant on the leaky bucket is the token bucket. The bucket is filled with tokens at a certain rate. A packet must grab and destroy a token to leave the bucket. Packets are never lost, they just have to wait for an available token.

* Why the leaky bucket algorithm allow just one packet per tick, irrespective of packet size ?

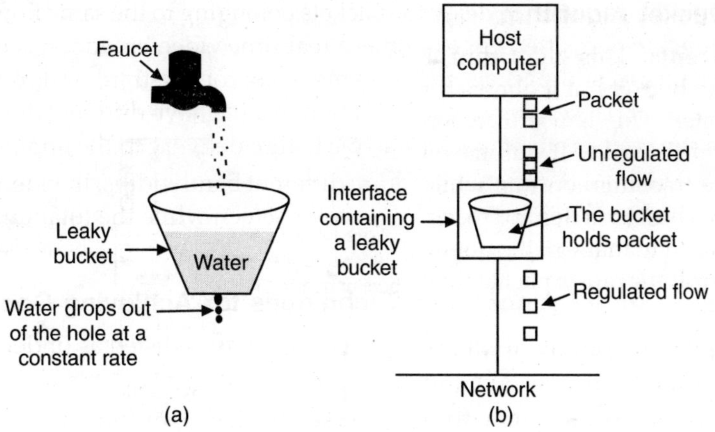

Fig. 7.21. Leaky bucket algorithm

2. Leaky Bucket Implementation: Figure 7.22 shows the implementation of leaky bucket principle. A FIFO queue is used for holding the packets. The implementation of figure 7.22 can be discussed under two different operating conditions, namely :

(i) For packets of fixed size.

(ii) For packets of variable size.

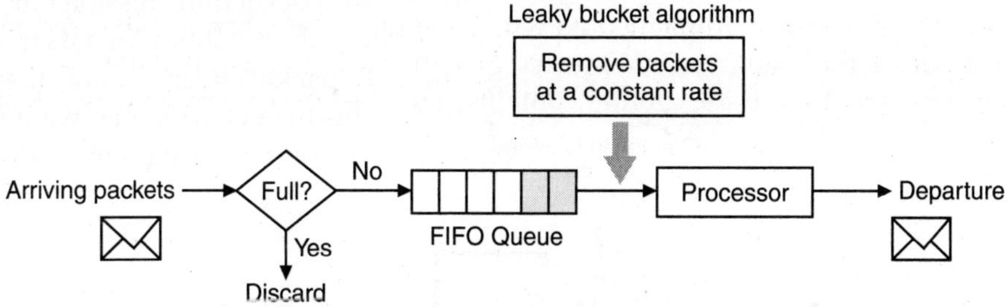

Fig. 7.22. Implementation of leaky bucket

(i) Fixed Size Packets: If the arriving packets are of fixed size (e.g., cells in ATM networks), then the process of figure 7.22 removes a fixed number of packets from the queue at each tick of the clock.

(ii) Packets of Variable Size: If the arriving packets are of different size, then the fixed output rate will not be based on the number of departing packets. Instead, it will be based on the number of departing bytes or bits. The algorithm for variable length packets is as follows :

3. Algorithm

(i) Initialize a counter to n at the tick of the clock.

(ii) If n is greater than the packet size, then send the packet and decrement the counter by the packet size.

(iii) Repeat step 2 until n becomes smaller than the packet size.

(iv) Reset the counter and go back to step 1.

Important Point :

A leaky bucket algorithm shapes the bursty traffic into a fixed rate traffic. It does so by averaging the data rate. It drops the packets if the bucket is full.

7.13.4. Token Bucket Algorithm*

1. Basic Concepts: This algorithm is similar to the leaky bucket but it allows for varying output rates. This is useful when larger burst of traffic arrive. It enforces a long-term average transmission rate while permitting bounded bursts. In this approach, a token bucket is used to manage the queue regulator that controls the rate of packet flow into the network.

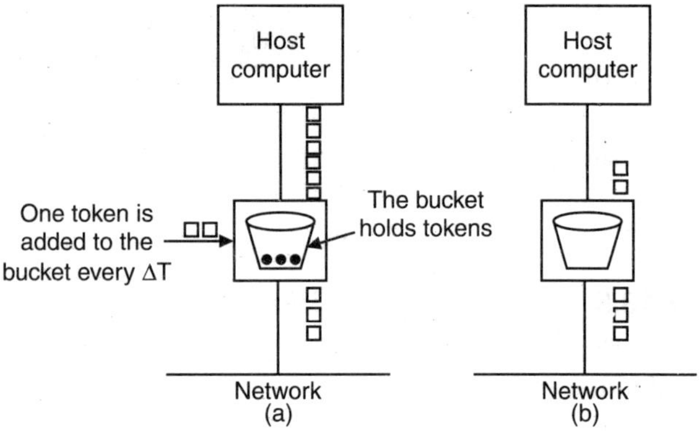

Fig. 7.23. Token bucket algorithm

A token generator constantly produces token at a rate of R tokens per second and places them into a token bucket with a depth of D tokens. Assuming that each token grants the ability to transmit a fixed number of bytes, if the token bucket fills, newly generated tokens are discarded. At the same time, an unregulated stream of packets arrive and are placed into a packet queue that has a maximum length of L. If the flow delivers more packets than the queue can store, the excess packets are discarded.

2. Implementation of Token Bucket: Figure 7.24 shows the implementation of token bucket. The token bucket can be easily implemented with a counter. The token is initialized to zero. Each time a token is added, the counter is incremented by 1 and each time a unit of data is dispatched, the counter is decremented by 1. If the counter contains zero, the host cannot send any data.

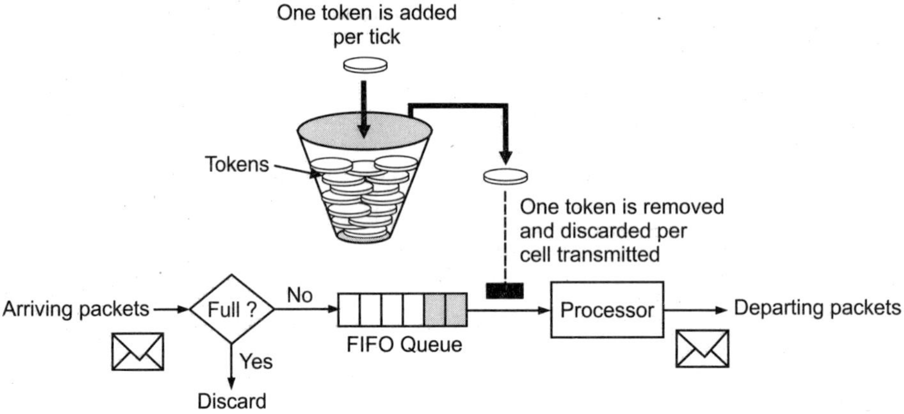

Fig. 7.24. Implementation of token bucket

Important Point:
The token bucket allows the bursty traffic at maximum possible rate.

3. Token Bucket Performance

Let s = burst length (seconds),
c = bucket capacity (bytes),
ρ = token arrival rate (bytes/second),
and m = maximum source rate (bytes/second).

* Explain token bucket algorithm. What problems of leaky bucket algorithm are addressed by it ?

What is the Duration of a Maximum-rate Burst Through a Token Bucket ?

(i) Maximum bytes sent from the token bucket during a burst is

$$c + \rho . s$$

(ii) Maximum bytes the source can send during a burst is

$$m . s$$

(iii) Setting the two equal and solving for s

$$s = \frac{c}{m - \rho}$$

7.13.5. Combination of Token Bucket and Leaky Bucket

The token bucket and leaky bucket techniques may be combined to obtain the following advantages :

(i) To credit an idle host

(ii) To regulate the traffic.

The token bucket is used prior to the leaky bucket. The rate of leaky bucket needs to be higher than the rate of tokens dropped in the bucket.

7.13.6. Resource Reservation

The data flow is dependent upon the following resources :

(i) Buffer

(ii) Bandwidth and

(iii) CPU time.

The QoS can be improved by reserving these resources. The QoS model called integrated services depends heavily on the principle of resource reservation, for improvement in QoS.

7.13.7. Admission Control

Admission control technique is used by a router or a switch. They use this mechanism to accept or reject the data flow based on predefined parameters called flow specifications. Before accepting a flow for processing, a router checks the flow specifications to judge if it can take up and handle this new data flow. It does this by checking its capacity in terms of bandwidth, buffer size, CPU speed etc. and its commitments to other flow.

7.14. QUEUEING DISCIPLINES

Each router must implement some queueing discipline which governs how the packets are buffered when they are waiting to get transmitted. There are also algorithms commonly used in packet switching networks namely,

(i) First in first out (FIFO)

(ii) Fair queueing (FQ).

7.14.1. FIFO Queueing

The FIFO Queueing is also called as first come first served (FCFS) Queueing. Its idea is simple and illustrated in figure 7.25.

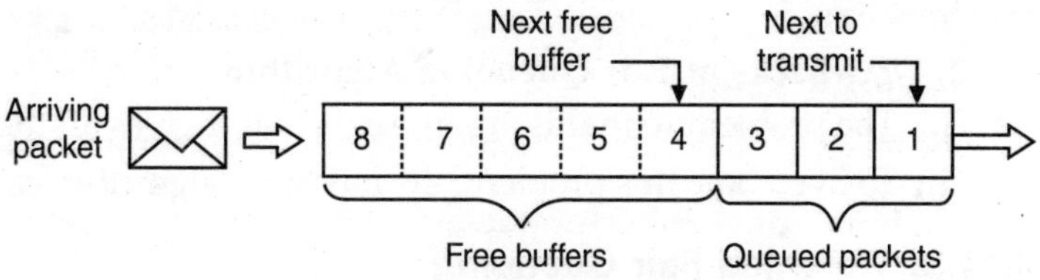

Fig. 7.25. FIFO queueing

The principle of FIFO queueing is that the first packet that arrives at a router is the first packet which is transmitted. In this sense, it is first out (FIFO) queueing. figure 7.25 shows a FIFO with slots to hold upto eight packets.

Discarding of a packet

The amount of buffer space at any router is finite. So, if a packet arrives when the queue (buffer space) is full, then the router discards that packet as shown in figure 7.26.

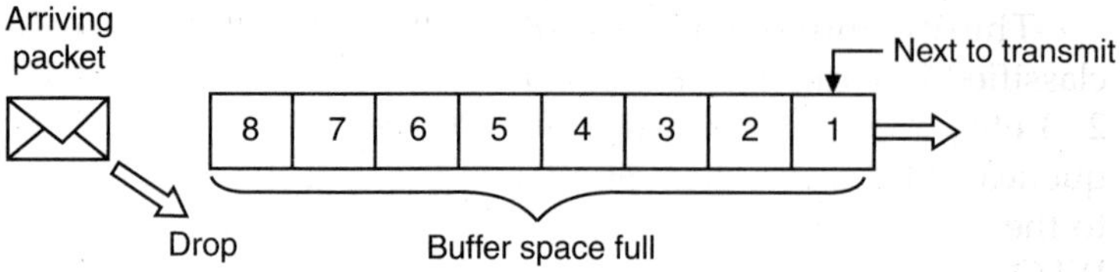

Fig. 7.26. Dropping of a packet

The packet is dropped (discarded) irrespective of which flow the packets belongs to or how important the packet is. This is sometimes called as tail drop because packets which arrive at the tail end of the FIFO are dropped. FIFO with tail drop, is the simplest queueing algorithm and so it is the most widely used one. A simple variation in basic FIFO queueing is priority queueing. In this, each packet is marked with a priority. The router then implements multiple FIFO queues, one for each priority class. The router will always transmit packets out of highest priority queues if that queue is non empty, before moving on to next priority queue.

7.14.2. Fair Queueing

1. Basic Concepts: The disadvantage of FIFO queueing is that there is absolutely no discrimination between the packets being received from different streams. Because the entire congestion control mechanism is implemented at the sources and the FIFO queueing does not provide any means of policing how well the source adhere to this mechanism, it is possible that an ill-behaved source may capture an arbitrary large fraction of the network capacity. In FQ, a separate queue is maintained for each flow currently being handled by the router.

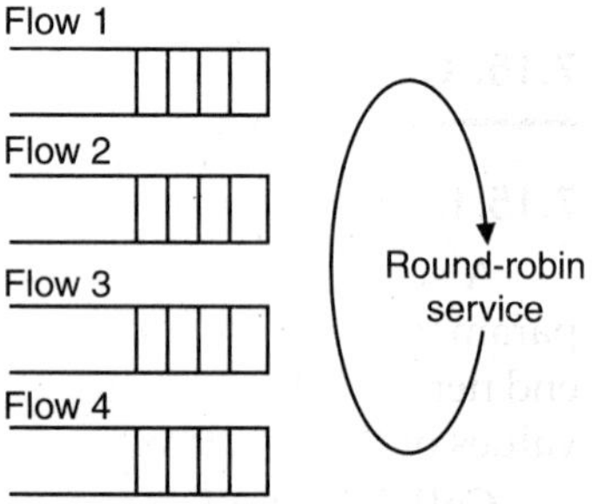

Fig. 7.27. Concept of fair Queueing (FQ) at the router

Fair queueing (FQ) is an algorithm which is proposed to solve this problem. The router then services these queues in the round robin manner as shown in figure 7.27. When a flow sends packets too quickly, then its queue is filled up. When a queue reaches a particular length, additional packets from that flow are discarded. This ensures that any source can not arbitrarily increase its share of the network's capacity. FQ is designed to be used in conjunction with an end-to-end congestion control mechanism. It simply segregates traffic so that an ill-behaved traffic sources do not interfere with the network. The main complication is that the packets being processed at a router are not of some length.

Ideally a bit by bit round is expected *i.e.* one bit from flow-1 is transmitted, then one bit from flow-2 and so on. But, practically, it is not possible. The FQ mechanism therefore simulates this behaviour by first determining when a given packet would finish being transmitted if it were being sent using bit by bit round robin, and then using this finishing time to sequence the packets for transmission.

2. Drawbacks of Fair Queueing Algorithm:

(i) The problem with this algorithm is that it gives all the hosts are given the same priority.

(ii) To overcome this problem, the modified algorithm called weighted fair queueing is used.

7.14.3. Weighted Fair Queueing

The disadvantage of choke packet technique can be overcome with the help of the weighted queueing technique. The queueing algorithm was proposed first in 1987. In this algorithm, it is proposed that the routers have a number of queues for each output line, with one queue for each source. Let us consider figure 7.28 to understand weighted fair queueing.

The incoming packets are classified into different classes (1, 2, 3 etc.) and stored in separate queues (*W*1, *W*2, *W*3 etc.). Similar to the round robin technique, the WFO scheduler, will output the packets from *W*1, *W*2, and *W*3 in a circular manner. If a queue is empty, the WFO scheduler will move immediately to the next queue. It will always keep the link occupied.

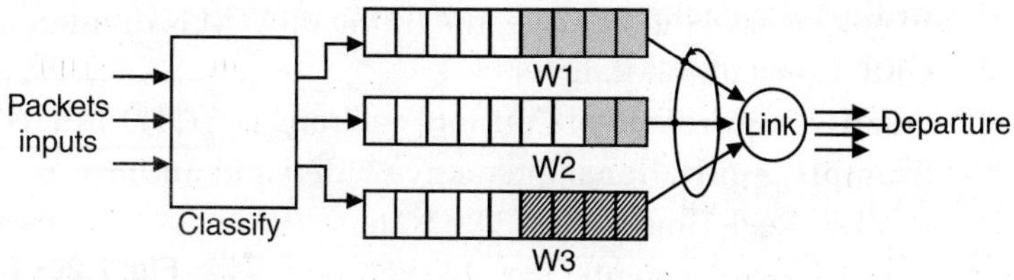

Fig. 7.28. Weighted fair queueing (WFQ)

Different Between Round Robin and WFQ

In round robin, the service received by each class is same but in WFQ, each class may receive a differential amount of service in any interval of time, depending on the weightage associated to it.

7.15. QoS IN ATM

7.15.1. Flow Characteristics in ATM

A set of parameters are negotiated when a connection is set up on ATM networks. These parameters are used to measure the Quality of Service (QoS) of a connection and quantify end-to-end network performance at ATM layer. The network should guarantee the QoS by meeting certain values of these parameters.

Cell Transfer Delay (CTD):

The delay experienced by a cell between the first bit of the cell is transmitted by the source and the last bit of the cell is received by the destination. Maximum Cell Transfer Delay (Max CTD) and Mean Cell Transfer Delay (Mean CTD) are used.

Peak-to-Peak Cell Delay Variation (CDV): The difference of the maximum and minimum CTD experienced during the connection. Peak-to-peak CDV and Instantaneous CDV are used.

Cell Loss Ratio (CLR): The percentage of cells that are lost in the network due to error or congestion and are not received by the destination.

7.15.2. Usage Parameters

Another set of parameters are also negotiated when a connection is set up. These parameters discipline the behavior of the user. The network only provide the QoS for the cells that do not violate these specifications.

Peak Cell Rate (PCR): The maximum instantaneous rate at which the user will transmit.

Sustained Cell Rate (SCR): The average rate as measured over a long interval.

Burst Tolerance (BT): The maximum burst size that can be sent at the peak rate.

Maximum Burst Size (MBS): The maximum number of back-to-back cells that can be sent at the peak cell rate. BT and MBS are related as follows:

Burst Tolerance = (MBS – 1)(1/SCR – 1/PCR)

Minimum Cell Rate (MCR)

The minimum cell rate desired by a user.

7.15.3. Service Categories

Providing desired QoS for different applications is very complex. For example, voice is delay-sensitive but not loss-sensitive, data is loss-sensitive but not delay sensitive, while some other applications may be both delay-sensitive and loss-sensitive.

It make it easier to manage, the traffic in ATM is divided into five service classes.

CBR: Constant Bit Rate

Quality requirements: Constant cell rate, i.e., CTD and CDV are tightly constrained; low CLR.

Example applications: interactive video and audio

rt-VBR: Real-Time Variable Bit Rate

Quality requirements: variable cell rate, with CTD and CDV are tightly constrained; a small nonzero random cell loss is possible as the result of using statistical multiplexing.

Example applications: interactive compressed video

nrt-VBR : Non-Real-Time Variable Bit Rate

Quality requirements : variable cell rate, with only CTD are tightly constrained; a small nonzero random cell loss is possible as the result of using statistical multiplexing.

Example applications: response time critical transaction processing.

UBR : Unspecified Bit Rate

Quality requirements : using any left-over capacity, no CTD or CDV or CLR constrained.

Example applications: email and news feed.

ABR: Available Bit Rate

Quality requirements: using the capacity of the network when available and controlling the source rate by feedback to minimize CTD, CDV and CLR.

Example applications: critical data transfer, remote proedure call and distributed file service. These service categories relate traffic characteristics and QoS requirements to network behavior. The QoS requirement for each class is different. The traffic management policy for them are different, too.

7.15.4. Important Points

(i) The CLR may be unspecified for CLP = 1.

(ii) Minimized for sources that adjust cell flow in response to control information.

(iii) May not be subject to CAC and UPC procedures.

(iv) Represents the maximum rate at which the source can send as controlled by the control information.

(v) These parameters are either explicitly or implicitly specified for PVCs or SVCs.

(vi) Different values of CDVT may be specified for SCR and PCR.

Among these service classes, ABR is commonly used for data transmissions which require a guaranteed QoS, such as low probability of loss and error. Small delay is also required for some application, but is not as strict as the requirement of loss and error. Due to the burstiness, unpredictability and huge amount of the data traffic, congestion control of this class is the most needed and is also the most studied.

ATM Forum Technical Committee specified the feedback mechanism for ABR flow control.

7.16. INTERNETWORKING

1. Basic Concepts: An internet is a general term for connection of different types of networks together. When we have different networks we have to deal with different protocols. Each machine will not use same protocol in each layer. A variety of protocols such as TCP/IP, SNA, DEC, NCP/IPX and specialized protocols used for satellites and cellular networks are around. All the networks being connected to each other may not use the same technology. The technology used for ATM and wireless networks is entirely different so the protocols for such networks can be completely different.

2. Types of Networks: The types of networks that are interconnected are LANs, MANs, different types of WANs etc.

3. Connection Between two Networks: The black boxes used for connecting two networks to each other depend on the layer that does the work. Table 7.2 shows the layer, corresponding name of the connector and its function.

Table 7.2.

Layer number	Connector	Function
Layer 1	Repeater segments.	Copy individual bits between cable
Layer 2	Bridges	Store and forward data link frames.
Layer 3	Routers	Multiprotocol routers forward packets between dissimilar networks.
Layer 4	Transport gateways	They connect byte streams in the transport layer.
Layer 5 and above	Application gateways	Allow internetworking above layer 4.

We have already discussed **repeaters and bridges Multi-protocol routers** are conceptually similar to bridges but they are network layer devices. **Transport gateways** are used for making a connection between two networks at the transport layer. **Application gateways** are used to connect two parts of an application in the application layer. Practically, an application gateway is rippled into two parts. Each one is called a **half gateway** and the two half gateways are connected by a wire.

7.16.1. Differing Factors in Various Networks

Networks can differ from each other in many different ways. Following are some of the differences in the network layer.

1. **Service offered :** It can be connection oriented or connectionless service.
2. **Protocol used :** Some of the possible protocols are IP, CLNP, DEC net etc.
3. **Addressing :** It can be flat or hierarchical.
4. **Multicasting :** Can be present or absent.
5. **Packet size :** It depends on the network.
6. **Error handling :** We can have reliable, ordered or unordered delivery.
7. **Flow control :** We can get sliding window, rate control etc.
8. **Congestion control :** Different algorithms such as leaky bucket, choke packets are available.
9. In addition to this, the networks differ in terms of security, parameters, accounting and quality of service.

While carrying out internetworking, one has to take into account all these differing factors.

7.16.2. Styles of Internetworking

The two commonly used styles of internetworking are shown in figure 7.29.

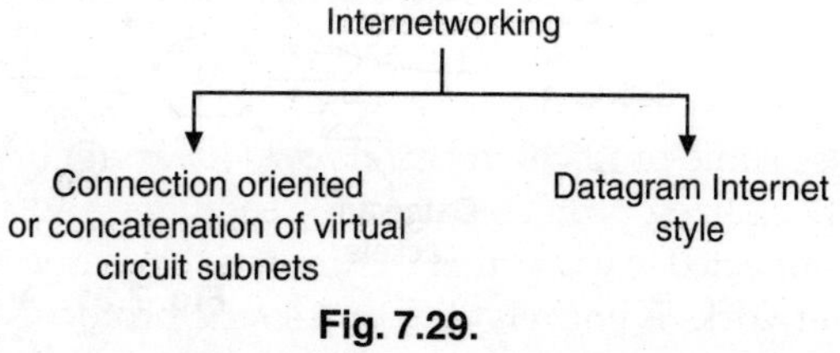

Fig. 7.29.

7.16.3. Concatenated Virtual Circuits

A concatenated virtual circuit model is shown in figure 7.30.

Process of Virtual Circuit Establishment: The subnet shows that the type of destination is remote destination and builds a virtual circuit to the router nearest the destination network. Next, it constructs a virtual circuit from that router to an external gateway (multiprotocol router). This gateway notes down the existence of this virtual circuit in its table and builds another virtual circuit to a router which is in the next subnet. This process continues until the destination host has been reached.

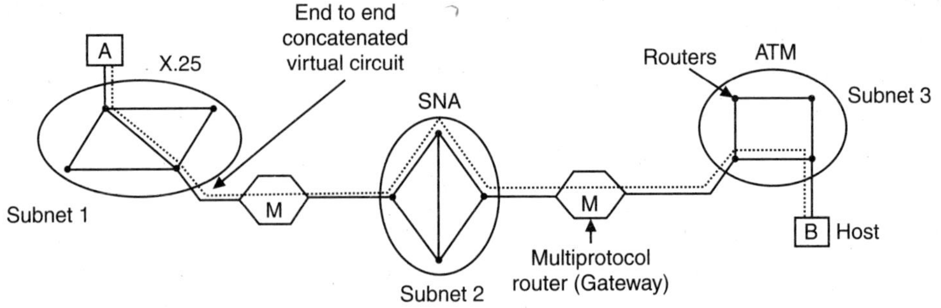

Fig. 7.30. Concatenated virtual circuit

After building the virtual circuit, the data packets will begin to flow along the path, via various gateways. All the data packets should traverse the same sequence of gateways and so arrive in order. In the concatenated virtual circuit approach, a sequence of virtual circuits is set up from the source through one or more gateways to destination. Each gateway maintains tables which tell about the virtual circuits passing through it, their routing path and the number given to different virtual circuits. This style of internetworking gives better results if all the networks have approximately the same properties, such as reliable delivery or non-reliable delivery.

7.16.4. Connectionless Interworking

> **DO YOU KNOW?**
>
> The IP packet, called the datagram, consists of a variable header and a variable data field.

1. Basic Concepts: This is the second style of internetworking which uses the diagram model shown in figure 7.31. This is connectionless internetworking. The network layer injects the datagrams into the subnet. This is the only service offered by the network layer to the transport layer. All the packets may or may not traverse the same sequence of gateways. Figure 7.31 shows the datagram for host A to host B, showing different paths taken by the datagram packets. The packets travel individually and the routing decision is made separately for each packet.

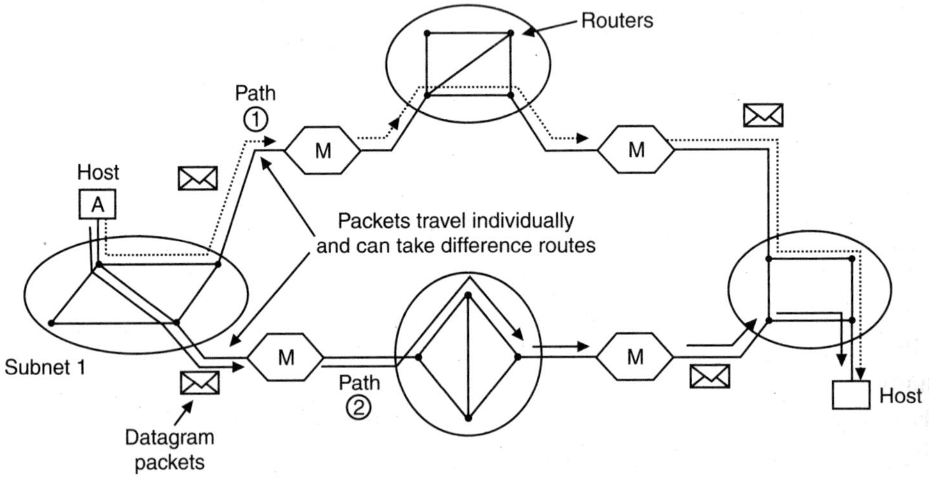

Fig. 7.31. A connectionless Internet

2. Advantage:

As this strategy uses multiple routes, it can achieve higher bandwidth than the concatenated virtual circuit model.

3. Drawbacks:

(i) Since the packets are taking different routes, there is no guarantee that the packets arrive at the destination in order.

(ii) Another disadvantage is that the model of figure 7.31 cannot work easily if each network has its own network layer protocol because their formats are different.

(iii) A more serious problem is addressing.

4. Remedy: One of the remedies to the problems mentioned earlier is to maintain a database mapping everything to everything. Another remedy is to design a universal Internet packet so that all routers can recognize it.

7.16.5. Performance Comparison of two Styles of Internetworking

(A) Advantages and disadvantages of concatenated virtual circuit model

1. The advantages of concatenated virtual circuit model are as under :
   (i) Buffers can be reserved in advance.
   (ii) Sequencing can be guaranteed.
   (iii) Short headers can be used.
   (iv) The problems caused by delayed duplicate packets can be avoided.
2. The disadvantages of concatenated virtual circuit model are as under :
   (i) Space for each open connection has to be kept in the router tables.
   (ii) There is no alternate route available to avoid the congestion areas.
   (iii) This scheme is vulnerable to router failure along a path.

(B) Advantages and disadvantages of datagram model

1. Advantages of datagram model are as under:
   (i) It is possible to adapt to the congestion problem in a better way.
   (ii) It is more robust in the event of router failure.
   (iii) Various adaptive routing algorithms are possible to be used.
   (iv) It can be used over the subnets which do not use the virtual circuits inside.
2. Disadvantages of the datagram model are as under :
   (i) There is more potential for congestion.
   (ii) Longer headers are required.

7.17. TUNNELING

1. Basic Concepts: A technique of Internetworking called tunneling is used when source and destination networks of same type are to be connected through a network of different type. For example, let us consider an Ethernet to be connected to another Ethernet through a WAN as shown in figure 7.32.

The task is send on IP packet from host A of Ethernet-1 to the host B of Ethernet-2 via a WAN.

2. Sequence of Events:

(i) Host A construct a packet containing the IP address of host B.

(ii) It then inserts this IP packet into an Ethernet frame. This frame is addressed to the multiprotocol router M_1.

(iii) Host A then puts this frame on Ethernet.

(iv) When M_1 receives this frame, it removes the IP packet, inserts it in the payload packet of the WAN network layer packet and addresses the WAN packet to M_2. The multiprotocol router M_2 removes the IP packet and sends it to host B in an Ethernet frame.

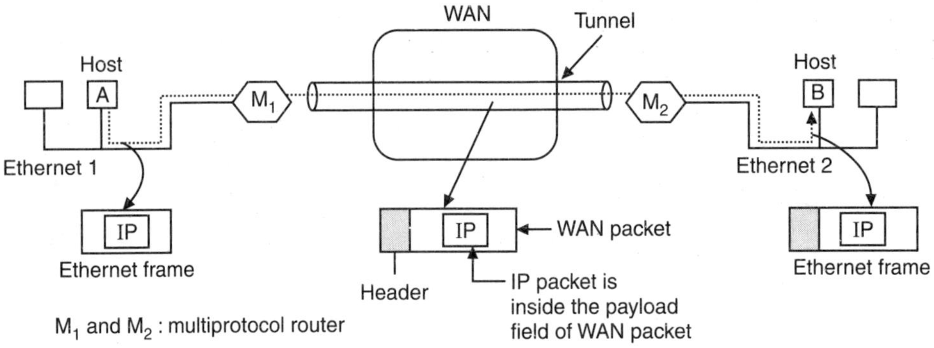

Fig. 7.32. Illustration of tunneling

3. Why is this Technique Called Tunneling ?: In this particular example, the IP packets do not have to deal with WAN. The hosts A and B also do not have to deal with the WAN. The multiprotocol routers M_1 and M_2 will have to understand about IP and WAN packets. Therefore, the WAN can be imagined to be equivalent to a big tunnel extending between the multiprotocol routers M_1 and M_2 and the technique is called as **tunneling**.

7.18. INTERNETWORK ROUTING

Internetwork routing means routing through Internetwork. It is similar to routing within a single subnet. However, it is bit more complicated. Figure 7.33 (a) shows an Internetwork with five networks. They are connected to each other by six multiprotocol routers M_1 to M_6. The graph of this network has been shown in figure 7.33(b).

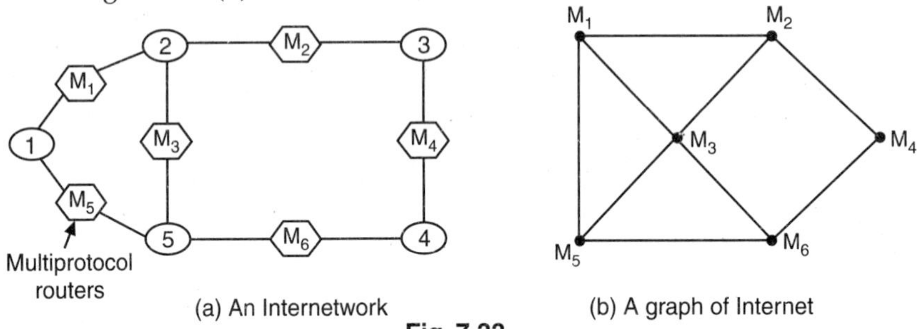

Fig. 7.33.

First step is to draw the graph. Second step is to apply a routing algorithm such a distance vector or link state algorithms to set the multiprotocol routers. For Internetwork routing, a two level routing algorithm is used. One of them is within each network (an interior gateway protocol) and the other is between the networks (an exterior gateway protocol). Note that gateway is an old term used for router.

All the network in an Internetwork are independent from each other and therefore called as Autonomous system. They may all use different protocols. An Internet packet first goes to the local multiprotocol router (in the MAC sublayer header). The network layer code decides which router is to be used to forward this packet. If it is possible for the packet to reach this router using the packet's native network protocol, then it is forwarded directly to the router.

Otherwise the packet is forwarded by tunneling technique. This process is repeated until the packet reaches the destination. As the Internetwork routing often crosses the international boundaries various international laws come into picture.

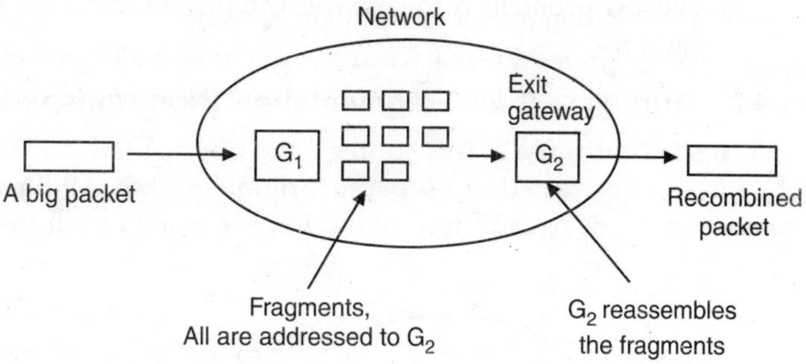

Fig. 7.34. Strategy - 1 for fragmentation

In the Internetwork routing, the costs associated with different possible routes to reach the same destination can differ. The quality of service offered by different networks may be different. In large Internetworks, the selection of best route is a time consuming job.

7.19. FRAGMENTATION

1. Basic Concepts: The network designers are not free to choose any size of the packet. The maximum packet size varies network to network and the factors which decide the maximum packet size may be listed as under :

(i) Width of the TDM transmission slot. (ii) Protocol used.

(iii) Type of operating system. (iv) International standards.

(v) Efforts to reduce retransmission.

(vi) Desire to prevent open packet from occupying the channel too long.

All these factors put a limit on the maximum packet size. The maximum payload size ranges from 48 bytes for an ATM cell to 65, 515 bytes for an IP packet. When a large packet wants to travel over a network whose maximum packet size is very small, we have a problem. The solution to this problem is to avoid this situation in the first place by using a routing algorithm which will avoid sending packets through the networks that cannot handle them. But, this solution cannot be exercised every time. The real solution to this problem is **Fragmentation**.

2. Fragmentation: In this technique, the gateways break up large packets into smaller ones called as fragments. Then each fragment is sent as a separate Internet packet. But, the reverse process of putting the fragments together is considerably difficult.

3. Recombination of Fragments: The recombination of fragments can be done by using one of the two strategies. Let us discuss two strategies inthe subsections to follow.

7.19.1. Strategy – 1 for Fragmentation (Transparent Strategy)

1. Basic Concepts: In this strategy, the fragmentation caused by a small packet network is made transparent to any subsequent network through which the packets will pass. When a large packet arrives at a gateway, it breaks the packet into fragments. Each fragment is then addressed to the same exit gateway. The exit gateway (G_2) recombines all these fragments. This strategy is illustrated in figure 7.34.

The subsequent networks are not even aware that fragmentation has taken place. Fragmentation in ATM networks is called segmentation, but the concept is same.

2. Drawbacks: The drawbacks of transport fragmentation are as under:

(i) The first problem with transport fragmentation is that the exit gateway G_2 has to know when it has received all the pieces. For this, a count field or an end of packet bit has to be included in each packet.

(ii) Another important factor is that all the packets should exit via the same gateway.

(iii) The last problem is the overhead required to repeatedly fragment and reassemble a large packet.

7.19.2. Strategy – 2 for Fragmentation (Non-transparent Strategy)

1. Basic Concepts: In this strategy, the fragmented packets are not reassembled at all. Instead, each fragment is treated as a separate original packet. All these packets are passed through the exit gateway or gateways and recombination is carried out at the destination host as shown in figure 7.35.

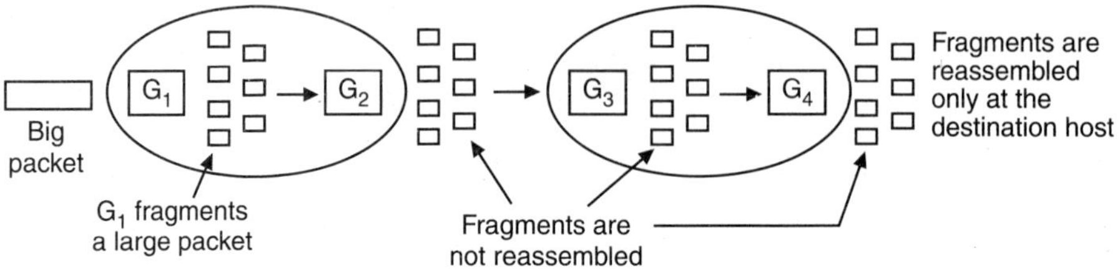

Fig. 7.35. Strategy - 2 for fragmentation

This is called as a non-transparent fragmentation.

2. Drawbacks: The drawbacks of non-transparent fragmentation may be listed as under:

(i) Every host must be capable of reassembling the fragments.

(ii) The total overhead increases due to fragmentation since each fragment has to have a header.

(iii) When a packet is fragmented, the fragments will have to be numbered in such a way that the original data stream can be reconstructed.

3. Advantage: The advantage of non-transparent strategy is that multiple exit gateways can be used to achieve higher performance.

7.20. FIREWALLS

Due to Internetworking, each computer is concerned to many other computers and networks. But this is problematic for the corporate companies in two ways. Most companies put a large amount of confidential information on line. Such an information should not get disclosed to the unauthorised persons. Secondly, there is a danger of viruses, worms and other digital pests leaking in. They can destroy valuable data and waste valuable administrator time. In order to overcome both these problems, we must have a mechanism to keep good bits in and bad bits out. One method is to use the encryption. This approach will protect the good bits from leaking out but it cannot keep the digital pests and hackers out. To accomplish this goal, we can use a technique called **Firewalls**. In this technique, a company can have many LANs connected in arbitrary ways, but all the traffic to and from the company is forced to flow through an electronic drawbridge called firewall. This has been shown in figure 7.36.

The firewall consists of following two components :

(i) Two routers for packet filtering.

(ii) An application gateway.

Therefore, every packet has to travel through two packet filtering routers and an application gateway while going in or coming out because there is no additional route existing.

7.20.1. Packet Filter

A packet filter is basically a standard router and some additional facilities. The additional facilities will make it possible to impact each incoming and outgoing packet. The packets satisfying certain criterias only are allowed to pass through. Those who fail to satisfy these conditions are dropped.

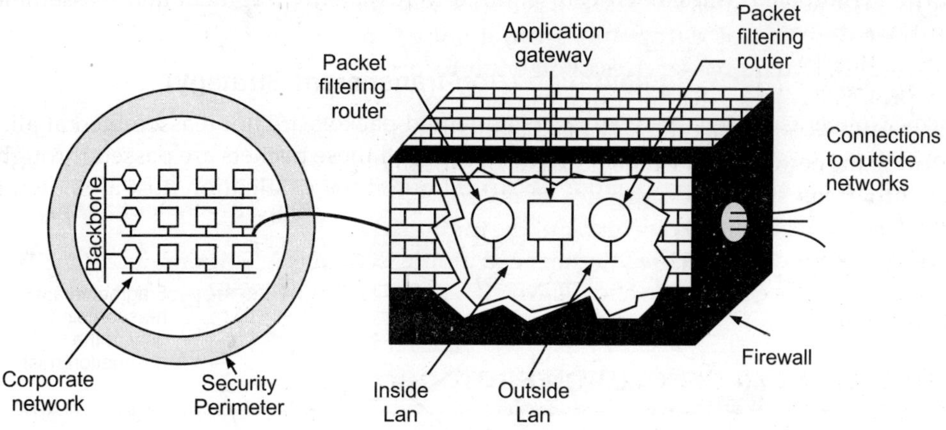

Fig. 7.36. Illustration of Firewall

The packet filter on the input side (inside LAN) checks the outgoing packets and that on the output side (outside LAN) checks the incoming packets. The application gateway makes further examination of the packets which have reached it through the router. Thus, every packet going in or coming out has to pass through the application gateway. However, the firewalls do not provide any protection if some of the machines are wireless and use radio communication because it passes over the firewalls in both the directions. The operation of packet filters is dependent on tables configured by system administrator. Such tables contain the list of sources and destinations which are acceptable and sources and destinations to be blocked. For example, in case of Unix setting, a source or destination consists of an IP address and a port. Ports indicate the desired service, like port 23 is for Telnet, port 79 for finger, port 119 for USENET etc. It is possible for a company to block the incoming packets for all the IP addresses combined with one of the ports mentioned above. If done so, no one outside the company could log in via Telnet or look up people using the finger daemon. The USENET news also will not be allowed.

Blocking Outgoing Packets

Till now we have only discussed how to block the incoming packets. Now let us see how to block the outgoing ones. In fact, it is more difficult to block the outgoing packets because it is not possible to force the sites to stick to the standard port naming conventions. For some services, like FTP (file transfer protocol) port numbers are assigned dynamically. On top of that if blocking the TCP connections is difficult then blocking a UDP packet is still more difficult, because we know a very little in advance about what they will do. Therefore, many packet filters simply ban the UDP traffic altogether.

7.20.2. Application Gateway

Application gateway is the second half of the firewall mechanism. The application gateway does not just look at the raw packets but it operates at the application level. It is possible to set up a mail gateway to examine each message going in or coming out.

The message is transmitted or discarded based on header fields, message size or the content (certain words in the message will result in discarding that message). It is possible to set up one or more application gateways for a specific application.

7.21. VARIOUS NETWORK LAYER PROTOCOLS

We have discussed the TCP/IP model in chapter 1. In the Internet model, or the TCP/IP suite, there are five main network layer protocols : ARP, RARP, IP ICMP and IGMP as shown in figure 7.37.

The main protocol in this layer is IP. It is responsible for host to host delivery of datagrams from a source to destination. But, IP requires services of other protocols. IP needs a protocol called ARP in order to find the MAC (physical) address of the next hop. IP needs the services of JCMP during the delivery of the datagram packets to handle unusual situations such as occurrence of an error. IP is basically designed for unicast delivery. But the multimedia and some new applications in the Internet need multicast delivery. So for multicasting, IP has to use the services of another protocol called IGMP. IPv4 is the current version of IP whereas IPv6 is the latest version of IP.

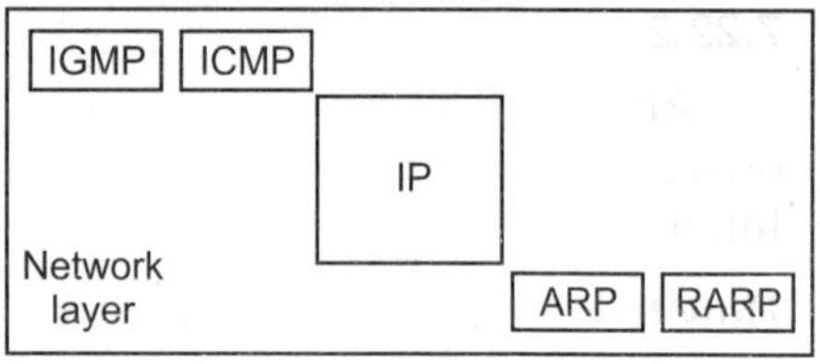

Fig. 7.37. Protocols at network layer

7.22. ARP (ADDRESS RESOLUTION PROTOCOL)

An Internet consists of various types of networks and the connecting devices like routers. A packets starts from the source host, passes through many physical networks and finally reaches the destination host. At the network level, the hosts and routers are recognised by their IP addresses.

1. IP Address: An IP address is an Internetwork address. It is a universally unique address. Every protocol involved in Internetworking requires IP addresses.

2. MAC Address: The packets from source to destination hosts pass through physical networks. At the physical level, the IP address is not useful but the hosts and routers are recognized by their MAC addresses. A MAC address is a local address. It is unique locally but it is not unique universally. The IP and MAC address are two different identifiers and both of them are needed, because a physical network can have two different protocols at the network layer at the same time. Similarly, a packet may pass through different physical networks.

Therefore, to deliver a packet to a host or a router, we require two levels of addressing namely IP addressing and MAC addressing. Most importantly, we should be able to map the IP address into a corresponding MAC address.

7.22.1. Mapping of IP Address into a MAC Address

We have seen the need of mapping an IP address into a MAC address. Such a mapping can be of following two types :

(i) Static mapping

(ii) Dynamic mapping.

1. Static mapping: In static mapping, a table is created and stored in each machine. This table associates an IP address with a MAC address. If a machine knows the IP address of another machine then it can search for the corresponding MAC address in its table. The limitation of static mapping is that the MAC address can change. To implement static mapping, the static mapping table needs to be updated periodically.

2. Dynamic mapping: In dynamic mapping technique, a protocol is used for finding the other address one type of address is known. There are two protocols designed to perform the dynamic mapping. They are as under:

(i) Address Resolution Protocol (ARP).

(ii) Reverse Address Resolution Protocol (RARP).

The ARP maps an IP address to a MAP address whereas the RARP maps a MAC address to an IP address.

DO YOU KNOW?

The address resolution protocol (ARP) associates an IP address with the physical address. On a typical physical network, such as a LAN, each device on a link is identified by a physical or station address usually imprinted on the network interface card (NIC).

7.22.2. ARP Operation

ARP is used for association an IP address to its MAC address. For a LAN, each device has its own physical or station address as its identification. This address is imprinted on the NIC (Network Interface Card).

How to Find the MAC Address?

When a router or a host needs to find the MAC address of another host or network, the sequence of events taking place is as follows:

The router or host A who wants to find the MAC address of some other router, sends as ARP request packet. This packets consists of IP and MAC addresses of the sender A and the IP address of the receiver (B). This request packet is broadcasted over the network as shown in figure 7.38.

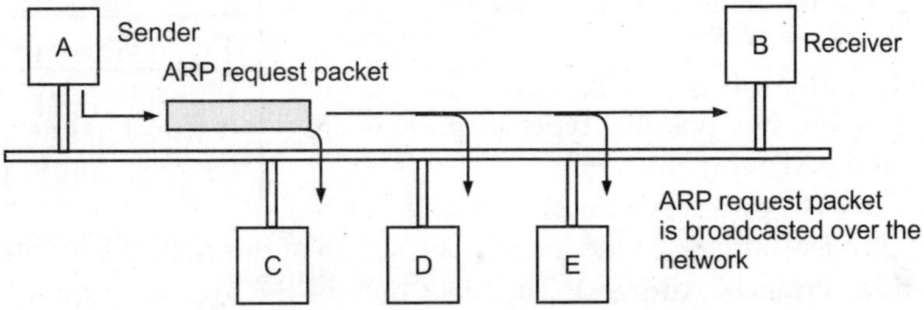

Fig. 7.38. ARP request is broadcasted

Every host and router on the network receives and processes the ARP request packet. But only the intended receiver (B) recognizes its IP address in the request packet and sends back an ARP response packet. The ARP response packet contains the IP and physical addresses of the receiver (B). This packet is delivered only to A (unicast) using A's physical address in the ARP request packet. This has been shown in figure 7.39.

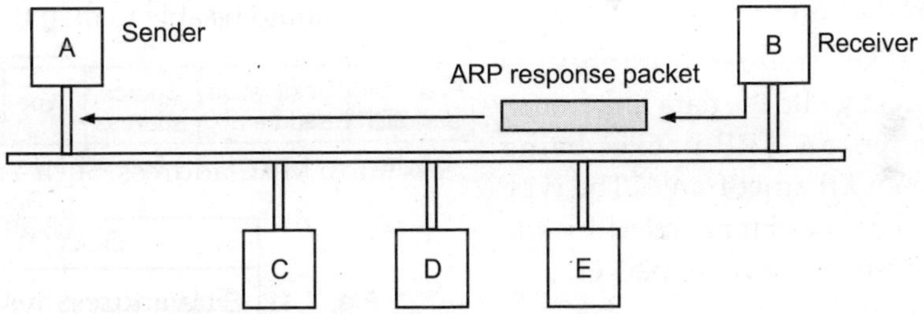

Fig. 7.39. ARP response unicast

7.22.3. ARP Packet Format

The ARP packet format is as shown in figure 7.40. The various fields in it are as under :

DO YOU KNOW?

ARP is used to find the physical address of the node when its Internet address is known.

1. HTYPE (Hardware type): This 16 bit field defines the type of network on which is ARP is being run. ARP can run on any physical network.

2. PTYPE (Protocol type): This 16 bit field is used to define the protocol using ARP. Note that ARP can be used with any higher-level protocol such as IPv4.

3. HLEN (Hardware length): It is an 8 bit field which is used for defining the length of the physical address in bytes. For example, this value is 6 for Ethernet.

Hardware Type (16 bits)		Protocol type (16 bits)
Hardware length	Protocol length	Operation request 1, Reply 2
Sender hardware address		
Sender protocol address		
Target hardware address		
Target protocol address		

Fig. 7.40. ARP frame format

4. PLEN (Protocol length): This field is 8 bit long and it defines the length of the IP address in bytes. For IPv4, this value is 4.

5. OPER (Operation): It is a 16 bit field which defines the type of packet. The two possible types of packets are : ARP request (1) and ARP reply (2).

DO YOU KNOW?

The reverse address resolution protocol (RARP) allows a host to discover its internet address when it knows only its physical address.

6. SHA (Sender Hardware Address): The field is used for defining the physical address of the sender. The length of this field is variable.

7. SPA (Sender Protocol Address): This field defines the logical address of the sender. The length of this field is variable.

8. THA (Target Hardware Length): It defines the physical address of the target. It is a variable length field. For the ARP request packet, the field contains all zeros because the sender does not know the receivers physical address.

9. TPA (Target Protocol Address): This field defines the logical address of the target. It is a variable length field.

7.22.4. Encapsulation

An ARP packet (request or reply) is encapsulated directly into the data link frame. Figure 7.41 shows an ARP packet being encapsulated in an Ethernet frame. The type field indicates that the data carried by the frame is an ARP request or reply packet.

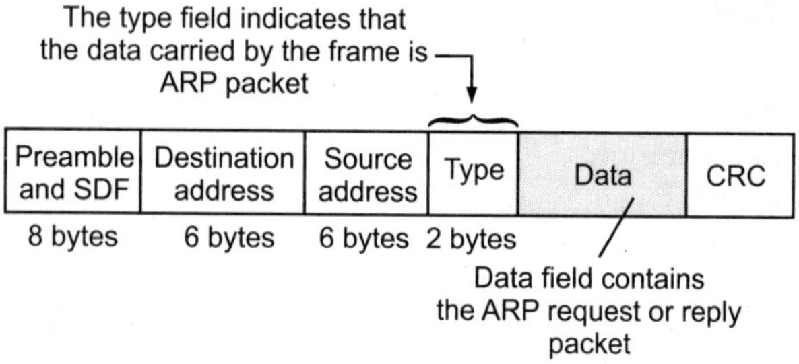

Fig. 7.41. Encapsulation of ARP packet

7.22.5. Operation of ARP on Internet

The services of ARP can be used under the following working conditions when it is being operated on Internet.

(i) The sender is a host and wants to send a packet to another host on the same net.

(ii) The sender is a host and wants to send a packet to another host on another network.

(iii) The sender is a router which has received a datagram destined for a host on another network.

(iv) The sender is a router that has received a datagram destined for a host in the same network.

Now, let us see how ARP works on the Internet.

Operation: The sender (host or router) knows the IP address of the target. IP orders ARP to create an ARP request message. The request packet consists of senders physical and IP addresses plus the IP address of the target. This ARP request packet is sent to the data link layer. Here, it is encapsulated in a frame. Every router or host receives this frame because it is broadcast. All the machines except the target drop this packet. The target machine replies back with an ARP reply packet which contains the target's physical address. This reply is unicast. The sender receives the reply packet. It now knows the physical address of the target. The IP datagram carrying data for the target machine is encapsulated in a frame and the frame is unicast to the destination.

7.23. REVERSE ADDRESS RESOLUTION PROTOCOL (RARP)

RARP (Reverse Address Resolution Protocol) is part of the TCP/IP protocol suite. It allows a computer, particularly a diskless workstation, to obtain an IP address from a server. When a diskless TCP/IP workstation is booted on a network, it broadcasts a RARP request packet on the local network. This address packet is broadcast on the network for all to receive because the workstation does not know the IP address of the server that can supply it with an address. It includes its own physical network address (the MAC address) in the request so the server will know where to return a reply. The server that receives the request looks in a table and matches the MAC address with an IP address, and then returns the IP address to the diskless workstation.

7.24. NETWORK LAYER IN THE INTERNET: INTERNET PROTOCOL (IP)* : IPv4 PROTOCOL

(RGTU, Bhopal, Sem. Exam., June, 2014)

This is the host to host network layer delivery protocol designed for the Internet. IP is a connectionless datagram protocol with no guarantee of reliability. It is an unreliable protocol because it does not provide any error control or flow control. IP can only detect the error and discards the packet if it is corrupted. If IP is to be made more reliable, then it must be paired with a reliable protocol such as TCP at the transport layer. Each IP datagram is handled independently and each one can follow a different route to the destination. Therefore, there is a responsibility of receiving out of order packets at the destination. Some packets may even be lost or corrupted. IP relies on a higher level protocol to take care of all these problems.

7.24.1. Datagram

Packets in IP layer are called datagrams. Figure 7.42 shows the IP datagram format. A datagram is a variable length with two parts namely the header and data. The header is 20 to 60 bytes in length. It contains the information essential for routing and delivery. The other part of the datagram is the data field which is of variable length.

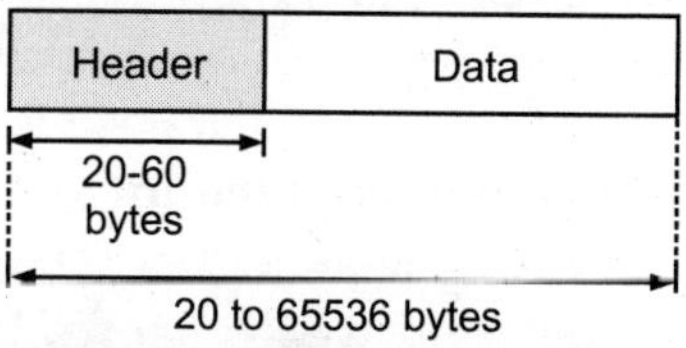

Fig. 7.42. IP datagram format

7.24.2. Structure of IP Frame Header

The IP frame header contains routing information associated with datagram delivery. The header structure is as shown in figure 7.43.

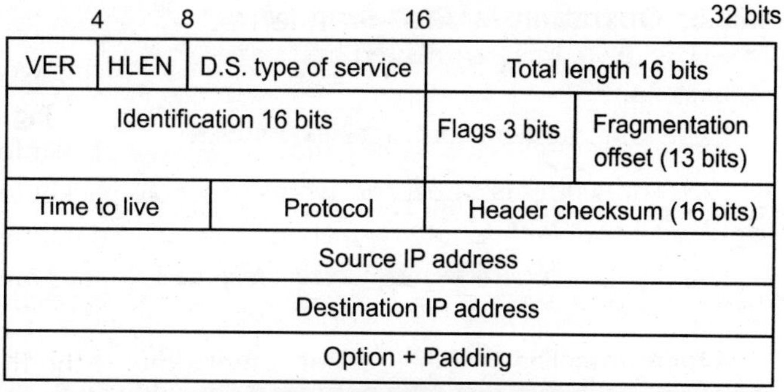

Fig. 7.43. IP header structure

Various fields in the IP header are as under :

1. VER (Version): This field define the version of IP. Current version of IP is IPv4 and the latest version of IP is IPv6. It is a four bit long field.

2. HLEN (Header length): This field defines the length of the datagram header in 4-byte word. Its value must be multiplicated by 4 to give the length in bytes.

* Sketch the IP header neatly and explain the function of each field. List major differences between IPv4 and IPv6.

3. Differential Service (DS): This field defines the class of the datagram for quality of service purpose. Networks may offer service precedence, meaning that they accept traffic only above certain precedence at times of high load. There is a three way trade off between low delay, high reliability and throughput.

4. Total length: The field defines the total length of the IP datagram. The total length includes the length of header as well as the data field. The field length of this fields is 16 bits so the total length of the IP datagram is restricted to $(2^{16} - 1) = 65535$ bytes of which 20 to 60 bytes are the header and the remaining are data. This field allows the length of a datagram to be upto 65,535 bytes, although such long datagrams are impractical for most hosts and networks. All hosts must be prepared to accept datagram of upto 576 bytes, regardless of whether they arrive whole or in fragments. It is recommended that hosts and datagram larger than 576 bytes only if the destination is prepared to accept larger datagram.

5. Identification, Flag and Offset

(i) Identification: This field identifies the datagram originating from the source host. When a datagram is fragmented, the value in the identification field is copied into all fragments. The identification number helps the destination in reassembling the fragments of the datagram.

(ii) Flags: This is a three bit field. The 3 bits are as shown in figure 7.44.

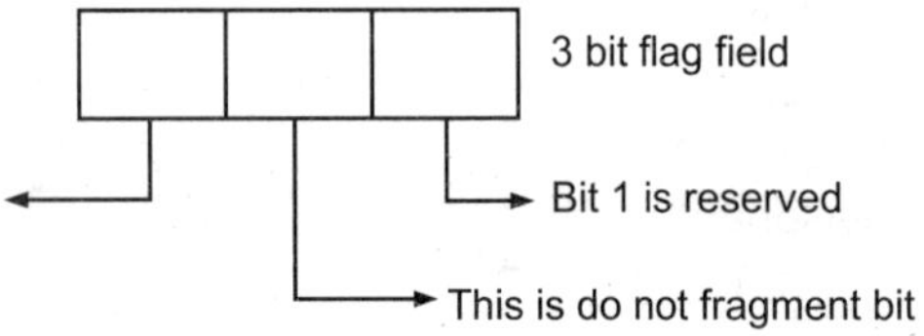

Fig. 7.44. Flag bits

First bit is reserved, and it should be 0. The second bit is called Do Not Fragment bit. If this bit is 1 then machine should not fragment the datagram. But if the value of this bit is 0 then the machine should fragment the datagram if only if necessary. The third bit is called as More Fragment Bit. If it is 1 it means that the datagram is not the last fragment but if its value is 0 it shows that this is the last or the only fragment.

(iii) Fragmentation Offset: This is a 13 bit field which shows the relative position of this fragment with respect to the whole datagram. It is the offset of the data in the original datagram measured in units of 8 bytes. To understand this let us consider figure 7.45. The original IP packet (datagram) contains 4000 bytes numbered from 0 to 3999. It is fragmented into three fragments.

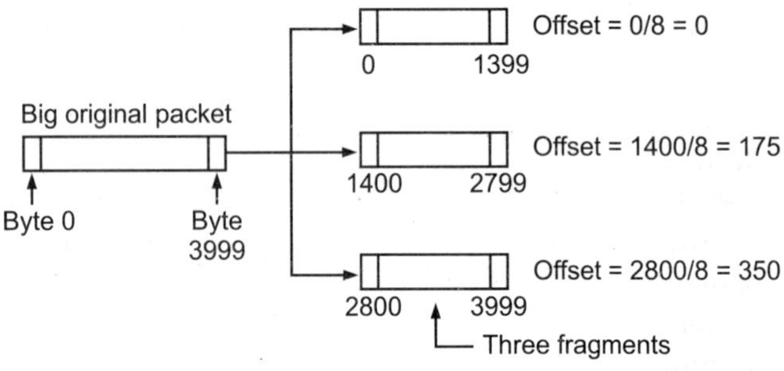

Fig. 7.45. Example of fragmentation

The first fragment contains 1400 bytes numbered from 0 to 1399. The offset for data fragment is $\frac{0}{8} = 0$. Similarly, the offsets for the other two fragments are $\frac{1400}{8} = 175$ and $\frac{2800}{8} = 350$ respectively as shown in figure 7.45. The offset is measured in units of 8 bytes. Because the length of the offset field is 13 bits. So, the fragments should be of size such that first byte number is divisible by 8.

6. Time to Live: This is an 8 bit long field which controls the maximum number of routers visited by the datagram.

7. Protocol: This field defines the higher-level protocol which uses the services of the IP Layer. An IP datagram can encapsulate data from various higher level protocols such as TCP, UDP, ICMP and IGMP. The protocol field specifies the final destination protocol to which the IP datagram should be delivered. Since IP multiplexes and demultiplexes data from different higher level protocols, the value of protocol field helps in demultiplexing at the final destination.

8. Header Checksum: A checksum in IP packet covers on the header only. Since some header fields change, this field is recomputed and verified at each point that the internet header is processed.

9. Source Address: This field is used for defining the IP address of the source.

10. Destination Address: This field is used for defining the IP address of the destination.

11. Options: Options are not required for every datagram. They are used for network testing and debugging. IP provides several optional features, allowing a packet's sender to set requirements on the path it takes through the network (source routing), trace the route a packet takes (record route), and label packets with security features.

7.24.3. Services Provided

IP provides following services:

1. Addressing: IP headers contain 32-bit addresses which identify the sending and receiving hosts. These addresses are used by intermediate routers to select a path through the network for the packet.

2. Fragmentation: IP packets may be split, or fragmented, into smaller packets. This permits a large packet to travel across a network which can only handle smaller packets. IP fragments and reasonable packets transparently.

3. Packet Timeout: Each IP packet contains a Time To Live (TTL) field, which is decremented every time a router handles the packet. If TTL reaches zero, the packet is discarded, preventing packets from running in circles forever and flooding a network.

4. Types of Service: IP supports traffic prioritization by allowing packets to be labelled with an abstract type of service.

7.24.4. IP Addresses

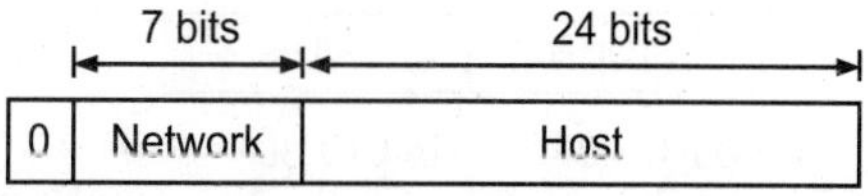

Fig. 7.46. (a) Class A IP address formats

As already stated every host and router on the internet has a unique IP address. All the IP addresses are 32 bit long and they are used in the source address and destination address fields of the IP header. The formats used for IP address are as shown in figure 7.46. The IP address for class A networks is shown in figure 7.46(*a*).

The IP numbers (addresses) for the hosts are assigned by the network administrator. For a public network on the Internet, we have to obtain a network number assigned by the network information center. An IP address consists of two parts. The first part of the address, called the network number, identifies a network on the Internet, the remainder, called the host ID, identifies an individual host on that network.

1. Class A address: The network field is 7 bit long as shown in figure 7.46 (a) and the host field is of 24 bit length. So the network field can have numbers between 1 to 126. But the host numbers will range from 0.0.0.0 to 127.255.255.255. Thus, in class A, there can be 126 types of networks and 17 million hosts. The 0 in the first field identifies that it is a class A network address.

2. Class B format: The class *B* address format is shown in figure 7.46 (b). The first two fields identify the network, and the number in the first field must be in the range 128 - 191.

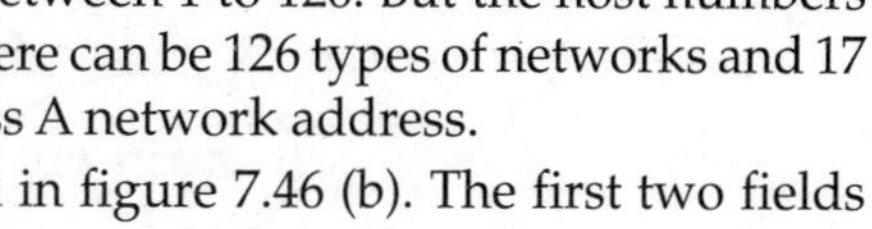

Fig. 7.46. (b) Class B format

Class *B* networks are large. Host numbers 0.0 and 255.255 are reserved, so, there can be upto 65,534 (216 - 2) hosts in a class *B* network. Most of the 16,382 class B addresses have been allocated. The first block covers address from 128.0.0.0 to 128.255.255.255 and the last block covers from 191.255.0.0 to 191.255.255.255. **Example :** 128.89.0.26, for host 0.26 on net 128.89.

3. Class C Format: The class C address format is shown in figure 7.46 (c).

The first block in class C covers addresses from 192.0.0.0 to 192.0.0.255 and the last block covers addresses from 223.255.255.0 to 223.255.255.255.

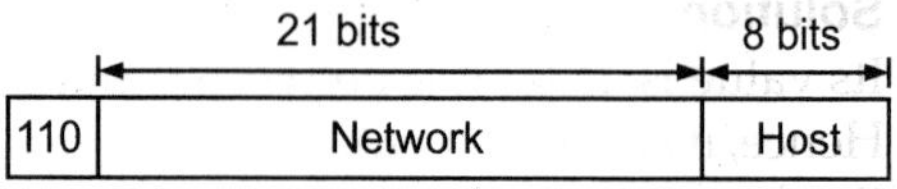

Fig. 7.46. (c) Class C Format

4. Class D Format: The class D address format is shown in figure 7.46 (d).

The class format allow for upto 2 million networks with upto 254 hosts each and class D format allows the multicast in which a datagram is directed to multiple hosts.

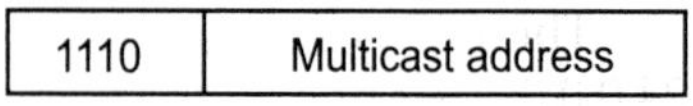

Fig. 7.46. (d) Class D format

5. Class E Address Format: Figure 7.47, shows the address format for a class *E* address. This address begins with 11110 which shows that it is reserved for the future use.

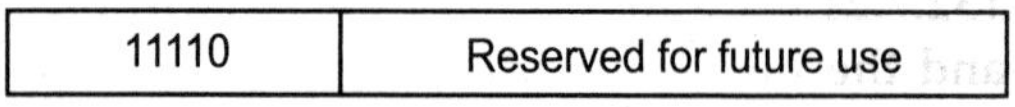

Fig. 7.47. IP address for class E network

The 32 bit (4 byte) network addresses are usually written in dotted decimal notation. In this notation each of the 4-bytes is written in decimal from 0 to 255. Therefore, the lowest IP address is 0.0.0.0 i.e. all the 32 bits are zero and the highest IP address is 255.255.255.255.

7.24.5. Network Address

The network address is an address that defines the network itself. It cannot be assigned to a host. Figure 7.48 shows the examples of network address for different classes.

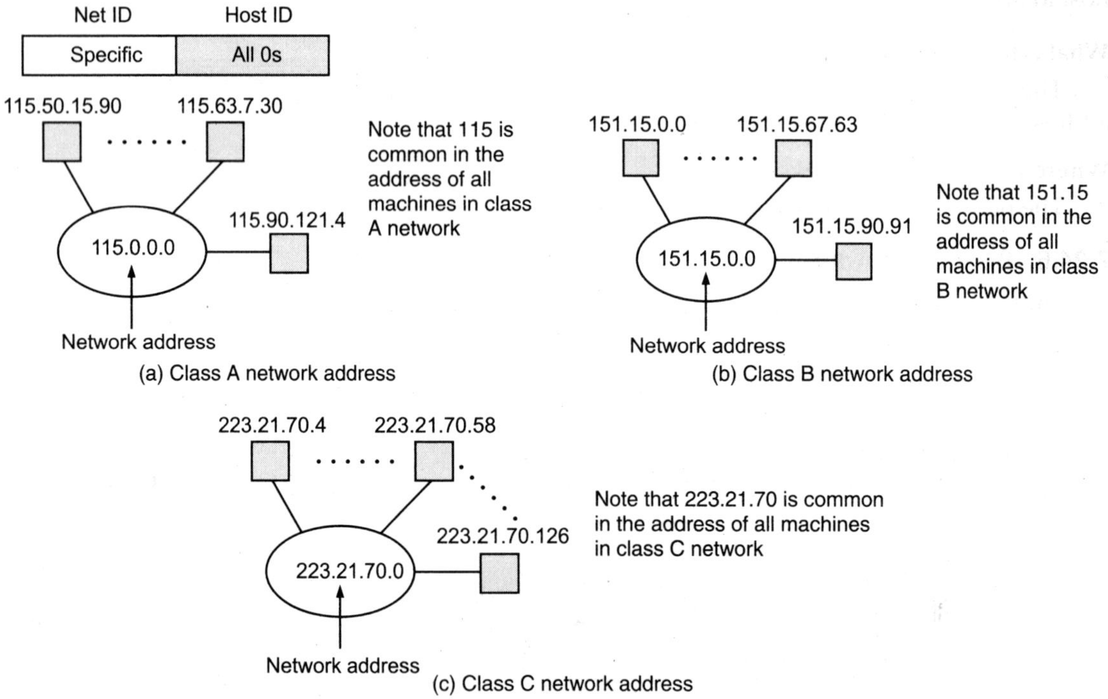

Fig. 7.48.

The following examples will enable you to find the network address.

EXAMPLE 7.1. For the address 24.46.8.95, identify the type of network and find the network address.

Solution: Let us examine the first byte. Its value is 24 *i.e.* it is between 0 and 127. Hence, it is a class *A* network. So, only the first byte defines the Net id. So we can find the network address by replacing the host id with 0s.

The process of obtaining the network address is shown in figure 7.49.

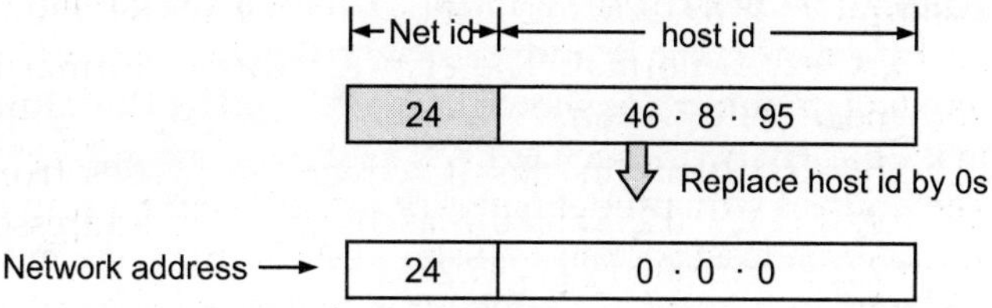

Fig. 7.49.

Therefore, the network address is 24.0.0.0. **Ans.**

EXAMPLE 7.2. For the address 132.7.21.84 find the type of network and the network address.

Solution: Let us examine the first byte. It is 132 *i.e.* between 128 and 192. So, it is a class *B* network. So, the first two bytes define the net id. Replace the host id with 0's to get the network address as shown in figure 7.50.

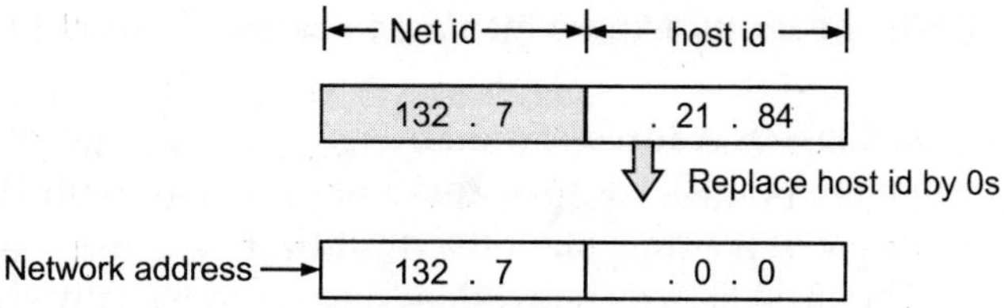

Fig. 7.50.

So, the network address is 132.7.0.0. **Ans.**

EXAMPLE 7.3. Find the class of the network if the address is 221.46.75.64.

Solution: The first byte is 221 *i.e.* between 192 and 255. So, this is a class C network. The net id and host id are as shown in figure 7.51.

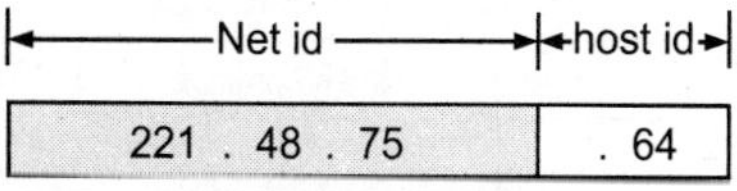

Fig. 7.51.

What is the Difference Between net id and Network Address ?

The network address is different from a net id. A network address has both net id and host id, with 0s for the host id.

Where to Use the Network Address?

The network address is used to route the packets to the desired location.

7.24.6. Special IP Addresses

Figure 7.52 shows some special IP addresses.

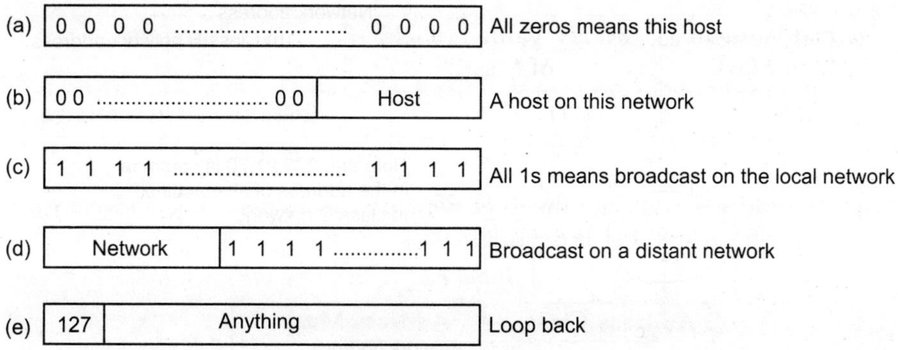

Fig. 7.52. Special IP addresses

All zeros means this host or this network and all 1s means broadcast address to all hosts on the indicated network. The IP address 0.0.0.0 is used by the hosts when they are being booted but not used afterward. The IP addresses with 0 as the network number refer to their own network without knowing its number as shown in figure 7.52 (*b*). The address having all ones is used for broadcasting on the local network such as LAN as shown in Fig. 7.52 (*c*). Now, let us consider figure 7.52 (*d*). This is an address with proper network number and all 1s in the host field. This address allow machines to send broadcast packets to distant LANs anywhere in the Internet. If the address is *127. Anything* as shown in figure 7.52 (*e*) then it is a reserved address **loopback testing**. This feature is also used for debugging network software.

7.24.7. Address Masks (Default Masks)

An address mask determines which portion of an IP address identifies the network and which portion identifies the host. Like the IP address, the mask is represented by four octets. (An octet is an 8-bit binary number equivalent to a decimal number in the range 0 - 255). If a given bit of the mask is 1, the corresponding bit of the IP address is in the network portion of the address, and if a given bit of the mask is 0, the corresponding bit of the IP address is in the host portion. As an example, let us consider a class C address 192.15.28.16. This is shown in figure 7.53. Here, it may be noted that 192.15.28 corresponds to the network part and 16 corresponds to the host part.

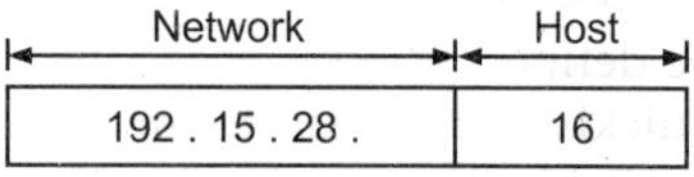

Fig. 7.53.

Hence, to differentiate the network and host parts, we have to use a mask 255.255.255.0. Table 7.3 shows the mask 255.255.255.0 in both decimal and binary form, aligned with the class C address 192.15.28.16, also in both decimal and binary form :

Table 7.3. IP address (In decimal and binary form)

Element	Network			Host
Mask	255	.255	.255	0
	11111111	11111111	11111111	00000000
Address	192	.15	.28	.16
	11000000	00001111	00011100	00010000

If a field of the network address is entirely used for the network number, the corresponding field of the mask has the decimal value 255 (binary 11111111), and if an address field is entirely used for the host ID, the corresponding field of the mask has the decimal value 0.

Table 7.4.

Decimal Value in Field of Mask	Binary Value in Field of Mask	Function
255	11111111	Identify network number
0	0000000	Identify host ID

Accordingly, the address masks for the three network classes described above are as shown in table 7.5. These masks are also called as default masks.

Table 7.5.

Address Class	Address Mask
A	255.0.0.0
B	255.255.0.0
C	255.255.255.0

Which IP Protocol Version is Being Used Currently ?

The network protocol in the Internet is currently IPv4. It was first introduced in 1970's. After that the world of data communication has grown beyond imaginations. Even though IPv4 is a well designed protocol, it has some limitations.

7.24.8. Limitations of IPv4

The most obvious limitation of IPv4 is its address field. IP relies on network layer addresses to identify end-points on networks, and each networked device has a unique IP address. IPv4 uses a 32-bit addressing scheme, which gives it 4 billion possible addresses. With the proliferation of networked devices including PCs, cell phones, wireless devices, etc., unique IP addresses are becoming scarce, and the world could theoretically run out of IP addresses. If a network has slightly more number of hosts than a particular class, then it needs either two IP addresses of that class or the next class of IP address. For example, let us say a network has 300 hosts, this network needs either a single class BIP address or two class C IP addresses. If class *B* address is allocated to this network, as the number of hosts that can be defined in a class *B* network is $(2 \wedge 16 - 2)$, a large number of host IP addresses are wasted. If two class C IP addresses are allocated, as the number of networks that can be defined using a class C address is only $(2 \wedge 21)$, the number of available class C networks will quickly exhaust. Because of the above two reasons, a lot of IP addresses are wasted and also the available IP address space is rapidly reduced. Other identified limitations of the IPv4 protocol are : Complex host and router configuration, non-hierarchical addressing, difficulty in re-numbering addresses, large routing tables, non-trivial implementations in providing security, QoS (Quality of Service), mobility and multi-homing, multicasting etc. To overcome these problems, the Internet protocol version 6 (IPv6) which is also known as internet protocol, next generation (IPng) was proposed. In IPv6, the Internet protocol was extensively modified for accommodating the unforeseen growth of the Internet. The format and length of the IP addresses has been changed and the packet format also is changed.

DO YOU KNOW?

Subnetting allows an additional level of hierarchy in IP addressing.

7.24.9. Subnetting in IP

All the hosts in a network must have the same network number. But, this property of IP addressing can be problematic as the network size increases. For example a company initially may have only the LAN but as the time passes by it might end up with many LANs each one having its own router and each one with its own class C network number. With increase in the number of distinct local networks, their management becomes a problem. Everytime a new network gets installed, the system administrator has to contact NIC to get a new network number and then this number is to be announced worldwide. Another problem is that if a machine is to be moved from one LAN to the other, then its IP address needs to be changed. This will require modification in its configurational files and its modified IP number needs to be announced to the world. The solution to this problem is that the network is split into several smaller networks internally but it acts like a single network to the outside world. The smaller parts of a network are

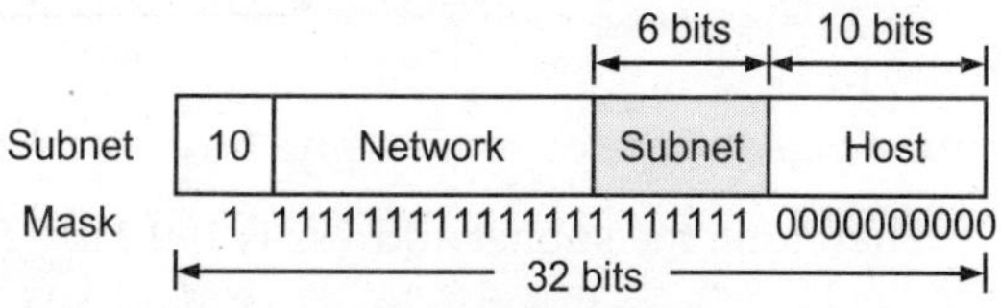

Fig. 7.54. One of the ways to subnet a class B network

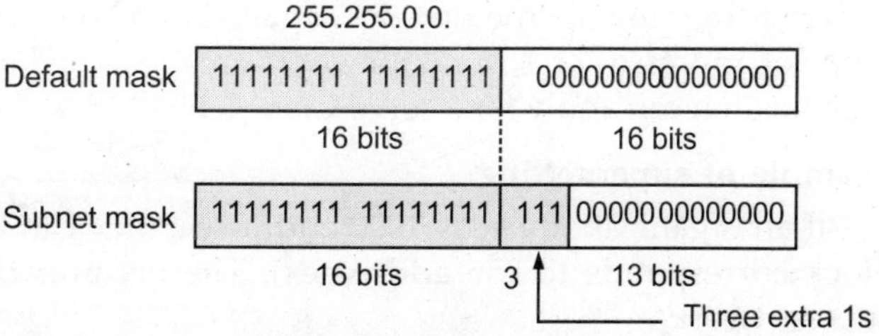

Fig. 7.55. Subnet mask

called as **subnets**. Now, let us continue with the same example taken at the beginning of this subsection. The growing company should start up with class B address instead of class C address and if it can number the hosts from 1 to 254. When a second LAN is to be installed it can split the 16 bit host number into a 6-bit subnet number and 10 bit host number as shown in figure 7.54.

Due to this split, it is possible to connect 62 LANs (0 and – 1 are reserved) and each one can contain upto 1022 hosts. Outside the network, the subnetting is not visible. Hence, even if a new subnet is created, it is not essential to contact NIC or change any database.

7.24.10. Subnet Mask

The number of 1's in the subnet mask is more than the number of 1s in the corresponding default mask. In subnet mask we change some of the left most 0s in the default mask to make a subnet mask. Figure 7.55 shows the difference between a class B default mask and subnet mask for the same block.

Number of subnets: This number of subnet is determined by the number of extra 1s. For 3 extra 1s, the number of subnets will be $2^3 = 8$. For n extra 1s, the number of subnets is 2^{11}.

EXAMPLE 7.4. A router inside an organization receives the same packet with a destination address 190.240.34.95. If the subnet mask is /19 (first 19 bits are 1s and following bits are 0s), find the subnet address.

Solution: To find the subnet address, AND the destination address with the subnet mask as shown in figure 7.56.

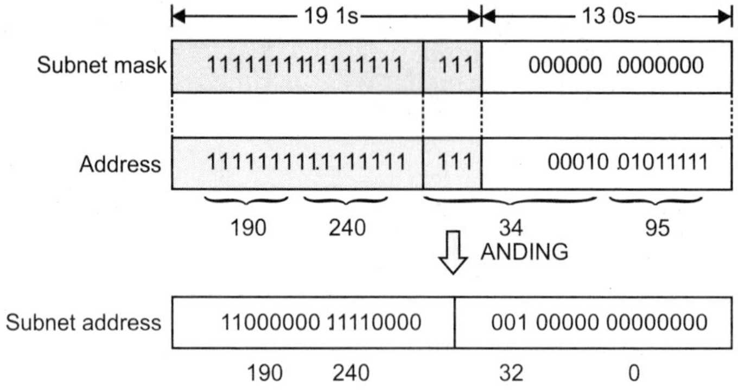

Fig. 7.56.

Therefore, the subnet address is 190.240.32.0.

7.24.11. Supernetting

The class A and class B addresses are almost depleted. But, class C addresses are still available. But, the size of class C address with a maximum number of 256 addresses does not satisfy the requirement of an organization. More addresses will be required. The solution to this problem is **supernetting**. In supernetting, an organization combines several class C blocks to create a large range of addresses *i.e.* several networks are combined to create a supernetwork. By doing this, the organization can apply for a set of Class C blocks instead of just one.

Example of supernetting

If an organization needs 1000 addresses, they can be obtained by using four C blocks (one C block corresponds to 256 addresses). The organization can then use these addresses in one supernetwork.

7.25. IPv6 PROTOCOL*

(RGTU, Bhopal, Sem. Exam., June, 2014)

IPv6 is the next generation Internet Protocol designed as a successor to the IP version 4. IPv6 was designed to enable high-performance, scalable Internet. This was achieved by overcoming many of the weaknesses of IPv4 protocol and by adding several new features.

Advantages of IPv6

1. **Larger address space:** IPv6 has 128-bit address space, which is 4 times wider in bits in compared to IPv4's 32-bit address space. So, there is huge increase in the address space.

2. **Buffer Header Format:** IPv6 has a better header format. In its header format, the options are separated from the base header. The options are inserted when needed, between the base header and upper layer data. This helps in speeding up the routing process.

3. **New Options:** New options have been added in IPv6 to increase the functionality.

4. **Possibility of Extension:** IPv6 has been designed in such a way that there is a possibility of extension of protocol if required.

5. **More Security:** IPv6 includes security in the basic specification. It includes encryption of packets (ESP: Encapsulated Security Payload) and authentication of the sender of packets (AH : Authentication Header).

6. **Support to Resource Allocation:** To implement better support for real time traffic (such as video conference), IPv6 includes flow label in the specification. With flow label mechanism, routers can recognize to which end-to-end flow the packets belongs.

7. **Plug and Play:** IPv6 includes plug and play in the standard specification. It therefore must be easier for novice users to connect their machines to the network, it will be done automatically.

8. **Clearer Specification and Optimization :** IPv6 follows good practices of IPv4, and rejects minor flaws/obsolete items of IPv4.

7.25.1. IPv6 Addresses

An IPv6 address consists of 16 bytes (octets) i.e. it is 128 bits long as shown in figure 7.57.

Hexadecimal Colon Notation

IPv6 uses a special notation called hexadecimal colon notation. In this, the 128 bits are divided into 8 sections, each one is 2 bytes long. 2 bytes correspond to 16 bits. So, in hexadecimal notation, we will require four hexadecimal digits. Hence, the IPv6 address consists of 32 hex digits and every group of 4 digits is separated by a colon as shown in figure 7.57. IPv6 uses 128-bit addresses. Only about 15% of the address space is initially allocated, the remaining 85% being reserved for future use. This remainder may be used in the future for expanding the address spaces of existing address types or for totally new uses.

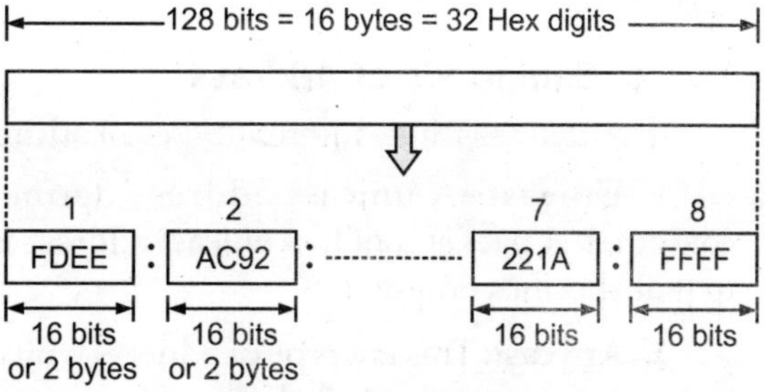

Fig. 7.57. IPv6 address

7.25.2. Abbreviation

The IPv6 address, even in hexadecimal format is very long. But in this address, there are many of the zero digits in it. In such a case, we can abbreviate the address. The leading zeros of a section (four digits between two colons) can be omitted. Note that only the leading zeros can be dropped but the trailing zeros can not be dropped. This has been illustrated in figure 7.58.

---

* List the advantages of IPv6. *(RGTU, Bhopal, Sem. Exam., June, 2014)*

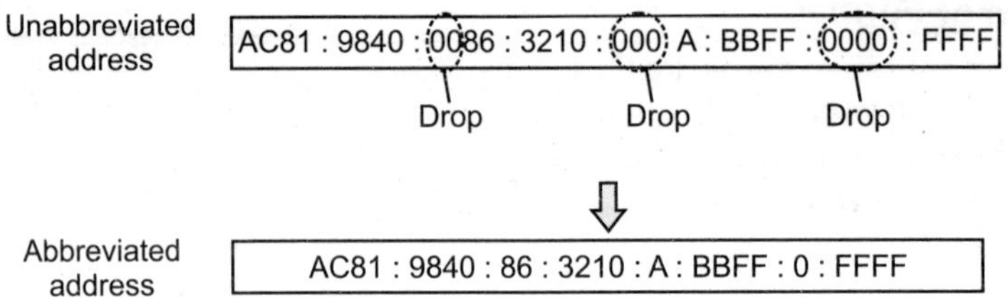

Fig. 7.58. Abbreviated address

Further Abbreviation: Further abbreviation are possible if there are consecutive sections consisting of only zeros. We can remove the zeros completely and replace them with double semicolon as shown in figure 7.59.

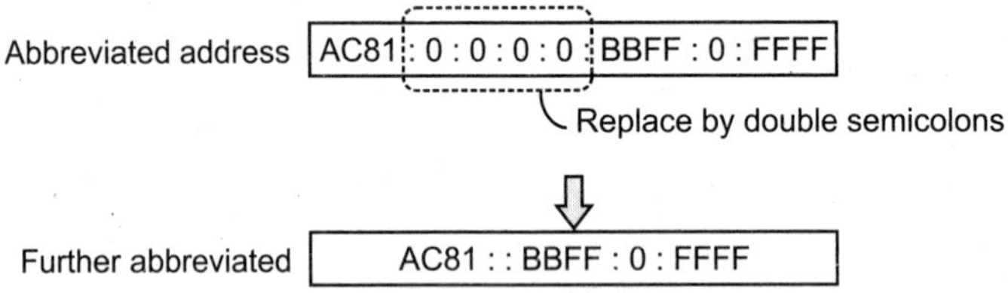

Fig. 7.59. Further abbreviation

It is important to note that abbreviation is allowed only once per address. Also note that if there are two runs of zero sections, then only one of them can be abbreviated.

7.25.3. CIDR Notation

IPv6 protocol allows classless addressing and CIDR notation. Figure 7.60 shows how to define a prefix of 60 bits using CIDR.

FDEC : 0 : 0 : 0 : 0 : BBFF : 0 : FFFF/60

Fig. 7.60. CIDR address

7.25.4. Categories of Address

IPv6 defines three different types of addresses.

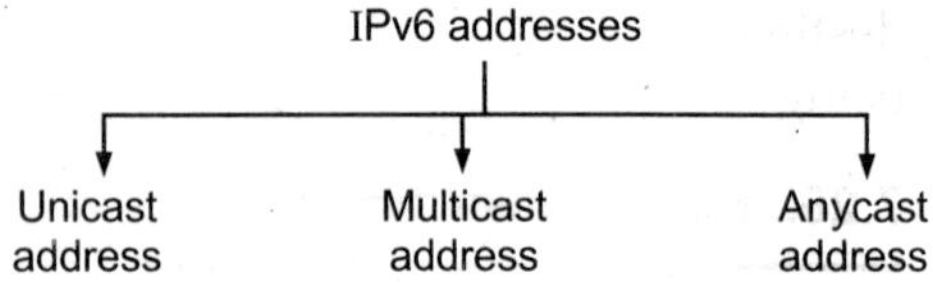

1. Unicast: A unicast address defines a single computer. A packet sent to a unicast address is delivered to that specific computer.

2. Anycast: This is a type of address which defines a group of computers with addresses which have the same prefix. A packet sent to an anycast address must be delivered to exactly one of the member of the group which is the closest of the most easily accessible.

3. Multicast Addresses: A multicast address defines a group of computers which may or may not share the same prefix and may or may not be connected to the same physical network. A packet sent to a multicast address must be delivered to each member of the set. There are no broadcast addresses in IPv6, because multicast addresses can perform the same function. The type of address is determined by the leading bits. Multicast addresses all start with FF (1111 1111) and all other addresses are unicast addresses. Anycast addresses are assigned from the unicast address space and they do not differ syntactically from unicast addresses. Anycast addressing is a rather new concept and there is little experience with the widespread use of anycast addresses. Therefore, some restrictions apply to anycast addressing in IPv6 until more experience is gained. An analyst address may not be used as the Source Address of an IPv6 packet and anycast addresses may not be assigned to hosts but to routers only.

7.25.5. IPv6 Packet Format

The IPv6 packet is shown in figure 7.61. Each packet consists of a base header which is mandatory followed by the payload. The payload is made up of two parts.

(i) Optional extension headers and

(ii) Data from an upper layer.

The base header is 40 byte length whereas the extension header and the data from upper layer contain upto 65,535 bytes of information.

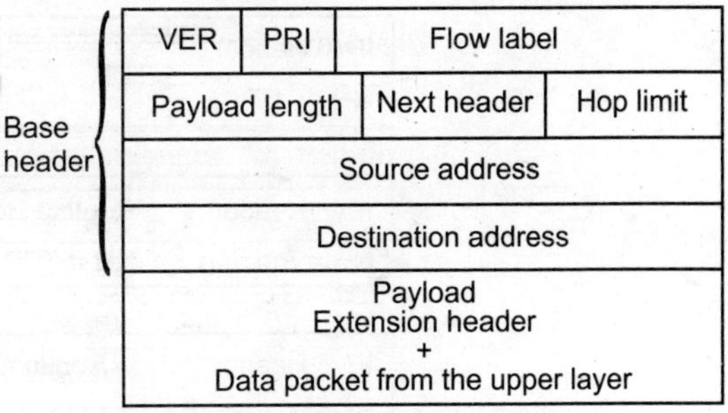

Fig. 7.61. Format of an IPv6 datagram

7.25.5.1. Base Header

Figure 7.61 shows the base header. It has eight fields. These fields are as follows:

1. Version (VER): It is a 4 bit field which defines the version of IP such as IPv4 or IPv6. For IPv6, the value of this field is 6.

2. Priority : It is a 4 bit field which defines the priority of the packet which is important in connection with the traffic congestion.

3. Flow Label: It is a 24 bit (3 byte) field which is designed for providing special handling for a particular flow of data.

4. Payload Length: This is a 2 byte length field which is used to define the total length of the IP datagram excluding the base header.

5. Next Header: It is an 8 bit field which defines the header which follows the base header in the datagram.

6. Hop Limit: This is an 8 bit field which has the same purpose as TTL (time to live) in IPv4.

7. Source Address: It is a 16 byte (128 bit) Internet address which identifies the original source of datagram.

8. Destination Address: This is a 16 byte (128 bit) Internet address which identifies the final destination of datagram. But this field will contain the address of the next router if source routing is being used.

7.26. EXTENSION HEADERS

The length of the base header is fixed at 40 bytes. But in IPv6, the base header can be followed by upto six extension headers. This is to give more functionality to the IP datagram. Many of the extension headers are options in IPv4. The IPv4 header has space for some optional fields requiring a particular processing of packets. These optional fields are not used often, and they can deteriorate router performance remarkably because their presence must be checked for each packet. IPv6 replaces optional fields by *extension headers.* In IPv6, optional Internet-layer information is encoded in separate headers that may be placed between the IPv6 header and the upper-layer header in a packet (see figure 7.62). There are a small number of such extension headers, each identified by a distinct Next Header value. An IPv6 packet may carry zero, one or more extension headers, each identified by the Next Header field of the preceding header. There are seven kinds of extension header.

Therefore, extension headers must be processed strictly in the order they appear in the packet ; a receiver must not, for example, scan through a packet looking for a particular kind of extension header and process that header prior to processing all preceding ones. Each extension header has a length equal to a multiple of 8 octets (64 bits). A full implementation of IPv6 must include support for the following extension headers.

IPv6 Header NH = TCP	TCP

IPv6 Header NH = Routing	Routing Header NH = TCP	TCP

IPv6 Header NH = Routing	Routing Header NH = Frag. Hdr	Fragment Header NH = TCP	TCP

Fig. 7.62. Examples of headers chain

When more than one extension header is used in the same packet, it is recommended that those headers appear in the following order :

1. IPv6 header.
2. Hop-by-Hop Options header.
3. Destination Options header.
4. Routing header.
5. Fragment header.
6. Authentication header.
7. Encapsulating Security Payload header.
8. Destination Options header.
9. Upper-layer header.

7.26.1. Fragmentation

Conceptually, the fragmentation is same as that in IPv4 but the place where fragmentation takes place is different. In IPv4, the fragmentation is done by the source or router, but in IPv6 only the original source can fragment.

7.26.2. Authentication and Privacy

IPv6 provides authentication and privacy using options in the extension header.

7.26.3. Migrating to IPv6 (Compatibility to IPv4)

It was IPv4's success that made an upgrade necessary, which means that there is a significant installed base of users to upgrade. Keeping the transition orderly was a major objective of the entire IPng program, and there are no plans for a cut over date when IPv6 would be turned on and IPv4 turned off. The strategy chosen for the upgrade is to deploy the IPv6 protocol stack in parallel with IPv4. In other words, hosts that upgrade to IPv6 will continue to exist as IPv4 hosts at the same time. An experimental IPv6 backbone, or 6bone, has been set up to handle IPv6. Internet traffic in parallel with the regular Internet. Such hosts will continue to have 32-bit IPv4 addresses but will add 128-bit IPv6 addresses. By 1999, hundreds of networks were linked to the 6bone. The transition can be achieved through two approaches: protocol tunnelling or IPv4/IPv6 dual stack.

7.26.4. Performance Comparison Between IPv4 and IPv6

S.No.	IPv4	IPv6
1.	In IPv4, there are only 2^{32} possible ways how to represent the address (about 4 billion possible addresses).	In IPv6, there are 2^{128} possible ways (about 3, 4*10^{38} possible addresses).
2.	The IPv4 address is written by dotted-decimal notation, e.g. 121.2.8.12.	IPv6 is written in hexadecimal and consists of 8 groups, containing 4 hexadecimal digits or 8 groups of 16 bits each e.g., FABC : AC77: 7834:2222:FACB: AB98 : 5432:4567.
3.	The basic length of the IPv4 header comprises a minimum of 20 bytes (without option fields). The maximum total length of the IPv4 header is 60 bytes (with option fields), and it uses 13 fields to identify various control settings.	The IPv6 header is a static header of 40 bytes in length, and has only 8 fields. Option information is carried by the extension header, which is placed after the IPv6 header.
4.	IPv4 header has a checksum, which must be computed by each router.	IPv6 has no header checksum because checksum are, for example, above the TCP/IP protocol suite, and above the Token Ring, Ethernet, etc.
5.	IPv4 contains an 8-bit field called Service Type. The Service Type field is composed of a TOS (Type of Service) field and a procedure field.	The IPv6 header contains an 8-bit field called the Traffic Class Field. This field allows the traffic source to identify the desired delivery priority of its packets.
6.	The IPv4 node has only Stateful auto-configuration.	The IPv6 node has both a stateful and a stateless address autoconfiguration mechanism.
7.	Security in IPv4 networks is limited to tunneling between two networks.	IPv6 has been designed to satisfy the growing and expanded need for network security.
8.	Source and destination addresses are 32 bits (4 bytes) in length.	Source and destination addresses are 128 bits (16 bytes) in length.
9.	IPsec support is optional.	IPsec support is required.
10.	No identification of packet flow for QoS handling by routers is present within the IPv4 header.	Packet flow identification for QoS handling by routers is included in the IPv6 header using the Flow Label field.
11.	Address Resolution Protocol (ARP) uses broadcast ARP Request frames to resolve an IPv4 address to a link layer address.	ARP Request frames are replaced with multicast Neighbour Solicitation messages.
12.	Must be configured either manually or through DHCP.	Does not require manual configuration or DHCP.
13.	ICMP Router Discovery is used to determine the IPv4 address of the best default gateway and is optional.	ICMP Router Discovery is replaced with ICMPv6 Router Solicitation and Router Advertisement messages and is required.
14.	Header includes options.	All optional data is moved to IPv6 extension headers.

EXAMPLE 7.5. Show by calculations how many network each IP address class can have with one example.

Solution: Number of networks in different IP address.

Class A address

The format of class A address is shown in figure 7.63 (a). Here, one byte defines the network ID and three bits define the host ID.

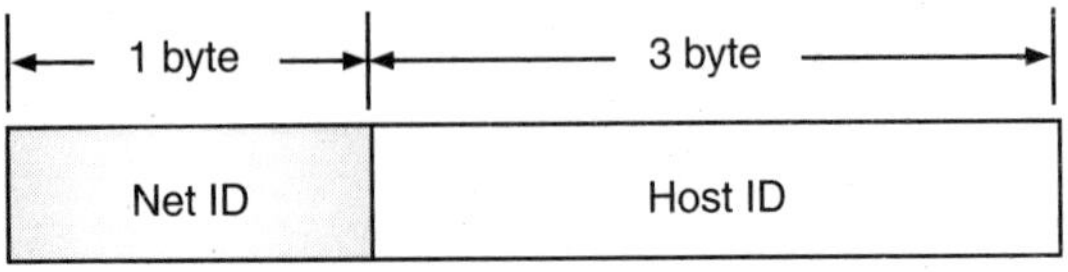

Fig. 7.63 (a). Class A address

The MSB in the network field is reserved. Hence, actually there are only 7 bits in the network field.

Therefore, the number of networks in class A address will be 128.

Class B address

The format of class B address is shown in figure 7.63(b). Here ,2 bytes are reserved for network field and remaining two bytes are for the host field.

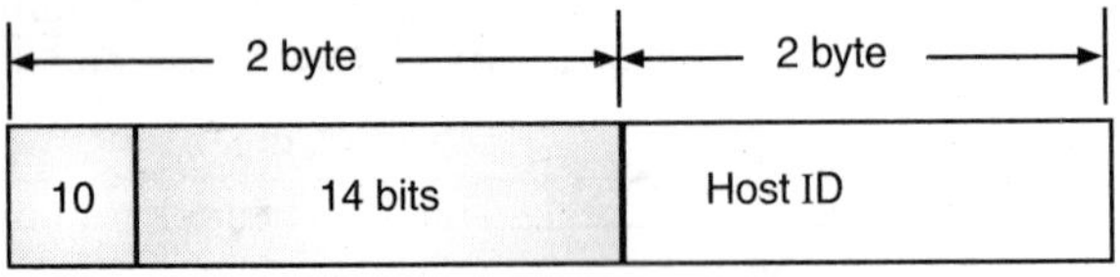

Fig. 7.63 (b). Class A address

Out of 16 bits in the network field, the first two bits (MSBs) are reserved. Hence, actually 14 bits are available in the network field. Thus, the number of networks in class B address is 2^{14} = 16, 368.

Class C address

The format of class C is shown in figure 7.63(c). Here, 3 bytes are reserved for network field and only one byte for the host field. Out of 24 bits in the network field, 3 bits are again reserved. Thus, actually only 21 bits are available.

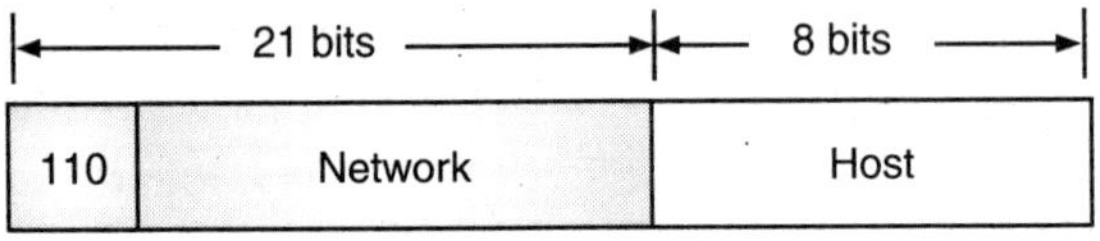

Fig. 7.63 (c). Class C address

Thus, the number of networks in class C addresses is 2, 097, 152.

EXAMPLE 7.6. How many host per network in each IP address class can exist, show with example.

Solution: Number of hosts in different IP addresses.

Class A

There are 3-bytes (24 bits) in the host. Hence, the number of hosts in class A address will be 2^{24} = 16, 7772, 16.

Class B

There are 2-bytes (16 bits) in the host field. Hence, the number of hosts in class B address will be 65536 i.e., 2^{16} per network.

Class C

There is 1-byte (8 bits) in the host field. Thus, the number of hosts in class C address will be 2^8 = 256 per network.

EXAMPLE 7.7. Convert the IP address whose hexadecimal representation is C22F15B2 to dotted decimal notation.

Solution:

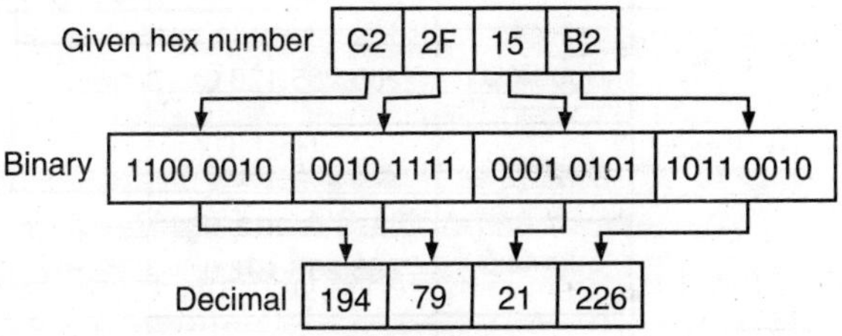

Fig. 7.64.

Hence, the IP address in the dotted decimal notation is as follows:

194.79.21.226

EXAMPLE 7.8. A class B network on internet has a subnet mask of 255.255.240.0. What is the maximum number of hosts per subnet?

Solution: The structure of class B address is as shown in figure 7.65 (a).

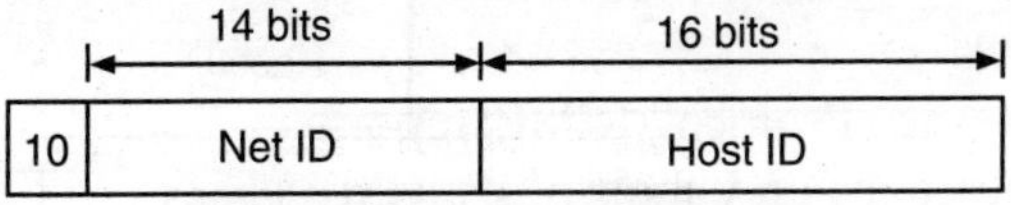

Fig. 7.65. (a). Class B address

The given subnet mask is 255.255.2400. Hence, it is as shown in figure 7.65 (b).

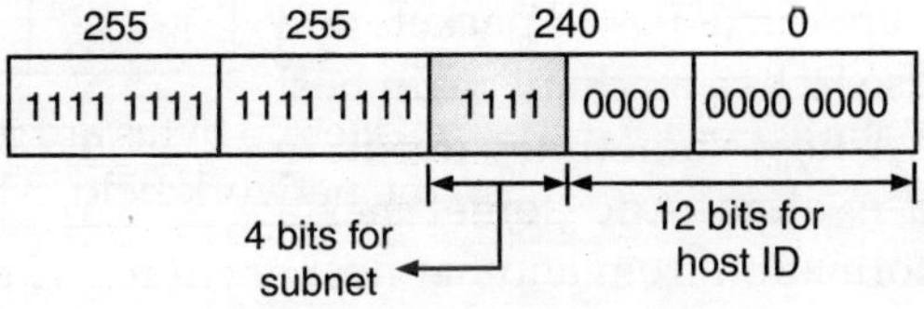

Fig. 7.65 (b). Subnet mask

Thus, there are 4 extra 1s as shown in figure 7.65 (b). Hence, there will be 16 subnets and each subnet can have 2^{12} = 4096 hosts.

EXAMPLE 7.9. Perform the subnetting of the following IP address 160.111. X.X.

Original subnet mask 255.255.0.0

Number of subnets 6(six).

Solution: The original subnet mask indicates that we are dealing with a class B address. In order to have six subnets, we need to use 3 extra bits from the bits that are reserved for host ID. Thus, the subnet mask is as shown in figure 7.66.

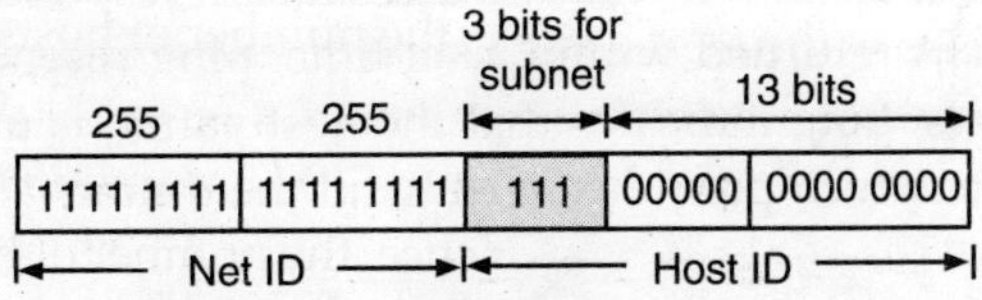

Fig. 7.66.

The 3-bits reserved for subnetting will have 8 combinations from 000 to 111 out of which any six combinations can be used for 6 subnets. Let us decide that the combinations 000 to 001 are not to be used. Then, the subnet masks for the 6 possible subnets will have the following addresses.

Subnet 1	255.255.64.0
Subnet 2	255.255.96.0
Subnet 3	255.255.128.0
Subnet 4	255.255.160.0
Subnet 5	255.255.192.0
Subnet 6	255.255.224.0

7.27. INTERNET CONTROL PROTOCOLS

The IP provides unreliable and connectionless datagram delivery. It was designed this way to make efficient use of network resources. IP is a best-effort delivery service that delivers a datagram from its original source to its final destination. However, it has two deficiencies : lack of error control and lack of assistance mechanisms.

The Internet Control Message Protocol (ICMP) reports errors and sends control messages on behalf of IP. ICMP does not attempt to make IP a reliable protocol. It simply attempts to report errors and provide feedback on specific conditions. ICMP messages are carried as IP packets and are therefore unreliable. IP also lacks a mechanism for host and management queries. A host sometimes needs to determine if a router host is alive. And sometimes a network manager needs information from another host or router. The Internet Control Message Protocol (ICMP) has been designed to compensate for the above two deficiencies. It is a companion to the IP. ICMP itself is a network layer protocol. However, its messages are not passed directly to the data link layer as would be expected. Instead, the messages are first encapsulated inside IP datagrams before going to the lower layer. (The value of the protocol field in the IP datagram is 1 to indicate that the IP data are an ICMP message.) This has been shown in figure 7.67.

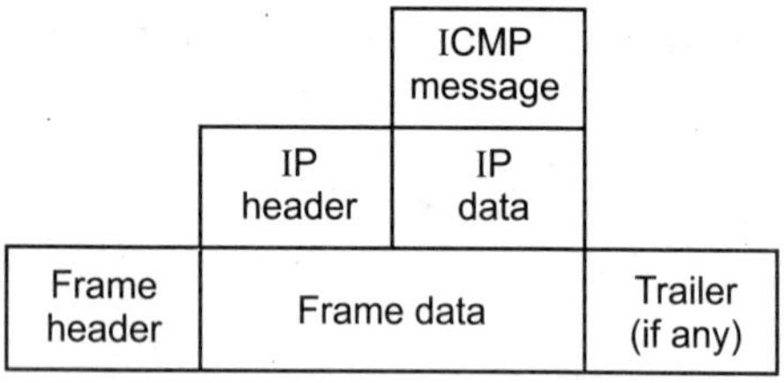

Fig. 7.67. ICMP encapsulation

The ping command uses ICMF as a probe to test whether a station is reachable. Ping packages an ICMP echo request message in a datagram and sends it to a selected destination. The user chooses the destination by specifying its IP address or name on the command line in a form such as :

ping 100.50.25.1

When the destination receives the echo request messages, it responds by sending an ICMP echo reply message. If a reply is not returned within a set time, ping resends the echo request several more times. If no reply arrives, ping indicates that the destination is unreachable. Another utility that uses ICMP is trace route, which provides a list of all the routers along the path to a specified destination.

7.27.1. Types of Messages

ICMP messages are broadly divided into following two categories as under:

(i) Error reporting messages

(ii) Query messages.

Error Reporting Messages

One of the important responsibilities of ICMP is error reporting. IP is an unreliable protocol. So error checking and control are not concerns of IP. Hence, ICMP was designed. But ICMP does not correct the errors. It simply reports them and error correction is left to the higher level protocols. ICMP always sends the error reporting messages back to the original source. ICMP can handle five types of errors. Figure 7.68 shows different types of error reporting messages.

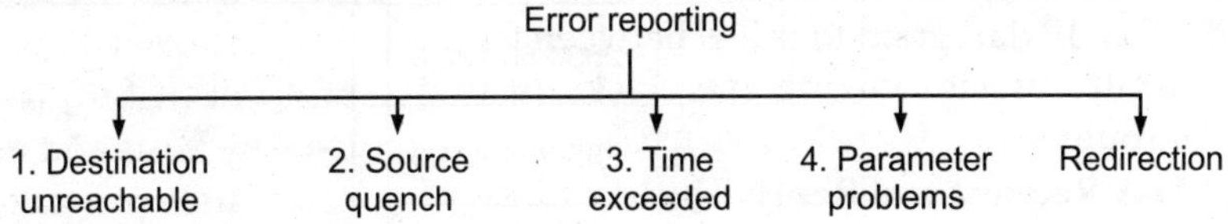

Fig. 7.68. Error reporting messages

1. Destination reachable: When a router cannot forward or deliver an IP packet, it sends a destination unreachable ICMP message back to the original source.

2. Source Quench Message: A host or router uses source quench messages to report congestion to the original source and to request it to reduce its current rate of packet transmission. There is no flow control or congestion control mechanism in IP. So, the source quench message is ICMP which is designed to add a kind of flow control and congestion control to IP.

This message serves following two purposes :

(i) It tells the source that the datagram has been discarded, and

(ii) It gives a warning to the source that the source should slow down (quench) because congestion has taken place somewhere.

3. Time Exceeded Message: This message is generated in two cases :

(i) If a router receives a datagram with a 0 in the TTL field then it discards that datagram and send a time exceeded message back to the original source.

(ii) If all the fragments which make up a message do not arrive at the destination host within a certain time limit then exceeded message is sent back.

4. Parameter Problem Message: There should not be any ambiguity in the header part of the datagram. If a router or destination host finds any such ambiguity or missing value in any field of the datagram then it discards that datagram and sends the parameter problem message back to the source.

5. Redirection Message: If a router or host wants to send a packet to another network then it should know the IP address of the next router. The routers and hosts must have a routing table to find the address of the next router and the routing table has to be updated constantly. For such an updating, the ICMP sends a redirection message back to its host.

7.27.2. Query

The ICMP can diagnose some of the network problems and such a diagnosis is accomplished through the query messages. The query messages is a group of four different pairs of messages as shown in figure 7.69.

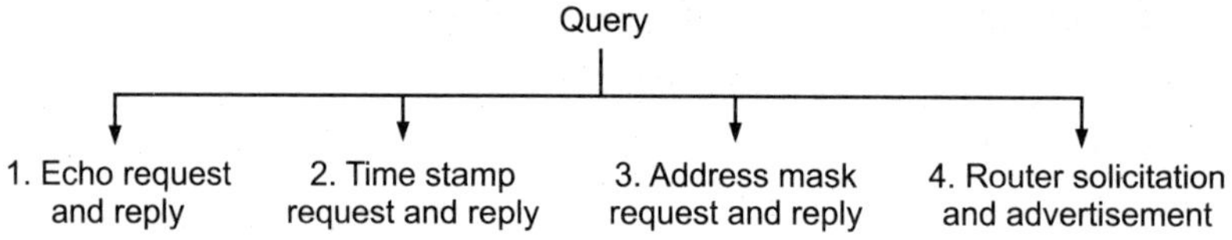

Fig. 7.69. Query messages

1. Echo Request and Reply: This is a pair of two messages namely echo request and echo reply messages and they are designed for diagnosis purpose. This pair of messages determines whether two systems (hosts or routers) can communicate with each other.

2. Time Stamp Request and Reply: This pair of messages can be used by the hosts and routers to determine the round trip time needed for an IP datagram to travel between them. It can also be used for synchronizing the clocks in two machines (hosts or routers).

3. Address Mask Request and Reply: The IP address of a host contains a network address, subnet address and host identifier. A host may know its full IP address but may not know its bifurcation. So, it can send an address mask request message to the router. The router then sends back the address mask reply message.

> **DO YOU KNOW?**
>
> The internet control message protocol (ICMP) is a mechanism used by hosts and routers to send notification of datagram problems back to the sender.

4. Router Solicitation and Advertisement: A host that wants to send data to a host or another network must know the address of routers connected to its own network. In such situations, the router solicitation and advertisement messages can help. A host can broadcast or multicast a router solicitation message. The routers receiving this message can broadcast their routing information using the router advertisement message.

7.28. MOBILE IP

Mobile IP is the underlying technology for support of various mobile data and wireless networking applications. With the advent of packet based mobile data applications and the increase of wireless computing, there is a corresponding need for the ability for seamless communication between the mobile node device and the packet data network (PDN) such as the Internet.

As PDAs and the next generation of data-ready cellular phones become more widely deployed, a greater degree of connectivity is almost becoming a necessity for the business user on the go.

Data connectivity solutions for this group of users are a very different requirement than it is for the fixed dialup user or the stationary-wired LAN user. Solutions here need to deal with the challenge of movement during a data session or conversation.

Cellular service providers and network administrators wanting to deploy wireless LAN technologies need to have a solution, which will grant this greater freedom.

7.29. TERMS USED IN MOBILE IP

Agent Advertisement : An advertisement message constructed by attaching a special Extension to a router advertisement message.

Binding : The association of the home address of a Mobile Node with a care-of address for that Mobile Node, along with the remaining lifetime of that association.

Care-of address : An IP address associated with a Mobile Node while visiting a foreign link; the subnet prefix of this IP address is a foreign subnet prefix. Among the multiple care-of addresses that a Mobile Node may have at a time (e.g., with different subnet prefixes), the one registered with the Mobile Node's Home Agent is called its *primary* care-of address.

Correspondent Node: A peer with which a Mobile Node is communicating. A correspondent node may be either mobile or stationary.

Foreign agent: A router on a Mobile Node's visited network, which co-operates with the Home Agent to complete the delivery of datagrams to the Mobile Node while it is away from home.

Foreign network: Any network other than the Mobile Node's Home Network.

Foreign subnet prefix : Any IP subnet prefix other than the Mobile Node's home subnet prefix.

Foreign link : Any link other than the Mobile Node's home link.

Home address : An IP address that is assigned for an extended period of time to a Mobile Node. It remains unchanged regardless of where the node is attached to the Internet.

Home network : A network, possibly virtual, is having a network prefix matching that of a Mobile Node's home address. Note that standard IP routing mechanisms will deliver datagrams destined to a Mobile Node's Home Address to the Mobile Node's Home Network.

Home address : An IP address assigned to a Mobile Node within its home link.

Home link : The link on which a Mobile Node's home subnet prefix is defined. Standard IP routing mechanisms will deliver packets destined for a Mobile Node's home address to its home link.

Home agent : A router on a Mobile Node's home link with which the Mobile Node has registered its current care-of address. While the Mobile Node is away from home, the Home Agent intercepts packets on the home link destined to the Mobile Node's address, encapsulates them, and tunnels them to the Mobile Node's registered care-of address.

Host: Any node that is not a router.

Mobile node : A node that can change its point of attachment from one link to another, while still being reachable via its home address.

Movement : A change in a Mobile Node's point of attachment to the Internet such that it is no longer connected to the same link as it was previously. If a Mobile Node is not currently attached to its home link, the Mobile Node is said to be *away from home*.

Tunnel : The path followed by a datagram while it is encapsulated. The model is that, while it is encapsulated, a datagram is routed to a knowledgeable decapsulating agent, which decapsulates the datagram and then correctly delivers it to its ultimate destination.

7.30. MOBILE IP ROUTING OPERATION

1. To accomplish this, mobile IP established the visited network as a foreign node and the home network as the home node.
2. Mobile IP uses a tunneling protocol to allow messages from the PDN to be directed to the mobile node's IP address. This is accomplished by way of routing messages to the foreign node for delivery via tunneling the original IP address inside a packet destined for the temporary IP address assigned to the mobile node by the foreign node.
3. The Home Agent and Foreign Agent continuously advertise their services on the network through an Agent Discovery process, enabling the Home Agent to recognize when a new Foreign Agent is acquired and allowing the Mobile Node to register a new Care of Address.

4. This method allows for seamless communications between the mobile node and applications residing on the PDN, allowing for seamless, always-on connectivity for mobile data applications and wireless computing.

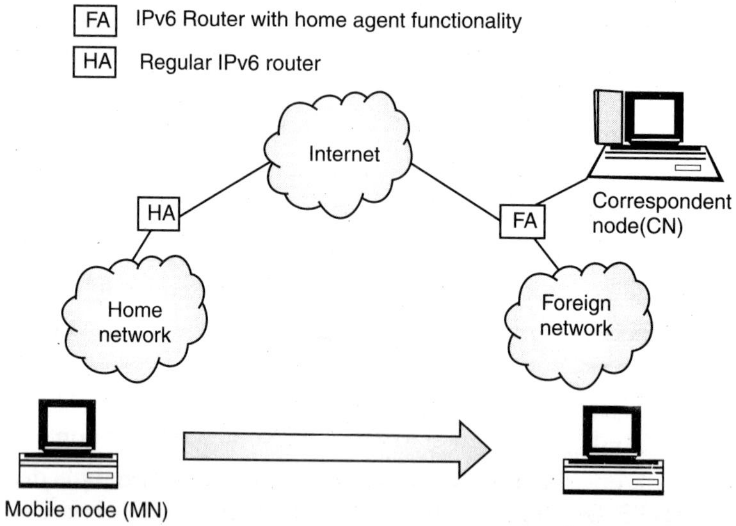

Fig. 7.70. Mobile IP routing operation

5. Mobile node is a host equipped with Mobile IPv6 stack the correspondent node can be a host with regular IPv6 stack but in this case, we assume it has Mobile-IPv6 stack.
6. Home Agent is the register for mobile terminals. Terminals register the care-of-addresses (i.e. where they really are today) to the home agent.
7. Unlike in Mobile IPv4, in Mobile IPv6, there are not foreign agents in the foreign network, just the IPv6 router.
8. Design requirements for Mobile IP:
   (i) (Detection) the mobile node must detect it has moved.
   (ii) (Address Configuration) the mobile node must discover or configure its care-of-address.
   (iii) (Binding Update) the mobile node must inform the home agent about its new care-of address.
   (iv) (Tunneling) the home agent must forward packets from the home network to the mobile node's care-of address.

7.31. FUNCTIONALITY IN THE HOME NETWORK

1. The home router sends router advertisements
2. The mobile node receives its home address from the home network via neighbor discovery protocol.
3. The mobile node can listen to the Router Advertisements (ICMPv6) or it can send Router Solicitation request to get or ensure the network prefix.
4. The mobile node starts for example a video call with correspondent node; the line signifies the route of the traffic.

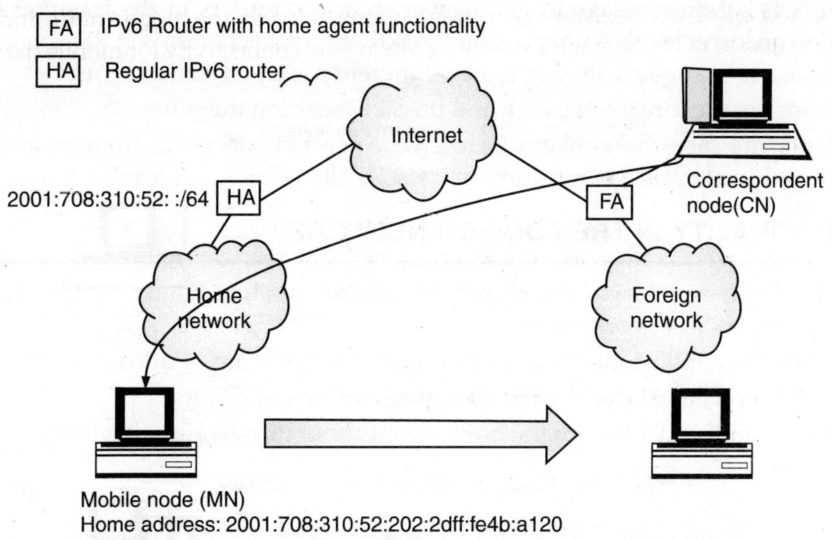

Fig. 7.71. Functionality in the home network

7.32. MOVING TO FOREIGN NETWORK

1. When moving to the foreign network, the mobile node is unreachable depending how long it takes to get IPv6 connectivity again.
2. The traffic between mobile node and correspondent node is stalled.
3. Home agent may know that the mobile node has already left the network and can for example buffer some of the traffic coming destined to the mobile node.

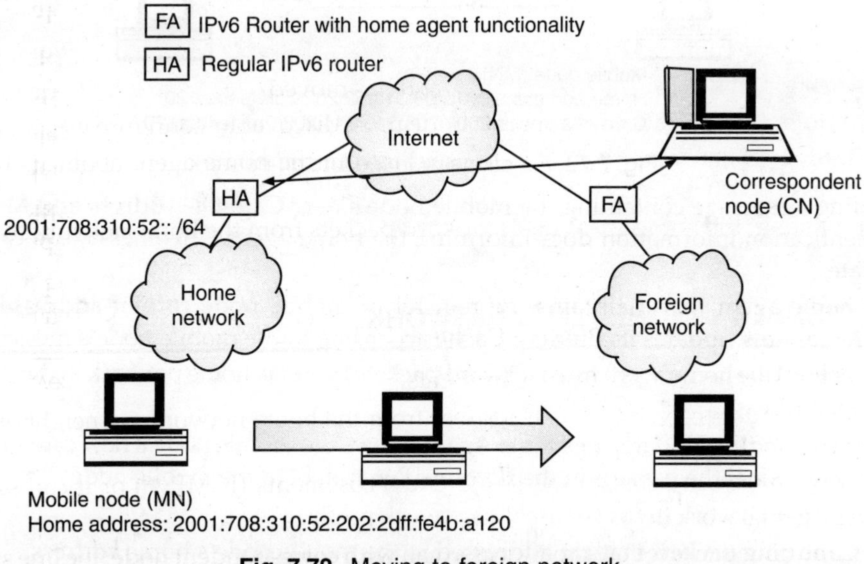

Fig. 7.72. Moving to foreign network.

4. There is a lot of research going on which purpose is to minimize the handover/configuration time so that this delay would be minimal.
5. It is recommended that mobile nodes use lower layer mechanisms for detecting the movement, for example, the signal/noise level/ratio in WLAN to detect movement.

6. Router Solicitation message to get answer from the routers in the foreign network. If the detection on lower levels is not possible, then there are mechanisms in RFC2461 for movement detection. For example, listening to new different router advertisements.
7. After entering the foreign network and possibly sending the Router.
8. Solicitation message, the mobile node receives a new network prefix from the foreign network's routers and constructs a new address for itself called Care-of-Address.

7.33. FUNCTIONALITY IN THE FOREIGN NETWORK

1. In the foreign network, we have regular IPv6 router sending Router Advertisements with the foreign network's network prefix
2. Now, also the mobile node has auto configured its new care-of-address.
3. The mobile node must discover or configure its care-of-address.
4. The mobile node must inform the home agent about its new care-of address

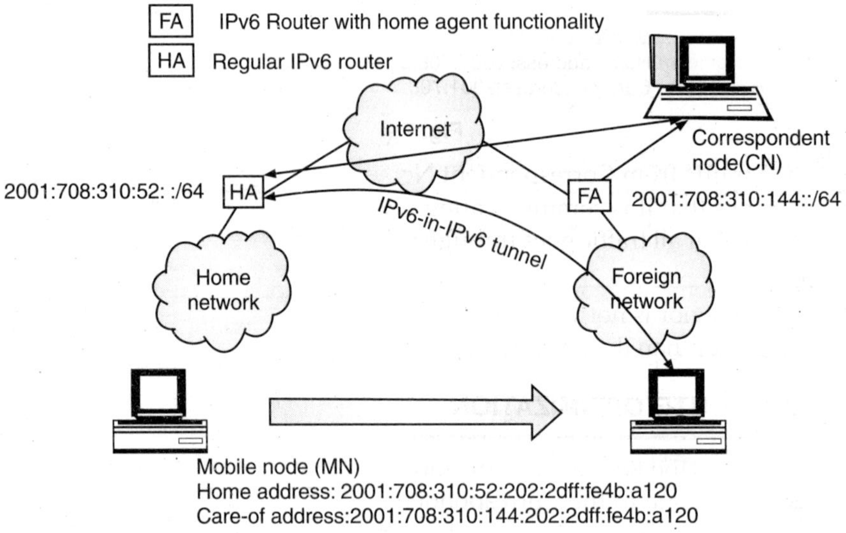

Fig. 7.73. Functionality in the foreign network.

5. Sending a message containing the mobile node's new Care-Of-Address and Mobile Node's authentication information does informing the home agent. This message is called binding update.
6. The home agent that maintains the register of mobile node care-of-addresses and home addresses now updates its Binding Cache according to the mobile node's message.
7. (Tunneling) the home agent must forward packets from the home network to the mobile node's care-of address
8. After the Binding Update, the Home Agent knows the mobile node's new care of address and is able to tunnel the packets in the IPv6-in-IPv6 tunnel to the mobile node
9. Home Agent's work items :
   (i) Capturing packets that are addressed to the mobile node's home address
   (ii) Participating in Duplicate Address Detection to protect the mobile node's addresses on the home network against reallocation to another network node.
   (iii) Redirecting the packets that it captures when they are forwarded to the home address
   (iv) Other tasks to inform the mobile node about renumbering events on the home network

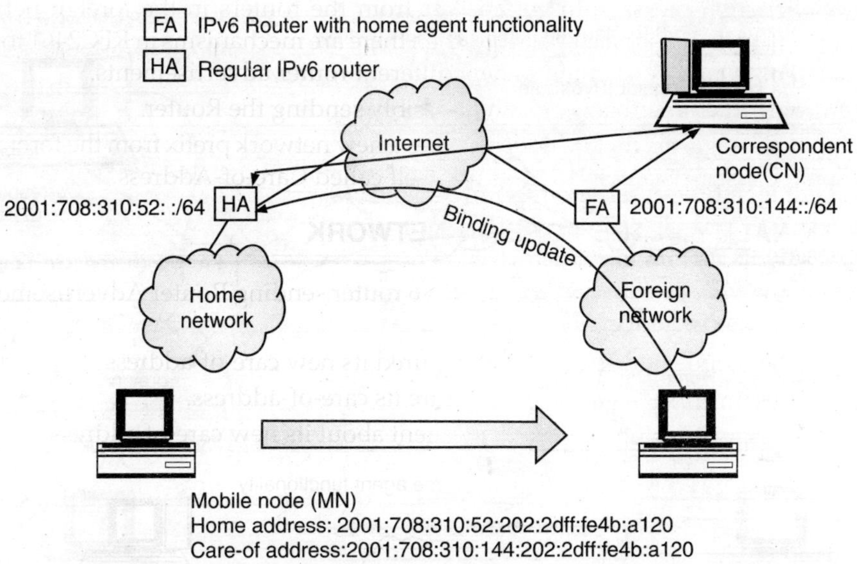

Fig. 7.74.

10. This way the traffic from Correspondent Node reaches the mobile node. The Correspondent Node still thinks that it is communicating with the mobile node that is in its home network and because of that all traffic goes through the home agent and tunneling before reaching the mobile node.
11. This kind of behavior is not very effective when the Correspondent Node and Mobile Node are closer together than the home network ⇒ Route Optimization is needed.

7.34. MOBILE IP ROUTE OPTIMIZATION

1. Binding Update to the Rescue, now the mobile node recycles the idea of binding update and sends Binding Update to the Correspondent Node that verifies it (key exchange protocols once again).
2. Once verified the Correspondent Node does not need to send Binding Update Acknowledgement to Mobile Node, instead it just creates an IPv6-in-IPv6 tunnel to the mobile node's Care-of-Address and redirects the traffic to the tunnel.
3. The advantages of this kind of route optimization are:
   (i) Reduces network utilization, since fewer links are traversed. This could mean much less congestion in all the routing fabric of the Internet.
   (ii) Reduces the load on the home agent, and the potential for complete isolation of the mobile node due to loss of the home agent.
   (iii) Reduced round-trip times for application endpoints.
   (iv) Better predictability of routing paths, which could be crucial for QoS Management.
   (v) More symmetric delay characteristics for incoming and outgoing traffic between mobile node and correspondent node.

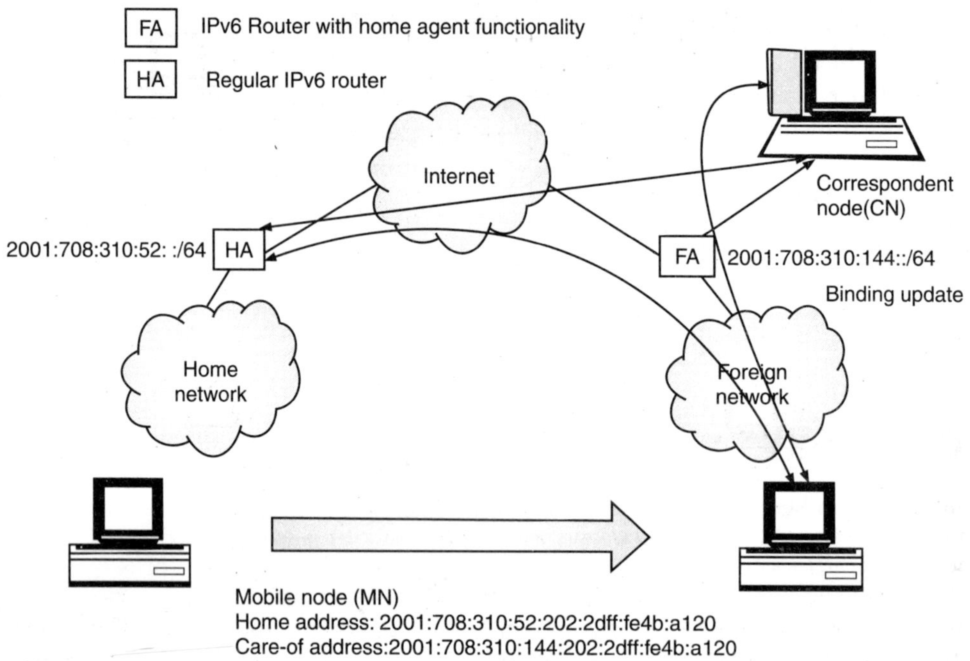

Fig. 7.75. Mobile IP route optimization.

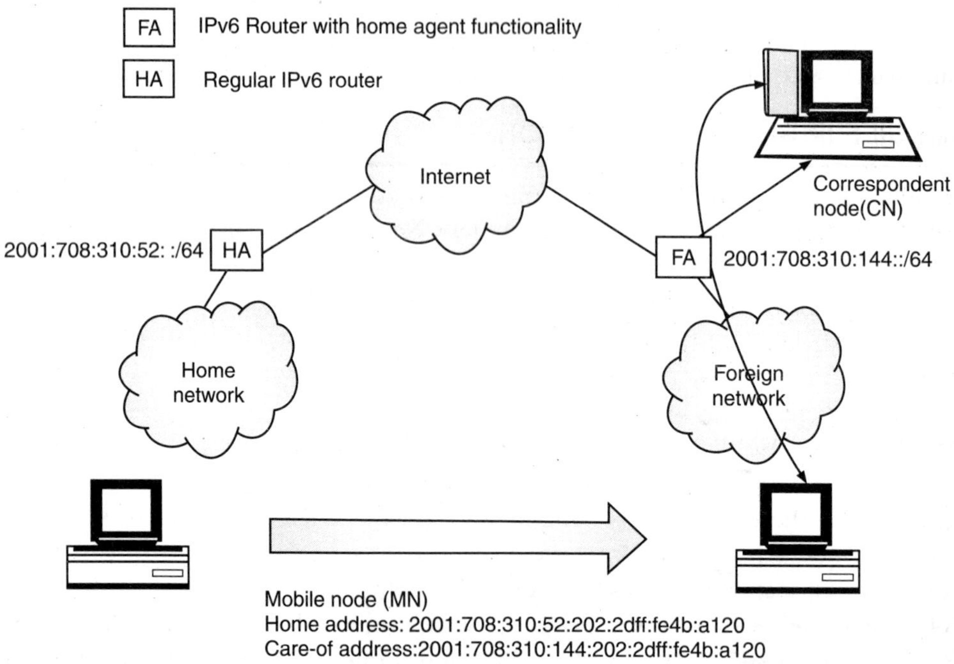

Fig. 7.76.

7.35. LIMITATIONS OF MOBILE IP

1. It is important to realize that Mobile IP has certain limitations in its usefulness in a wireless

environment. It was designed to handle mobility of devices, but only relatively infrequent mobility. This is due to the work involved with each change.

2. This overhead is not a big deal when you move a computer once a week, a day or even an hour. It can be an issue for *real-time* mobility such as roaming in a wireless network, where hand-off functions operating at the data link layer may be more suitable.
3. Mobile IP was designed under the specific assumption that the attachment point would not change more than once per second.
4. It should also point out that Mobile IP is intended to be used with devices that maintain a static IP configuration.
5. Since the device needs to be able to always know the identity of its home network and normal IP address, it is much more difficult to use it with a device that obtains an IP address dynamically, using something like DHCP.

GLOSSARY

- **Connectionless Service** A type of service in which messages are partitioned into *packets* and routed through the network. Each packet is independent of the other packets that carry parts of the message, and each packet carries a destination address. Unlike *connection-oriented service*, no physical link is established between sending and receiving nodes before data transmission.
- **Connection-Oriented Service** A type of service in which prior to the transfer of data a physical (and virtual) link is set up between the transmitting and receiving nodes. This link remains in effect for the duration of the session. After the session is completed, the link is removed. Characteristics of a connection-oriented service include wasted bandwidth (in fact, link must remain established even during idle periods of a transmission), a high potential for a hung network (there is always a possibility that a link will not be terminated), and guaranteed sequential arrival of packets at the destination node.
- **Addressing** A network concept which describes the process of assigning unique identification numbers (also called *addresses*) to a networked device.
- **Congestion** A situation when a network is consumed with excessive network traffic (i.e., lots of packets), resulting in performance degradation. Congestion occurs when routers are quite slow and thus causing queues to lengthen, or when routers are quite fast, causing queues to build up whenever input traffic is greater than the capacity of output lines. The ultimate level of congenstion is called as deadlock, which occurs when one router cannot proceed until a second router does somethings, and the second router cannot proceed since it is waiting for the first router to do something. Congestion control is provided by layer 3 of the *OSI model*.
- **Datagram** A grouping of bits organized as a logical unit of data at the network layer. IP datagram serve as the Internet's primary unit of information. In the OSI model, a datagram is usually referred to as a *packet*.
- **Distance-Vector Algorithm** A routing algorithm which determines the distance between source and destination nodes by evaluating the number of router hops a packet traverses and route from the source network to the destination network. An example of a distance-vector algorithm is the Bellman-Ford algorithm.
- **IP Address** A network address assigned to a node's network interface and used to uniquely identify (locate) the node within the Internet. Two versions are currently implemented i.e., IPv4 and IPv6.
- **IPv4 Address** An *IP address* based on IPv4. These addresses consists of 32 bits (0 though 31)partitioned into four groups of eight bits each (called octets) and organized into five classes (A through E) based on the values of bits 0 through 3.
- **IPv6** It is an acronym for *Internet protocol version* 6, which is an evolutionary replacement to IPv4. IPv6 maintains most IPv4 functions, relegates certain functions that either were not working or were rarely used in IPv4 as optimal and adds new functionality that is missing from IPv4. It is sometimes called IPng (for next generation).

- **IPv6 Address** An *IP address* based on *IPv6*. As *IPv6* address consists of 128 bits and is 4 billion × 4 billion times size of the IPv4 address space (2^{128} vs. 2^{32}). Unlike IPv4 addresses, IPv6 addresses use a colon as their delimiter (instead of a *dot* notation), and they are written as eight 16-bit integers expressed in hexadecimal force.
- **Multicast** A data transmission which is destined to a group of recipients.
- **Subnet Mask** A special network address that is used to identify a specific subnetwork. Using a unique bit combination, a mask partitions an address into a network ID and a host ID.
- **Subnetting** The partitioning of a network address space into separate, autonomous *subnetworks*. It is key to subnetting is a network's *subnet mask.*
- **Subnetwork** A network segment, commonly abbreviated *subnet.*
- **Unicast** A data transmission which is destined to a single recipient.

SUMMARY

- The network layer is responsible for carrying the packet from the source all the way to destination. In short, it is responsible for host-to-host delivery. The network layer has a higher responsibility than the data link layer, because the data link layer is only supposed to move the frames from one end of the wire to the other end. Thus, network layer is the lowest layer that deals with the end to end transmission.
- Addressing is necessary to identify each device on the Internet uniquely. This is similar to a telephone system. The addresses used in the network layer should uniquely and universally define the connection of a computer.
- In a network, there are multiple roots available from a source to a destination and one of them is to be chosen. The network layer decides the root to be taken. This is called as routing and it depends on various criterions.
- The other issues which are not directly related to the duties of network layer but need to be discussed are as under:
  (i) Address resolution. (ii) Multicasting.
  (iii) Routing protocols.
- There are certain desirable properties of a routing algorithm as under :
  (i) Correctness. (ii) Robustness.
  (iii) Stability. (iv) Fairness and
  (v) Optimality.
- Routing algorithms can be divided into two groups as under:
  (i) Non-adaptive algorithms (ii) Adaptive algorithms.
- The examples of static algorithms may be listed as under :
  (i) Shortest path routing. (ii) Flooding.
  (iii) Flow based routing.
- For certain applications, widely separated processes need to work together in groups. A process has to send a message to all other processes in the group. For a small group, it is possible to send a point-to-point message. But, this is expensive if the group is large.
- An important issue in a packet switching network is congestion. When too many packets are present in a part of a subnet, the performance degrades. This situation known as **congestion**.
- Jitter may be defined as the variation in delay for the packets belonging to the same flow. The real time audio and video cannot tolerate jitter on the other hand the jitter does not matter if the packets are carrying an information contained in a file.
- Some of the techniques useful in achieving good QoS may be listed as under :
  (i) Buffering. (ii) Traffic shaping.
  (iii) Leaky bucket algorithm. (iv) Token bucket algorithm.
  (v) Resource reservation. (vi) Admission control.
  (vii) Proportional routing. (viii) Packet scheduling.

SHORT QUESTIONS WITH ANSWERS

Q.1. Write the Serviced Provided to the Transport Layer by the Network Layer.

Ans. The network layer services are designed to achieve the following goals :

(i) The services should be independent of the subnet technology.

(ii) Transport layer should be shielded from the number, type and topology of the subnet.

(iii) The network address which is made available to the transport layer must use a uniform nubmerging plan.

Q.2. Write the desired properties of a Routing Algorithm.

Ans. There are certain desirable properties of a routing algorithm as under :

(i) Correctness. (ii) Robustness.
(iii) Stability. (iv) Fairness and
(v) Optimality.

Q.3. Write the types of Routing Algorithms.

Ans. Routing algorithms can be divided into two groups as under:

1. Nonadaptive algorithms 2. Adaptive algorithms.

Q.4. Write the examples of Static Algorithms.

Ans. The examples of static algorithms may be listed as under :

(i) Shortest path routing. (ii) Flooding. (iii) Flow based routing.

Q.5. What is need of congestion Control ?

Ans. It is not possible to completely avoid the congestion but it is necessary to control it. Congestion will lead to a large queue length, which results in buffer overflow and loss of packets. Therefore, congestion control is necessary to ensure that the user gets the negotiated QoS (quality of service).

Q.6. What is the difference between Congestion Control and Flow Control ?

Ans. Congestion control makes it sure that the subnet is able to carry the offered traffic. Congestion control is dependent on the behaviour of all the hosts, all the routers and other factors which reduce the carrying capacity of a subnet. On the contrary, the flow control is related to point to point traffic. Flow control ensures that a fast sender does not send data at a rate faster than the rate at which the receiver can receive it. Flow control involves some kind of feedback.

Q.7. What is Open Loop Control ?

Ans. Open loop solutions try to solve the problems by excellent design to prevent the congestion from happening. Open loop control is exercised by using the tools such as deciding when to accept the new packets, when to discard the packets, which packets are to be discarded and making the scheduling decisions at various points.

Q.8. Which layer is responsible for the physical connection between the source and the destination?

Ans. Network layer.

Q.9. What are Unicast Routing Protocols ?

Ans. Various unicast routing protocols are shown in figure 7.77. The popular interior protocols are RIP (Routing Information Protocol) and OSPF (Open Shortest Path First). Whereas the exterior protocol used popularly is BGP (Border Gateway Protocol).

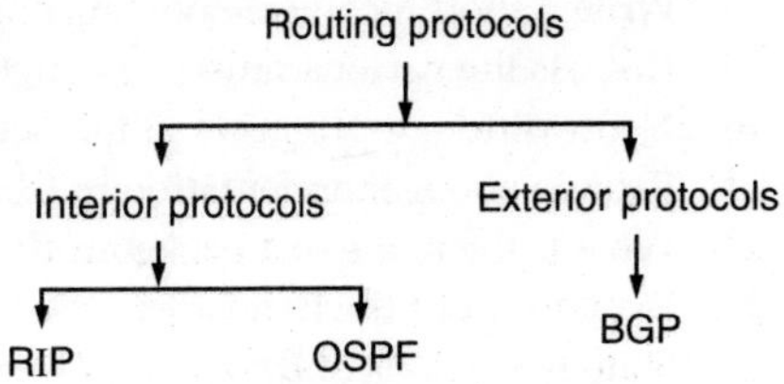

Fig. 7.77. Unicast routing protocols

Q.10. Explain broadcast address.

Ans. A station uses the broadcast address in the destination field of the frame to send the frame to all stations on the network. The destination address field for broadcast is set to all ones.

Q.11. List the functions of a router.

Ans. A router accepts a frame from its port and retransmits the frame according to its routing table. A router can also change the frame format, for example: IEEE 802.3 to IEEE 803.5.

Q.12. Explain dynamic routers.

Ans. A dynamic router automatically generates a routing table.

Q.13. Explain the function of IP.
Ans. Delivering packets to a destination.

Q.14. What is the function of a router?
Ans. The function of a router is to determine the path for transporting information through internetworking (routing).

Q.15. Explain static routing and dynamic routing.
Ans. A static routing table is configured manually but a dynamic routing table is configured automatically.

Q.16. What is the function of a firewall?
Ans. A firewall is a system for preventing unauthorized users access to private networks.

REVIEW QUESTIONS

1. Explain about the connection oriented and connectionless services.
2. What are the network design issues involved in designing a typical network and what are the supporting design tools available to make this design as a good design. Explain how these design tools help to address design issues.
3. Why modern computer use dynamic routing ? Explain with an example how distance vector routing is used to route the packet and why count-to-identify problem arises and how does to get solved.
4. Why leaky bucket algorithm should allow 1 packet tick independent of how large the packet is ?
5. A computer on 6 Mbps network is regulated by token bucket. The token bucket is filled at a rate of 1 Mbps. It is initially filled to capacity with 8 megabits. How long can be computer transmit at the full 6 Mbps.
6. What is Firewall ? What is difference between packet filtering and proxy server gateways?
7. What do you mean by fragmentation ? Explain.
8. Write a short technical note on leaky bucket algorithm.
9. Write a short technical note on Congestion control.
10. Enlist and discuss various design layer issues.
11. What is meant by fragmentation ? Is fragmentation needed in concentrated virtual circuit internets, or any in datagram system ?
12. Write a short technical note on Hierarchical routing.
13. Give an efficient algorithm for finding the shortest parts between all pairs of nods in a tree. What is the complexity of the algorithm ?
14. Write a short technical note on: (i) Tunneling, (ii) Jitter control, (iii) mobile IP, (iv) network design issues.
15. What is the difference between flow control and congestion control.
16. What is the difference between end to end delay and packet differ ? What are the causes of packet Jitter ?
17. What is transparent and non transparent fragmentation ?
18. Write a short technical note on congestion prevention policies.
19. Discuss the various causes of congestion in subnet.
20. Name different protocols in the network layer.
21. Explain about fragmentation in IP.
22. What is the name of a packet in IP ? Explain.
23. Explain about the IP header.
24. State limitations of IPv4.
25. How many types of addresses IPv6 allows ?
26. Given an IP address, how will you extract its net id and host id.
27. How can you compare IPv4 and IPv6 ?
28. A network on the Internet has a subnet mask of 255.255.240.0. Write is the maximum number of hosts it can handle ?
29. What is the difference between net id and network address ?
30. Which IP Protocol version is being used currently ?

UNIT – IV

SYLLABUS:

The Transport Layer: Design issues and services, Transport protocols, congestion control, UDP and TCP protocols, performance issues.

The above syllabus has been covered in following chapter:

Chapter 8 : The Transport Layer

Chapter 8

The Transport Layer

Syllabus

The Transport Layer: Design issues and services, Transport protocols, congestion control, UDP and TCP protocols, performance issues.

Chapter Outline

- Introduction
- Transport Layer Design Issues
- Duties of Transport Layer
- The Transport Layer Services
- Transport Service Primitives
- Sockets
- Transfer Protocols
- Addressing
- Connection-Oriented Service
- Simple Transport Protocol
- The Internet Transport Protocols (TCP and UDP)
- Connection Management from Client's Point of View
- TCP Transmission Policy
- TCP Congestion Control
- TCP Timer Management
- User Datagram Protocol (UDP)
- Wireless TCP and UDP
- Performance Issues

8.1. INTRODUCTION

The transport layer is the core of the Internet model. The application layer programs interact with each other using the services of the transport layer. Transport layer provides services to the application layer and takes services from the network layer. Figure 8.1 shows the position of the transport layer in the 5-layer Internet model. The transport layer is fourth layer in this model.

Fig. 8.1. Position of transport layer

8.2. TRANSPORT LAYER DESIGN ISSUES

Transport layer is meant for the process to process delivery and it is achieved by performing a number of functions.

Figure 8.2 lists the design issues of a transport layer.

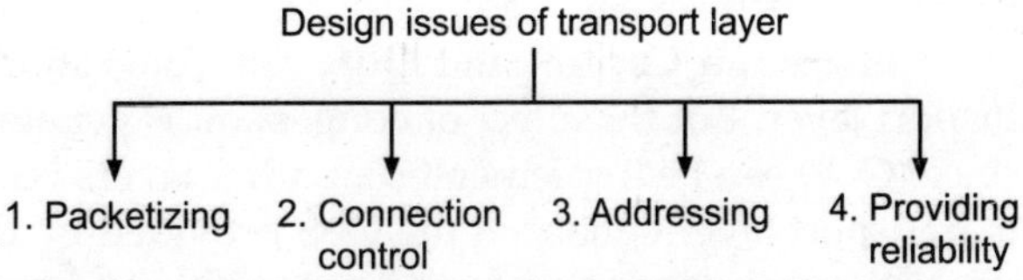

Fig. 8.2. Design issues of transport layer

Pioneer in Communications : Robert M. Metcalfe (1946—)

Robert (Bob) Metcalfe was born in 1946 in Brooklyn, New York, but grew up on Long Island. He got his graduation from Massachusetts Institute of Technology (MIT) in 1969 with BAs in Electrical engineering and business management. He had also earned an MA in Applied Mathematics from Harvard University in 1970, as well as a Ph.D in computer science in 1973. His research at PARC involved discovering a technique to interconnect prototypes of a new personal computer known as the Alto, and in 1973, he and his colleague, D. R. Boggs, discovered the local area networking technology Ethernet.

1. Packetizing: The transport layer creates packet out of the messages received from the application layer. Packetizing is a process of dividing a long message into smaller ones. These packets are then encapsulated into the data field of the transport layer packet and headers are added. The length of the message which is to be divided can vary from several lines (e-mail) to several pages. But the size of the message can become a problem. The message size can be larger than the maximum size that can be handled by the lower layer protocols. Hence, the messages must be divided into smaller sections. Each small section is then encapsulated into a separate packet. Then a header is added to each packet to allow the transport layer to perform its other functions.

DO YOU KNOW?

The transport layer is responsible for end to end delivery, segmentation, and concatenation.

2. Connection Control: Transport layer protocols may be divided into following two categories:

(i) Connection oriented.

(ii) Connectionless.

(i) Connection oriented delivery: A connection oriented transport layer protocol establishes a connection *i.e.*, virtual path between sender and receiver. This is a virtual connection. The packet may travel out of order. The packets are numbered consecutively and communication is bidirectional.

(ii) Connectionless delivery: A connectionless transport protocol will treat each packet independently. There is no connection between them. Each packet can take its own different route.

3. Addressing: The client needs the address of the remote computer it wants to communicate with. Such a remote computer have a unique address so that it can be distinguished from the other computers.

4. Providing reliability: For high reliability, the flow control and error control should be incorporated.

5. Flow control: We know that data link layer can provide the flow control. Similarly, transport layer also can provide flow control. But this flow control is performed end to end rather than across a single link.

6. Error control: The transport layer can provide error control as well. But error control at transport layer is performed end to end rather than across a single link. Error connection is generally achieved through retransmission.

7. Congestion Control and QoS: The congestion can take place in the data link, network or transport layer. But the effect of congestion is generally evident in the transport layer. Quality of service (QoS) can be implemented in other layers but its actual effect is felt in the transport layer. The transport layer enhanced the QoS provided by the network layer. The transport service may allow the user to specify preferred, acceptable and minimum values of various service parameters at the time of setting up a connection.

The typical QoS parameters for transport layer are as under:

1. Connection establishment delay
2. Connection establishment failure probability
3. Throughput
4. Transit delay
5. Protection
6. Resident error ratio
7. Priority
8. Resilence

8.3. THE TRANSPORT LAYER SERVICES* (RGTU, Bhopal, Sem. Exam; June, 2014)

The task of transport layer is to provide reliable, cost effective transport of data from source machine to destination machine. To achieve this goal, the transport layer makes use of the services provided by the network layer.

Transport entity

The hardware and/or software within the transport layer which does the work of making use of the services provided by the network layer is called as transport entity. The relationship (logical) between the network, transport and application layers is shown in figure 8.3.

The transport services are of following two types :

(i) Connection oriented.

(ii) Connectionless.

In the connection oriented transport service, there are three phases namely : establishment, data transfer and release. Addressing and flow control in transport layer is similar to that in the network layer. The connectionless transport service also is very similar to connectionless network layer service. Eventhough the network and transport layers provide the similar services, the transport layer can improve the quality of service which a network layer cannot do. If a transport entity is told that its network connection is abruptly terminated, then it can set up a new network connection to the remote transport entity. Through this new connection, it can ask about which data has already arrived and which is yet to arrive. In other words, the transport layer makes the transport service more reliable than the network layer service. Lost packets and damaged data can be detected and compensated for by the transport layer. The transport service primitives can be designed to be independent of the network service primitives. The transport layer makes it possible for application layer programs to be written using a standard non-varying set of primitives (network layer service primitives vary a lot). These programs can work on a wide variety of networks due to transport layer. Thus, the transport layer isolates the upper layers from the technology, design and imperfections of the subnet. Hence the layers of reference model are divided into two groups. Layers 1 to 4 from the first group called transport service provider and the layers above 4 are in the second group called transport service users.

DO YOU KNOW?
The transport layer needs ports or service access points.

* Discuss the various types of services provided by the transport layer.
(RGTU, Bhopal, Sem. Exam; June, 2014)

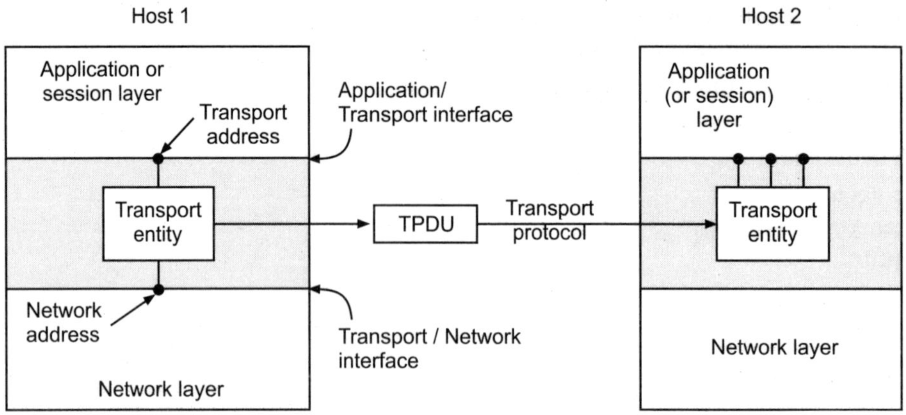

Fig. 8.3. Relation between network, transport and application layers

8.3.1. Quality of Service (Qos)

As mentioned earlier, the QoS parameters are as under :

1. Connection establishment delay: The time difference between the instant at which a transport connection is requested and the instant at which it is confirmed is called as **connection establishment delay**. The shorter the delay, the better the service.

2. Connection establishment failure probability: It is the probability that connection is not established even after the maximum connection establishment delay. This can be due to network congestion, lack of table space or some other problems.

3. Throughput: It measures the number of bytes of user data transferred per second, measured over some time interval. It is measured separately for each direction.

4. Transit delay: It is the time between a message being sent by the transport user on the source machine and its being received by the transport user on the destination machine.

5. Residual error ratio: It measures the number of lost or garbled messages as a function of the total messages sent. Ideally, the value of this ratio should be zero and practically, it should be small as possible.

6. Protection: This parameter provides a way to protect the transmitted data from being read or modified by some unauthorized parties.

7. Priority: This parameter provides a way for the user to show that some of its connection are more important (have higher priority) than the other ones. This is important while handling the congestions. Because the higher prirority connections should get service before the low priority connections.

8. Resilence: Due to internal problem or congestion, the transport layer spontaneously terminates a connection. The resilence parameter gives the probability of such a termination.

8.4. TRANSPORT SERVICE PRIMITIVES

The transport service primitives allow the transport user such as application programs to access the transport service. Each transport service has its own access primitives. The transport service is similar to network service but there are some important differences. The main difference is that the connection-oriented transport service is reliable. The second difference between the network service and transport service is whom the services are intended for. The transport primitives are seen by many programs and programmers. Hence, the transport service is convenient and easy to use. We can get the idea about the transport services by referring to Table 8.1 which lists the five primitives.

The transport interface allows the application programs to establish, use and release connections. Let us see how these primitives are used in actual applications.

(i) The server executes a LISTEN primitive. This will make a system call to block the server until a client turns up.

(ii) When a client wants to talk to the server it executes the CONNECT primitive.

(iii) In response, the transport entity blocks the caller and sends a packet to the server. The transport layer message is encapsulated in the payload of this packet for the server's transport entity.

Table 8.1. Primitives for a simple transport service

S. No.	Primitive	TPDU sent	Meaning
1.	LISTEN	None	Block until some process tries to connect
2.	CONNECT	Connection request	Actively attempt to establish a connection
3.	SEND	Data	Send data
4.	RECEIVE	None	Block until a data TPDU arrives
5.	DISCONNECT	Disconnection request	Replace the connection.

TPDU: The message sent from transport entity to transport entity is called as transport protocol data unit or TPDU.

8.4.1. Nesting of TPDUs, Packets and Frames

The TPDUs which are exchanged by the transport layer are contained in the packets that are exchanged by the network layer. These packets are in turn contained in the frames which are exchanged by the data link layer. When a frame arrives, the data link layer processes the frame, header and passes the contents of the frame payload field to the network entity.

The network entity processes the packet header and passes the contents of the packet payload to the transport entity. This is called as Nesting and is illustrated in figure 8.5.

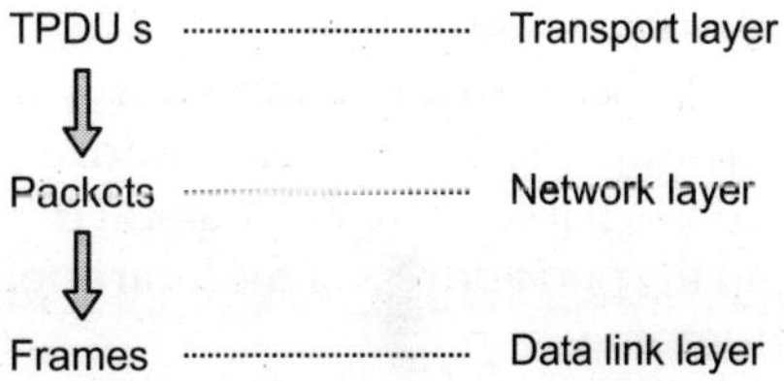

Fig. 8.4.

Connect primitive: If a client gives the CONNECT call, then a connection request TPDU is sent to the server. When this TPDU arrives, the transport entity checks if the server is blocked on a LISTEN. It then unblocks the server and sends a connection accepted TPDU back to the client. On arrival of this TPDU, the client is unblocked and connection is established.

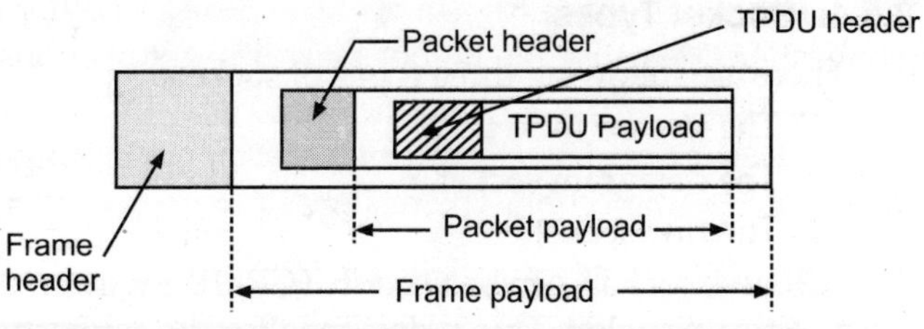

Fig. 8.5. Nesting of TPDUs, packets and frames

SEND and RECEIVE Primitives: The SEND and RECEIVE primitives can be used for exchange of data. The data exchange at the network layer is more complicated than that at the transport layer. In transport layer data exchange, every data packet is eventually acknowledged. The packets carrying control TPDUs are also acknowledged. All these acknowledgements are managed by the transport entities using the network layer protocols. The transport entities have to take care of issues like timers and retransmission. The transport layer connection acts as a reliable bit pipe through which the bits sent by a sender come out from the other side of type.

Connection Release: A connection should be released when it is no longer needed. This is essential in order to free up the table space within the two transport entities.

Disconnection can be of following two types :

(i) Asymmetric

(ii) Symmetric

8.5. SOCKETS

The socket interface was originally based on UNIX. It defines a set of system calls or procedure. The communication structure that we need in socket programming is called as a **socket**. A socket acts as an end point. Two processes can communicate if and only if they have a socket at each end.

Socket structure: Figure 8.6. shows a simplified socket structure.

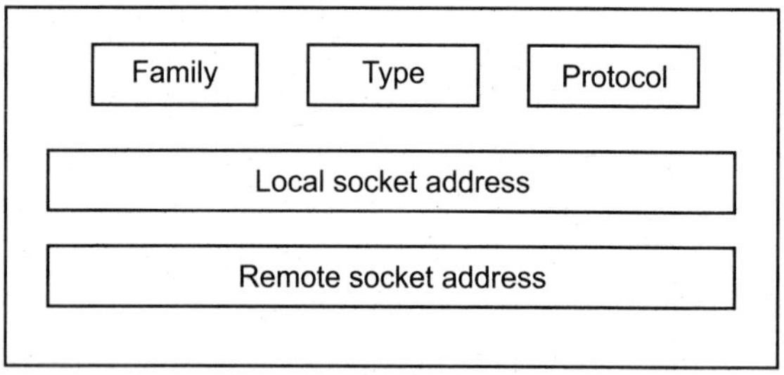

Fig. 8.6. Socket structure

Various fields in the socket structure are as under :

1. Family: This field is used for defining the protocol group such as IPv4 or IPv6, UNIX domain protocol etc.

2. Type: This field is used for defining the type of socket such as stream socket, packet socket or raw socket.

3. Protocol: This field is usually set to zero for TCP and UDP.

4. Local socket address: It is used for defining the local socket address. This address is a combination of local IP address and the port address of the local application program.

5. Remote socket address: It is used for defining the remote socket address which is a combination of remote IP address and the port address of the remote application program.

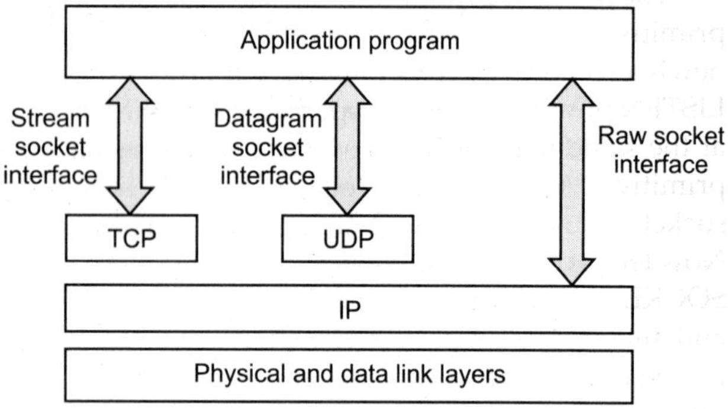

Fig. 8.7. Type of sockets

8.5.1. Socket Types

There are following three types of sockets:

1. The stream socket.
2. The datagram socket.
3. The raw socket.

All these sockets can be used in TCP/IP environment. Let us discuss them one by one.

1. Stream socket: This is designed for the connection oriented protocol such as TCP. The TCP uses a pair of stream sockets for connecting one application program to the other across the Internet.

2. Datagram socket: This type of socket is designed for the connectionless protocol such as UDP. UDP uses a pair of datagram sockets for sending a message from one application program to another across the Internet.

3. Raw socket: Raw sockets are designed for the protocols like ICMP or OSPF, because these protocols do not use either stream packets or datagram sockets.

8.5.2. Berkeley Sockets

1. Various Transport Primitives: Table 8.2 lists various transport primitives used in Berkeley UNIX for TCP.

Table 8.2.

S. No.	Primitive	Meaning
1.	SOCKET	Create a new communication end point.
2.	BIND	Provide a local address to a socket.
3.	LISTEN	Show willingness to accept connections.
4.	ACCEPT	Block the caller as long as a connection attempt.
		does not arrive.
5.	CONNECT	Attempt to establish a connection.
6.	SEND	Send data.
7.	RECEIVE	Receive data.
8.	CLOSE	Release the connection.

The first four primitives in the table 8.2 are executed in the same order by the server. The SOCKET primitive creates a new end point and allocates table space for its within the transport entity. The newly created sockets do not have addresses. These are assigned using the BIND primitive. The LISTEN primitive allocates space to queue the incoming calls in case if several clients wish to connect at the same time. To block waiting for an incoming connection, the server executes an ACCEPT primitive. When a TPDU requesting for a connection arrives, the transport entity creates a new socket and returns a file descriptor for it. These were the primitives corresponding to server side. Now let us consider the client side. On the client side also a socket needs to be created first using the SOCKET primitive, however, the BIND is not required. The CONNECT primitive blocks the caller and initiates the connection process. When it completes (which is indicated by an appropriate TPDU received from the server), the client process is unblocked and the connection is established. After this, both the sides can use SEND and RECEIVE primitives to send and receive data. In order to release the connection, both sides have to execute a CLOSE primitive.

2. Steps followed for Socket Programming: The steps followed for the socket programming are as under:

Server side

(i) Server creates a socket and checks for errors using SOCKET.

(ii) Assign address to the newly created socket using BIND.

(iii) Use the LISTEN to allocate space to queue the incoming calls.

(iv) Execute an ACCEPT for blocking the waiting incoming connections.

Client side

(i) Create a socket using Socket.

(ii) Use CONNECT to initiate connection process.

(iii) Establish the connection.

8.5.3. Connectionless Iterative Server

Let us now discuss connectionless, iterative client-server communication using UDP and datagram sockets. The server that uses UDP is usually connectionless iterative. So, the server one request at a time. A server gets the request received in a packet from UDP, it processes the request and gives the response to UDP to send it to the client. The server does not pay any attention to the other packets. The other packets are stored in a queue waiting for the service. They are processed one by one. The server uses one single port for this purpose, the well known port. All the packets arriving at this port will wait in line to be served.

Server functions: The server performs the following functions :

1. **Create a socket:** The server asks the operating system to create a socket.

2. **Bind:** The server asks the operating system to enter information in the socket related to the server. This is called as binding the server socket.

3. **Repeat:** The server repeats the following steps for infinite number of times.

(a) Receive a request.

(b) **Process:** The request is processed by the server.

(c) **Send:** The response is sent to the client.

Client functions: The client performs the following functions:

1. **Create a socket:** The client asks the operating system to create a socket. There is no need of binding.

2. **Repeat:** The client repeats the following steps as long as it has requests.

(a) Send: Client asks the operating system to send a request.

(b) Receive: Client asks the operating system to wait for the response and deliver it, when it has arrived.

3. **Destroy:** When the client does not have any more requests, it asks the operating system to destroy the socket.

8.5.4. Connection Oriented Concurrent Server

The connection oriented concurrent client server communication uses TCP and stream socket. The servers using TCP are normally concurrent. This means that a server is serving many clients at the same time. The type of communication is connection oriented. Once a connection is established, it remains open until entire stream of bytes is processed. After that the connection is terminated. The server must have one buffer for each connection. The bytes from the client are stored in buffers and handled concurrently by the server. In order to provide this type of server, the concept of parents and child server is used.

DO YOU KNOW?

The transport layer, by hiding all of the manipulations necessary to move a message from source to destination, makes data transmission transparent to the upper layers.

Parent server: A parent server is the server running infinitely and accepting connections from clients. It uses the well known port. After establishing a connection, the parent server creates a child server and an ephemeral port to allow the child server to handle the client.

Server function: The server performs the following functions

1. **Create a socket:** The server asks the operating system to create a socket.

2. **Bind:** The server asks the operating system to enter information in the socket.

3. **Listen:** The server asks the operating system to be passive and listen to the client which needs to be connected to this server. This is because TCP is a connection oriented protocol so a connection needs to be made before transferring the data.

4. **Repeat:** The server repeats the steps given below infinitely:

(a) Create a child: When a child requests a connection, the operating system creates a temporary child process and assigns the duty of serving the client to the child. The parent process is then free for listening to new clients.

(b) Create a new socket: A new socket is created which is to be used by the child process.

(c) Repeating: The child repeats the following steps as long as it has requests from the client. (i) Read; (ii) Process; (iii) Write; (iv) Destroy socket.

Client functions: The client performs the following functions :

(i) Create a socket. (ii) Connect.

(iii) Repeat the write and read operations. (iv) Destroy : Close the connection.

Client and Server Program: Client-server programs are written in the languages such as C, C + +, Java. It requires advanced knowledge of the particular language.

8.6. TRANSPORT PROTOCOLS

In order to implement the transport layer services between the two transport entities, we have to use a transport protocol.

The transport protocols have to deal with the following tasks.

(i) Error control

(ii) Sequencing and

(iii) Flow control.

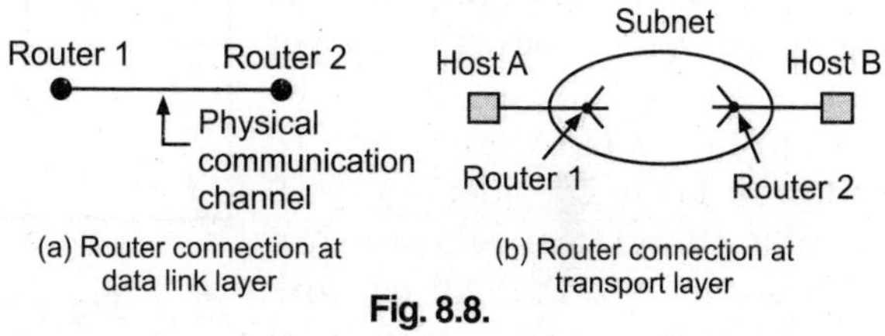

Fig. 8.8.

The transport protocols are similar to the data link protocols in many ways but there are some dissimilarities as well. At the data link layer, two router communicate directly via a physical channel as shown in figure 8.8 (a), whereas at the transport layer, the physical channel is replaced by the entire subnet as shown in figure 8.8 (b).

The difference between data link and transport layer communication is as under:

S.No.	Data link layer	Transport layer
1.	Communication is through a physical channel.	Communication is through a subnet.
2.	It is not necessary to specify the destination router.	Explicit addressing of destination is essential.
3.	Establishing a connection is simple.	Initial connection establishment is more complicated.
4.	No storage capacity.	There is some storage capacity in the subnet.
5.	No additional delay.	Delay is introduced due to the storing capacity of subnets.
6.	Different approaches are used for buffering and flow control by the two layers.	

Elements of transport protocols

Following are some of the important elements of transport protocols :

(i) Addressing (ii) Establishing a connection

(iii) Releasing a connection (iv) Flow control and buffering

(v) Multiplexing (vi) Crash recovery.

In the following sections, we will discuss all these elements.

8.7. ADDRESSING

If a remote process wants to get connected to another remote process, then it has to specify the one to which connection is to be established. So, a transport address is to be defined. This is an address to which processes can listen for connection requests. In the Internet, these end points are IP address local port pairs whereas in ATM networks, the AAL-SAP pairs are used. Note that end points and addresses carry the same meaning. In case of transport layer communication, let us use a general term TSAP (Transport Service Access Point) for the same. The analogous end points in the network layer *i.e.* the network layer addresses are called as NSAPs. The examples of NSAPs are the IP address. Let us consider figure 8.9 which shows the relationship between the NSAP, TSAP, network connection and transport connection for the connection oriented subnet such as an ATM.

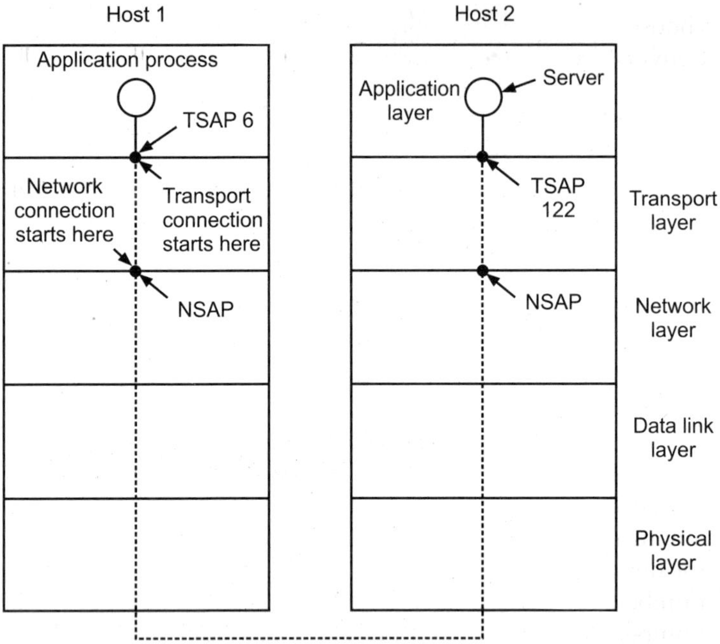

Fig. 8.9. Relationship between NSAP, TSAP, network connection and transport connection

With reference to figure 8.9, we can explain the connection scenario for a transport connection over a connection oriented network layer is as under:

1. The time-of-day server process on host 2 gets connected to TSAP 122 so as to wait for an incoming call. For this, it can use the LISTEN call.
2. An application process on host 1 wants to find out the time of day. So, it issues a CONNECT request that specifies TSAP 6 as the source and TSAP 122 as the destination.
3. The transport entity on host 1 then sets up a network connection between the two hosts. Such a network layer connection is not required with a connectionless subnet. Using this network connection, the host 1's transport entity can communicate with transport entity on host 2.
4. Transport entity on host 1 then request a connection between TSAP 6 and TSAP 122 to transport entity on host 2.
5. The transport entity on host 2 then asks the time of day server at TSAP 122 if it is willing to accept a new connection. If it says yes, then a new transport connection is established.

Important Point: This shows that the transport connection goes from TSAP to TSAP whereas a network connection goes only from NSAP to NSAP.

The question here is that how does the user process on host 1 know that the time- of-day server is attached to TSAP 122.

If this server is attached to TSAP 122 for a long time, then it is possible. But this is not always true. Hence, a scheme called *initial connection protocol is used.*

8.7.1. Addressing in Transport Layer

At the data link layer, we need a MAC address, at the network layer, we need an IP address to choose on host among millions. A datagram in the network layer needs a destination IP address for delivery and source IP address for the destination's reply.

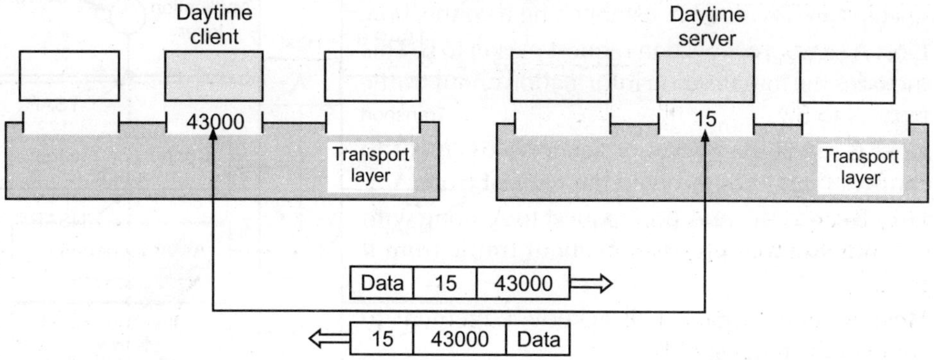

Fig. 8.10. Port numbers

At the transport layer, a transport layer address called a port number is required to choose among multiple processes running on the destination host. The destination port number is required for delivery and the source port number is needed for the reply. In the Internet model, the port numbers are 16 bit integers between 0 and 65,535. The client program defines itself with a port number which is chosen randomly. This number is called as **ephemeral port number**. The server process should also define itself with a port number but this port number can not be chosen randomly. The Internet uses universal port numbers for servers and these numbers are called as **well known port numbers**. Every client process knows the well-known port numbers of the corresponding server process. For example, a Day time client process can use an ephemeral (temporary) port number 43000 for identifying itself, the Day time server process must use the well known (permanent) port number 15. This is illustrated in figure 8.10.

8.7.2. Difference between IP Addresses and Port Numbers?

The IP addresses and port numbers play different roles in selecting the final destination of data. The destination IP address defines the host among the millions of hosts in the world. After a particular host is selected, the port number defines one of the processes on this selected host.

8.7.3. Socket Address

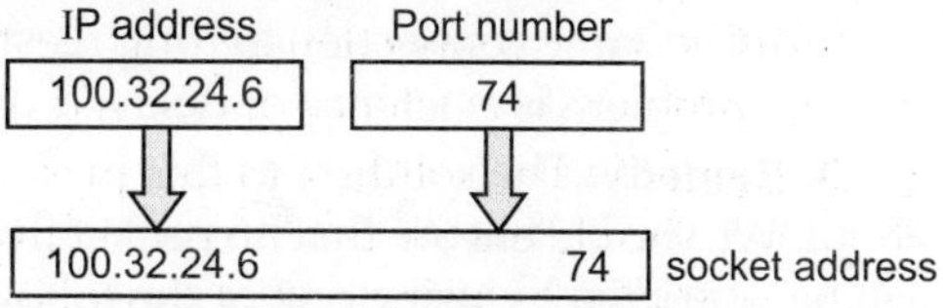

Fig. 8.11. Socket address

Process to process delivery (transport layer communication) needs two identifiers, one is IP address and the other is port number at each end to make a connection. Socket address is the combination of IP address and port number as shown in figure 8.11. The client socket address defines the client process uniquely whereas the server socket address defines the server process uniquely. A transport layer protocol requires the client socket address as well as the server socket address. These two addresses contain four pieces. These four pieces go into the IP header and the transport layer protocol header.

The IP header contains the IP addresses while the UDP and TCP headers contain the port numbers.

8.8. CONNECTION-ORIENTED SERVICE

In a connection oriented service, a connection is established between source and destination. Then the data is transferred and at the end, the connection is released.

8.8.1. Connection Establishment

1. Basic concepts: Let us consider figure 8.12 to understand the connection establishment. Following steps are involved in establishing a connection.

(i) Host A sends a connection request packet to B. This includes the initialisation information about traffic from A to B.

(ii) Host B sends the packet of acknowledgement to confirm that it has received the request from A.

(iii) Host B sends a connection request to A along with the initialisation information about traffic from B to A.

(iv) Host A sends a packet of acknowledgement to confirm the request of B.

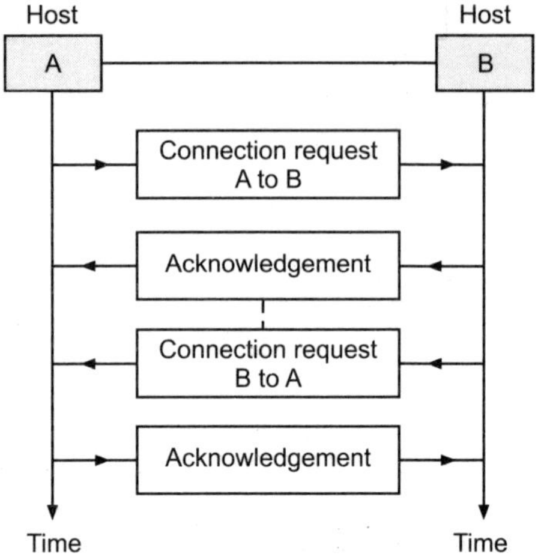

Fig. 8.12. Establishing a connection

It is possible to merge the steps (ii) and (iii). Note that each connection request must have a sequence number which is helpful in recovering from the loss or duplication of the packets. For the same reason, each acknowledgement also should have an acknowledgement number. The first sequence number in each direction should be random for each connection established. This is to ensure that a sender can not create more than one connection which starts with the same sequence number (e.g. 2). This is to recognise the duplication of packets. Since a sequence number is required for each connection, the receiver has to keep the history of sequence numbers for each remote host for a specific time.

2. Problems: Establishing a connection sounds easy. But, actually it is a very tricky job. The problem occurs when the network can lose store and duplicate packets. The problems can be elaborated as under :

(i) Due to congestion on a subnet, the acknowledgements do not get back in time. So retransmission of each packet takes place.

(ii) If the subnet uses datagrams inside and every packet travels on a different route. Some of the packets might get stuck in a traffic jam and take a long time to arrive.

(iii) The same connection getting reestablished due to duplication of packet. So, the crux of the problem is existence of delayed duplicate packets.

3. Remedy: The solution to this problem is to kill off the aged packets that are still wandering about. We should ensure that no packet lives longer than some predecided time. The packet lifetime can be restricted by using one of the following techniques:

(i) Restricted subnet design.

(ii) Putting a hop counter in each packet.

(iii) Time stamping each packet.

8.8.2. Three Way Handshake Technique

The delayed duplicate packet problem can be solved by using a technique called *three way handshake*. The principle of three way handshake is shown in figure 8.13.

1. Normal Operation

(i) Host 1 chooses a sequence number x and sends a connection request (CR) TPDU containing it to host 2.

(ii) Host 2 replies with a connection accepted TPDU to acknowledge x and to announce its own sequence number y.

(iii) Host I acknowledges host 2 and send the first data TPDU.

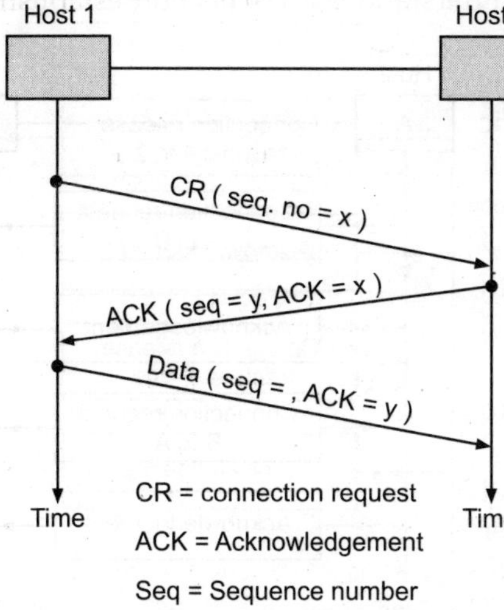

Fig. 8.13. Three way handshake technique

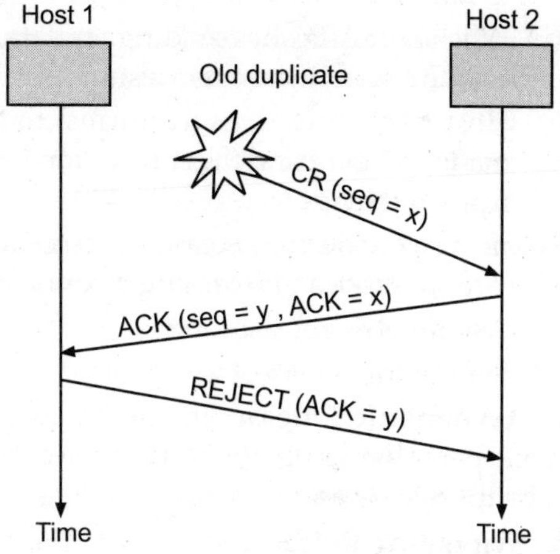

Fig. 8.14. Response to an OLD duplicate

2. Operation in the Abnormal Circumstances: Now, let us see how the three way handshake works in presence of delayed duplicate control TPDUs. Let us consider figure 8.14. The first TPDU is a delayed duplicate CONNECTION REQUEST from an old connection. The HOST 1 does not know about it.

Host 2 receives this TPDU and sends to host 1, a connection accepted TPDU. But, Host 1 is not trying to establish any connection so it sends a REJECT alongwith ACK = y. So, host 2 realizes that it was folded by a delayed duplicate and abandons the connection.

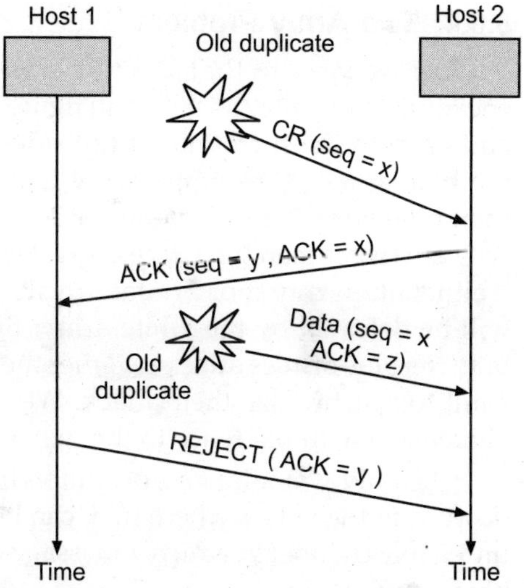

Fig. 8.15. Duplicate CR and duplicate ACK

3. Duplicate CR and Duplicate ACK: This is another abnormal situation. Let us consider figure 8.15 to understand this situation. This is the worst case in which delayed duplicates of connection request (CR) and acknowledgement (ACK) are floating around in the subnet. Host 2 gets a CR and it replies to it by sending ACK. Note that host 2 has proposed a connection with a sequence number y. When the second delayed TPDU (duplicate) arrives at host 2 it understands that z has been acknowledged and not y. So, it understands that this too is an OLD duplicate.

8.8.3. Connection Release

Any one of the two parties involved in data exchange can close the connection. But the problem is that, when connection is terminated from one end, the other party can continue to send data in the other direction. Hence, to ensure a proper connection release one has to follow the steps given below:

Procedure to Release a Connection: Let us consider figure 8.16 to understand this procedure.

(i) Host 1 sends a connection release request to host 2.

(ii) Host 2 sends an acknowledgement to confirm the request of host 1.

(iii) After this, the connection is closed in one direction but host 2 can continue to send data to host 1.

(iv) When host 2 finishes sending his data, it sends a connection release request to host 1.

(v) Host 1 acknowledges (confirms) the request made by host 2 and the connection is released from both ends.

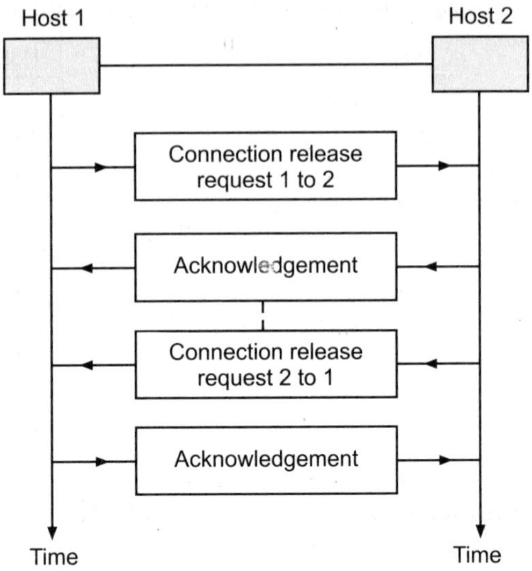

Fig. 8.16. Connection release

Releasing a connection is easier than establishing it. There are two styles of releasing a connection.

1. Asymmetric release, and
2. Symmetric release.

1. Asymmetric Release: In asymmetric release, when one party hangs up, the connection is broken. It is an abrupt release and may result in loss of data.

2. Symmetric Release: Symmetric release treats the connection as two separate unidirectional connections and requires each one to be released separately. Symmetric release is as shown in figure 8.16 and does not involve any loss of data.

8.8.4. Two Army Problem

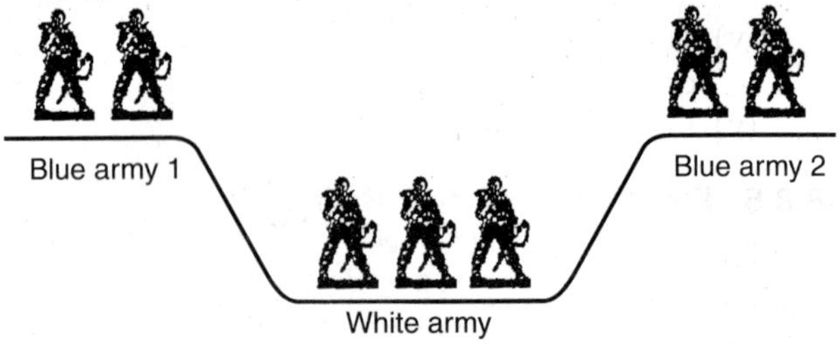

Fig. 8.17. Two army problem

Let us assume that a white army is encamped in a valley as shown in figure 8.17 and on both the surrounding hill sides are the blue armies. The white army is larger than blue army 1 and 2 alone but together they are larger than the white army. Hence, if either blue army (1 or 2) attack by itself, it will be defeated by the white army. But if both the blue armies attack together then they will win over the white army. Hence, the blue armies want to synchronize their attack. We will now discuss the principles of two way and three way handshake with reference to the two army problem.

The only way of communication between the two blue armies is to send messengers on foot down into the valley where they can be captured and the message is lost. This is equivalent to an unreliable channel in which a message can be lost. Now, let us formulate a protocol that allows the blue army to win.

Protocol 1 : Two Way Handshake

(i) Chief of blue army 1 sends a message to the chief of blue army 2 stating "I purpose we attack in the morning of January 1. Is it OK with you ?"

(ii) Suppose that the messenger reaches to blue army 2 and the chief of blue army 2 agrees with the idea.

(iii) He sends the message "yes" back to blue army 1 and assume that this message also gets through.

(iv) This process is called two way handshake and it is shown in figure 8.18.

(v) The question is will the attack take place ? The answer is probably not because the chief of blue army 2 does not know whether his reply got through.

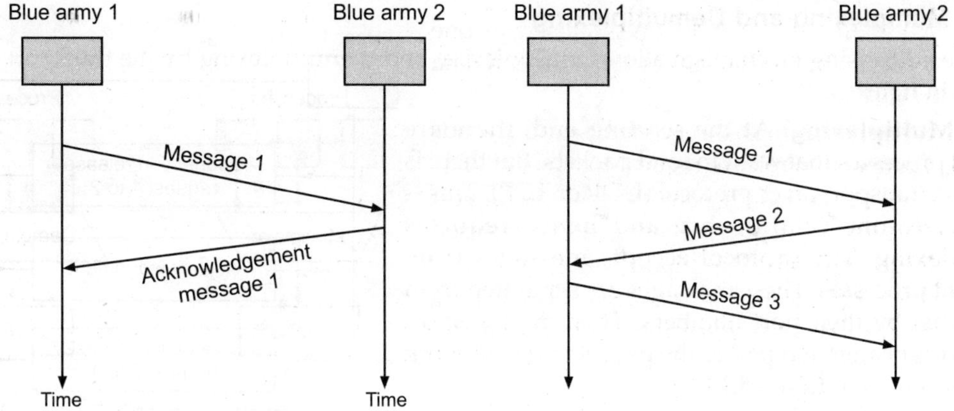

(a) Two way handshake (b) Three way handshake

Fig. 8.18.

Protocol 2 : Three way handshake

(i) Now, let us improve the two way handshake protocol by making it a three way handshake.

(ii) This protocol is shown in figure 8.18 (b). Assuming no messages lost, blue army – 2 will also get the acknowledgement.

(iii) But, now, the chief of blue army – 1 will hesitate because he does not know if the last message he has sent has got through or not.

(iv) Therefore, we can make a four way handshake. But it also does not help because in every protocol, the uncertainty after last handshake message always remains.

(v) In fact, we can prove that no protocol exists that works.

8.8.5. Flow Control and Buffering

For flow control, a sliding window is required on each connection to keep a fast transmitter from overrunning a slow receiver (the same as the data link layer). Since a host may have numerous (*e.g.*, 64) connections, it is impractical to implement the same data link buffering strategy (using dedicated buffers for each line). The sender should always buffer outgoing TPUDs until they are acknowledged. The receiver may not dedicate specific buffers to specific connections. Instead, a single buffer pool may be maintained for all connections. When a TPDU comes in, if there is a free buffer available, the TPDU is accepted, otherwise it is discarded. However, for high-bandwidth traffic (*e.g.*, file transfers), it is better if the receiver dedicate a full window of buffers, to allow the data to flow at maximum speed. How large the buffer size should be ? As the traffic pattern changes, the transport protocol should allow a sending host to request buffer space at the other end. Alternatively, the receiver could tell the sender "I have reserved buffers for you". A general way to manage dynamic buffer allocation is to decouple the buffering from the acknowledgements.

Dynamic buffer management in a variable sized window.

Initially, the sender requests a certain number of buffers, based on its perceived needs. The receiver then grants as many buffers as it can afford. Every time the sender transmits a TPDU, it must decrement its allocation, stopping altogether when the allocation reaches zero. The receiver then separately piggybacks both acknowledgements and buffer allocations onto reverse traffic. To prevent deadlock caused by loss of control TPDUs, each host should periodically and control TPDUs giving the acknowledgement and buffer status on each connection. The sender's window size could be dynamically adjusted not only by the availability of the buffers at the receiver side, but also by the capacity and traffic of the subnet. The bigger the subnet's carrying capacity and lighter the traffic, larger is the sender's window size.

8.8.6. Multiplexing and Demultiplexing

The addressing mechanism allows multiplexing and demultiplexing by the transport layer as shown in figure 8.19.

1. Multiplexing: At the sending end, there are several processes that want to send packets. But there is only one transport layer protocol (UDP or TCP). This is a many-to-one relationship and hence requires multiplexing. The protocol accepts messages from different processes. These messages are separated from each other by their port numbers. Then the transport layer adds header and passes the packet to the network layer as shown in figure 8.19.

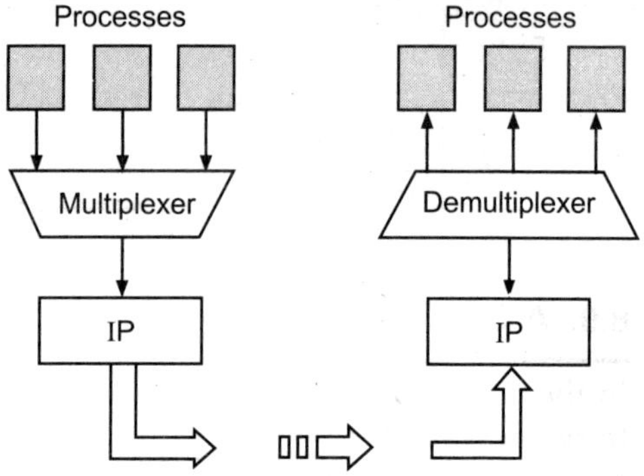

Fig. 8.19. Multiplexing and demultiplexing

2. Demultiplexing: At the receiving end, the relationship is one as to many. Therefore, we need a demultiplexer. First the transport layer receives datagrams from the network layer. The transport layer then checks for errors and drops the header to obtain the messages and delivers them to appropriate process based on the port number.

8.8.7. Crash Recovery

The host and routes are subnet to crashes, and the recovery from such crashes is essential. Such crashes will result in loss of data packets. If the transport entity is completely inside the hosts then the recovery from the network and router crashes is simple. This is as explained below.

In case of a **router crash**, the two transport entities must exchange information after the crash to determine which TPDUs were received and which were not. The crash can be recovered by retransmitting the lost ones. If the network layer provides connection oriented service, then the loss of a virtual circuit (crash) can be handled by establishing a new virtual circuit. Then the remote transport entity can be asked about which TPDUs it has received and which TPDUs are lost. Those lost can be retransmitted over the newly established virtual circuit. It is very difficult to recover from the host crash. Let us consider that the sender is sending a long file to the receiver using a simple stop-and-wait protocol. Part way through the transmission the receiver crashes. When the receiver comes back up it might send a broadcast TPDU to all other hosts, requesting the other hosts to inform if of the status of all open connections before the crash.

The sender can be in one of two states : one TPDU outstanding, or no TPDUs outstanding. What happens if the sender retransmits ? Think about the situation when the receiver first sends an ACK and then performs the wire (to the output stream), but crash occurs in the middle (lost TPDU). How about reversing the order of sending ACK and performing the write ? It does not help (duplicated TPDU). No matter how the sender and receiver are programmed, there are always situations where the protocol fails to recover properly.

Conclusion: If the crash takes place at layer N, then the crash recovery can be only done by the $N + 1$ layer. Only if the higher layer retain enough status information. This means that transport layer can recover from the failures in the network layer provided that each end of connection keep a track of where it is.

Recovery from IMP and Host Crashes: The strategy planned to recover from IMP and host crashes is as under :

1. Transport entity should exchange information after the crash to determine which TPDU's were received and which are lost.
2. Receiver may send broadcast TPDU's, to all other hosts announcing that it has just crashed. It requests the other hosts to inform about all the open connections.

3. Transport entity first send write and then sends acknowledgement.
4. Receiver can be programmed in one of the following two ways : acknowledgement first or write first.
5. Sender can be programmed in one of the following ways :
   (i) Always transmit the last TPDU
   (ii) Never transmit the last TPDU
   (iii) Retransmit only in the state SO
   (iv) Retransmit only in the state SI.

8.9. A SIMPLE TRANSPORT PROTOCOL

In the previous chapter, we have discussed many ideas related to the transport layer. To make these ideas more concrete, let us take an example.

8.9.1. The Example Service Primitives

The abstract service primitives are connection-oriented, as listed in table 8.3

Table 8.3. The Primitives for a Simple Transport Service

Primitive	Packet Sent	Meaning
LISTEN	(none)	Block until some process tries to connect.
CONNECT	CONNECTION REQ.	Actively attempt to establish a connection.
SEND	DATA	Send information.
RECEIVE	(none)	Block until a DATA packet arrives
DISCONNECT	DISCONNECTION REQ.	This side wants to release the connection.

The problem that we have to discuss is how to express these transport primitives correctly.

1. CONNECT: For CONNECT, we have to call the library procedure *connect* with necessary parameters to establish a connection. These parameters are the local and remote TSAPs. During the call, the caller is suspended (*i.e.* blocked) for a time when the transport layer is trying to set up the connection. If the connection is established, the caller is unblocked. Then it can start transmitting the data.

2. LISTEN: When a process wants to accept the incoming calls, it calls the library procedure *listen* which specifies a particular TSAP to listen to. The process then blocks itself until some remote process makes an attempt to establish a connection to its TSAP. This model is symmetric because one side is passive which executes *listen* and waits until something happens. The other side is active which initiates the connection. What happens if the active side begins first when the passive side is not ready ? To solve this problem, the receiving and holds the connection request for a certain time interval. If a process on that host call's *listen* before the timer goes off then the connection is established. Otherwise it is rejected and the caller is unblocked and given an error return.

3. DISCONNECT: The procedure *disconnect* is used to release a connection. When both the sides have disconnected, the connection is released. That means we use a symmetric disconnection model.

4. SEND and RECEIVE: Both these are related to the data transfer. But sending is active and receiving is passive. This problem is same as that faced with the connection establishment. Hence, we can use the same solution, an active call *send* for data transmission and a passive call *receive* that blacks until a TPDU arrives.

Therefore, our concrete service definition consists of five primitives namely :

(a) LISTEN (b) CONNECT

(c) SEND (d) RECEIVE

(e) DISCONNECT.

Each primitive corresponds exactly with a library procedure which executes the primitive. The parameters for the service primitives and library procedures are as under :

- connum = LISTEN (local)
- connum = CONNECT (local, remote)
- connum = SEND (connum, buffer, bytes)
- status = RECEIVER (connum, buffer, bytes)
- status = DISCONNECT (connum).

(a) LISTEN: This primitive discloses the caller's willingness towards accepting connection requests. The user of this primitive is blocked when the connection is being attempted. There is no time limit for this primitive.

(b) CONNECT: The CONNECT primitive takes two parameters, a local TSAP (*i.e.*, transport address), *local* and a remote TSAP *remote* and establishes a transport connection between the two. If it is successful, it returns in *connum* a non negative number used to identify the connection on subsequent call. But if it is not successful then the reason for failure is put in *connum* as a negative number. The possible reasons for such failure are : one of the transport address is currently in use, remote host down, illegal local address, and illegal remote address.

(c) SEND: This primitive transmits the contents of the buffer on the indicated transport connection. The errors are returned in *status*. The possible errors are : no connection, illegal buffer address or negative count.

(d) RECEIVE: The primitive shows that caller wants to accept data. The size of incoming message is placed in *bytes*. If the remote process has released the connection or if the buffer address is illegal, *status* is set to an error code which indicates the nature of the problem.

(e) DISCONNECT: This primitive terminates the transport the transport connection. The parameter *connum* tells which connection is to be terminated. The possible errors are *connum* belongs to other processes or *connum* is not a valid connection identifier. An error code (if an error exists) or a "0" indicating success is returned in *status*.

8.9.2. The Example Transport Entity

The transport layer makes use of the network service primitives to send and receives TPDUs. We have chosen to use a connection-oriented reliable network service, so that we can focus on transport issues, such as connection establishment, connection release and credit management, etc., that do not occur in the lower layers. The transport entity can be in :

(i) the host operating system kernel.

(ii) a separate user process.

(iii) a package of library routines running within the user's address space, or

(iv) a co-processor chip or network board plugged into the host's backplane.

In following example, the **transport entity** is not really a separate entity at all, but part of the user process. Therefore, when the user executes a primitive that blocks, such as LISTEN, the entire transport entity blocks as well.

The Interface to the Network Layer: The interface to the network layer is via the procedures "to_net" and "from_net"

to_net (int cid, int q, int m, pkt_type pt, unsigned char *p, int bytes);

from_net (int *cid, int *q, int *m, pkt_type *pt, unsigned char *p, int *bytes);

Each one has six parameters. First is the connection identifier which is required for mapping one-to-one onto the network virtual circuits. Next are the Q and M bits. If Q = 1 then it indicates a control message and if M = 1 then it indicates that more data from this message follows in the next packet. Net is the packet type. One of the six packet types given in table 8.4 is to be selected.

Table 8.4. The types of network Layer Packets

S.No.	Network layer	Meaning
1.	CALL REQUEST	With to establish a connection
2.	CALL ACCEPTED	Response to call request
3.	CLEAR REQUEST	Want to release the connection
4.	CLEAR CONFIRMATION	Response to clear request
5.	DATA	Transport data
6.	CREDIT	Control packet for managing the window.

On call to "to_net" the transport entity fill in all the parameters which are to be read by the network layer. On call to "from_net" the network layer dismembers an incoming packet for the transport entity. By passing information as procedure parameters rather than the actual outgoing or incoming packet itself, the transport layer is shielded from the details of the network layer protocol. If the transport entity should attempt to send a packet when the underlying virtual current's sliding window is full, it is suspended transparently within to_net until there is room in the window. In addition to this transparent suspension mechanism, there are also explicit "sleep" and "wakeup" procedures called by the transport entity to block/unblock itself :

- **sleep ()** is called when the transport entity logically needs to wait for an external event to happen. After sleep has been called, the (main stream of the) transport entity is blocked.
- **wakeup ()** is called by the event handling procedure (*i.e.*, packet_arrival ()) to unblock the sleeping (main stream of the) transport entity.

Connection States

Each connection is always in one of the following seven states.

1. IDLE – a connection is yet to be established.
2. WAITING – CONNECT has already been executed and CALL REQUEST has been sent.
3. QUEUED – A CALL REQUEST has arrived but no LISTEN yet.
4. ESTABLISHED – The connection has been made.
5. SENDING – The user waits for a permission to send his packet.
6. RECEIVING – A RECEIVE has been made.
7. DISCONNECTING – A connection has been locally disconnected.

An existing state of a connection will change if any one of the following events occurs.

(i) A primitive is executed.
(ii) A packet arrives.
(iii) The timer expires.

Most of procedures in the transport entity can be called directly by user programs. However, the following two procedures are effectively (software) interrupt routines and can be called only when the main stream of the transport entity is sleeping.

packet–arrival( ) is triggered by the packet arrival event (generated by the underlying network layer).

clock ( ) is triggered by the clock ticking event.

A flow control mechanism based on **credit** is used in the examples transport entity. When a user calls, RECEIVE, a special **credit message** is sent to the transport entity on the sending machine and is recorded in the conn array. When SEND is called, the transport entity checks to see if a credit has arrived on the specified connection. If so, the message is sent (in multiple packets if need be) and the credit is determined. If not, the transport entity puts itself to sleep until a credit arrives.

8.10. TCP AND UDP PROTOCOLS: THE INTERNET TRANSPORT PROTOCOLS (TCP AND UDP)

The internet has two main protocols in the transport layer. One of them is connection oriented and the other one supports the connectionless service. TCP (Transmission control protocol) is a connection oriented protocol and UDP (user's data protocol) is the connectionless protocol. UDP is basically just IP with an additional short header.

8.10.1. Transmission Control Protocol (TCP)

The TCP provides reliable transmission of data in an IP environment. TCP corresponds to the transport layer (Layer 4) of the OSI reference model. Among the services, TCP provides are stream data transfer reliability, efficient flow control, full-duplex operation, and multiplexing. With stream data transfer, TCP delivers an unstructured stream of bytes identified by sequence numbers. This service benefits applications because they do not have to chop data into blocks handling it off to TCP. Instead TCP groups bytes into segments and passes them to IP for delivery. TCP offers reliability by providing connection-oriented, end-to-end reliable packet delivery through an internetwork. It does this by sequencing bytes with a forwarding acknowledgement number that indicates to the destination the next byte the source expects to receive. Bytes not acknowledged within a specified time period are retransmitted. The reliability mechanism of TCP allows devices to deal with lost, delayed, duplicate, or misread packets. A time-out mechanism allows devices to detect lost packets and request retransmission. TCP offers efficient flow control, which means that, when sending acknowledgements back to the source, the receiving TCP process indicates the highest sequence number that it can receive without overflowing its internal buffers. Full-duplex operation means that TCP processess can both send and receive at the same time. Finally, TCP's multiplexing means that numerous simultaneous upper-layer conversations can be multiplexed over a single connection.

8.10.2. The TCP Service Model

For obtaining the TCP service, it is necessary for both sender and receiver to create end points called **sockets**. Each socket has a **socket number** or **socket address**. The socket address is made of two parts.

(i) An IP address of the host.

(ii) A port number which is a 16 bit local to host.

A port is the TCP name for TSAP. In order to obtain TCP service, it is necessary to establish a connection between the sockets on the sending and receiving machines. Various socket calls and their meanings are given in table 8.5.

Table 8.5. Socket calls

S.No.	Socket call	Meaning
1.	Socket	Create a new end point (socket).
2.	BIND	Give a local address to a socket.
3.	LISTEN	Show willingness to accept connections.
4.	ACCEPT	Block the caller until a connection attempt arrives.
5.	CONNECT	Attempt to make a connection.
6.	SEND	Send data over the connection.
7.	RECEIVE	Receiver data over the connection.
8.	CLOSE	Release the connection.

The same socket can be used for establishing more than one connection at a time. Connections are identified by the socket identifiers at both ends. No virtual circuit numbers or other identifiers are used.

Well known ports: The port numbers below 1024 are known as well known ports and they are reserved for standard circuits. All TCP connections are point to point and full duplex. This means that each connection has two endpoints and each connection is bi-directional. TCP does not support multicasting or broadcasting. A TCP connection is a byte stream and not a message stream.

PUSH Flag and Buffering: The message boundaries are not preserved end to end. When an application passes data to TCP, the TCP may send it immediately or may collect the data for some time and send at once (which is called buffering). But if an application wants the data to be sent immediately, then it can use the PUSH flag which will force the TCP to send data, without any delay.

Urgent Data: This is another important feature of the TCP, service. If the sending application puts some control information in the data stream and gives it to TCP alongwith the URGENT flag then the TCP will stop accumulating data and transmit everything it has for that connection immediately. When the urgent data reaches the destination, the receiving application is interrupted and the urgent data stream is presented to it. The end of urgent data is always indicated for the application to understand.

8.10.3. The TCP Protocol *(RGTU, Bhopal, Sem. Exam; June, 2014)*

Let us take a general overview of the TCP protocol. Every byte on a TCP connection has its own 32 bit sequence number. These numbers are used for both acknowledgement and for window mechanism.

1. Segments: The sending and receiving TCP entities exchange data in the form of segments. A segment consists of a fixed 20 byte header (plus and optional part) followed by zero or more data bytes.

2. Segment Size: The segment size is decided by the TCP software. Two limits restrict the segment size as under :

(i) Each segment including the TCP header, must fit in the 65535 byte IP payload.

(ii) Each segment must fit in the **MTU (maximum transfer unit) :** Each network has a maximum transfer unit. Practically an MTU which is a few thousand bytes defines the upper limit on the segment size.

3. Fragmentation: If a segment is too large, then it should be taken into small segments. Using fragmentation by a router, each new segment gets a new IP header. Therefore, the fragmentation by router will increase the overhead.

4. Timer: The basic protocol used by TCP entities is the sliding window protocol. A sender starts a timer as soon as a sender transmits a segment. When the segment is received by the destination, it sends back acknowledgement alongwith data if any. The acknowledgement number is equal to the next sequence number it expects to receive. If the timer at the sender goes out before the acknowledgement reaches back, they will retransmit that segment again.

5. Possible Problems: As the segments can be fragmented, a part of the transmitted segment only may reach the destination with the remaining part lost. Segments can arrive out of order. Segments can get delayed so much that timer is out and unnecessary retransmission will take place. If a retransmitted segment takes a different route than the original segment is fragmented then the fragments of original and retransmitted segments can reach the destination in a sporadic way. Hence, a careful administration is required to achieve reliable byte stream. There is possibility of congestion or broken network along the path. TCP should be able to solve these problems in an efficient manner.

8.10.4. The TCP Segment Header*

(RGTU, Bhopal, Sem. Exam; June, 2014)

Figure 8.20 shows the layout of a TCP segment. Every segment begins with a 20 byte fixed format header. The fixed header may be followed by header options. After the options, if any, upto 65535 – 20 – 20 = 65495 data bytes may follow. Here, it may be noted that the first 20 bytes correspond to the IP header and the next 20 correspond to the TCP header. The TCP segment without data are used for sending the acknowledgements and control messages.

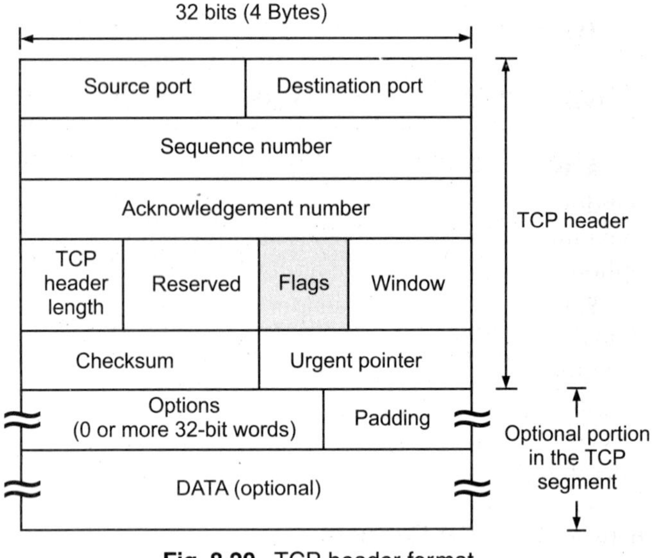

Fig. 8.20. TCP header format

1. Source Port: A 16-bit number identifying the application the TCP segment originated from within the sending host. The port numbers are divided into three ranges, well-known ports (0 through 1023), registered ports (1024 through 49151) and private ports (49152 through 65535). Port assignments are used by TCP as an interface to the application layer.

2. Destination port: A 16-bit number identifying the application the TCP segment is destined for on a receiving host. Destination ports use the same port number assignments as those set aside for source ports.

3. Sequence Number: A 32-bit number identifying the current position of the first data byte in the segment within the entire byte stream for the TCP connection. After reaching $2^{32} - 1$, this number will wrap around to 0.

4. Acknowledgement Number: A 32-bit number identifying the next data byte the sender expects from the receiver. Therefore, the number will be one greater than the most recently received data byte. This field is only used when the ACK control bit is turned on.

---

* Draw the diagram of TCP header and explain the use of following :
(i) Source and destination port addresses
(ii) Sequence and acknowledgement numbers.

5. Header Length or Offset: A 4-bit field that specifies the total TCP header length in 32-bit words (or in multiples of 4 bytes if you prefer). Without options, a TCP header is always 20 bytes in length. The largest a TCP header may be is 60 bytes. The field is required because the size of the options field(s) cannot be determined in advance. Note that this field is called *data offset* in the official TCP standard, but header length is more commonly used.

6. Reserved: A 6-bit field is currently unused and is reserved for future use.

7. Control Bits or Flags

(i) Urgent Pointer (URG) : If this bit field is set, the receiving TCP should interpret the urgent pointer field.

(ii) Acknowledgement (ACK) : If this bit field is set, the acknowledgement field described earlier is valid.

(iii) Push Function (PSH) : If this bit field is set, the receiver should deliver this segment to the receiving application as soon as possible.

(iv) Reset the Connection (RST) : If this bit is present, it signals the receiver that the sender is a right connection and all queued data and allocated buffers for the connection can be freely operated

(v) Synthesis (SYN) : When present, this bit field signifies that sender is attempting to achieve the sequence numbers.

(vi) No More Data from Sender (FIN) : If set, this bit field tells the receiver that the sender has received end of the stream for the current TCP connection.

8. Window: A 16-bit integer is used by TCP for flow control in the form of a data transmission window size. This number tells the sender how much data the receiver is willing to accept. The maximum value for this field would limit the window size to 65,535 bytes, however a *window-scale* option can be used to make use of even larger windows.

9. Checksum: A TCP sender computes a value based on the contents of the TCP header and data fields. This 16-bit value will be compared with the value the receiver generates using the same conaputation. If the values match, the receiver can be very confident that the segment arrived intact.

10. Urgent Pointer: In certain circumstances, it may be necessary for a TCP sender to notify the receiver of urgent data that should be processed by the receiving. Application as soon as possible. This 16-bit field tells the receiver when the last byte of urgent data in the segment ends.

11. Options: In order to provide additional functionality, several parameters may be used between a TCP sender and receiver. Depending on the option(s) used, the length of this field will vary in size, but it cannot be larger than 40 bytes due to the size of the header length field (4 bits). The most common option is the maximum-segment size (MSS) option. A TCP receiver tells the TCP sender the maximum segment size it is willing to accept through the use of this option. Other options are often used for various flow control and congestion control techniques.

12. Padding: Because options may vary in size, it may be necessary to pad the TCP header with zeros so that the segment ends on a 32-bit word boundary as defined by the standard.

13. Data: Although not used in some circumstances (*e.g.* acknowledgement segments with no data in the reverse direction), this variable length field carries the application data from TCP sender to receiver. This field coupled with the TCP header fields constitutes a TCP segment.

8.10.5. Checksum

A checksum is provided to ensure extreme reliability. It checksums the header, the data and the conceptual pseudo header shown in figure 8.21.

When the checksum is being computed, the TCP checksum field is set to zero, and the data field is padded out with an additional zero byte if its length is an odd-number. Then all the 16 bit words are added in 1's complement and then 1's complement of the sum is taken to get the checksum. When a receiver performs the calculation on the entire segment including the checksum field, the result has to be zero. The pseudo header contains the 32 bit IP address of the source and destination machines, the protocol number for TCP. *i.e.* 6 and the TCP segment length as shown in figure 8.21.

32 bits

Source address		
Destination address		
000 000 00	Protocol = 6	TCP segment length

Fig. 8.21. The psuedo header included in the TCP checksum

8.10.6. Protocol Operation

The TCP incorporates many of the features of the HDLC (High Level Data Link Control) protocol. It supports duplex message transfers and incorporates a go-back-N error control procedure with a type of sliding window flow control mechanism. All the PDUs that the TCP uses to set up a connection, transfer data, and clear a connection have a standard format and are known as segments. All segments are transferred between the TCP entities in the user data field of IP datagrams. As with a UDP datagram, the first two fields in the header are the source port and destination port addresses. They have the same meaning as in a UDP datagram except that with TCP the two addresses indicate the end points of the logical connection between the two application protocols. The sequence number has the same role as the send sequence number in the HDLC protocol, and the acknowledgement number the same role as the receive sequence number. The former thus relates to the flow of data in the direction of the sending TCP entity and the latter to the flow of the reverse direction. The presence of an option field in the segment header means that the header can (in theory) be of variable length. The header length indicates the number of 32-bit words in the header. The reserved field, as its name implies, is reserved for future use.

All segments (and hence PDU types) have the same header format and the validity of selected fields in the segment header is indicated by the setting of bits in the 6-bit codefield; if a bit is set (= 1), the corresponding field is valid. Note that multiple bits can be set in a single segment. The window field relates to the sliding window flow control scheme. It indicates the number of data octets relative to the octet number in the acknowledgement field that the source is prepared to accept. It is determined by the amount of buffer storage the source has available for this connection. As with a UDP datagram, the checksum relates to the complete segment, header plus contents. The checksum is the complement of the sum of all the 16-bit words in the segment added together using 1's complement arithmetic. When the URG flag is set in the code field, the urgent pointer indicates the amount of urgent (expedited) data in the segment. Normally, this is delivered by the receiving TCP entity immediately it is received. The default maximum number of octets in the data field of a segment is 536. This has been chosen on the assumption than WANs are present in the route since, in general, they have an inferior but error rate probability.

8.10.7. TCP Connection Establishment

To make the transport services reliable, TCP hosts must establish a connection-oriented session with one another. Connection establishment is performed by using a **three-way handshake** mechanism. A three-way handshake synchronizes both ends of a connection by allowing both sides to agree upon initial sequence numbers. This mechanism also guarantees that both sides are ready to transmit data and know that the other side is ready to transmit as well. This is necessary so that packets are not transmitted or retransmitted during session establishment or after session termination. Each host randomly chooses a sequence number used to track bytes within the stream it is sending and receiving. Then, the three-way handshake proceeds in the manner shown in figure 8.22.

The requesting end (HOST A) sends a SYN segment specifying the port number of the server that the client wants to get connected to, and the client's initial sequence number (x). The server (HOST B) responds with its own SYN segment containing the server's initial sequence number (y). The server also acknowledges the client's SYN by acknowledging the client's SYN plus one (x + 1). A SYN consumes one sequence number. The client must acknowledge this SYN from the server by acknowledging the server's SYN plus one. (SEQ = x + 1), ACK = y + 1). This is how a TCP connection is established.

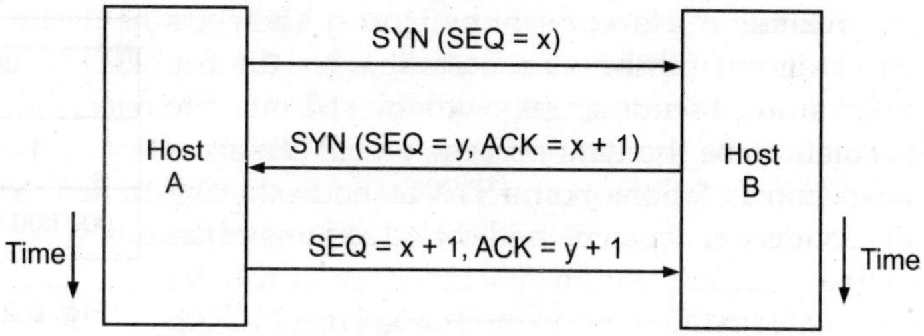

Fig. 8.22. TCP connection establishment (Three way)

8.10.8. Connection Termination Protocol (Connection Released)

While it takes three segments to establish a connection, it takes four to terminate a connection. Since a TCP connection is full-duplex (that is, data flows in each direction independently of the other direction), each direction must be shut down independently. The termination procedure of each host is shown in figure 8.23. The rule is that either end can send a FIN when it has finished sending data.

When a TCP receives a FIN, it must notify the application that the other end has terminated that direction of data flow. The sending of a FIN is normally the result of the application issuing a close. The receipt of a FIN only means there will be no more data flowing in that direction. A TCP can still send data after receiving a FIN. The end that first issues the close (*e.g.*, sends the first FIN) performs the active close and the other end (that receives this FIN) performs the passive close. When the server receives the FIN it sends back an ACK of the received sequence number plus one. A FIN consumes a sequence number, just like a SYN. At this point, the server's TCP also delivers an end-of-file to the application (the discard server). The server then closes its connection, causing its TCP to send a FIN, which the client TCP must and sends back an ACK by incrementing the received sequence number by one. Connections are normally initiated by the client, with the first SYN going from the client to the server. Either end can actively close the connection (*i.e.*, send the first FIN). Often, however, it is the client that determines when the connection should be terminated, since client processes are often driven by an interactive user, who enters something like quit to terminate. This is how the TCP connection is related.

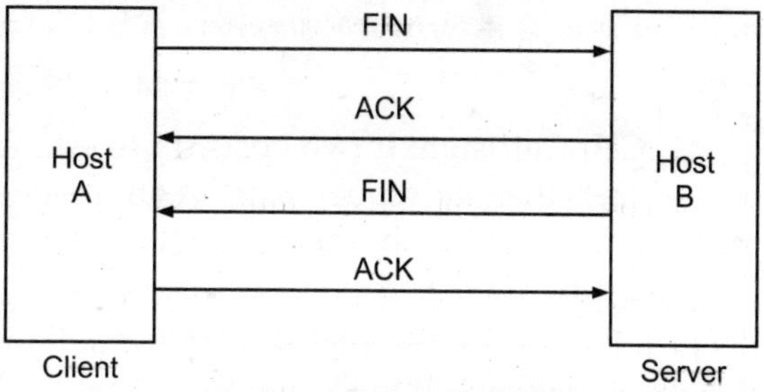

Fig. 8.23. TCP termination

8.10.9. TCP Connection Management

1. Basic concepts: Connections are established in TCP using the three way handshakly as discussed earlier. To establish a connection, one side, say the server, passive waits for an incoming connection by executing the LISTEN and ACCEPT primitives, either specifying a particular other side or nobody in particular. The other side executes a connect primitive, specifying the IP and port to which it wants to connect, the maximum TCP segment size, possible other options and optionally some user data (e.g., a password). The CONNECT primitive sends a TCP segment with the SYN bit on and the ACK bit off and waits for a response. The sequence of TCP segments sent in the normal case is shown in figure 8.24.

When the segment sent by Host – 1 reaches the destination i.e. host – 2, the receiving server checks to see if there is a process that has done a LISTEN on the port given in the destination port field. If not, it sends a reply with the RST bit on to reject the connection. Otherwise it gives the TCP segment to be listening process, which can accept or refuse (e.g., if it does not like the client) the connection. On acception, a SYN is send, otherwise a RST. Note that a SYN segment occupies 1 byte of sequence so it can be acknowledged unambiguously.

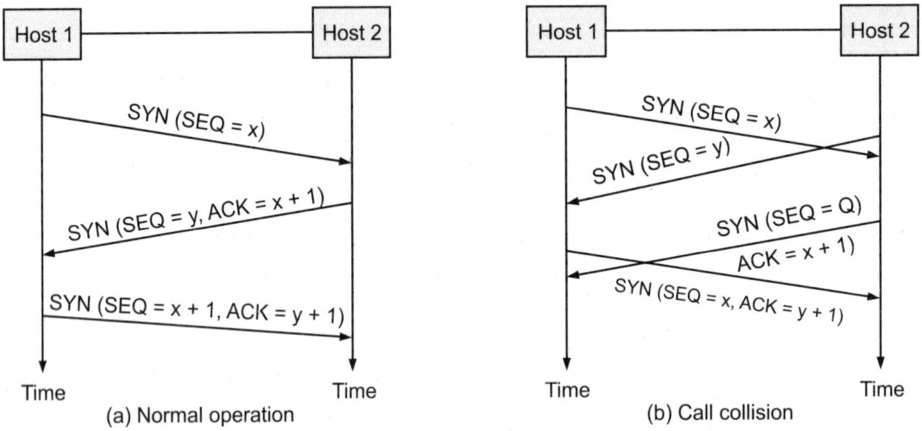

Fig. 8.24. TCP connection management

2. Call Collision: If two hosts try to establish a connection simultaneously between the same two sockets then the sequence of events is as shown in figure 8.24. Under such circumstances, only one connection is established. Both the connections can not be established because connection are identified by their end points. If the first set up results in a connection which is identified by (x, y) and second connection is also set up, then only one table entry will be made i.e. for (x, y). For the initial sequence number, a clock based scheme is used, with a clock pulse coming after every 4 μ sec. For ensuring an additional safety, when a host crashes, it may not reboot for 120 sec. which is maximum packet lifetime. This is to make sure that no packets from previous connections are roaming around.

3. TCP Connection Release: A TCP connection is actually a full duplex connection but to understand the connection release we can consider them to be a pair of simplex connections. We can then imagine that each simplex connection is being released independently. Releasing a TCP connection is symmetric. Either part can send a TCP segment with the FIN bit set, meaning it has no more data to send. When the FIN is acknowledged, that direction is shut down, but data may continue to flow indefinitely in the other direction. If a response to a FIN is not received within 2 maximum packet lifetimes, the sender of the FIN releases the connection. The receiver will eventually notice that it receives no more data and time-out as well. Normally four TCP segments are required to release a connection i.e. one FIN and one ACK in each direction.

However, the First ACK and Second FIN can be combined in the same segment.

4. Finite State Machine for Connection Establishment and Release: The steps involved in TCP connection establishment and release can be represented using a finite state machine. The 11 states in such a state machine are given in table 8.6.

Table 8.6. Different States in TCP Finite State Machine.

S.No.	State	Description
1.	CLOSED	No connection is active or pending.
2.	LISTEN	The server is waiting for an incoming call.
3.	SYN RECEIVED	A connection request has arrived; wait for ACK.
4.	SYN SENT	The application has started to open a connection.
5.	ESTABLISHED	The normal data transfer state.
6.	FIN WAIT 1	The application has said it is finished.
7.	FIN WAIT 2	The other side has agreed to release.
8.	TIMED WAIT	Wait for all packets to die off.
9.	CLOSING	Both sides have tried to close simultaneously.
10.	CLOSE WAIT	The other side has initiated a release.
11.	LAST ACK	Wait for ack of FIN of last close.

In each of the 11 states, shown in table 8.6, some specific events are legal. Corresponding to every legal event, some action may be taken, but if some other event happens, then error is reported. The finite state machine is shown in figure 8.25.

Each connection is always in the CLOSED state initially. It comes out of this state when it does either the passive open (LISTEN) or an active open (CONNECT). A connection is established, if the other side does the opposite and the state becomes ESTABLISHED. When both the sides initiate a connection release, the connection is terminated and the state returns to CLOSED state.

5. Various Types of Lines in the Finite State Machine: Various types of lines are used in the finite state machine of figure 8.25.

(i) Heavy solid lines : These lincs show a client actively connecting to a passive server.

(ii) Heavy dotted lines : The lines are used for the server.

(iii) The light faced lines : These are unusual event sequences.

Each line in figure 8.25 is marked by an event/action pair. The event can either be a user-initiated system call (CONNECT, LISTEN, SEND, or CLOSE), a segment arrival (SYN, FIN, ACK, or RST), or a time-out. For the TIMED WAIT state, the event can only by a time-cut of twice the maximum packet length. The action is the sending of a control segment (SYN, FIN or RST) or nothing. The time-outs to guard for lost packets (e.g. in the SYN SENT state) are not shown here. There are 11 states used in the TCP connection management finite state machine. Data can be sent in the ESTABLISHED and the CLOSE WATT states and rejected in the ESTABLISHED and FIN WAIT 1 states.

6. Explanation: To understand the finite state machine of figure 8.25, first follow the path of a client i.e., the heavy solid state. After that follow the path of the server (the heavy dashed line).

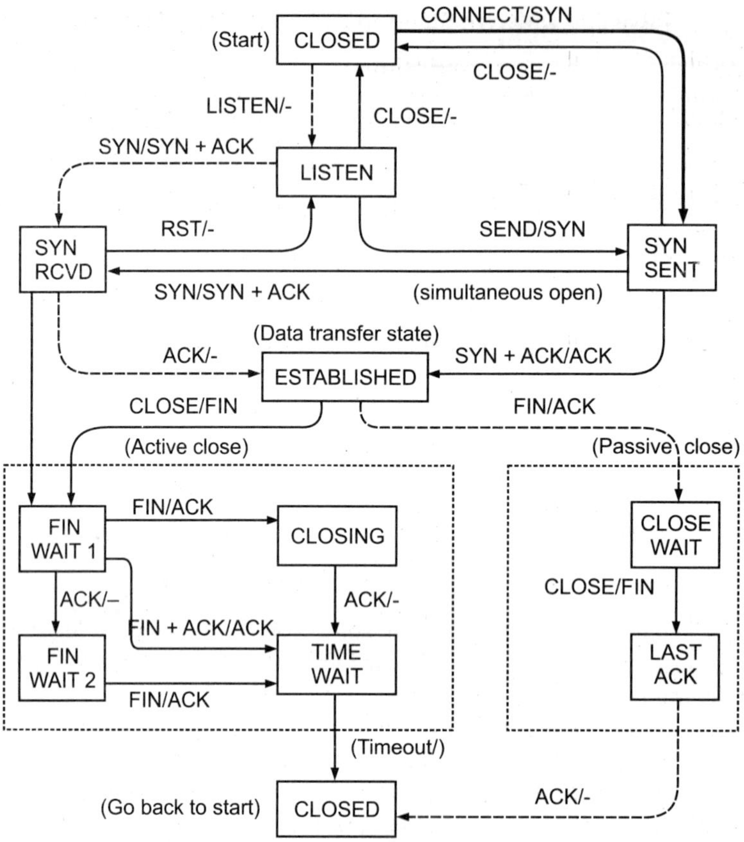

Fig. 8.25. TCP connection management finite state machine

8.11. CONNECTION MANAGEMENT FROM CLIENT'S POINT OF VIEW

An application on the client machine issues a CONNECT request, the head TCP entity marks it as SYN SENT state and sends a SYN segment. When the SYN + ACK arrives, the TCP sends the final ACK of the three way handshake and goes into the ESTABLISHED state. It is possible to now send and receive the data. When sending data is finished, an application executes a CLOSE primitive. This will cause the local TCP entity to send FIN segment and then wait for ACK (see dashed box marked active close). When the ACK is obtained, a transition is made to state FIN WAIT 2 to close the connection in one direction. When the other side also closes, a FIN comes in from that side which is acknowledged. Now, both sides are closed. But TCP waits for a time equal to the maximum packet lifetime to ensure that all the packets die out. When the timer goes out, TCP deletes the connection record.

8.11.1. Connection Management from Servers Point of View

The server does LISTEN and waits. When a SYN comes in, it is acknowledged and the server enters into SYN RCVD state. When the server SYN is acknowledged, the three way handshake is complete and the server goes to the ESTABLISHED state. Data can now be transferred. When the client does a CLOSED, a FIN is sent to the client. After receiving ACK from client, the server releases the connection and deletes the connection record.

8.12. TCP TRANSMISSION POLICY

Now, let us see how the window policy is used in transmission policy of TCP. Window management in TCP is normally decoupled from the acknowledgements. To understand the window management, let us consider figure 8.26.

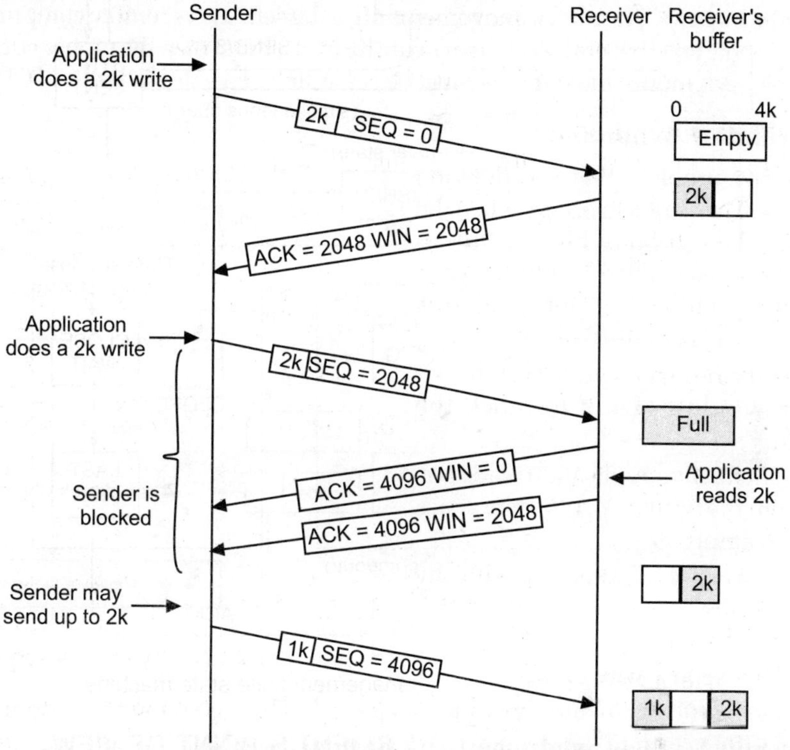

Fig. 8.26. Windows management in TCP

Explanation: Let the receiver in figure 8.26, has a 4 kbyte i.e. 4096 byte buffer space. The sender transmits a 2048 byte (2 kbyte) segment with a sequence number SEQ = 0. These bytes occupy half space of the receiver's buffer and the receiver will send back acknowledgement of this segment (ACK 2048, WIN = 2048). Here, WIN = 2048 is the window which advertises an empty buffer space of 2048. Now, the sender sends another 2k i.e. 2048 bytes segment (SEQ = 2048) which is acknowledged by the receiver ACK = 4096, WIN = 0) which shows that window = 0 because the receiver buffer space is 0. The sender must now be blocked until the application process on the receiver removes some data from the buffer. As soon as the application on the receiver side reads 2k bytes, the buffer becomes partially empty and an acknowledgement with a window of 2k (ACK = 4096%, WIN = 2048) is sent back to sender. The sender may send upto 2 kbytes. When the window = 0, the sender should not normally send any segment. But under two exceptional conditions, the sender will continue to send data.

(i) First, urgent data may be sent, e.g., to allow the user to kill the process running on the other machine.

(ii) Second, the sender may send a 1-byte segment to make the receiver reannounce the next byte expected and the window size. This is used to prevent deadlock if a window announcement ever gets lost.

Senders are not supposed to transmit data as it arrives from an application. The receivers also are not supposed to send acknowledgements as soon as possible. This is to reduce the usage of the system. A way to reduce the system usage, an algorithm called Nagle's algorithm is used. It can be stated as follows.

When data come into the sender one byte at a time (e.g., on a Talnet connection), just the first byte is send and the rest is buffered until the outstanding byte is acknowledged. Nagle's algorithm is widely used but sometime it is better not to use it. For example, let us consider an X-windows application being run over the Internet. The mouse movements are to be sent to the remote computer. If we gather the mouse movements together and send them in the form of a burst then the mouse cursor movements will be erratic. So, each mouse movements must be separately. This degrades the TCP performance.

8.12.1. Silly Window Syndrome

This is another problem that can degrade TCP performance. This problem occurs when the sender transmit data in large blocks, but an interactive application on the receiver side reads data 1 byte at a time. To understand this problem, let us consider figure 8.27.

(i) Initially the receiver's buffer is full so it send a window size 0 to block the sender.

(ii) But the interactive application reads one byte from the buffer. So, one byte space becomes empty.

(iii) The receiving TCP sends a window update to the sender informing that it can send 1 byte.

(iv) The sender send 1-new byte.

(v) The buffer is full again and the window size is 0. This behaviour can continue forever.

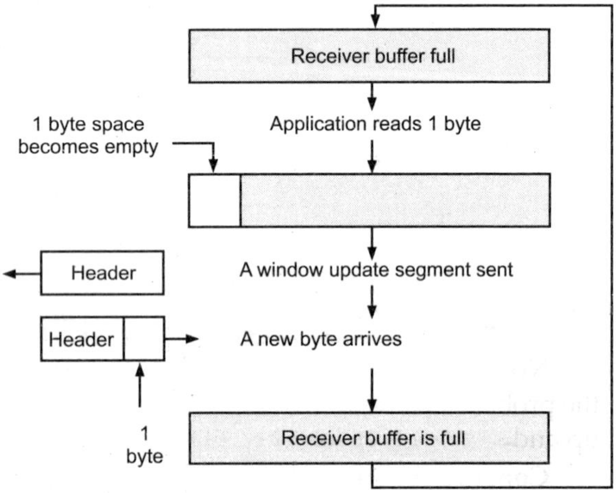

Fig. 8.27. Silly window syndrome

Solution to Silly Window Syndrome: Clark suggested a solution to silly window syndrome as under:

He suggested that the receiver should not send a window update for 1 byte. Instead, it must wait until it has a substantial amount of buffer space available and then send the window update. To be specific, the receiver should not send a window update until it can handle the maximum window size it has advertised at the time of establishing a connection or its buffer is half empty, whichever is smaller. The sender can also help to improve the situation. It should not send tiny segment. Instead, it must wait and send a full segment or at least one containing half of the receivers buffers size.

8.13. CONGESTION CONTROL: TCP CONGESTION CONTROL

We have already discussed the reasons of congestion in networks and the Internet is no exception. The network layer also tries to manage congestion but the TCP takes maximum responsibility in congestion control. This is because, the real solution to congestion control is to slow down the data rate.

Principle of Congestion Control: The basic principle is do not inject a new packet into the network until an old one is delivered. TCP tries to do this by dynamically manipulating the window size.

Step 1 : Detect the Congestion: This is the first step in congestion control. Now-a-days, packet loss due to transmission errors is very rare due to the use of optical fiber links. So, most transmission time-outs (loss of packets) are due to congestions. So, all the Internet TCP algorithms assume that time-outs are caused by congestion.

Step 2 : Try to Prevent Congestion: After establishing a connection, a suitable window size is to be chosen. The receiver window size is based on its buffer capacity. If the sender adjusts its transmission rate to suit for this capacity as shown in figure 8.28, the congestion due to buffer overflow will never take place.

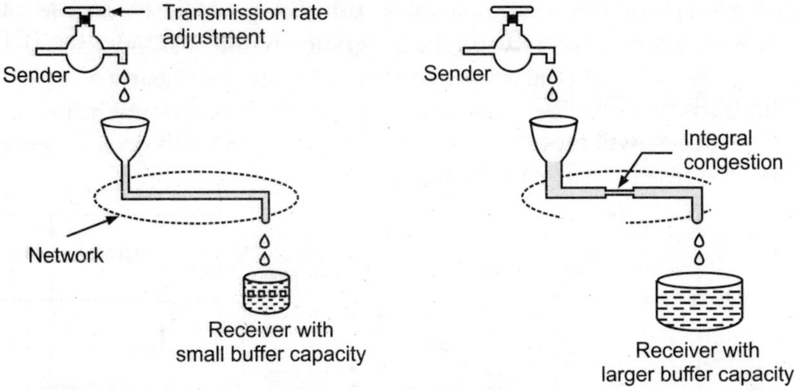

(a) No congestion (b) Internal

Fig. 8.28. Congestion

Now, let us consider figure 8.28. The sender is slow, the receiver has a large buffer capacity but the problem is internal carrying capacity of the network. If the sender is too fast, the water will back up and some will be lost (loss of packets) and congestion will take place.

Conclusion: To prevent congestion, TCP has to deal with two problems separately – receiver capacity and network capacity.

Solution: To deal with the two problems mentioned earlier each sender maintains two windows; the window the receiver has granted (indicating the receiver capacity) and the **congestion window** (indicating the network capacity). Each window reflects the number of bytes the sender may transmit. The number of bytes that may be sent by the sender is the minimum of the two windows. So the effective window is the minimum of what the sender and the receiver both think is all right.

Slow Start Algorithm: After establishing a connection, the sender initialises the congestion window to the size of maximum segment in use on the connection. It then sends one maximum segment. If this segment is acknowledged, then it adds bytes corresponding to one full segment to the congestion window. So, now the window is two maximum size segments. The sender then sends two segments. As each of these segments is acknowledged, the size of congestion window is increased by maximum segment size. This is shown in figure 8.29. When the congestion window is n segments, if all n are acknowledged before time-out takes place, the congestion window is increased by the byte count corresponding to n segments. But, there is a limit on the exponentially growing congestion window. The congestion window stops growing as soon as either the time-out occurs or the receiver's window size is reached. If the congestion window can grow to 1024 (1 kbyte) byte, 2048 byte, but a burst of 4096 bytes gives a time-out when we have to set the congestion window at 2048 in order to avoid congestion. Once this is done, no data bursts longer than 2048 bytes will be sent by the sender even if receiver grants a wider window. The name of this algorithm is slow algorithm and it is required to be supported by all the TCP implementations.

Internet Congestion Control Algorithm: Till now only two parameters have been used namely receiver window and congestion window. But in the algorithm we are going to discuss, a third parameter called threshold is used. Initially the threshold is set to 64 kbyte. When the time-out occurs, the threshold is set to half of the current congestion window and the congestion window is reset to one maximum segment. The slow start algorithm is then used to find what the network can handle. But the exponential growth of the congestion window is stopped as soon as it reaches the threshold. After this point, the congestion window grows linearly (and not exponentially) by one maximum segment for each burst instead of one per segment. This is illustrated in figure 8.29.

The maximum segment size here is 1024 *i.e.* 1 kbyte. Initial value of congestion windows was 64 k, but timeout occurs. Hence, threshold is set to 32 k and congestion window to 1 k at 0. Then the congestion window grows exponentially till the congestion window size reaches the threshold of 32 k. The threshold occurs at 32 k so the congestion windows grows linearly after 32 k. The time-out occurs as the 13th transmission. The threshold is set to half the current window and slow start is initiated again. The process will repeat thereafter. If no more time-outs occur, the size of congestion window continues to grow upto the size of the receiver window.

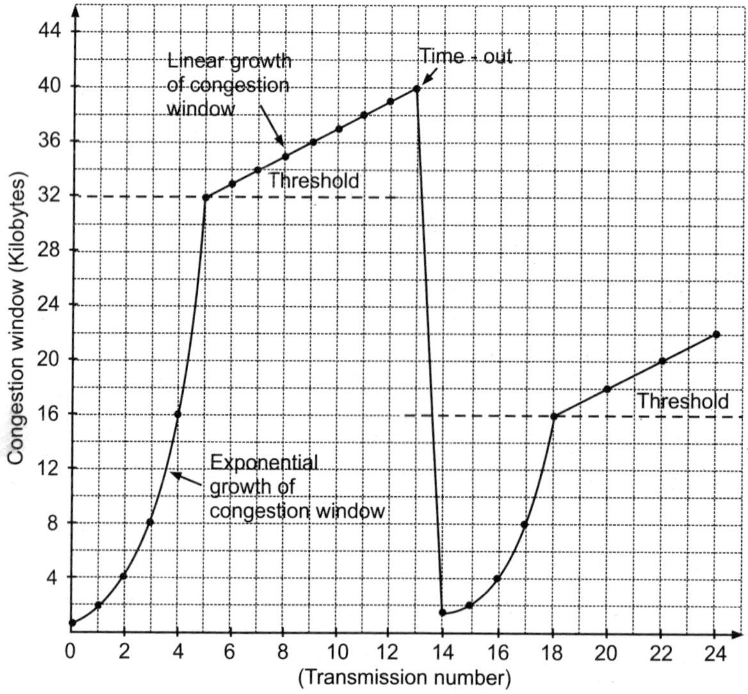

Fig. 8.29. Internet congestion control algorithm

8.14. TCP TIMER MANAGEMENT

The TCP, at least conceptually uses more than one timer. But the most important of them is the retransmission timer. This timer is started as soon as a segment is sent. The timer is stopped if the segment acknowledgement is received, before the timer expires. But if the timer expires before the arrival of an "ack" then the segment is retransmitted and the timer is started again.

What should be the time-out interval ?: The most important question about the retransmission timer is that how long should the time-out interval be ? The answer to this question is difficult in the transport layer as compared to that in the data link protocol. Figure 8.30 shows the probability density function for the time taken by data link and TCP segment acknowledgements. Determining the round trip time to destination is not simple and even if we know it, deciding the value of time-out is difficult.

Let us consider figure 8.30 (b). If the value of time-out is too small (T_1 for example) then unnecessary retransmission will take place. If time-out is too long say T_2, then the performance will degrade due to long retransmission delays whenever a packet is lost. The solution to this problem is to use a highly dynamic algorithm which adjusts the time-out interval constantly. This adjustment is based on continuous measurement of network performance.

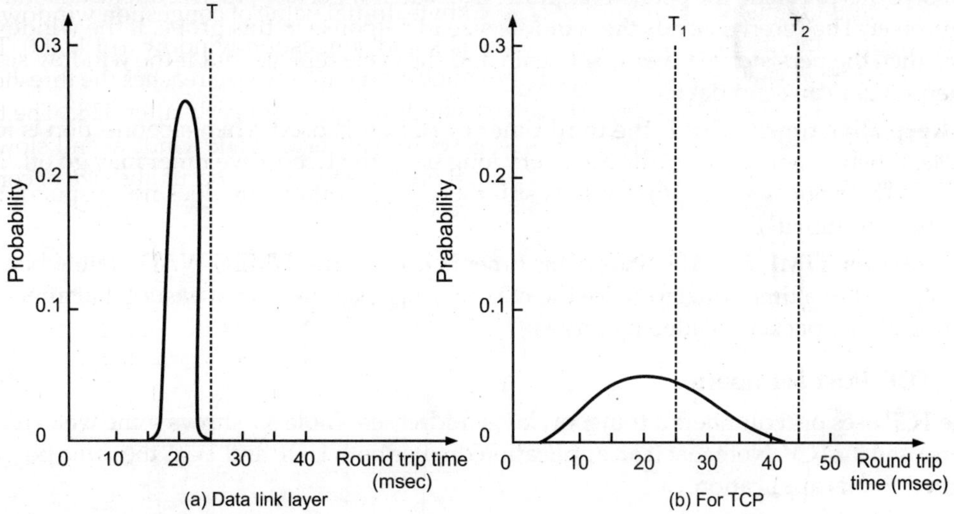

Fig. 8.30. Probability density of acknowledgement arrival times

8.14.1. Jacobson's Algorithm

This is the algorithm, generally used by the TCP. The RTT (round trip time) for each connection of TCP is variable. When a segment is sent, timer is started. This is to measure the time required to receive ACK and to trigger retransmission if ACK takes too long to come. If the acknowledgement returns back before timer goes out, then TCP measures the time taken by the ACK (say M) and adjusts RTT to a new value using the following equation:

$$RTT = \alpha\, RTT + (1 - \alpha)\, M \qquad \text{...(8.1)}$$

Here, α is called as smoothing factor.

$$\text{Typically,} \quad \alpha = \frac{7}{8}$$

Even if a good value of RTT is given, it is not easy to choose the time-out. Jacobson proposed a new smoothing factor D (deviation) which is given by

$$D = \alpha\, D + (1 - \alpha)\, RTT - M$$

And the time-out is calculated using the following equation.

$$\text{Time-out} = RTT + 4D$$

Problem: One problem with dynamic estimation of RTT is what should we do to RTT when a segment times out and has to be retransmitted ? When ACKs come, it is confusing to find whether the ACK correspond to the original transmission or retransmission.

Solution: The solution to this problem is Karn's algorithm, which is accepted by most TCP implementations. With karn's algorithm, RTT and D are not updated for retransmitted segments. Instead, the time-out is doubled on each failure until the segments get through the first time.

8.14.2. Other Timers in TCP

1. Persistence timer : The second timer in TCP is called **persistence timer.** It is designed to solve the following problem :

(i) The receiver sends an ACK with window size = 0. So the sender will wait.
(ii) After the receiver buffer becomes partially empty it sends a window update to the sender.
(iii) But the packet containing this window update is lost on its way to sender.
(iv) So both sender and receiver will be waiting for ever.

To solve this problem, the persistence timer is usual. If it goes off, then sender transmits a probe to the receiver. The receiver sends the window size in response to this probe. If the window size is still zero then the persistence timer is set again and the cycle repeats. But if the window size is non zero then sender can send data.

2. Keep alive timer : This is the third timer in TCP. It is used when a connection is idle for a long time. When a connection is idle for a very long time, the Keep alive timer may go off. This will cause one side to check if the other side is still there. If the other side does not respond, then the connection is terminated.

3. Timer for TIMED WAIT State: This timer is used in the TIMED WAIT state while closing. This timer is set to a time equal to twice the maximum packet lifetime to ensure that after closing a connection all the packets created by it die off.

8.14.3. TCP Port Numbers

The TCP uses port numbers a transport layer addresses. Table 8.7 shows some well known port numbers used by TCP. Note that if an application can use both UDP and TCP, the same port number is assigned to this application.

Table 8.7. Well Known Ports Used by TCP

Port	Protocol	Description
7	Echo	Echoes a received datagram back to sender
9	Discard	Discards any received datagram
11	Users	Active users
13	Daytime	Returns the data and the time
17	Quote	Returns a quote of the day
19	Chargen	Returns a string character
20	FTP, Data	File Transfer protocol (data connection)
21	FTP, Control	File Transfer protocol (control connection)
23	TELENT	Terminal network
25	SMTP	Simple Mail Transfer Protocol (SMTP)
53	DNS	Domain Name Server
67	BOOTP	Bootstrap Protocol
79	Finger	Finger
80	HTTP	Hypertext Transfer Protocol
111	RPC	Remote Procedure Call

8.14.4. TCP Services

Following are some of the services offered by TCP to the processes at the application layer.

1. Stream delivery service
2. Sending and receiving buffers.

3. Bytes and segments.
4. Full duplex service
5. Connection oriented service
6. Reliable service.

Let us discuss them one by one.

1. Stream Delivery Service

TCP is a stream oriented protocol. It allows the sending process to deliver data as a stream of bytes and the receiving process to obtain data as a stream of bytes. TCP creates a working environment in such a way that the sending and receiving processes seem to be connected by an imaginary *tube* as shown in figure 8.31. This is called as **stream delivery service**.

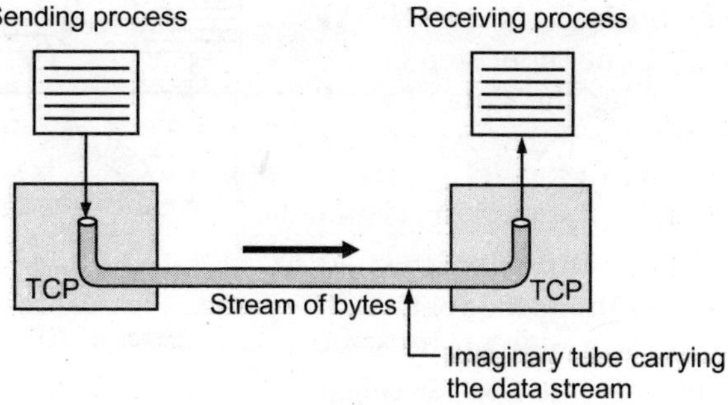

Fig. 8.31. Stream delivery service

2. Sending and Receiving Buffers: The sending and receiving processes may not produce and receive data at the same speed. Hence, TCP needs buffers for storage. There are two types of buffers used as each direction.

(i) Sending buffer
(ii) Receiving buffer.

A buffer can be implemented by using a circular array of 1 byte locations as shown in figure 8.32. Figure 8.32 shows the movement of data in one direction on the sending side the buffer has following three types of locations:

(i) Empty locations.
(ii) Locations containing the bytes which have been sent but not acknowledged. These bytes are kept in the buffer till an acknowledgement is received.
(iii) The locations containing the bytes to be sent by the sending TCP.

In practice, the TCP may be able to send only a part of data which is to be sent, due to slowness of the receiving process or congestion in the network. The buffer at the receiver is divided into following two parts :

(i) The part containing empty locations.
(ii) The part containing the received bytes which can be consumed by the sending process.

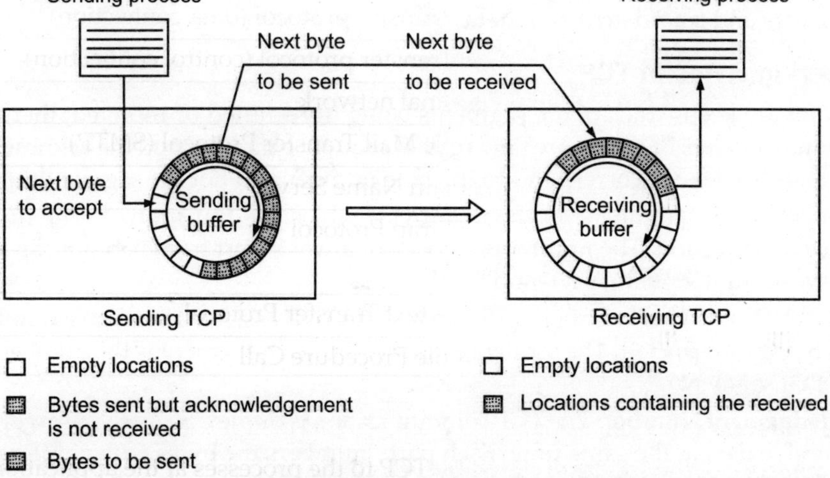

Fig. 8.32. Sending and receiving buffers

3. Bytes and Segments: Buffering is used to handle the difference between the speed of data transmission and data consumption. But only buffering is not enough. We need one more step before sending the data. The IP layer, as a service provider for TCP, needs to send data in the form of packets and not as a stream of bytes. At the transport layer, TCP groups a number of bytes together into a packet called a segment. A header is added to each segment for the purpose of exercising control. The segments are encapsulated in an IP datagram and then transmitted. The entire operation is transparent to the receiving process. The segments may be received out of order, lost or corrupted when it reaches the receiving end. Figure 8.33 shows how segments are created from the bytes in the buffers.

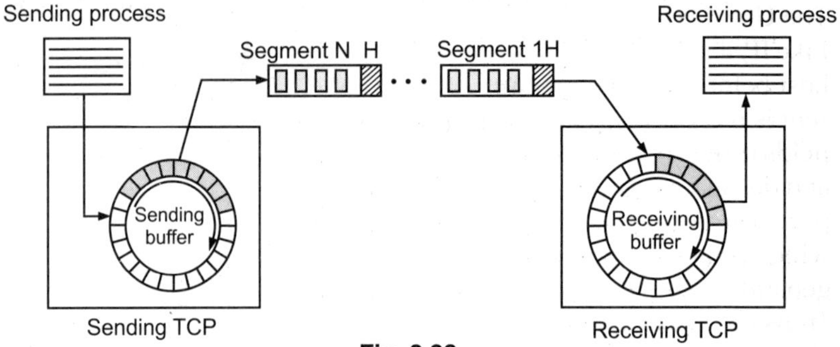

Fig. 8.33.

The segments are not of the same size. Each segment can carry hundred of bytes.

4. Full duplex service: TCP offers full duplex service where the data can flow in both the directions simultaneously. Each TCP will then have a sending buffer and receiving buffer. The TCP segments are sent in both the directions.

5. Connection oriented service: TCP is a connection oriented protocol. When process – 1 wants to communicate (send and receive) with another process (process – 2), the sequence of operations is as under :

(i) TCP of process – 1 informs TCP of process – 2 and gets its approval.

(ii) TCP of process – 1 and TCP of process – 2 exchange data in both the directions.

(iii) After completing the data exchange, when buffers on both sides are empty, the two TCPs destroy their buffers.

The type of connection in TCP is not physical, it is virtual. The TCP segment is encapsulated in an IP datagram and can be sent out of order. These segments can get lost or corrupted and may have to be resent. Each segment may take a different path to reach the destination.

6. Reliable service: TCP is a reliable transport protocol. It uses an acknowledgement mechanism for checking the safe and sound arrival of data.

8.14.5. Numbering Bytes in TCP

The TCP software keeps track of the segments being transmitted or received, there is no field for a segment number value. But there are fields called **sequence number** and the **acknowledgement number**. Note that these fields correspond to the byte number and not the segment number.

1. Byte numbers: TCP numbers all the data bytes which are transmitted. The numbering is independent in each direction. The numbering does not always start from 0, but it can start with a randomly generated number between 0 and $2^{32}–1$.

2. Sequence number: After numbering the bytes, the TCP assigns a sequence number to each segment that is being transmitted. The sequence number for each segment is the number of the first byte carried in that segment.

3. Acknowledgement Number: The TCP communication is duplex. So, both the involved parties can send and receive data at the same time. Each party numbers the bytes with a different starting byte number. The acknowledgement number is cumulative *i.e.* the receiver takes the number of the last byte received, adds 1 to it and announces this sum as the acknowledgement number.

8.15. USER DATAGRAM PROTOCOL (UDP) *(RGTU, Bhopal, Sem. Exam; June, 2014)*

The User Datagram Protocol is a very simple protocol. It adds little to the basic functionality of IP. Like IP, it is an unreliable, connectionless protocol. You do not need to establish a connection with a host before exchanging data with it using UDP, and there is no mechanism for ensuring that data sent is received. A unit of data sent using UDP is called a Datagram. UDP adds four 16-bit header fields (8 bytes) to whatever data is sent. These fields are : a length field, a checksum field, and source and destination port numbers. *Port number*, in this context, represents a software port, not a hardware port. The concept of port numbers is common to both UDP and TCP. The port numbers identify which protocol module sent (or is to receive) the data. Most protocols have standard ports that are generally used for this. For example, the Telnet protocol generally uses port 23. The Simple Mail Transfer Protocol (SMTP) uses port 25. The use of standard port numbers makes it possible for clients to communicate with a server without first having to establish which port to use. The port number and the protocol field in the IP header duplicate each other to some extent through the protocol field is not available to the higher level protocols. IP uses the protocol field to determine whether data should be passed to the UDP or TCP module.

UDP or TCP use the port number to determine which application protocol should receive the data. Although UDP is not reliable, it is still an appropriate choice for many applications. It is used in real-time applications like. Net audio and video where, if data is lost, it is better to do without it than send it again out of sequence. It is also used by protocols like the Simple Network Management Protocol (SNMP).

8.15.1. Purposes of UDP

User Datagram Protocol (UDP) provides a connectionless packet service that offers unreliable *best effort* delivery. This means that the arrival of packers is not guaranteed, nor is the correct sequencing of delivered packets. Applications that do not require an acknowledgement of receipt of data, for example, audio or video broadcasting uses UDP. UDP is also used by applications that typically transmit small amounts of data at one time, for example, the Simple Network Management Protocol (SNMP). UDP provides a mechanism that application programs use to send data to other application programs. UDP provides protocol port numbers used to distinguish between multiple programs executing on a single device. That is, in addition to the data sent, each UDP message contains both a destination port number and a source port number. This makes it possible for the UDP software at the destination to deliver the message to the correct application program, and for the application program to send a reply.

Source port	Destination port
Length	UDP checksum
DATA	

Fig. 8.34. UDP Datagram

The UDP header is divided into the following four 16-bit fields :

1. Source port: Source port is an optional field when meaningful, it indicates the port of the sending process, and may be assumed to be port to which a reply should be addressed in the absence of any other information. If not used, a value of zero is inserted.

2. Destination port: Destination port has a meaning within the context of a particular Internet destination address.

3. Length: This is the size in bytes of the UDP packet, including the header and data. The minimum length is 8 bytes, the length of the header alone.

4. UDP Checksum: This is used to verity the integrity of the UDP header. The checksum is performed on a pseudo *header* consisting of information obtained from the IP header (source and destination address) as well as the UDP header.

8.15.2. UDP Pseudo Header

The purpose of using a pseudo-header is to verify that the UDP packet has reached its correct destination. The correct destination consists of a specific machine and a specific protocol port number within that machine. The UDP header itself specifies only the protocol port number. Thus, to verify the destination, UDP on the sending machine computes a checksum that covers the destination IP address as well as the UDP packet. At the ultimate destination, UDP software verifies the checksum using the destination IP address obtained from the header of the IP packet that carried the UDP message. If the checksum agrees, then it must be true that the packet has reached the intended destination host as well as the correct protocol port within that host.

0 8 15 31

Source Address		
Destination Address		
Zero	Protocol	UDP Length

Fig. 8.35. UDP pseudo header

1. User interface: A user interface should allow the creation of new receive ports, receive operations on the receive ports that return the data octets and an indication of source port and source address, and an operation that allows a datagram to be sent, specifying the data, source and destination ports and addresses to be sent.

2. IP Interface: The UDP module must be able to determine the source and destination Internet addresses and the protocol field from the Internet header. One possible UDP/IP interface would return the whole Internet datagram including the entire Internet header in response to a receive operation. Such as interface would also allow the UDP to pass a full Internet datagram complete with header to the IP to send. The IP would verify certain fields for consistency and compute the Internet header checksum.

3. Protocol Application: The major uses of this protocol are the Internet Name Server, and the Trivial File Transfer.

4. Protocol Number: This is protocol 17 (21 octal) when used in the Internet Protocol.

8.15.3. Performance Comparison of UDP and TCP

S.No.	Characteristic/ Description	UDP	TCP
1.	General Description	Simple, high-speed low-functionality *wrapper* that interfaces applications to the network layer and does little else.	Full-featured protocol that allows applications to send data reliably without worrying about network layer issues.
2.	Protocol Connection Setup	Connectionless data is sent without setup.	Connection-oriented; connection must be established prior to transmission.
3.	Data Interface to Application	Message-based data is sent in discrete packages by the application.	Stream-based data is sent by the application with no particular structure.
4.	Reliability and Acknowledgements	Unreliable, best-effort delivery without acknowledgements.	Reliable delivery of messages; all data is acknowledged.
5.	Retransmissions	Not performed. Application must detect lost data and retransmit if needed.	Delivery of all data is managed, and lost data is retransmitted automatically.
6.	Features Provided to Manage Flow of Data	None.	Flow control using sliding windows; window size adjustment heuristics ; congestion avoidance algorithms.

7.	**Overhead**	Very low.	Low, but higher than UDP.
8.	**Transmission Speed**	Very high.	High, but not as high as UDP.
9.	**Data Quantity Suitability**	Small to moderate amounts of data (up to a few hundred bytes).	Small to very large amounts of data (up to gigabytes).
10.	**Types of Applications that Use the Protocol**	Applications where data delivery speed matters more than completeness, where small amounts of data are sent ; or where multicast/broadcast are used.	Most protocols and applications sending data that must be received reliably, including must file and message transfer protocols.
11.	**Well-known Applications and Protocols**	Multimedia applications, DNS BOOTP, DHCP, TFTP, SNMP, RIP, NFS (early versions).	FTP, Telnet, SMTP, DNS, HTTP, POP, NNTP, IMAP, BGP, IRC, NFS (later versions).

8.16. WIRELESS TCP AND UDP

In theory, transport protocols should be independent of the technology of the underlying network layer. In practice, most TCP implementations have been carefully optimised based on assumptions that are true for wired networks but not for wireless networks. Of course, they work correctly for wireless network, but the performance is low.

1. Problem: The main problem in application of TCP on wireless networks is the congestion control algorithm. As stated earlier, almost all the TCP implementations assume that time-outs are due to congestion and not by lost packets. So, when a timer times-out, the TCP assumes that congestion his take place and shows down the sender. This is to reduce the load on the network and avoid congestion. But the wireless links are not very reliable and they keep loosing packets all the time. So, if a packet is lost, then the sender should retransmit it as early as possible instead of slowing down.

2. Conclusion: The conclusion is when a packet is lost on the wired network, the sender should slow down. But when a packet is lost on a wireless network, the sender should try harder. The problem arises when the sender does not know the type of network (wired or wireless).

3. Practical difficulty: In practice, the path from sender to receiver is not homogeneous. A part of its can be over a wired network and the remaining may be wireless. Under such circumstances, the decision about time-out becomes more difficult.

8.16.1. Solution (Indirect TCP)

The solution suggested for this problem is to use indirect TCP in which the TCP connection is split into two separate connection as shown in figure 8.36.

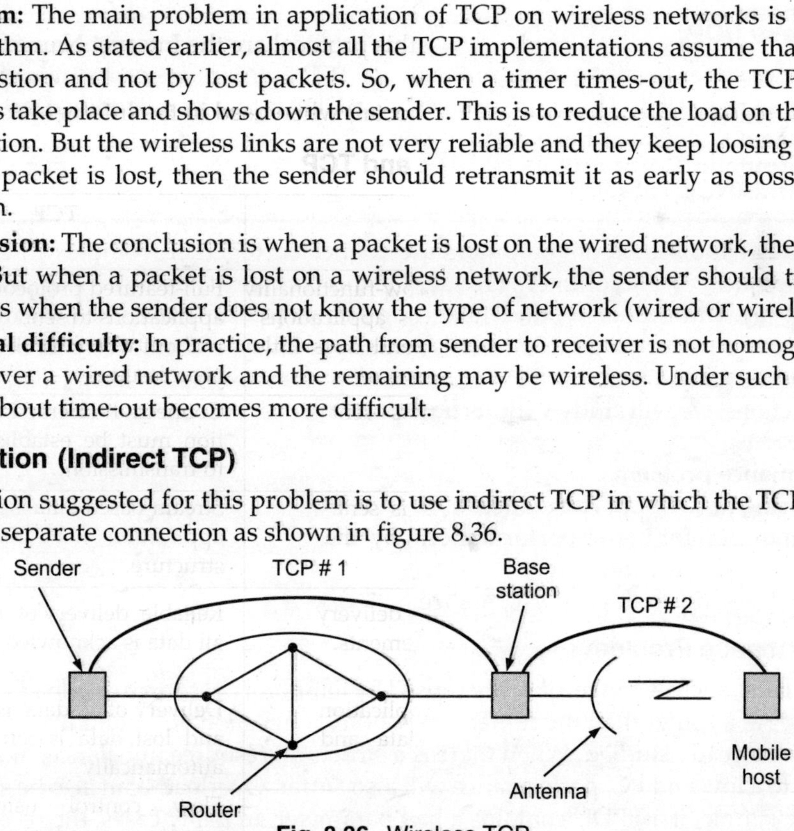

Fig. 8.36. Wireless TCP

The first connection is between sender and base station and the second one is between base station and receiver which is a mobile host. The base station only copies the packets between the connections in both the directions. The first connection is on the wires whereas the second one is wireless. The first connection is on the wires whereas the second one is wireless. The problem of non-homogeneous connections has been solved. The first connection can respond to the time-outs by slowing down the sender whereas the second one will speed up the sender in the event of time-outs.

8.16.2. Alternative Solution

In an alternative solution, the TCP is not broken instead modifications are made to network layer code in the base station. One of the important change is the addition of a **snooping agent.** It observe and caches TCP segments going out to the mobile host and the ACKs being received from it. The snooping agent has a timer, which is set every time a segment goes out to the mobile host. If no ACK is received before the expiry of the timer, then the snooping agent retransmits that segment without informing the source. A disadvantage of this scheme is that if the wireless link is too lossy, then the source may time-out without waiting for ACK and will invoke the congestion control algorithm. The problem will not be there with indirect TCP because there the congestion control will be exercised if and only if there is congestion in the wired part of the network. What if segments originating from the mobile host are lost ? The solution is when the base station notices a gap in the inbound sequence numbers, it sends a request for selective repeat. Thus the alternative solution proposed by Balakrishnan make the wireless link more reliable in both the direction without the source even knowing about it and without changing the semantics of TCP.

8.16.3. Wireless UDP

UDP does not suffer from the same problems of TCP but wireless communication introduces some other problems in UDP. The main problem with UDP is that its reliability decreases to a great extent as compared to that over the wired networks.

8.17. PERFORMANCE ISSUES

When hundreds or thousands of computers are connected together, complex interactions, with unforeseen consequences, are common. Frequently, this complexity leads to poor performance and no one knows why. Unfortunately, understanding network performance is more of an art than a science. There is little underlying theory that is actually of any use in practice. The best we can do is give rules of thumb gained from hard experience and present examples taken from the real world. In following sections, we will study various transport layer aspects of network performance. These are as follows :

1. Performance problems.
2. Measuring network performance.
3. System design for better performance.
4. Fast TPDU processing.
5. Protocols for future high-performance networks.

8.17.1. Performance Problems

Some problems, such as congestion, are caused by temporary resource overloads. If more traffic suddenly arrives at a router than the router can handle, congestion will build up and performance will suffer, as we already studied. When there is a structural resource imbalance, like a high speed line connected to a low end PC, performance will also suffer. Overloads can also be synchronously triggered. For example, if a TPDU contains a bad parameter, in many cases, the receiver will send back an error notification.

We consider a bad TPDU broadcasted to 10000 machines, the resulting **broadcast storm** of error messages could cripple the network. UDP suffered from this problem until the protocol was changed to not sending error messages for bad UDP TPDU's sent to broadcast addresses. Another example is what happens after an electrical power failure. When power comes back on, all the machines start rebooting, which might require going to a (RARP) server to learn ones' true identity and then to some file server to get a copy of the operating system. If hundreds of systems all do this at once, the server might collapse under the load.

Another tuning issue is setting time-outs correctly. For better quantity, we have to keep in mind the **bandwidth-delay product.** It is the capacity in bits of the pipe from the sender to the receiver and back. To achieve good performance, the receiver's window should be at least as large as this product. For a transatlantic gigabit line, with 40 msec round trip time, this window size is 40 megabit or 5 Mbytes for each connection. For time-critical applications, like audio and video, the jitter in transmission time is important as discussed earlier.

8.17.2. Measuring the Network Performance

For improving the network performance, first the network performance must be measured. The basic loop to improve network performance is :

1. Measure the relevant network parameters and performance.
2. Try to understand what is going on and where the bottleneck is.
3. Change one parameter

The most basic kind of measurement is to start a timer at the beginning of some activity to see how long it takes, e.g. round trip time. Other measurements are made with counters to record how often some event has happened, e.g. number of lost TPDU's. Finally, one is often interested in knowing the amount of something, e.g. the number of bytes processed in a given time interval. For measuring the network performance, the following factors should be carefully observed:

(i) Make sure the sample size is large enough (average a million RTT's).

(ii) Make sure that the samples are representative (no congestions at lunch time).

(iii) Be sure nothing unexpected is going on during the tests (no automatic backup program).

(iv) Caching can cause havoc (FTP of files still in the disk cache).

(v) Know what you are measuring (file transfer times depend on: network, operating systems used on both sides, disk or cache speed, drivers, etc.).

(vi) Be careful about extrapolating results (response time as function of the load).

8.17.3. System Design for Better Performance

Measuring and tuning can improve network performance considerably, but they cannot substitute for good design. Some thumb rules based on experience with many networks, can be given. They relate to system design, not just network design, since the software and operating system are often more important than the routers and interface boards.

1. CPU Speed is More Important than Network Speed: In nearly all the networks, operating system and protocol overhead dominates actual time on the wire. The biggest problem in running at 1 Gbps is getting the bits from the user's buffer out on the fiber fast enough and having the receiving CPU process them as fast as they come in.

2. Reduce Packet Count to Reduce Software Overhead

(i) Processing a TPDU has a certain amount of processing per byte (e.g. checksum) and per TPDU (e.g. header processing), which can be reduced by using larger TPDU size.

(ii) In addition to the TPDU overhead, there is also overhead in the lower layers. Each arriving packet causes an interrupt which is costly in modern RISC processors: breaking the CPU pipeline, interfering with the cache, changing the memory management context, forcing CPU register saving and restoring.

3. Minimize Context Switches: Context switches have the same bad properties as interrupts, the worst being a long series of initial misses.

4. Minimize Copying: It is not unusual for an incoming packet to be copied 3 to 4 times before the TPDU enclosed in it is delivered. Thus, pass *pointers* or *references* instead of data. Try to combine copying with checksum calculation, that can give a speed gain.

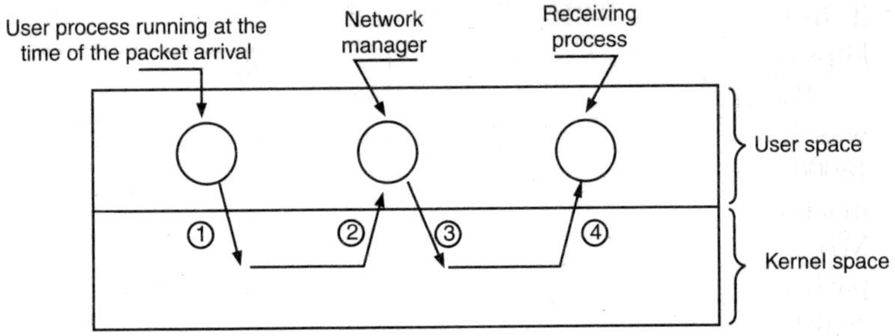

Fig. 8.37. System design for better performance

5. You can Buy More Bandwidth but not Lower Delay: Putting a second fiber next to the first one doubles the bandwidth but does nothing to reduce the delay. Making that shorter requires improving the protocol software, the operating software or the network interface, but the speed of light in the fiber remains a limiting factor.

6. Avoiding congestion is better than recovering from it.
7. Avoid time-outs.

8.17.4. Fast TPDU Processing

The key to fast TPDU processing is to separate out the normal case (data transfer in the ESTABLISHED state, no PSH or URG, enough window space) and handle it separately. On the receiver side, the connection record for an incoming TPDU has to be looked up. For TCP these can be stored in a hash table for which some simple function of the 2 IP and port numbers is a key. Trying first the last connection, gives a hit rate of more than 90%. Timer management should be optimized for the case of timers rarely expiring. In the normal case, the headers of consecutive data TPDU's and NPDU's are almost the same. They can largely be copied from a stored prototype. The calculation of a checksum is often combined with copying of the data. Copying and checksum calculation are so critical for fast TPDU processing that they are often coded in assembler.

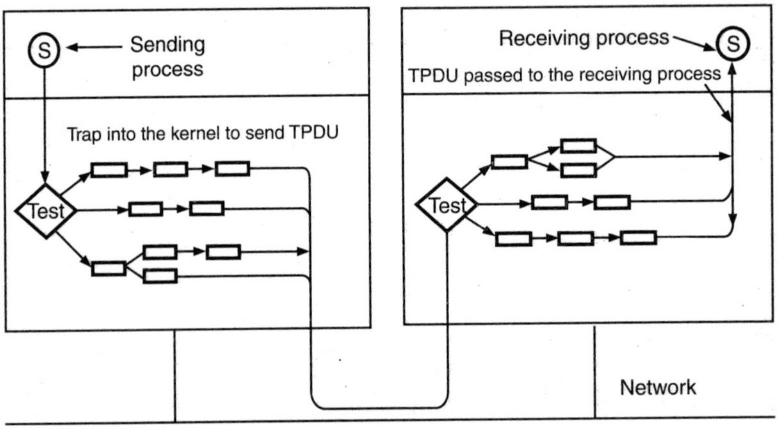

Fig. 8.38. TPDU processing

8.17.5. Protocols for Gigabit Networks

The first problem is the use of 16 or 32 bit sequence numbers. At a rate of 1 Gbps, it takes only 32 sec to send 2^{32} bytes, and in the Internet packets can live for 120 sec. The second problem is that communication speeds have improved much faster than computing speeds, for those who can pay it. In the 70's, the ARPANET ran at 56 kbps and had 1 MIPS computers.

Packets were 1008 bits, thus, there was 18 msec to process a packet, thus, 18000 CISC instructions (for a dedicated machine). Compare this to modern 100 MIPS computers exchanging 4 KB packets over a gigabit line. There are then 30 usec to process a packet, or 3000 RISC instructions, which are less powerful than CISC instructions.

The third problem is that the go back n protocol performs poorly on lines with a large bandwidth-delay product. The fourth problem is that gigabit lines are fundamentally different from megabit lines in that long ones are delay limited rather than bandwidth limited. Figure 8.39 shows the time to transfer and acknowledge a 1 megabit file over a 4000 km line.

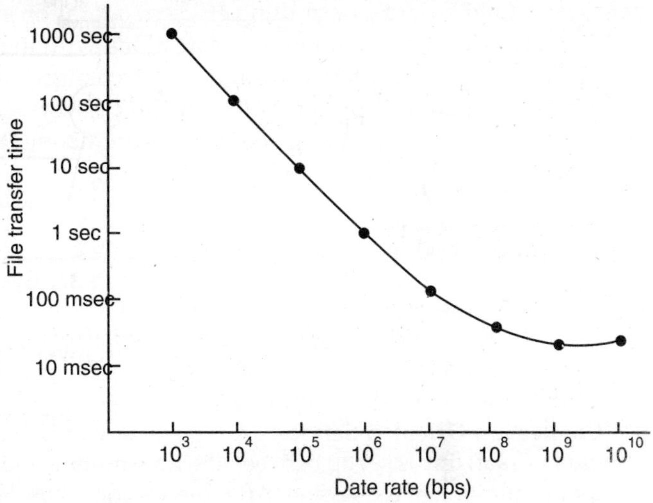

Fig. 8.39.

The fifth problem is due to new applications. For many gigabit applications, such as multimedia, the jitter in packet arrival time is as important as the mean delay itself. The basic principle that all gigabit network designers should learn by heart is: "design for speed, not for bandwidth optimization". Old protocols were often designed to minimize the number of bits on the wire, frequently by using small fields and packing them together into bytes and words. Now, not bandwidth but protocol processing is the problem, so protocols should be designed to minimize it. A way to go fast is to build fast network interfaces in hardware. The difficulty with this is that unless the protocol is very simple, hardware here just means a plug-in board with its own CPU and software. To avoid that, this coprocessor is as expensive as the main CPU, it is often a slower chip. As a consequence, many a times it happens that the fast CPU is waiting for the slow coprocessor to do the critical work. Furthermore, elaborate protocols are needed between the two CPU's to synchronize them correctly. Usually, the best strategy is to make the protocols simple and have the main CPU do the work of the higher layers. Due to the relatively long delay loop, feedback should be avoided; it takes too long for the receiver to signal the sender. One example is to avoid the delay inherent in the receiver sending windows updates to the sender, it is better to use a rate-based protocol. There the sender can send all it wants, provided it does not send faster than some rate (mean, peak or other parameters) the sender and receiver have agreed upon in advance.

Another example is the slow start algorithm of TCP, it makes multiple probes to see how much the network can handle, and with long delays, it takes a lot of time and wastes a large amount of bandwidth. It is better to reserve the needed resources in advance, making it also easier to reduce jitter. In short, going to high speeds pushes the design toward connection oriented operation, or something very close to it. The header should contain as few field as possible, to reduce processing time, and these fields should be big enough to do the job and be *word* aligned, adapted to the characteristics of modern CPU and memory buses. Big enough means that problems such as sequence numbers wrapping around while old packets still exists, receivers being unable to announce enough window space because the window field is to small, and so on, do not occur.

The header and the data should be separately checksummed. The header can than be checked before the data is copied to the user space allowing the data checksum to be done during copying. Otherwise, if there was an error in the header, the copying might be to a wrong process. The maximum data size should be large, permitting efficient operation even in the face of long delays. It should be possible to send a normal amount of data along with the connection request. This is important for connections were only a few messages are exchanged, like RFC or bank card operations. Design and implementations of protocols for gigabit networks should aim at minimizing the processing time for the case that everything goes right. Minimizing the processing time when an error occurs is secondary.

GLOSSARY

- **Connectionless Service** A type of service in which messages are partitioned into *packets* and routed through the network. Each packet is independent of the other packets that carry parts of the message, and each packet carries a destination address. Unlike *connection-oriented service*, no physical link is established between sending and receiving nodes before data transmission.
- **Connection-Oriented Service** A type of service in which prior to the transfer of data a physical (and virtual) link is set up between the transmitting and receiving nodes. This link remains in effect for the duration of the session. After the session is completed, the link is removed. Characteristics of a connection-oriented service include wasted bandwidth (in fact, link must remain established even during idle periods of a transmission), a high potential for a hung network (there is always a possibility that a link will not be terminated), and guaranteed sequential arrival of packets at the destination node.
- **User Datagram Protocol (UDP)** A connectionless protocol providing an unreliable datagram service. UDP does not furnish any end-to-end error detection or correction and it does not retransmit any data it did not receive.
- **Addressing** A network concept which describes the process of assigning unique identification numbers (also called *addresses*) to a networked device.
- **Congestion** A situation when a network is consumed with excessive network traffic (i.e., lots of packets), resulting in performance degradation. Congestion occurs when routers are quite slow and thus causing queues to lengthen, or when routers are quite fast, causing queues to build up whenever input traffic is greater than the capacity of output lines. The ultimate level of congenstion is called as deadlock, which occurs when one router cannot proceed until a second router does somethings, and the second router cannot proceed since it is waiting for the first router to do something. Congestion control is provided by layer 3 of the *OSI model*.

SUMMARY

- A connection oriented transport layer protocol establishes a connection i.e., virtual path between sender and receiver. This is a virtual connection. The packet may travel out of order. The packets are numbered consecutively and communication is bidirectional.
- The transport layer can provide error control as well. But error control at transport layer is performed end to end rather than across a single link. Error connection is generally achieved through retransmission.
- The typical QoS parameters for transport layer are as under:
  - (i) Connection establishment delay
  - (ii) Connection establishment failure probability
  - (iii) Throughput
  - (iv) Transit delay
  - (v) Protection
  - (vi) Resident error ratio
  - (vii) Priority
  - (viii) Resilence

- The task of transport layer is to provide reliable, cost effective transport of data from source machine to destination machine. To achieve this goal, the transport layer makes use of the services provided by the network layer.
- The connection oriented concurrent client server communication uses TCP and stream socket. The servers using TCP are normally concurrent. This means that a server is serving many clients at the same time.
- The IP addresses and port numbers play different roles in selecting the final destination of data. The destination IP address defines the host among the millions of hosts in the world. After a particular host is selected, the port number defines one of the processes on this selected host.
- The IP header contains the IP addresses while the UDP and TCP headers contain the port numbers.
- In a connection oriented service, a connection is established between source and destination. Then the data is transferred and at the end, the connection is released.
- Releasing a connection is easier than establishing it. There are two styles of releasing a connection.
  (i) Asymmetric release, and
  (ii) Symmetric release.
- The basic principle is do not inject a new packet into the network until an old one is delivered. TCP tries to do this by dynamically manipulating the window size.
- To prevent congestion, TCP has to deal with two problems separately – receiver capacity and network capacity.

SHORT QUESTIONS WITH ANSWERS

Q.1. What do you mean by Congestion Control and QoS?

Ans. The congestion can take place in the data link, network or transport layer. But the effect of congestion is generally evident in the transport layer. Quality of service (QoS) can be implemented in other layers but its actual effect is felt in the transport layer. The transport layer enhanced the QoS provided by the network layer. The transport service may allow the user to specify preferred, acceptable and minimum values of various service parameters at the time of setting up a connection.

Q.2. What is Transit Delay?

Ans. It is the time between a message being sent by the transport user on the source machine and its being received by the transport user on the destination machine.

Q.3. What do you mean by Sockets?

Ans. The socket interface was originally based on UNIX. It defines a set of system calls or procedure. The communication structure that we need in socket programming is called as a **socket**. A socket acts as an end point. Two processes can communicate if and only if they have a socket at each end.

Q.4. What are Socket Types?

Ans. There are following three types of sockets:
1. The stream socket.
2. The packet socket.
3. The raw socket.

Q.5. What are the Elements of transport protocols ?

Ans. Following are some of the important elements of transport protocols :
(i) Addressing
(ii) Establishing a connection
(iii) Releasing a connection
(iv) Flow control and buffering
(v) Multiplexing
(vi) Crash recovery.

In the following sections, we will discuss all these elements.

Q.6. What is the Difference between IP Addresses and Port Numbers ?

Ans. The IP addresses and port numbers play different roles in selecting the final destination of data. The destination IP address defines the host among the millions of hosts in the world. After a particular host is selected, the port number defines one of the processes on this selected host.

Q.7. What are the TCP Services ?

Ans. Following are some of the services offered by TCP to tne processes at the application layer.
1. Stream delivery service
2. Sending and receiving buffers.
3. Bytes and segments.
4. Full duplex service
5. Connection oriented service
6. Reliable service.

Q.8. List the protocols in the transport level of TCP/IP.

Ans. TCP (Transmission Control Protocol) and UDP (User Datagram Protocol).

Q.9. What is the size of a TCP header?

Ans. 20 bytes.

Q.10. What is the size of a UDP header?

Ans. 8 bytes.

Q.11. List two applications of UDP.

Ans. TFTP (Trivial File Transfer Protocol) and RCP (Remote Call Procedure).

REVIEW QUESTIONS

1. What do you mean by congestion control and QoS ?
2. What are the parameters of QoS ? Explain
3. Define the term, Socket.
4. List the types of Socket.
5. What are the steps used for socket programming ? Explain.
6. What are the elements of transport layer ? Explain.
7. What is difference between IP addresses and port number ?
8. What are the functions of client and server ? Explain.
9. What problems will occur in establishing a connection ? Explain.
10. What is meant by TCP and UDP ?
11. State threshold condition in congestion.
12. Explain the significant of listen call. Does it apply to all sockets ? Explain.
13. What parameters are specified by its various arguments ? Explain.
14. Explain in detail how TCP provides flow control.
15. Define a term silly window syndrome and possible solution to overcome its effect.
16. What are the techniques used to improve QoS ? Explain.
17. What is meant by fair queueing ?
18. What are the drawbacks of fair queueing ? Explain.
19. What are the duties of transport layer ? Explain in brief.
20. Draw and explain the relation between network layer, transport layer and application layer.
21. What are the transport serves primitives ? Explain.
22. Draw and explain the various fields of socket structure.
23. Explain about connection oriented concurrent server.
24. Writes a short technical note on Addressing transport layer.
25. Writes a short technical note on Flow control and buffering.
26. Explain about multiplexing and demultiplexing used in transport layer.
27. Explain about the TCP transmission policy, congestion control.
28. Explain about the following issues of transport protocol :
    (i) Establishing a connection; (ii) Releasing a connection.
29. Explain about the structure of UDP header.
30. Explain about the TCP header and working of the TCP protocol.
31. Explain about the various fields of TCP header with the help of neat diagram.
32. Explain about the various steps that are followed in releasing a TCP connection.

UNIT – V

SYLLABUS:

The Application Layer: The Domain Name System, E-mail, World Wide Web, streaming audio and video, content delivery.

The above syllabus has been covered in following chapter:

Chapter 9 : The Application Layer

CHAPTER 9

THE APPLICATION LAYER

Syllabus

The Application Layer: The Domain Name System, E-mail, World Wide Web, streaming audio and video, content delivery.

CHAPTER OUTLINE

- Introduction
- Domain Name System (DNS)
- Caching
- Electronic Mail (E-mail)
- Multipurpose Internet Mail Extensions (MIME)
- Message Transfer
- Simple Mail Transfer Protocol (SMTP)
- E-mail Gateways
- Final Delivery (Delivery Protocols)
- Mail Server
- World Wide Web (WWW)
- Uniform Resource Locator (URL)
- Hypertext Transfer Protocol (http)
- Browser Architecture
- Structure of a Web Page
- Dynamic Document
- Performance Enhancement
- Introduction to Multimedia
- Examples of Multimedia Applications
- Problems for Multimedia in Today's Internet
- Expectations of Multimedia from the Internet
- Audio and Video Compression
- Content Delivery Network?

9.1. INTRODUCTION

As a matter of fact, the application layer is the topmost (fifth layer) of the Internet model. This is layer where all the interesting applications are found. The application layer allows people to use the Internet. The layers below the application layer provide reliable transport but they do not do any real work for the users. In other words, the other four layers are created so that people can use the various application program.

Figure 9.1 shows the position of application layer in the five layer Internet model. The application layer provides services to the users. The users can be humans or software. It enables the user to access the network.

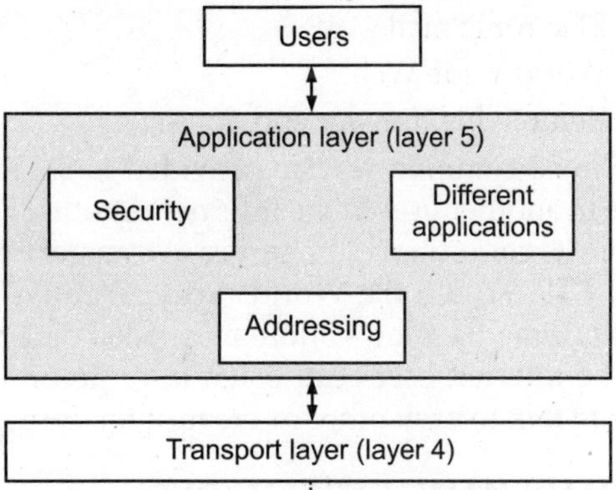

Fig. 9.1. Position of application layer in five-layer Internet

The application layer receives services from the transport layer. For the real applications in the application layer to function, there is a requirement of support protocols.

Pioneer in Communication: Radia Perlman (1951-)

Radia Perlman's work has had a great impact upon the field of networking, and she has been considered as specialist in areas of bridging and routing, as well as sabotage-proof networks. Best known as the inventor of spanning tree algorithm used in bridges and switches, Perlman is also the inventor of several important algorithms that make modern link-state routing protocols robust and quite efficient, including self-stabilizing flooding of routing updates and designated routers. This technology is the core of all modern link-state protocols, including IS-IS, OSPF, and PNNI.

The three areas or protocols required for such support may be listed as under :

(i) Network security.

(ii) Domain Name Service (DNS)

(iii) Network management.

Basically, security is not a single protocol but it consists of a large number of concepts and protocols used for providing privacy. DNS is used to handle naming or addressing within the Internet. The third support protocol is network management.

9.1.1. Client-Server Paradigm

The application layer programs are based upon the concept of clients and servers.

9.1.2. Addressing

For communication between client and server, addressing is needed. When a client requests a service from the server, it has to include the servers address as destination address and its own address as source address. When the server responds, it reverses the addresses. DNS is used for addressing as well be discussed later on in the chapter.

9.1.3. Applications

Some of the important applications may be listed as under :

(i) Electronic mail.

(ii) USENET (net news).

(iii) World Wide Web.

(iv) Multimedia.

(v) Remote file transfer and access.

The most common service provided is SMTP or electronic mail. It allows the user to send a message to another user in the Internet. Another common service is file transfer. A user can transfer a file from its computer to the server or transfer a file from a server to its computer. This application is called FTP. To use the World Wide Web (WWW), we need a simple transfer protocol called HTTP. Multimedia allows audio and video data to be attached with the computer for the Internet users. The Internet users can listen to radio or TV or music that is streamed from a server. It is possible to talk to each other or create a teleconferencing environment.

9.2 DOMAIN NAME SYSTEM (DNS) **(RGTU, Bhopal, Sem. Exam; June 2014)**

Addressing: For communication to take place successfully, the sender and receiver both should have addresses and they should be known to each other. The addressing in application program is different from that in the other layers. Each program will have its own address format. For example, an e-mail address is like sanjaysharma @ vsnl.net where as the address to access a web page is like http://www.google.com/

> **DO YOU KNOW?**
>
> The TCP/IP application layer corresponds to the combined session, presentation, and application layers of the OSI model.

It is important to note that there is an alias name for the address of remote host. The application program uses an alias name instead of an IP address. This type of address is very convenient for the human beings to remember and use. But it is not suitable for the IP protocol.

Therefore, the alias address has to be mapped to the IP address. For this, an application program requires service of another entity. This entity is an application program called DNS. It may be noted that DNS is not used directly by the user. It is used by another application programs for carrying out the mapping.

9.2.1. DNS Working

To map a name onto an IP address, an application program calls a library procedure called the resolver. The name is passed on to the resolver as a parameter. The resolver sends a UDP packet to a local DNS server which looks up the name and returns the corresponding IP address to the resolver. The resolver then sends this address to the caller. Then, the program can establish a TCP connection with the destination or sends in the UDP packets.

9.2.2. The DNS Name Space

Conceptually, the Internet has been divided into hundreds of top level domains. Each domain covers many hosts. Each domain is divided into several subdomains and they are further partitioned and so on. These domains can be represented by a tree as shown in figure 9.2.

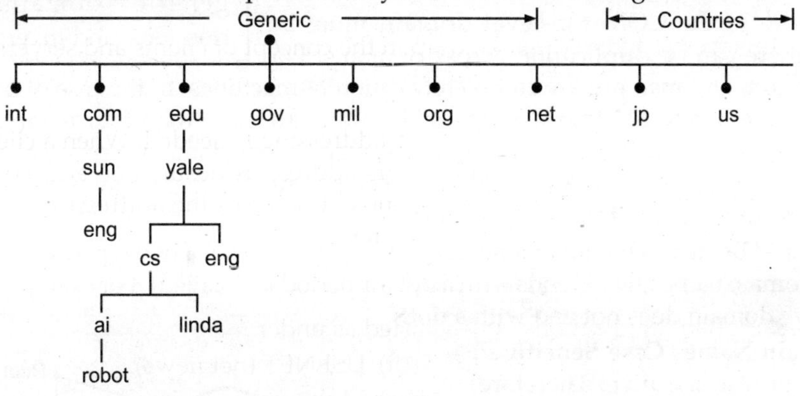

Fig. 9.2. A portion of Internet domain name space

The top level domains are of two types namely generic and countries.

1. Generic Domains: The generic domains are com (commercial), edu (educational institutions), gov (government), int (some international organizations), mil (military), net (network providers) and org (non-profit organizations). The country domains include one entry for every country. Each domain is named by following an upward path. The components are separated by dots *e.g.*, **eng.sum.com.** This is called hierarchical naming. Another example of hierarchical naming is shown in figure 9.3. The upward followed path has been shown by an arrow.

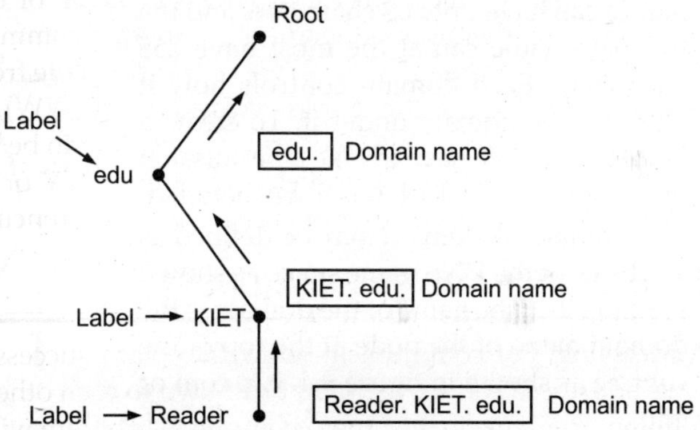

Fig. 9.3. Domain names, labels and hierarchical naming

2. Label: Each node in the tree has a label (or component) and it can be specified using upto 63 characters. If we had to remember the IP addresses of all of the Web sites we visit every day, we would all go nuts. Human beings just are not that good at remembering strings of numbers. We are good at remembering words, however, and that is where domain names come in. You probably have hundreds of domain names stored in your head. For example :

- www.yahoo.com - the world's best-known name
- www.kiet.edu - a popular EDU name
- encarta.msn.com - a Web server that does not start with www
- www.bbc.co.uk - a name using four parts rather than three
- ftp.microsoft.com - an FTP server rather than a Web server

The COM, EDU and UK portions of these domain names are called the **top-level domain** or **first-level domain.** There are several hundred top-level domain names, including COM, EDU, GOV, MIL, NET, ORG and INT, as well as unique **two-letter combinations for every country**.

Within every top-level domain there is a huge list of **second-level domains.** For example, in the COM first-level domain, you have got :

- yahoo
- msn
- microsoft
- plus millions of ohers...

DO YOU KNOW?

The domain name space is divided into three sections: generic domains, country domains, and inverse domain.

Every name in the COM top-level domain must be **unique**, but there can be duplication across domains. For example, **msn.com** and **msn.org** are completely different machines. In the case of bbc.co.uk, it is a third-level domain. Up to **127 levels** are possible, although more than four is rare. The left-most word, such as **www** or **encarta**, is the **host name.** It specifies the name of a specific machine (with a specific IP address) in a domain. A given domain can potentially contain millions of host names as long as they are all unique within that domain.

Absolute and Relative Domain Names: Domain names can be of two types : absolute or relative. An absolute domain name always ends with a dot (or period as it called). For example, eng.sun.com. But the relative domain does not end with a dot.

Are Domain Names Case Sensitive?: No they are not case sensitive. Therefore, com and COM means the same thing.

How Many Characters?: Component names can have upto 63 characters and the full path name can at the most have 255 characters. Each domain controls how it allocates the domain under it. To create a new domain, we have to take a permission of the domain in which it is to be included.

Domain: A domain may be defined as a subtree of the DNS name space as shown in figure 9.4. The name of the domain is the domain name of the node at the top of the subtree as shown in figure 9.4. e.g. com or edu.

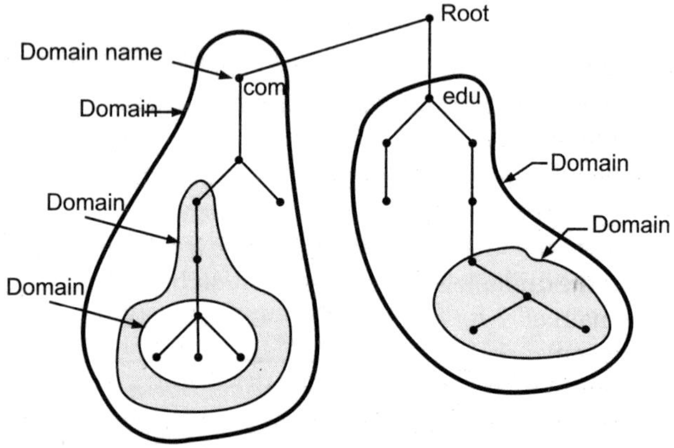

Fig. 9.4. Illustration of domains

A domain may be divided into subdomains as shown as shown in figure 9.4. It may be noted that the naming follows organizational boundaries, not physical networks. This means that even if two different departments are located in the same building, they can have distinct domains. But the computers belonging to the same department kept in two different buildings will not have different domains.

9.2.3. Name Servers

1. Basic concepts: Name server consists of the DNS database *i.e.* the various names and their corresponding IP addresses. Theoretically, a single name server could contain the entire DNS database. But, practically to store such a huge information at one place is inefficient and unreliable.

Such a server will be soon overloaded and be useless and worst things is if it ever goes down the entire Internet will go down. The solution to this problem is to distribute the information among many computers called **DNS servers.** Then we have to use a hierarchy of the Name servers as shown in figure 9.5. First, the whole space is divided into many first level domains. The root server stands alone and can create as many first level domains as needed. The first level domains are further divided into smaller subdomains called second level domains. They can be further divided as shown in figure 9.5.

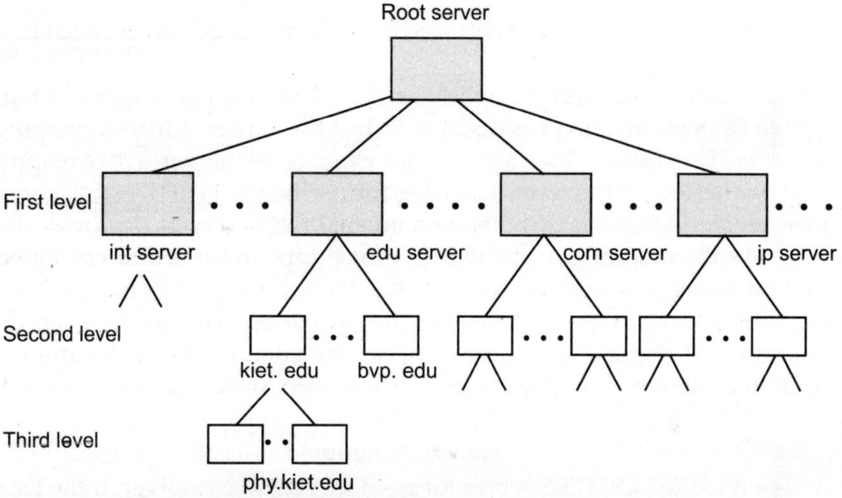

Fig. 9.5. Hierarchy of name servers

Each server can be responsible (authoritative) to either a large or small domain. It may be noted that the hierarchy of servers is similar to the hierarchy of names. The whole DNS name space is divided up into non overlapping **zones.** The concept of zones may be explained as under:

2. Zones: What a server is responsible for or has authority over is called as a zone. If a server is appointed for a domain and the domain is not further divided into subdomains then the domain and zone will be the same as shown in figure 9.6. The server makes a database called a zone file. It keeps all information about every node under that zone. But if a server divides its domains into subdomains and delegates a part of its authority to other servers than domains and zone will be different from each other. This has been shown in figure 9.6. The information about the nodes in the subdomains is stored in the servers at the lower levels. The higher level and original server keeps some sort of reference of these lower level servers.

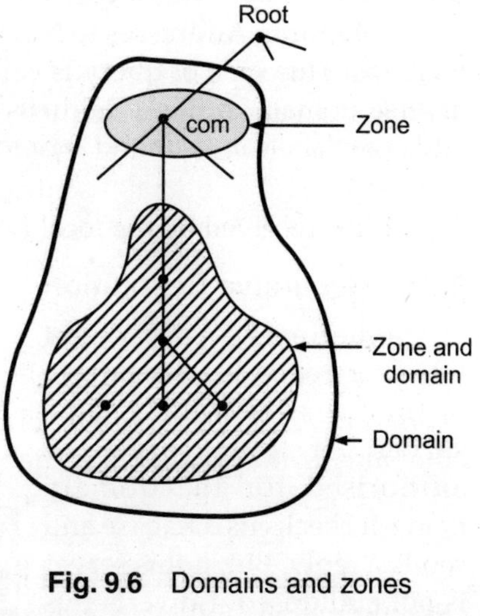

Fig. 9.6 Domains and zones

3. Root Server : A root server is a server whose zone consists of the whole DNS tree. It does not store any information about domains but delegates the authority to other servers. It only keeps the reference of these servers. There are more than 13 root servers and they are distributed all around the world.

4. Primary and Secondary Servers : DNS defines two types of servers namely the primary servers and the secondary servers.

5. Primary Server: It is a server which stores a file about its zone. It is authorized to create, maintain and update the zone file. It stores the zone file on a local disk.

6. Secondary Server: This server transfers complete information about a zone from another server which may be primary or secondary server. The secondary server is not authorized to create or update a zone file. If its zone file is to be updated, then it is to be done by the primary server.

Important Point: A primary server loads all the information from the disk file, whereas the secondary server loads all information from the primary server.

9.2.4. Resolution

The process of mapping a name to an address or an address to a name is called as name address resolution.

1. Resolver: DNS is the client server application. A host which wants to map a name to address or vice versa calls a DNS client named as resolver. When the name « address mapping is necessary a host calls a resolver. The resolver then accesses the closest DNS server with a mapping request. If this server has the requested information, it satisfies the resolver but if it does not have the requested information, then it refers the resolver to other servers or asks other servers to provide the information. The resolver receives the mapping. It then checks for errors and if found error free delivers the mapping to the requesting process.

2. Mapping Names to Addresses: The resolver gives a domain name to the server and requests for the corresponding IP address. The server checks the generic or country domains to get the mapping. If the domain name is from the generic domain section then the resolver receives a domain name such as

xxx.yyy.zzz.edu

The query is sent to the local DNS server for resolution by the resolver. If the local server does not resolve the query, it will refer the resolver to other servers or asks them directly. The same procedure is followed for a name from country domain.

3. Mapping Addresses to Names: Here, a client sends an IP address to a server and requests for its name. This type of query is called as PTR query. To answer the PTR query, the DNS uses the inverse domain. If the IP address is 142.36.48.118 then the resolver first inverts the address and adds two labels *in_addr* and *arpa* to it. Therefore, the domain name sent is :

118.48.36.142.in_addr.arpa.

This is received by the local DNS and resolved.

9.2.5. Recursive Resolution

Sometimes, a client (resolver) requests for recursive of final answer from a name server. If this server is authorised for the domain name, it checks its database and sends a reply. But, if this server is not authorised it diverts this

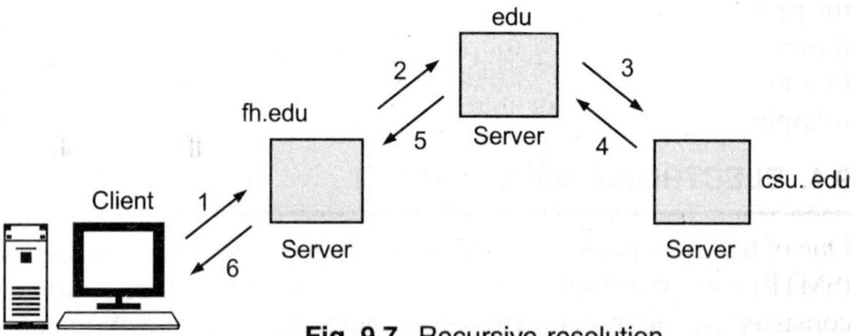

Fig. 9.7. Recursive resolution

request to another server (usually the parent server) and waits for the response. If the patent is an authority, it responds, otherwise it sends the query to another sever.

When the query is solved, the response is returned back to the requesting client. Such a query is called as recursive query and the process is called recursive resolution. It has been illustrated in figure 9.7.

9.2.6. Iterative Resolution

This type of mapping can be done if the client does not ask for recursive answer. In iterative resolution, if the server has authority for the name it will send the answer. But if it does not have the authority then it returns to the client IP address of the server that it thinks has the answer to the query. The client has to repeat the query to this server. If this server also cannot answer it returns the IP address of a new server to the client. Now, the client should send the query to this third server. This process is called as iterative resolution because client sends the same query to different servers. Figure 9.8 illustrates the iterative resolution.

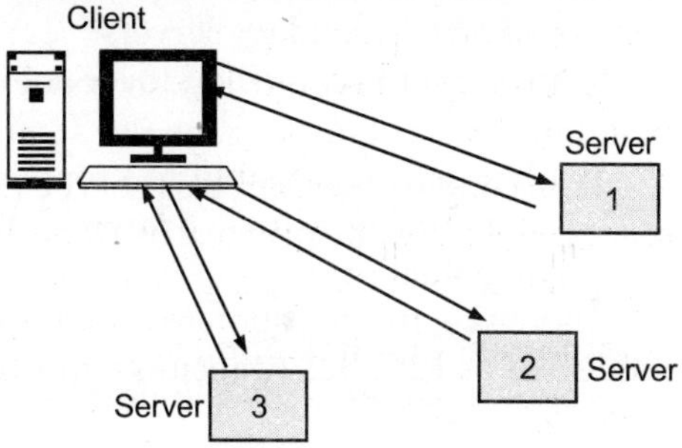

Fig. 9.8. Illustration of iterative resolution

DNS Examples: The DNS system is a database, and no other database on the planet gets this many requests. No other database on the planet has millions of people changing it everyday, either. This is what makes the DNS system so unique.

For example

www.yahoo.com - the world's best-known name

www.kiet.edu - a popular EDU name

encarta.msn.com - a Web server that does not start with www

www.bbc.co.uk - a name using four parts rather than three

ftp.microsoft.com - an FTP server rather than a Web server

www.spce.ac.in - Server in India 'in' domain.

The COM, EDU and UK portions of these domains names are called the top-level domain or first-level domain. There are several hundred top-level domain names, including COM, EDU, GOV, MIL, NET, ORG and INT, as well as unique two-letter combinations for every country.

9.3. CACHING

Every time a query is asked, the server has to spend time in searching the corresponding IP address. If this searching time is reduced then efficiency would go up. The searching time can be reduced by using a technique called caching. When a server asks for a mapping from another server and receives the response, it stores this information in its cache memory before sending it to the client. If the same or other client request for the same mapping, it can check its cache memory and resolve the problem. But the problem with caching is that, if a server caches a mapping for a long time then it may send an outdated mapping to the client. This problem can be solved by adding the time to live information (TTL) to the mapping and each server is asked to keep a TTL counter for each mapping it caches.

9.4. ELECTRONIC MAIL (E-MAIL) (RGTU, Bhopal, Sem. Exam; June, 2014)

One of the most popular network services is electronic mail (e-mail). Simple Mail Transfer Protocol (SMTP) is the standard mechanism for electronic mail in the Internet. The first email systems simply consisted by file transfer protocols. But some of the limitations of this system were as under :

(i) Sending a message to a group of people was inconvenient.
(ii) Message did not have any internal structure. Therefore, its computer processing was difficult.
(iii) The sender never used to know if a message arrived or not.
(iv) It was not easy to handover one's e-mails to someone else.
(v) It was not possible to create and send messages containing a text, drawing, facsimile and voice together.

DO YOU KNOW?

In an electronic mail system, the user deposits an electronic message with an electronic mail delivery system. The delivery system cooperates with other systems to transfer the message to the mailbox of the intended receiver.

Therefore, more elaborate e-mail systems were proposed, ARPANET e-mail proposals were published as RFC 821 (transmission protocol) and RFC 822 (message format). These are used in Internet.

9.4.1. E-mail Architecture and Services

An e-mail system consists of two subsystems as under:
(i) User agents and
(ii) Message transfer agents.

User agents : They allow the people to read and send e-mail.

Message transfer agents : They move the messages from the source to the destination.

Basic Functions: E-mail systems support five basic systems which may be listed as under :

1. Composition
2. Transfer
3. Reporting
4. Displaying and
5. Disposition.

1. **Composition:** The process of creating messages and to answer them is known as composition. The system can also provide assistance with addressing and a number of header fields attached to each message.
2. **Transfer:** It is the process of moving messages from the sender to the recipient. This includes establishment of a connection from sender to destination or some intermediate machine, outputting the message, and releasing the connection.
3. **Reporting:** This is to tell the sender about whether the message was delivered or rejected or lost.
4. **Displaying :** It is the process of displaying the incoming messages. For this purpose, simple conversions and formatings are required to be done.
5. **Disposition:** This is concerned with what the recipient does with the message after receiving it. Some of the possibilities are as under :
   (i) Throw after reading.
   (ii) Throw before reading.
   (iii) Save messages.
   (iv) Forward messages.
   (v) Process messages in some other way.

Advanced Features of E-mail Systems: Some of the advanced features included in addition to the basic functions may be listed as under:

1. Forwarding an e-mail to a person away from his computer.
2. Creating and destroying mailboxes to store incoming e-mail

3. Inspecting contents of mailbox, insert and delete messages from the mailboxes.
4. Sending a message to a large group of people using the idea of mail list.
5. To provide registered e-mail.
6. Automatic notification of undelivered e-mails.
7. Carbon copies.
8. High priority e-mail.
9. Secret (encrypted e-mail)
10. Alternative recipient.

E-mail Envelope: In the modern e-mail systems, there is a distinction made between the e-mail and its contents. An e-mail envelope contains the message, destination address, priority, security level etc. The message transport agents use this envelope for routing.

Message: The actual message inside the envelope is made of following two parts:

(i) Header and

(ii) Body.

Header carries the control information while body contains the message contents. Envelopes and messages are shown in figure 9.9.

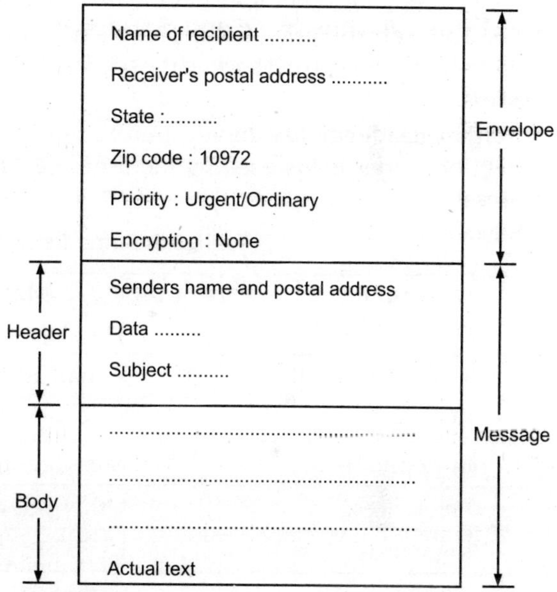

Fig. 9.9. Illustration of envelope and message

9.4.2. Message Formats

Now, let us discuss the e-mail message formats.

RFC 822: As a matter of fact, messages consist of a primitive envelope, some number of header fields, a blank line and then the message body. Each header field logically consists of a single line of ASCII test which contains the field name, a colon and a field. It may be noted that RFC 822 is an old standard. Normally, the user agent builds a message and passes it to the message transfer agent which uses some header fields for construction of an envelope. Table 9.1 shows the principle header fields related to the message transport. Let us discuss them one by one.

Table 9.1. RFC 822 header fields related to message transport

Header Name	Meaning
To :	E-mail address of primary recipients
Cc :	E-mail address of secondary recipients
Bcc :	E-mail address for blind carbon copies
Form :	Person who created the message
Sender :	E-mail address of the actual sender
Received :	Line address by each transfer agent along the route
Return – Path :	Can be used to identify the path back to the sender.

1. The To : field: This field gives the DNS address of the primary recipient. It is allowed to have multiple recipients.

2. The Cc : field: This field gives the addresses of any secondary recipients. Cc stands for carbon copy.

3. The Bcc : field: The long form of Bcc is blind carbon copy. This field is like Cc field, except that this is deleted from all the copies sent to the primary and secondary recipients. This feature allows the people to send copies to third parties without primary and secondary recipients knowing this.

4. From : and Sender : fields: These fields tell about who wrote the message and who actually sent the message respectively because the person who creates the messages and the person who sends it can be different. The From : field is required but the Sender : field can be omitted, if it is same as the From : field. These fields are required in case the message is undeliverable and is to be returned to the sender.

5. Received : field: A line containing Received : is added by each message transfer agent along the way. This line carries the agent's identity, date and time at which the message was received. It also contains some other information that can be used to find bugs in the routing system.

6. The Return-Path : field: This field is added by the final message transfer agent and it is intended to tell how to get back to the sender. This information can be gathered from all the Received : headers.

Other header-fields: In addition to the fields of table 9.2, RFC 822 messages may contain a variety of header fields used by the user agents or human recipients some of them are shown in table 9.2.

Table 9.2. Some fields in RFC 822 message header

Header	Meaning
Date :	The date and time of the message.
Reply-To :	E-mail address to which the reply is to be sent.
Message-Id :	Message identifying number.
In-Reply-To :	Message-Id of the message to which this is a reply.
References:	Other relevant message identifying numbers.
Keywords :	Keywords chosen by user.
Subject :	Summary of the message for the one line display.

The RFC 822 allows the users to invert new headers for their own private use but it is essential that these headers start with the string X- for example X-Event of the week.

Message Body: The message body comes after the header. The users can put whatever they want to send, in the message body. It is possible to terminate the messages with ASCII cartoons, quotations, political statements etc.

9.5 MULTIPURPOSE INTERNET MAIL EXTENSIONS (MIME)

In the early days, the e-mail used to consist of only the text messages in English and expressed in ASCII. RS 822 was sufficient for this environment. But in the worldwide Internet environment, this approach is not adequate. Some problems are encountered in sending and receiving the following types of messages:

(i) Messages in the languages having accents such as French or Germans.

(ii) Messages which do not contain text e.g. audio and video.

(iii) Messages in the languages which do not have alphabets (e.g. Chinese and Japanese).

(iv) Messages in non Latin alphabets such as Russian or Hebrew.

The solution to these problems was MIME *i.e.* Multipurpose Internet Mail Extensions. It was proposed in RFC 1341 and then updated in RFC 1521.

9.5.1. Principle of MIME

MIME uses the same RFC 822 format but it adds structure to the message body (In RFC 822 there is no structure to the message body). In addition to this, MIME defines encoding rules for non ASCII message. MIME messages can be sent using the existing mail programs and protocols. The sending and receiving programs need to be changed which users can do themselves.

New Message Headers: Five new message headers are defined for MIME. They are listed in table 9.3.

1. MIME-Version: It tells the user agent that the message is a MIME message and it also specifies the version of MIME being used.

Table 9.3. New headers in MIME

S. No.	Header Name	Meaning
1.	MIME – Version :	Identifies the MIME version
2.	Content – Description :	Tells what is in the message
3.	Content – Id :	Identifier
4.	Content – Transfer – Encoding :	How is the body wrapped for transmission
5.	Content – Type :	Nature of the message.

2. Content-Description: This field tells with the message is. It is in the form of ASCII string. This header is required because the recipient will know whether it is worth decoding and reading the message.

3. Content-Id: This field identifies the contents. Its format is same as the format of standard Message–Id header.

4. Content-Transfer-Encoding: This field tells how the body is wrapped for transmission through a network. The network can object to most characters except for the letters, numbers and punctuation marks. Five schemes are provided in addition to an escape to new schemes. These schemes are as under :

(i) The first scheme simply uses ASCII text. Each ASCII character uses 7 bits and can be carried directly if each line contains at the most 1000 characters.

(ii) The second scheme is similar to the first scheme except for one change. Instead of 7 bit characters, new 8 bit characters are used. The maximum line length remains unchanged.

(iii) Third is the messages using binary encoding. The problems with these messages are that they do not always use all the 8 bit and do not respect the 1000 line per character limit. There is no guarantee that these messages arrive error free.

(iv) The fourth type is encoded binary messages using base 64 encoding which is also called as ASCII armor.

> **DO YOU KNOW?**
>
> Multipurpose Internet Mail extension (MIME) is an extension of SMTP that allows the transfer of multimedia messages.

(v) The fifth type is the messages with a special type of encoding called **quoted-printable encoding.**

5. Content-Type: This is last header in table 9.4 and it is used to specify the type of the message body. RFC 1521 defines seven types with each one having one or more subtypes. The type and subtype are separated by a slash such as,

Content– Type : Video/mpeg.

The subtype must be given in the header. Table 9.4 gives the initial list of types and subtypes specified in RFC 1521.

Table 9.4. The MIME Types and Subtypes in RFC 1521

Type	Subtype	Description
Text	Plain Rich text	Text in the unformatted way Text including simple formatting commands.
Image	Gif Jpeg	Still pictures in GIF format Still pictures in JPEG format
Audio	Basic	Audible sound
Video	Mpeg	Movie in MPEG format
Applications	Octet-stream Post script	Byte sequence in uninterpreted form A printable document in Post script
Message	RFC 822 Partial External body	A MIME RFC 822 message Split message for transmission Message itself should be fetched over the net
Multipart	Mixed Alternative Parallel Digest	Independent parts in the specified order. Same message in different formats Parts must be viewed simultaneously Each part is a complete RFC 822 message

It may be noted that many new types have been added to the basic list of table 9.4 and the addition is still being made.

(a) Text: The text type is for the straight text. There are two subtypes namely plain and richtext. The text/plain combination represents the original messages without any encoding or further processing. The text/richtext allows simple formatting in the text. It allows the text with boldface, italics, small and large point sizes, indentation, subscripts , page layout etc.

(b) Image: This MIME type is used for transmitting still pictures. There are many formats used for storing and transmitting images with or without compression. The two sub types are GIF and JPEG.

(c) Audio and Video: The audio type is for sound and video is for moving pictures. The video does not include a soundtrack. Only one video format defined is MPEG which is decided by the Moving Picture Experts Group (MPEG.)

(d) Applications: This type is used for format which require external proceeding and which is not covered by any other type. The octet stream is a sequence of uninterpreted bytes. When it is received, a user agent should display it and suggest the user to copy it in a file for further processing. The other subtype is post script. It corresponds to the post script language produced by Abode systems which is used for describing printed pages.

(e) Message: This type allows one message to be fully encapsulated inside the other message. This is useful forwarding e-mail. The partial subtype allows to break an encapsulated message into pieces and send them separately. The external body subtype can be used for very long messages such as video films.

(f) Multipart: This is the last type of multipart. It allows a message to contain more than one parts. The beginning and end of each part is clearly delimited. There are four subtypes. The mixed subtype allows each part to be different. In the alternative subtype, each part should contain the same message expressed in a different medium or encoding.

The alternative subtype can be used for multiple languages as well. The parallel subtype is used for viewing all parts simultaneously *e.g.*, audio and video parts of a movie. The fourth subtype is digest. It is used when many messages are packed together to form a composite message.

9.6. MESSAGE TRANSFER

The message transfer system is used to relay the messages from originator (sender) to the recipient. This can be achieved by establishing a transport connection between the source machine and destination machine. However, in certain situations, this does not work and something else needs to be done for this.

9.7 SIMPLE MAIL TRANSFER PROTOCOL (SMTP)

In Internet the source machine establishes a connection to port 25 of the destination machine so as to deliver an e-mail. An e-mail daemon which speaks SMTP is listening to this port. This daemon is supposed to perform the following tasks :

1. Accept the incoming connections, and copy messages from them into appropriate mailboxes.
2. Return an error message to the sender, if a message is not delivered.

> **DO YOU KNOW?**
>
> The TCP/IP protocol that supports e-mail on the Internet is called Simple Mail Transfer Protocol (SMTP).

SMTP is a simple ASCII protocol. Once a TCP connection between a sender and port 25 of the receiver is established, the sending machine operates as a client and the receiving machine acts as a server. The client then waits for the server to talk first. The server sends a line of text to give it identity and announces its willingness/unwillingness to receive mail. If the server is not prepared, the client releases the connection and tries again later. But if the server is willing to accept e-mail, then the client announces the sender of e-mail and its recipient. If such a recipient exists at the destination, then the server tells the client to send the message. The client, then sends the message and the server acknowledges it. No checksums are generally required because TCP provides a reliable byte stream. If there is more e-mail, it can now be sent.

After exchanging all the e-mail, the connection is released. SMTP uses numerical codes. The lines sent by the client are marked C : ; and those sent by the server are marked S : ;

Some of the commands, useful for communication are

HELO, RCTP, DATA, QUIT etc.

RCTP represents recipient. If only one command is used then the message is being sent to only one recipient. If the command is used many times, then the message is sent to more than one recipient. In such a case, each message is individually acknowledged or rejected. The syntax of four commands for the clients are rigidly specified but the syntax for the replies are not that rigid. The SMTP protocol is well defined by RFC 821 but some problems are still present.

Problems in SMTP: Some of the problems in SMTP may be listed as under:

(i) Some older implementations are not capable of handling messages longer than 64 kB.

(ii) If client and server have different time-outs, then one of them may give up when the other is still busy. This will terminate the connection unexpectedly.

(iii) In rate situations, infinite mailstorms can be triggered.

Extended SMTP (ESMTP): Some of these problems can be solved by using the extended SMTP (ESMTP) defined in RFC 1425.

9.8 E-MAIL GATEWAYS

(RGTU, Bhopal, Sem. Exam., June, 2014)

The E-mail using SMTP can work properly if both the sender and the receiver are on the Internet and can support TCP connections between sender and receiver. But there can be many machines which are not on the Internet still want to send and receive e-mail. This is made possible by using the application layer e-mail gateways, as shown in figure 9.10.

In figure 9.10, host *A* speaks only TCP/IP and RFC 822 whereas host *B* speaks only OSI TP4 and X.400. Therefore, without the e-mail gateway they cannot exchange e-mails. But e-mail gateway allows them to exchange e-mails.

Host *A* first establishes a TCP connection to the gateway. Then it uses SMTP and transfer message (1) to the gateway message buffer. Then the gateway daemon establishes a TP4 connection (OSI equivalent of TCP) with the destination host *B*, and message (2) is transferred using the OSI equivalent of SMTP. What a gateway process has to do is extract incoming messages from one queue and deposit them in the other.

Problems

(i) The first problem is that the Internet addresses are totally different from the X.400 addresses so an elaborate mapping between them is required.

(ii) The envelopes and header fields of one system are different from those of the other.

(iii) The body parts are incompatible. For simple unstructured text messages in ASCII, gatewaying is a good solution.

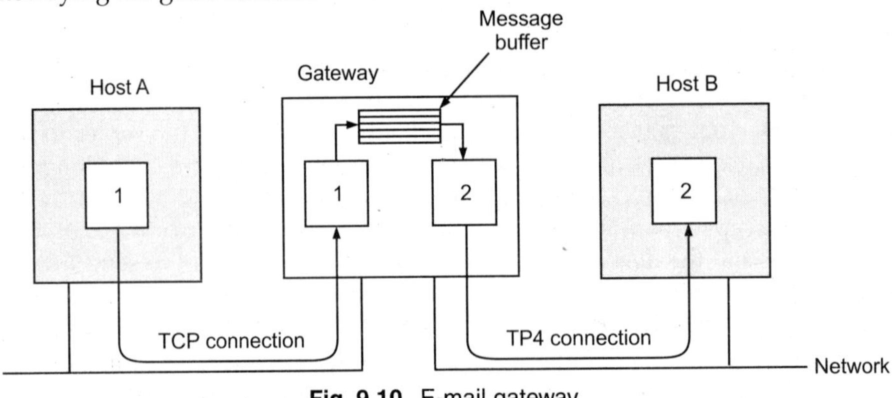

Fig. 9.10. E-mail gateway

9.9. FINAL DELIVERY (DELIVERY PROTOCOLS)

In many companies, the users work at desktop PCs which are not on the Internet and they are not capable of sending or receiving e-mail. To send or receive e-mail, the company uses one or more e-mail servers. A PC talks to an e-mail server using some kind of delivery protocol. Some of the important delivery protocols are :

1. Post office protocol (POP3)
2. Interactive Mail Access Protocol (IMAP)
3. Distribution Mail System Protocol (DMSP)

> **DO YOU KNOW?**
>
> Post office protocol (POP) is a protocol used by a mail server in conjunction with SMTP to receive and hold mail for hosts.

1. Post Office Protocol (POP 3): This is a simple protocol used for fetching e-mail from a remote mailbox. It is defined in RFC 1225. It provides the commands for the user to log in, log out, fetch messages and delete messages. POP 3 consists of ASCII text. It fetches e-mail from the remote mailbox and stores it on the user's local machine.

2. Interactive Mail Access Protocol (IMAP): IMAP is more sophisticated than POP3 and it is defined in RFC 1064. IMAP is ideal for a user having multiple computers such as a laptop on the road, PC at home and a workstation in office. IMAP maintains a central repository which can be accessed from any machine. Therefore, IMAP does not copy e-mail to the user's personal machine. An important feature of IMAP is ability to address mail not by arrival number but by using attributes. This means that the mailbox is like a relational database system than a linear sequence of messages.

IMAP provides the following extra functions :

(i) A user can check the e-mail header before downloading.
(ii) A user can search the contents of e-mail.
(iii) A user can partially download e-mail.
(iv) A user can create, delete or rename mailboxes on the mail server.
(v) A user can create a hierarchy of mailboxes in a folder for e-mail storage.

3. Distributed Mail System Protocol (DMSP): This is the third delivery protocol. It is a part of the PCMAIL system and described in RFC 1056. DMSP allows user to download e-mail from the server to a laptop or PC or workstation and then disconnect. It is possible to read and answer the e-mail while disconnected. After reconnecting, e-mail is transferred.

Additional Processing Facilities: Many systems provide additional processing facilities to the incoming e-mail. One of such important tool is the ability to set up filters. Filters are some rules which are checked when e-mail comes in. Each rule specifies a condition and corresponding action. Another facility is the ability to forward incoming e-mail to different addresses. One more feature is the ability to install a vacation daemon. It is a program which examines each incoming message and returns the sender a reply such as : I am out of station. I will be back on 1st January. It is possible to add some other information such as whom to contact in case of an emergency etc.

9.9.1. Mailing List

Mailing list is the list of e-mail addresses. When a message is sent to the mailing list, copies of the message are delivered to everyone who is on the list. For example, if the message such as *Happy New Year* can be sent to a number of persons, using the facility of mailing list. The mailing list contains e-mail-id of each person along with his name. Sender edits the message addressed to the list using his own mail user agent (MUA). This MUA is passed to mail transport agent (MTA), which transfers it to the host MTA which sends to subscriber MTA and local subscriber's mail box.

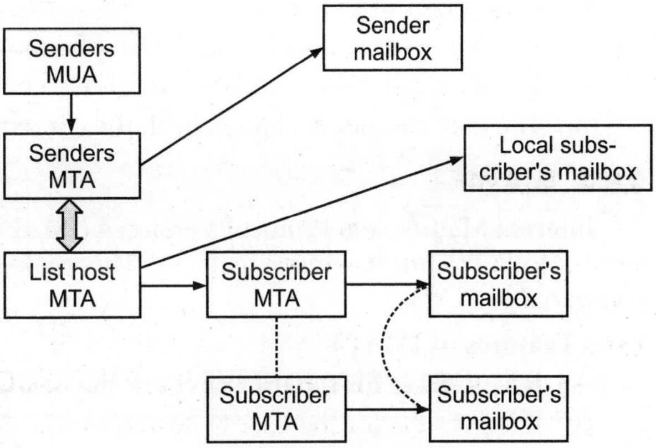

Fig. 9.11. Mailing list

The block diagram of mailing list is shown in figure 9.11.

9.9.2. Mail Access Protocols

The E-mail delivery takes place in three stages. The first and second stages use SMTP but the third stage does not use SMTP because it is a push protocol. This means that SMTP will push the messages from sender to receiver even if the receiver does not want it. The third stage requires a pull protocol. Therefore, the third stage uses a mail access protocol. There are two mail access protocols being used as under:

1. Post office protocol (POP) version 3 *i.e.* POP 3.

2. Internet Mail Access Protocol Version 4 *i.e.* IMAP4.

9.9.3. POP 3

The POP3 consists of client POP3 software and server POP3 software. Out of these, the client POP3 software is installed on the receiving computer whereas the server POP3 software is installed on the mail server. When the user wants to download e-mail from the mailbox on the emial server, the events take place in the following sequence, (figure 9.12):

(i) The client (user) opens a connection with the server on TCP port 110.

(ii) It sends it user name and password to the server so as to access the mailbox.

(iii) The user is then allowed to list and get the mail messages one by one.

This is called as downloading. It is illustrated in figure 9.12.

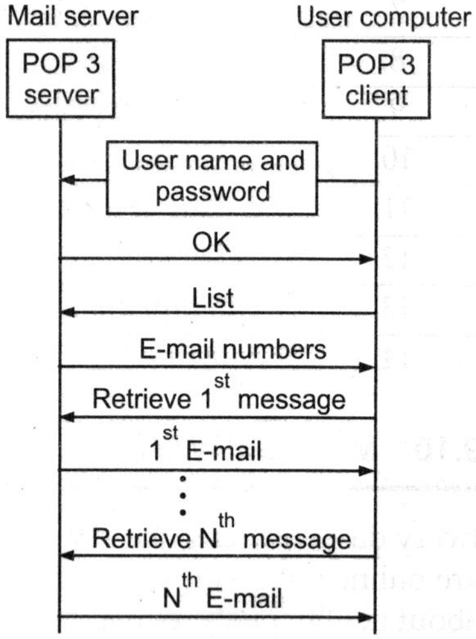

Fig. 9.12. Downloading in POP3

Modes of POP 3: POP3 has following two modes :

(i) Delete mode, and

(ii) Keep mode.

Delete Mode: In this mode, the mail is deleted from the mailbox after each retrieval. This mode is used when the user is working at his permanent computer.

Keep Mode: If operated in this mode, the mail remains in the mailbox after retrieval. This mode is used when the user accesses mail away from the primary computer.

Disadvantages of POP3

(i) POP3 does not allow organization of e-mail on the server.

(ii) The user can not have different folders on the server.

(iii) The user can not partially check the contents of E mail before downloading.

9.9.4. IMAP4

Internet Mail Access Protocol Version 4 (IMAP4) is another mail access protocol which is very similar to POP3 but has more features. This makes IMAP4 more powerful but more complex as compared to POP3.

Extra Features of IMAP4

(i) It is possible for the user to check the header before downloading.

(ii) It is possible for the user to search for the contents of E mail before downloading.

(iii) It is possible to partially download E mail.

(iv) It is possible for the user to create, rename or delete mailboxes on the mail server.

(v) It is possible for the user to create a hierarchy of mailboxes in a folder for e-mail storage.

9.9.5. Comparison of IMAP and POP 3

S.No.	Parameter	POP 3	IMAP
1.	Protocol is defined at	RFC 1939	RFC 2060
2.	TCP port used	110	143
3.	e-mail is stored at	User's PC	Server
4.	e-mail is read	Off lin3e	On line

5.	Connect time required	Small	Long
6.	Use of server resources	Minimal	Extensive
7.	Multiple mail boxes	No	Yes
8.	Who backs up mailboxes	User	ISP
9.	For mobile users	Not good	Good
10.	User control over download	Little	Great
11.	Partial message downloads	No	Yes
12.	Disk quotas	No problem	Could be
13.	Simplicity in implementation	Yes	No
14.	Widespread support	Yes	Increasing

9.10 MAIL SERVER

Every day, Netizens, the citizens of the Internet send each other billions of e-mail messages. If you are online a lot, you yourself may send a dozen or more e-mails each day without even thinking about it. Obviously, e-mail has become an extremely popular communication tool.

9.10.1. An E-mail Message

The first e-mail message was sent in 1971 by an engineer named Ray Tomlinson. Prior to this, you could only send messages to users on a single machine. Tomlinson's breakthrough was the ability to send messages to other machines on the Internet, using the @ sign to designate the receiving machine. An e-mail message has always been nothing more than a simple **text message** - a piece of text sent to a recipient.

In the beginning and even today, e-mail messages tend to be short pieces of text, although the ability to add attachments now makes many e-mail messages quite long. Even with attachments, however, e-mail messages continue to be text messages – we will see why, when we get to the section on attachments.

9.10.2. E-mail Clients

You have probably already received several e-mail messages today. To look at them, you use some sort of **e-mail client**. Many people use well-known stand-alone clients like **Microsoft Outlook**. Outlook Express, Eudora on Pegasus. People who subscribe to free e-mail services like Hotmail or Yahoo use an e-mail client that appears in a **Web page.**

If you are an AOL, customer, you use AOL's e-mail reader. No matter which type of client you are using, it generally does four things :

(i) It shows a list of all of the messages in your mailbox by displaying the **message headers**. The header shows you who has sent the mail, the subject of the mail and may also show the time and date of the message and the message size.

(ii) It lets you select a message header and read the body of the e-mail message.

(iii) It lets you create new messages and send them. You type in the e-mail address of the recipient and the subject for the message, and then type the body of the message.

(iv) Most e-mail clients also let you add attachments to messages you send and save the attachments from messages you receive.

Sophisticated e-mail clients may have all sorts of bells and whistles, but at the core, this is all that an e-mail client does.

9.10.3. A Simple E-mail Server

Given that you have an e-mail client on your machine, you are ready to send and receive e-mail. All that you need is an **e-mail server** for the client to connect to. Let us imagine what the simplest possible e-mail server would look like in order to get a basic understanding of the process. Then we will look at the real thing. There are Web servers, FTP servers, talent servers and e-mail servers running on millions of machines on the Internet right now. These applications run all the time on the server machine and they listen to specific **ports,** waiting for people or programs to attach to the port. The simplest possible e-mail server would work something like this :

It would have a list of e-mail accounts, with one account for each person who can receive e-mail on the server. My account name might be **mbrain,** John Smith's might be **jsmith**, and so on. It would have a text file for each account in the list. So the server would have a text file in its directory named MBRAIN.TXT, another named JSMITH.TXT, and so on. If someone wanted to send me a message, the person would compose a text message ("Marshall, Can we have lunch on Monday ? John") in an e-mail client, and indicate that the message should go to mbrain. When the person presses the Send button, the e-mail client would connect to the e-mail server and pass to the server the name of the recipient (mbrain), the name of the sender (jsmith) and the body of the message. The server would format those pieces of information and append them to the bottom of the MBRAIN.TXT file. The entry in the file might look like this :

- From : jsmith
- To : mbrain
- Marshal,
- Can we have lunch on Monday ?
- John.

There are several other pieces of information that the server might save into the file, like the time and date of receipt and a subject line; but overall, you can see that this is an extremely simple process. As other people sent mail to mbrain, the server would simply append those messages to the bottom of the file in the order that they arrived. The text file would accumulate a series of five or 10 messages, and eventually I would log in to read them. When I wanted to look at my e-mail client would connect to the server machine. In the simplest possible system, it would :

1. Ask the server to send a copy of the MBRAIN.TXT file
2. Ask the server to erase and reset the MBRAIN.TXT file
3. Save the MBRAIN.TXT file on my local machine
4. Pass the file into the separate messages (using the word "Form": as the separator)
5. Show me all of the message headers in a list.

You have to admit that this is a VERY simple system. Surprisingly, the real-e-mail system that you use every day is not much more complicated than this !

9.10.4. The Real E-mail System

For the vast majority of people right now, the real e-mail system consists of two different servers running on a server machine. One is called the SMTP Server, where SMTP stands for Simple Mail Transfer Protocol. The SMTP server handles outgoing mail. The other is a POP3 Server, where POP stands for Post Office Protocol. The POP3 server handles incoming mail. A typical e-mail server is shown in figure 9.13. The SMTP server listens on well-known port number 25, while POP3 listens on port 110.

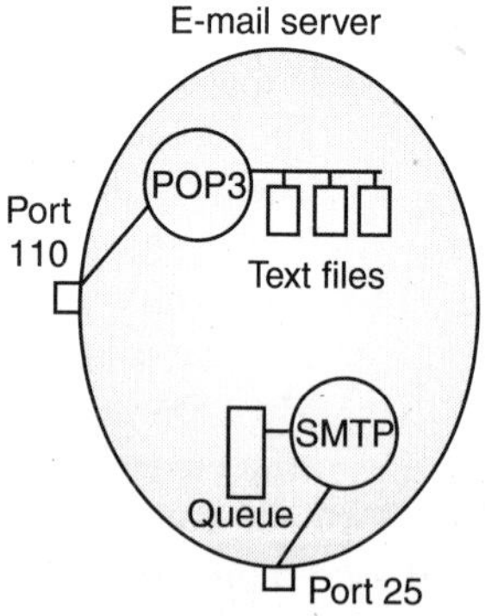

Fig. 9.13. Typical e-mail server.

9.10.5. The SMTP Server

Whenever you send a piece of e-mail, your e-mail client interacts with the SMTP server to handle the sending. The SMTP server on your host may have conversations with other SMTP servers to actually deliver the e-mail. Let us assume that you want to send a piece of e-mail. Your e-mail ID is **brain**, and you have your account on howstuffworks.com. You want to send e-mail to jsmith@mindspring.com. You are using a stand- alone e-mail client like Outlook Express. When you set up your account at howstuffworks, you have told Outlook Express the name of the mail server – mail. howstuffworks.com. When you compose a message and press the Send button, here is what happens :

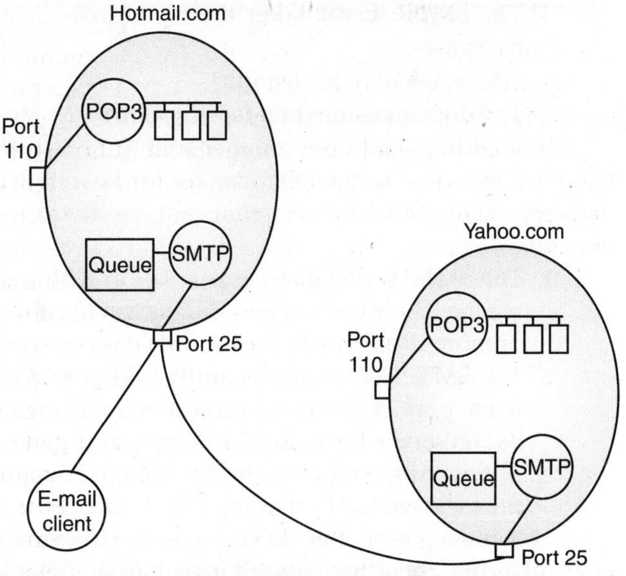

Fig. 9.14. Working of a typical mail server.

1. Outlook Express connects to the SMTP server at mail. howstuffworks. com using **port 25.**

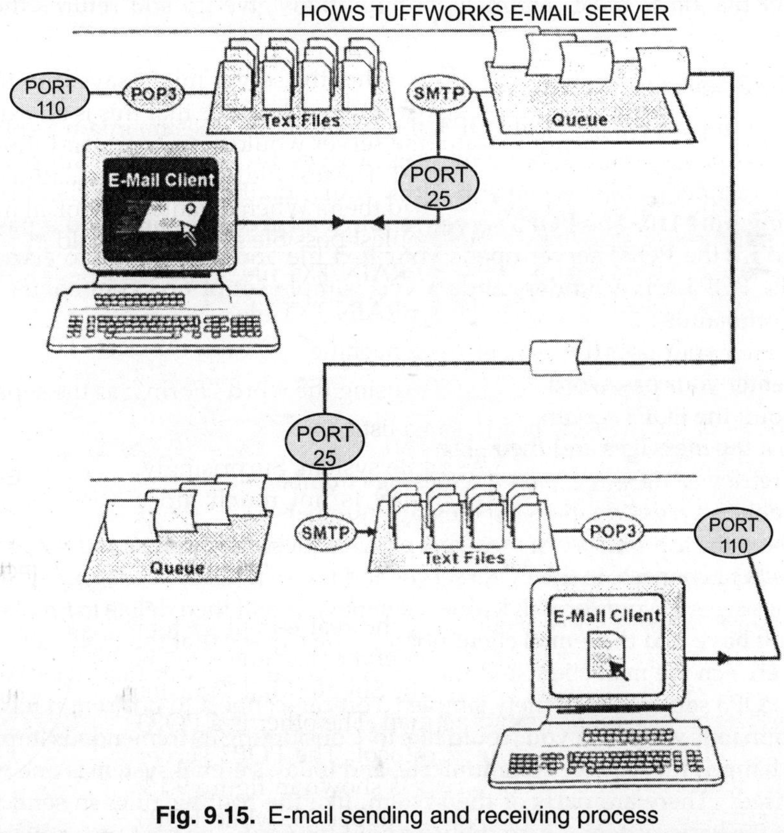

Fig. 9.15. E-mail sending and receiving process

2. Outlook Express has a conversion with the SMTP server, telling the SMTP server the address of the sender and address of the recipient, as well as the body of the message.

3. The SMTP server takes the to address (jsmith @ mindspring.com) and breaks it into two parts :
   The recipient name (jsmith)
   The domain name (mindspring.com)

If the address had been another user at howstuffworks.com, the SMTP server would simply hand the message to the POP3 server for howstuffworks.com (using a little program called the **delivery agent**). Since the recipient is at another domain, SMTP needs to communicate with that domain.

1. The SMTP server has a conversation with a Domain Name Server. It says, "Can you give me the IP address of the SMTP server for mindspring.com?" The DNS replies with the one or more IP addresses for the SMTP server(s) that Mindspring operates.
2. The SMTP server at howstuffworks.com connects with the SMTP server at Mindspring using port 25. It has the same simple text conversation that my e-mail client had with the SMTP server for HowStuffWorks, and gives the message to the Mindspring server. The Mindspring server recognizes that the domain name for jsmith is at Mindspring so it hands the message to Mindspring's POP3 server, which puts the message in jsmith's mailbox.

If, for some reason, the SMTP server at HowStuffWorks cannot connect with the SMTP server at Mindspring, then the message goes into a queue. The SMTP server on most machines uses a program called **sendmail** to do the actual sending, so this queue is called the **sendmail queue**. Sendmail will periodically try to resend the messages in its queue. For example, it might retry every 15 minutes. After four hours, it will usually send you a piece of mail that tells you there is some sort of problem. After five days, most sendmail configurations give up and returns the mail to you undelivered.

9.10.6. The POP3 Server

In the simplest implementations of POP3, the server really does maintain a collection of text files, one for each e-mail account. When a message arrives, the POP3 server simply appends it to the bottom of the recipient's file ! When you check your e-mail, your e-mail client connects to the POP3 server using **port 110**. The POP3 server requires an **account name** and a **password**. Once you have logged in, the POP3 server opens your text file and allows you to access it. Like the SMTP server, the POP3 server understands a very simple set of text commands. Here are the most common commands :

- **USER :** enter your user ID
- **PASS :** enter your password
- **QUIT :** quit the POP3 server
- **LIST :** list the messages and their size
- **RETR :** retrieve a message, pass it a message number
- **DELE :** delete a message, pass it a message number
- **TOP :** show the top x lines of a message, pass it a message number and the number of lines.

Your e-mail client connects to the POP3 server and issues a series of commands to bring copies of your e-mail messages to your local machine. Generally, it will then delete the messages from the server (unless you have told the e-mail client not to). You can see that the POP3 server simply acts as an interface between the mail client and the text file containing your messages. And again, you can see that the POP3 server is extremely simple ! You can connect to it through telnet at port 110 and issue the commands yourself if you would like to. Considering its tremendous impact on society, having forever changed the way we communicate, and today's e-mail system is one of the simplest things ever devised! There are parts of the system, like the routing rules in send mail, that get complicated, but the basic system is incredibly straight forward. The next time you send an e-mail, you know exactly how it is getting to its destination!

9.11. WORLD WIDE WEB (WWW)

(RGTU, Bhopal, Sem. Exam., June, 2014)

People have become aware of the power of Internet through WWW. HTTP is a file transfer protocol which is specifically designed to facilitate access to the WWW. The World Wide Web is an architectural framework for accessing documents which are spread out over a number of machines over Internet. It has a colourful graphical interface which is easy for the beginners to use. It provides information on almost every subject. The web began in 1989 at CERN the European center for nuclear research. The web was designed primarily to get collaboration of scientists stationed all over the world.

The web is basically a client-server system. The web pages are written in the languages HTML and Java. The growth of the World-Wide Web (WWW or simply Web) today is simply phenomenal. Each day, thousands or more people gain access to the Internet (above 100 million users at recent estimates). Easy retrieval of electronic information in conjunction with the multimedia capabilities of Web browsers (like Mosaic or Netscape) is what started this explosion. This topic provides some basic information behind some of this technology used in accessing the World-Wide Web.

Difference between Web and Internet : The Web and the Internet are not the same thing. The Web is a collection of standard protocols or instructions, sent back and forth over the Internet to gain access to information. The Internet, on the other hand, is a **network of networks** — a more physical entity.

9.11.1 Web from the Users Side

The user (client) looks at the web as a collection of vast world wide collection of documents called pages in short.

Links or Pointers : Each page may contain links or pointers to it, related pages, anywhere in the world. A user can follow a link by clicking on it. This will take him to the pages pointed by the links. This process can be repeated indefinitely.

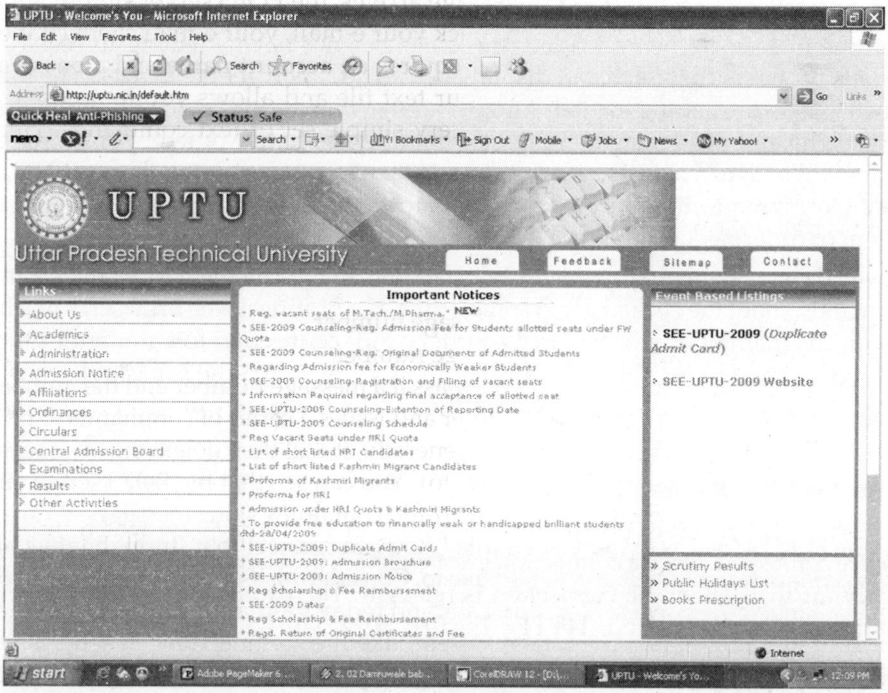

Fig. 9.16. A web page.

Hypertext : Pages which point to the other pages are said to use hypertext.

Browser : The program used for viewing pages is called as a browser. The job of a browser is to fetch the page requested by the user, interprets the text and formatting commands which it contains, display the page with proper format on the screen. An example of a web page is shown in figure 9.16.

A web page starts with a title and contains the following

(i) Some information

(ii) Strings of text, linked to other pages

(iii) E-mail address of the page's maintainer.

Hyperlinks : Strings of text that are links to other pages are called hyperlinks. They are highlightened by underlining, using special colour or both. In order to follow a link, the user has to place the cursor on the highlightened area using the mouse or arrow keys and select it by clicking or hitting ENTER. The browsers can be of two types, namely the graphical browsers and non-graphical browsers. But, the graphical browsers are more popular. Voice based browsers are also being developed.

Most browsers have numerous buttons and features that make the navigation on web easier. There can be a button to back to the previous page or a button for going forward to the next page. Some browsers can provide a facility of having a button or menu item to set a book mark on a given page and another one to display the list of bookmarks. This makes it possible to revisit any of them with a single click on mouse. It is also possible to save pages or print them. Lot of options are available to control the screen layout and setting various preferences of the users. The web pages can also contain line drawings, icons, maps, photographs etc., and they can be linked (if required) to another page.

Hypermedia: All pages may not be viewable in the conventional way, some pages may contain audio tracks, video clips or both. If the hypertext pages are mixed with other media, the result of such a mixing is called as **hypermedia**. Some browsers are capable of displaying all kinds of hypermedia but others cannot do so. Many web pages contain large images that take a long time to load. When the images are being loaded, the user does not have anything to see. To solve this problem, some browsers first fetch and display the text and then get the images. The user can read the text when images are getting loaded. Another strategy can be to provide an option to disable the automatic fetching and displaying of images. One more alternative opted by some page writers is to display the full image in a coarse resolution and then to fill up the details gradually. Some web pages contain forms requesting the user to fill up information. This is to search a database for a user supplied item or ordering a product etc.

Some web pages contain maps which allow the users to click on them to get the zooming facility or get information about the clicked geographical area. For hosting a web browser, a machine should be directly on the Internet or at least have a SLIP or PPP connection to a router or other machine which is directly on Internet. This is because of the way in which the browser fetches a page. To fetch a page, it has to establish a TCP connection to the machine from where the page is to be fetched.

9.11.2. Web from the Servers Side

Every website has a server process. It is listening to TCP port 80 on which incoming clients (browsers) are connected. Once a connection is established, the client sends a request and the server sends a reply for that. Then, the connection is released. The protocol used for defining the legal request and replies is called HTTP. Figure 9.17 shows various parts of the web model.

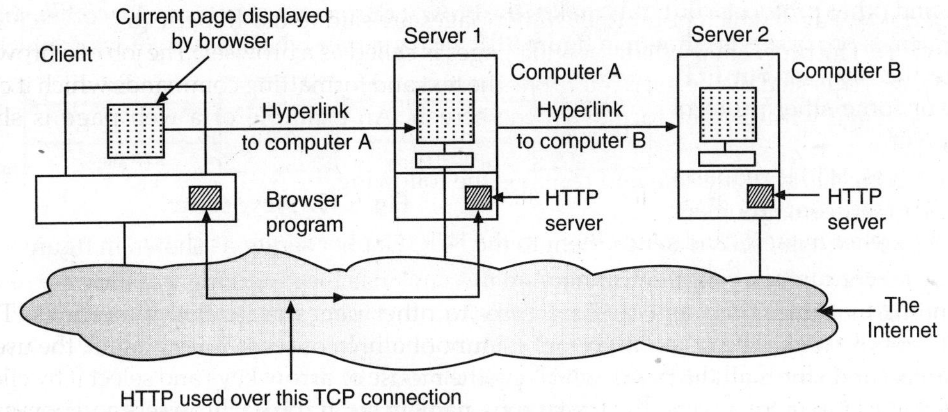

Fig. 9.17. Web model

9.12. UNIFORM RESOURCE LOCATOR (URL)

URL has following three parts :

(i) Name of the protocol (e.g. http)

(ii) Name of the machine where the page is located.

(iii) The name of the file containing the page.

The example of a URL is as under :

http:// www.w4.org / hypertext / WWW/Project. html.

Protocol Machine name. File name containing the page

Steps between user's click and page being displayed

(i) The browser determine the URL.

(ii) Browser asks DNS for the IP address of the specified machine (www.w4.org).

(iii) DNS sends the IP address say 19.24.0.22.

(iv) Browser establishes a TCP connection to port 80 on the machine having IP address 19.24.0.22.

(v) Browser then sends a command reading

GET / hypertext / WWW / Project. html.

(vi) The specified machine sends the file 'Project html'.

(vii) The TCP connection is released.

(viii) The browser displays all the text in 'Project.html' file.

(ix) The browser fetches and displays all images in the 'Project.html' file.

(x) The browsers display the step that they are currently executing.

(xi) For each in-line image such as drawing or photo, the browser establishes a new TCP connection to the appropriate server to fetch images.

(xii) HTTP is an ASCII protocol. Hence, it is easy for a person at a terminal to directly talk to the web servers. All that is required is a TCP connection to port 80 on the server and to get such a connection, we have to use the Telnet program.

9.12.1 Proxy Server

All the servers cannot speak HTTP, some of them use the FTP, Gopher or some other protocols. A large information is available on FTP and Gopher servers, hence, it must be made available to web users. To do so, one solution can be to have a browser which can use the HTTP as well as FTP,

Gopher and other protocols. But, this makes the browser unnecessarily large. The other solution to this problem is proxy server, shown in figure 9.18. Proxy server is basically a gateway which speaks HTTP to the browser but FTP, Gopher or some other protocol to the server.

It receives HTTP requests from a browser, converts them in FTP or Gopher requests and sends them to the FTP/Gopher server as shown in figure 9.18.

Fig. 9.18. Proxy server

Proxy server can be a program running on the same machine working as a browser or can be a free standing machine. The users can configure their browsers with proxies for protocols that the browser cannot speak. The other important feature of a proxy server is caching. A caching proxy server collect and stores all the pages which pass through it.

When a user asks for a page, the proxy server has to see if it has the page stored with it. If the page is there then it will see if the page is upto date. If the page is current, it passes the page to the user otherwise a new copy of page is fetched. A proxy server can be put inside a firewall. The user can access the web but it is not allowed the full Internet access. In such situation, the user talks to the proxy server and the server contacts different sites and fetches pages on behalf of the user.

9.13. HYPERTEXT TRANSFER PROTOCOL (HTTP)

HTTP is used mainly to access data on WWW. This protocol transfers data in the form of plaintext, hypertext, audio, video etc. The function of HTTP is like a combination of FTP and SMTP. It uses services of TCP. It uses only one TCP connection (port 80). There is no separate control connection. Only the data transfer takes place between the client and server. The data transfer in HTTP is similar to SMTP. The format of the messages is controlled by MIME like headers.

9.13.1 Principle of HTTP Operation

The principle of HTTP is simple. A client sends a request. The server sends a response. The request and response messages carry data in the form of a letter with a MIME like format. Figure 9.24 shows the HTTP transactions between client and server. The client initializes the transaction by sending a request message and the server replies it by sending a response.

9.13.2 The Web and HTTP

As a matter of fact, HTTP is the Web's application layer protocol. Actually, it is the heart of the Web.

Basically, HTTP socket can be implemented in following two programs :

1. A client's program
2. A server's program

These programs are executed on different systems and talk to each other by following exchanging HTTP messages. HTTP defines how Web clients such as browsers request Web pages from Web servers and how servers transfer Web pages to clients. HTTP uses TCP as its under lying transport protocol (rather than using UDP). The HTTP client first initiates a TCP connection with the server. After establishing connection, the browser and the server processes access TCP through their socket interface. TCP provides a reliable data transfer service to HTTP. This means that each HTTP request message, transmitted by a client will eventually arrive intact at the server.

In a similar manner, each HTTP response message transmitted by the server will eventually arrive intact at the client, due to the reliable TCP connection. Due to this kind of layered architecture HTTP need not have to worry about the lost data or about the details of how TCP deals with the loss and retransmission of data. It is managed by TCP.

Statelessness

In HTTP, the server sends the files requested to the client without storing any information about the client. Hence, it may happen that the same client may ask the information repeatedly to the server and the server would not even understand it. Therefore, it will keep resending those files. As the HTTP servers does not maintain any information about the state of client, it is called as a stateless protocol.

9.13.3. Types of HTTP Connections

As a matter of fact, there are following two types of connections in HTTP :

(i) Non-persistent connections

(ii) Persistent connections

HTTP use persistent connections in its default mode. But, it is possible to configure HTTP clients and servers to use non-persistent connections.

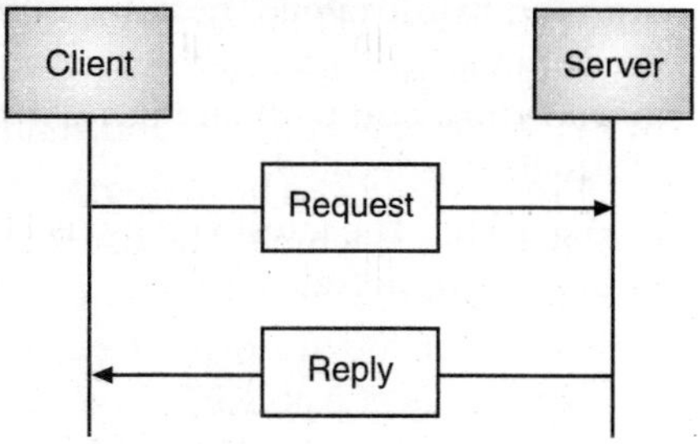

Fig. 9.19 HTTP transaction.

The non-persistent connections have some problems. First is that everytime a new connection requires to be established and maintained. Second problem is that each object suffers a delivery delay of two RTTs (Round Trip Time) — one RTT corresponding to TCP connection establishment and the other is request and receive an object.

If the persistent types of connection is used, then the server will leave the TCP connection open after sending a response. Hence, the same connection can be utilized sending request and response between a pair of client and server.

Further, the persistent connections can be of following two versions :

(i) Without pipelining

(ii) With pipelining

The default mode of HTTP uses persistent connections with pipelining.

9.13.4 HTTP Messages

Basically, the HTTP messages are of following two types :

(i) Request message

(ii) Response message

The format of both these messages is almost the same.

9.13.5 Request Message

Figure 9.20 shows the format of the request message. It consists of a request line, headers and sometimes a body.

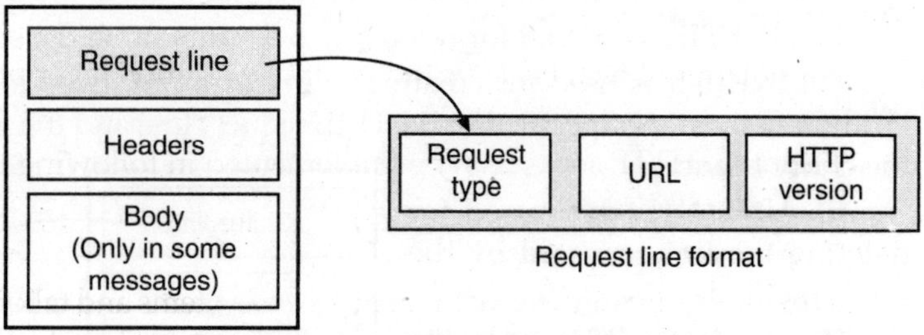

Fig. 9.20. HTTP request message

Request Line: The request line is used for defining the request type, resource (URL) and HTTP version as shown in figure 9.20.

Request type : Several request types are defined.

Uniform Resource Locator (URL) : The client accessing a web page needs an address. The HTTP uses the URL to facilitate the access of any document distribution over the world. The URL defines four thing as shown in figure 9.21. They are as under :

(i) Method (ii) Host computer (iii) Port (iv) Path

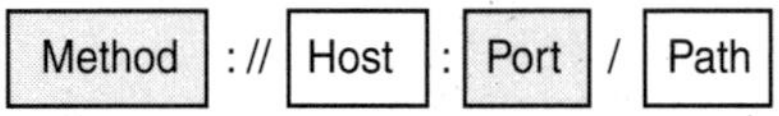

Fig. 9.21. URL

Method is the protocol used such as FTP, HTTP. Host is the computer where the required information is located. The name of the computer begins with www but this is not mandatory. URL can optionally contain the server's port number. If the port is included then it must be inserted between host and path and it must be separated by a colon. Path is the name of the file where the information is located.

Version : The latest version is HTTP is 1.1, but, the versions 0.9 and 1 are used. The example of URL is shown in figure 9.22.

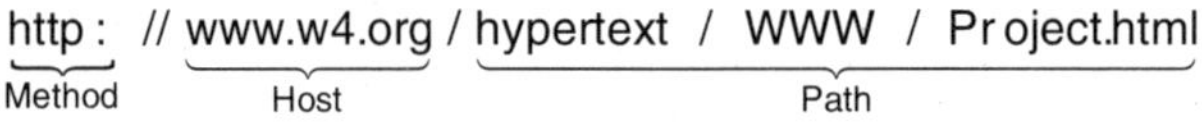

Fig. 9.22. Example of URL

9.13.6. Methods (Request Type)

This is one of the fields in the request line format. It defines different kinds of messages referred to a request types or methods. The request method is a command or request issued by the client to the server. Following are some of the important methods (request types).

1. GET : This method is used when the client wants to retrieve a document from the server. The address from where this document is to be obtained is defined in the URL.

2. HEAD : This method is used when the client wants some information about a document but not the document itself.

3. POST : This is used by the client to provide some information to the server.

4. PUT : This is used by client for providing a new or replacement document to be stored on the server.

5. PATCH : This is similar to PUT. But, there is one change. The request contains a list of differences which should be implemented in the existing file.

6. COPY : This method is used to copy a file to another location.

7. MOVE : This method is used for moving a file to another location.

8. DELETE : It is used for removing a document on the server.

9. LINK : It is used for creating a link or a link from a document to another location. The location of the file is specified in the URL request line and the location of destination is specified in the entity header.

10. UNLINK : It is used for deleting the links created by the LINK method.

11. OPTION : It is used by the client to ask the server about various available options.

9.13.7. Response Message

Figure 9.23(a) shows the format of the response message. It consists of a status line, a header and sometimes a body.

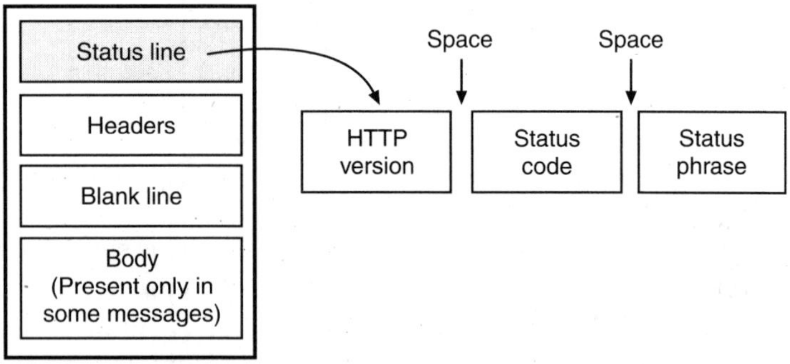

(a) Response message (b) Status line format

Fig. 9.23.

Status Line: Basically, the status line is used for defining the status of the response message. As shown in figure 9.23 (b), it consists of HTTP version, status code and status phrases with spaces in between.

HTTP Version : This field is same as the corresponding field in the request line.
Status Code : This field is similar to those in FTP and SMTP protocols. It contains three digits.
Status Phrase : It explains the status code in the text form.

9.13.8. Headers

Headers in the response message are used for exchanging additional information between the client and server. The header can be of one line or more lines. The format of a header line is shown in figure 9.24 which shows that it consists of a header name, a colon, a space and a header value.

An header line can be one of the following four types :

1. General header
2. Request header
3. Response header, and
4. Entity header

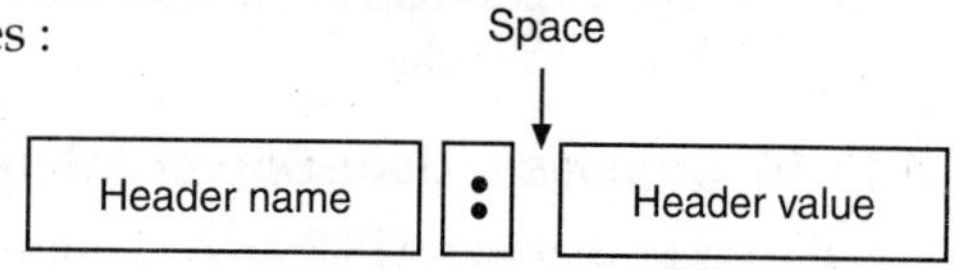

Fig. 9.24. Header format

A request message can contain only general, request and entity headers, but, a response message can contain only general, response and entity headers.

The comparison of request message and response message is shown in figure 9.25.

1. General Header: It gives general information about the message. It is present request message as well as response message.

2. Request Header : It can be present only in the request message. It is used to specify the client's configuration and the client's preferred file format.

3. Response Header : It can be present only in the response message. It is used for specifying the server's configuration.

4. Entity Header: It is used for giving the information about the body of the document. It can be present in the request message as well as the response message.

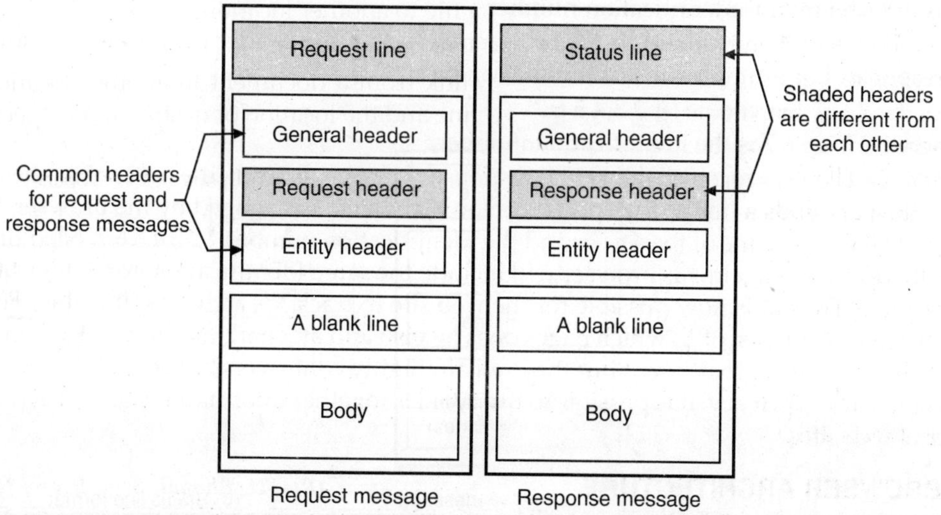

Fig. 9.25. Comparison of request message and response message.

9.13.9 Performance Comparison of HTTP and SMTP

S.No.	SMTP	HTTP
1.	Message is transferred from client server.	Message transfer is from client to server or the other way round.
2.	Uses TCP.	Uses TCP.
3.	Uses port 25 for transmission.	Uses port 80 for transmission.
4.	SMTP messages are to be read by humans.	HTTP messages are to be read and understood by the HTTP servers and HTTP clients.
5.	These messages are first stored and then forwarded.	These messages are immediately delivered.

9.13.10 Cookies : User-Server Interaction

We know that the HTTP servers are stateless. The drawbacks of being stateless is that the server cannot identify the client. However, the advantage of statelessness is that the server design is simplified to a great extent and it permits the engineers to develop high performance web servers which can handle thousands of simultaneous TCP connections. However, many a times, it is necessary for a web site to identify users. In such cases, HTTP uses cookies. Cookies are identified in RFC 2109 and they allow sites to keep track of users. Cookies are not used by all the sites but some of the prominent sites that use cookies are : Yahoo, Amazon etc.

Components of Cookie Technology: Following are the four components of the cookie technology:

(i) A cookie header line in HTTP response message.
(ii) A cookie header line in the HTTP request message.
(iii) A cookie file kept on the user's end system and managed by user's browser.
(iv) A back end database at web site.

Operating Principle: If a new user *X* contacts a site (that uses cookies) for the first time, then the web site creates a unique identification number and then creates an entry in its back end database which is indexed by this identification number.

The server will then respond to *X*'s browser including the header set-cookies : header, in the HTTP response. For example the header line can be :

Set-cookie : 1 2 3 4 5 6 7

Where 1 2 3 4 5 6 7 is the identification number.

When *X*'s browser receives the HTTP response message, it reads the set cookie : header. The browser then appends a line to the special cookie file which is managed by the browser. This line will include the host name of the server and the identification number 1 2 3 4 5 6 7. Next time when *X* visits this same site again, his browser will include the same identification number in each of his HTTP request. Thus, it is now possible for the web site to track X's activities. It is then possible to know the areas of interest of *X*, which pages does he visit and at what time etc. Cookies simplify the Internet shopping to a great extent but they remain highly controversial because they are thought as invasion in users privacy. It is possible to use cookies together information about *X* across a large number of web sites.

9.14. BROWSER ARCHITECTURE *(RGTU, Bhopal, Sem. Exam; June, 2014)*

Even though a number of browsers are available around, the browser architecture is nearly the same. Each browser consists of the following parts :

(i) A controller
(ii) Client Program
(iii) Interpreters

Figure 9.26 shows the general architecture of a browser.

The controller receives input from the keyboard or mouse. It then uses the client programs like HTTP, FTP etc., to access the document. After accessing the document, the controller makes use of an interpreter such a HTML or Java (depending on type of document) and displays the accessed document on the screen.

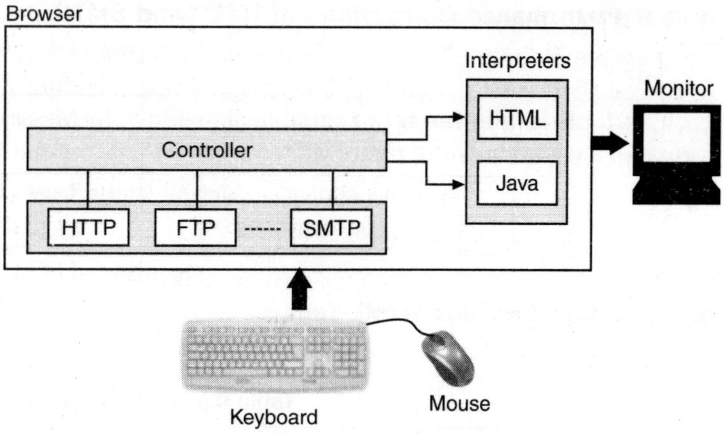

Fig. 9.26. Browser architecture

9.14.1. Web Documents

The web documents may be classified into three categories as shown in figure 9.27

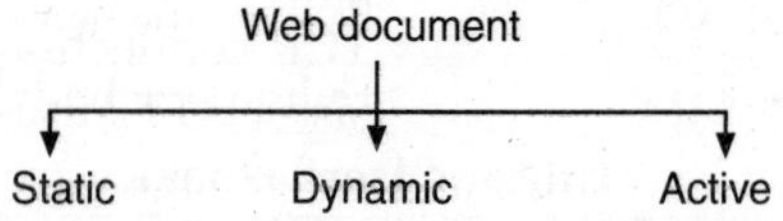

Fig. 9.27. Categories of web documents.

Static Documents: The contents of static documents are fixed. They are created and stored in a server. If required the client can get a copy of static document. The contents of the static document are determined when it is created. The contents cannot changed when the static document is being used. It is possible to change the contents of static document at the server but the user cannot change them.

9.14.2 Hypertext Markup Language (HTML)

HTML is a language used for creating the web pages. It uses certain marks to format the text. For example, if a part of text is required to be *boldface* then we can use the beginning and ending bold face tags (marks) in the text as shown below :

<B> – Beginning of boldface
</B> – End of boldface

Here, <B> and </B> are the instructions for the browser. The browser will make the part of the text between these tags bold. HTML lets the user to use only ASCII characters for the main text as well as for formatting instructions. Hence, every computer can receive the whole document as an ASCII document. The formatting instructions are used by the browser to format the data.

9.15. STRUCTURE OF A WEB PAGE

Basically, a web page may be divided into following two parts :

1. Head
2. Body

1. Head : This is the first page of a web page. The head carries the title of the web page and some other parameters that the browser will use.

2. Body : The body carries actual contents of a web page. It includes the text and tags. Text is the actual information carrying matter and the tags decide the appearance of the document.

9.15.1 HTML Tags

Tags are the marks which are embedded into the text. A tag is always enclosed in two signs (< and >). They usually come in pairs as beginning and ending tags. A tag will have a list of attributes. Each attribute is followed by an equal to sign and value associated with the attribute. The general format of a tag is shown in figure 9.28.

<Tag Name Attribute = Value Attribute = Value ... >	< / Tag Name >
(a) Beginning tag	(b) Ending tag

Fig. 9.28.

9.15.2. Some Common HTML Tags

Table 9.5 shows some common HTML tags.

Table 9.5. Common Tags.

Beginning Tag	Ending Tag	Meaning
	Skeletal Tags	
<HTML>	</HTML>	Defines an HTML document
<HEAD>	</HEAD>	Defines the head of the document
<BODY>	</BODY>	Defines the body of the document
	Title and Header Tags	
<TITLE>	</TITLE>	Defines the title of the document
<Hn>	</Hn>	Defines different headers (n is an integer)
	Text Formatting Tags	
<B>	</B>	Boldface
<I>	</I>	Italic
<U>	</U>	Underlined
_		Subscript
[	]	Superscript
	Data Flow Tags	
<CENTER>	</CENTER>	Centered
 		Line break
	List Tags	
<OL>	</OL>	Ordered list
<UL>	</UL>	Unordered list
<LI>	</LL>	An item in a list
	Image Tags	
<IMG>		Defines an image
	Hyperlink Tags	
<A>	</A>	Defines an address (hyperlink)
	Executable Contents	
<APPLET>	</APPLET>	The document is an object

9.16. DYNAMIC DOCUMENT

The dynamic documents are not present in a predefined format, like static documents. A dynamic document is created by a web server whenever a browser requests for the document. Let us consider figure 9.29 to understand how a dynamic document is created and passed on to the client. First the client sends a request to the web server. After receiving this request, the web server will execute an application program to create a dynamic document. The server returns the dynamic document as a response of the request to the client.

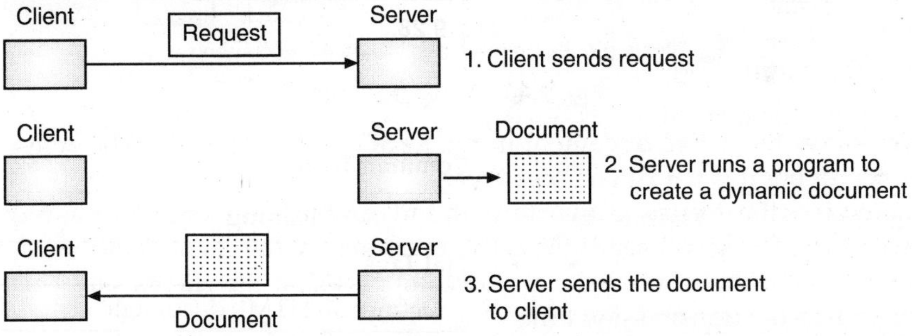

Fig. 9.29. Dynamic document.

The contents of a dynamic document will be different corresponding to every request. A simple example of a dynamic document is to get time and data from the server. A server follows the following steps given below to handle dynamic documents :

(i) The server checks the URL in order to find if it has defined a dynamic document.

(ii) If the URL has defined the dynamic document, then the server executes the program.

(iii) The output of this program is the dynamic document. It is returned back to the client.

9.16.1. Common Gateway Interface (CGI)

Basically, CGI is the name of a technology which creates the dynamic documents and handle them too. CGI is in fact a set of standards. It defines the way in which a dynamic document should be written, the way in which input data be supplied to the program and how the output result be used. Note that CGI is not a new language. It allows the user to use the existing languages such as C, C++, Perl etc. However CGI defines rules and terms which are to be followed by the programmers.

The word **common** in CGI shows that this standard defines some rules which are common to any language.

The word **gateway** indicates that a CGI program is a gateway which can be used to access other resources such as database and graphic packages.

Lastly the word **interface** in CGI indicates the presence of a set of terms, calls and variables which can be used in any CGI program.

CGI Program: It is a code which is written in one of the languages that supports CGI (such as C, C++, etc.).

9.16.2. Active Documents

Active documents can be defined as the program, that is needed to be run at the client side. The examples of active documents are the programs creating animated graphics on the screen or the one which helps interaction with the user. Let us consider figure 9.30. It shows that whenever a browser requests for an active document, the server will send a copy of document in the form of byte code. The active document will then be run at the browser (client) site.

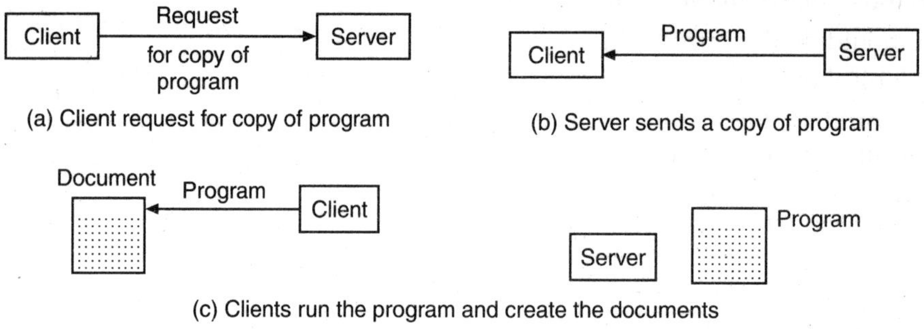

Fig. 9.30. Active document

The server stores the active document in the form of a binary code. The active document is stored on the server but it is not run on the server.

The client receives the document and stores it, and can run it as many times as required without repeating the request. The server sends the active document to the client in the binary form. Hence, it can be compressed at the server's site and decompressed at the clients site. This will save the bandwidth as well as the transmission time.

Steps in Creation of an Active Document: Let us consider figure 9.30 to understand the creation, compilation and execution of an active document.

(i) At the server, a program is written in source code and stored in a file.

(ii) Then, the program is compiled and binary code is created and stored in a file at the server's site.

(iii) A client (browser) requests for a copy of program as shown in figure 9.30 (a). This program is transported from the server to the client in the compressed form.

(iv) The client converts the received program from binary code into executable code using its own software.

(v) The client runs the program to create a result which can include animation or interaction with the user.

9.17. PERFORMANCE ENHANCEMENT

Because of ever increasing number of users of world wide web, the servers, routers and lines are frequently overloaded and the user has to wait a lot. To reduce this endless delay, researchers have developed various techniques for improving the performance. Some of these techniques may be listed as under :

(i) Caching

(ii) Server replication

(iii) Content delivery network

9.17.1. Caching in Web Browser

This is a very simple technique used in which pages which have requested are saved (stored) in case they are used again. This technique is specially useful if the pages being requested are being used a great deal such as www.yahoo.com.

The usual procedure is for some process called proxy to maintain the cache. To use caching, a browser can be so configured that all page requests are made to a proxy instead of to the page's real server. If the proxy has the requested page stored with it then it return the page immediately. But, the proxy does not have the requested page, then it fetches the page from the server, add it to the cache for future use and returns the page to the requesting client.

The two most important questions related to caching are as under :

1. Who should do the caching ?
2. How long should the pages be cached ?

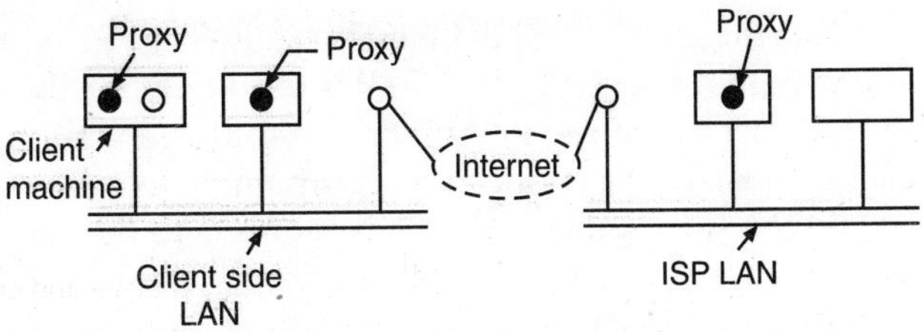

Fig. 9.31. Hierarchical caching using three proxies.

1. Who should do the caching ?: There can be several answers to this question. The first type of user doing caching is individual PC. Individual PCs run proxies so that they can quickly look up pages previously visited. On a LAN, the machine shared by all the machines is often the proxy. Often all these caches operate at the same time, so requests first go to the local proxy. If the local proxy does not have the requested page, then it queries the LAN proxy. If it fails, then the LAN proxy queries the ISP proxy. The ISP proxy must succeed either from its cache or higher level cache or from the server itself. This scheme is called hierarchical caching and its implementation is shown in figure 9.31.

2. How long should the pages be cached ?: Some pages must not be cached at all. Some pages containing a fast change information such as share prices need to be updated. Hence, the most important factor which determines when to evict a page from cache is how much staleness the users can put up with. If the proxy updates the cache pages it will rarely return a stale page, but here effectiveness is reduced. Note that caching is a client side technique of improving the performance. Web pages containing dynamic content should never be cached, because parameters may be different next time. Another approach to improve performance is *proactive caching*. When a proxy fetches a page from a server, it can inspect the page to see if there are any hyperlinks on it. If they are, then it issues request to the relevant servers preload the cache with the pages pointed to, just in case they are needed. However disadvantage of this approach is that the reduction in access time subsequent requests is at the expense of flooding of communication lines with pages that are never required.

9.17.2. Server Replication

Caching is a technique used on the client side to improve the performance but server replication is the technique used for improving performance on the server side. This technique is also called as mirroring. In this technique, servers replicate the contents at multiple, widely-separated locations. The performance degradation also takes place due to a phenomenon called as flash crowds. This happens if a website which is not very known and normally handling a small traffic suddenly becomes known and gets flooded with a heavy traffic. Hence, there has to be a way for a website which suddenly notices a massive increase traffic to automatically clone (replicate) itself at as many locations as needed and known those sites operational as long as heavy traffic continues to pour in. Another strategy is to create dynamic replicas on a per page basis depending on when the traffic is coming from.

9.17.3. Content Delivery Networks

Companies called CDNs (content delivery networks) talk to content provides such a music sites, newspapers etc. and offer to deliver them to end users by charging fees. After signing the contract the content owner gives the CDN the contents of its website for preprocessing and distribution. After this, the CDN talks to large number of ISPs and offers to pay them if they permit to place a remotely managed server having valuable contents on their LANs. The contents replicated at various sites world wide, the performance is likely to improve to a great extent.

9.17.4. Telnet

As a matter of fact, Telnet is age old terminal emulation program to login into remote systems. It uses TCP/IP protocol and underlying communication can take place through PSTN, Satellites etc.

One thing we can note that telnet allows us to login in system for any operation (OS commands and program execution) and FTP is used only for file transfer use. Telnet is an application used on the Internet to connect to a remote computer, enabling access to the computer and its resources. The action consists of a connection, comparable to a connection via telephone, except the link remains within the Internet. Our terminal connects to the server on the other computer and is able to give actions (input), and receive data in return (output). Telnet is used for a number of activities such as telneting to a site, or checking e-mail at another account, other on-line services, or perhaps an Internet BBS, in addition to many other possibilities.

Telnet is another example of cyberspace extension, or cybertravel, which enables the user to travel all across the internet accessing machines or databases that may offer different services or information. In addition to machines across the net that are accessible, each machine has a number of ports that offer different types of access to the system. As with any other type of travel, the key part of beginning our adventure is to know something about where we are going or want to go. To use Telnet, we need a specific address, however, now it is possible to access telnet connections through hyperlinks on the web, this means that browsing and stumbling across something that we are interested in is possible.

9.17.5. Logging into Telnet

All uses of telnet begin with the designation of a destination; an address that we want to go to; a computer to be accessed. Although most would automatically take the direct route by telneting directly to their intended destination, one should always consider the option of not taking a direct route. As mentioned elsewhere, the Internet has very little provisions regarding privacy. When we telnet to a site, the host (server) that accepts our entry, it also records where we have telneted from. If this is a concern of ours, it can be possible to telnet along several sites to our destination to cloud our trail and original departure point. With an address, it may sometimes be beneficial to also have a port to telnet. If we will not have one we will be sent to the default telnet port, and generally, this is the right thing to do.

However, sometimes with M.U.D or irc, a special port number is required. Similarly, with other on-line services, a special port number is required. Other options in participating with a system are available through a number of other open ports that will allow telnet access. There are thousands of ports on a machine and there's no telling what some of them could be set up to do. Generally, the command for telneting to a site is :

- telnet presence.lglobal.com

  or with port numbers

- telnet presence.lglobal.com 23

Basically, we start the command, then give the address with the option of including the port number. Sometimes, this command is also referred to as open. If we have the correct address and the appropriate port number is required, we will be then prompted to login. If we do not have instructions on what to login, or a login name of our own, well we are probably out of luck.

It is always worth trying guest or new, sometimes this is the method of getting an account of our own. If we have a working login, we will then usually be prompted for a password. Again, if we do not have one, we are probably out of luck. If we have successfully passed these two points, we are then in the system, at which point we are usually on our own. Each system tends to have its own uniqueness, and our telneting for a service, we will be automatically in some sort of environment that probably user-friendly and menu driven.

An important keyboard command to know is :

Ctrl] or ^].

It will always remove us you from the environment of the system that we are on, and return us to the telnet environment where we will see a telnet prompt and can type quit or close if we want out.

Otherwise, just press a clear return and we will return to the system we were on. When we have done our session on the machine, type logout or exit to return to our original destination. Remember, if we have taken a roundabout route, we successively logouts to get home.

9.18. INTRODUCTION TO MULTIMEDIA

The meaning of multimedia is two or more media. But, in practice, when we use the word multimedia, it means the combination of two or more **continuous media**. Usually, the two media are audio and video. Multimedia is a system, which is a combination of following components such as :

(i) Text
(ii) Images or Graphics
(iii) Animation
(iv) Sound
(v) Video

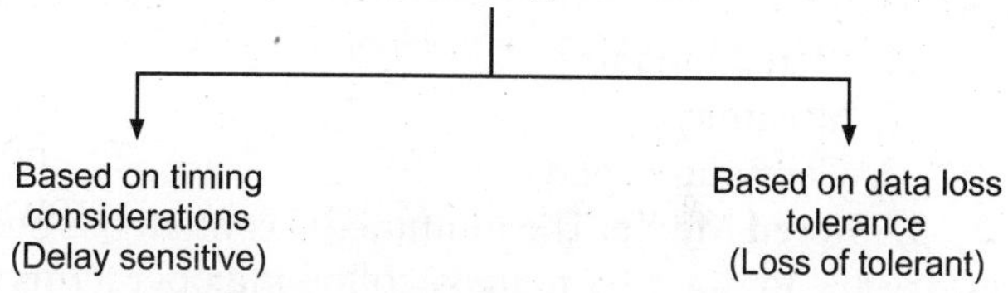

Fig. 9.32. Classification of multimedia applications

In Multimedia, all these five media are used to represent the information to the users via the computer. In practice, many multimedia packages comprise of only two or three medias of the above media, as sometimes the use of all media types may not be appropriate for the information exchange. Typically, all computers have inbuilt sound cards. Playing of notes, music is possible. We can classify the multimedia application in two major categories as shown in figure 9.32.

Delay Sensitive: These applications are sensitive to timing delays. Many multimedia applications are useless if the sender to receiver delay is more than a few hundred milliseconds.

Loss of Tolerant: Most of the multimedia applications are loss tolerant. This means that they allow the loss of data to some extent without compromising with the quality. Therefore, the multimedia applications are delay sensitive but loss tolerant, which are different from the elastic applications such as E-mail, FTP and Telnet.

9.19. EXAMPLES OF MULTIMEDIA APPLICATIONS *(RGTU, Bhopal, Sem. Exam; June, 2014)*

The multimedia applications can be broadly categorized into three classes as under :

1. Streaming stored audio/video
2. Streaming live audio/video
3. Interactive audio/video.

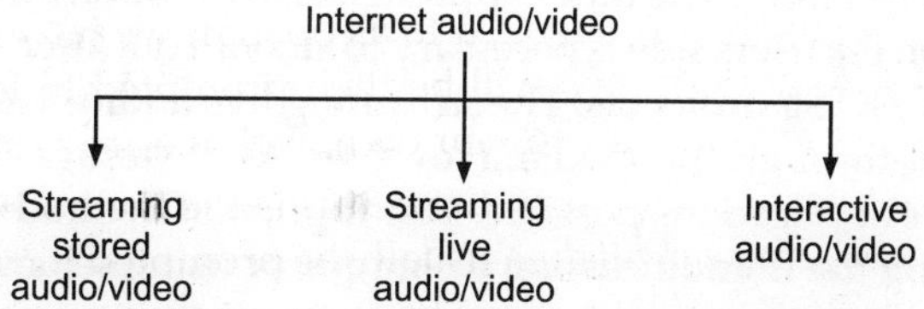

1. Streaming Stored Audio/Video: A client can download the files through internet which are in compressed form and stored on server.

e.g. : Stored audio songs, famous lectures.

Stored video : Movies, T.V. shows and music clips.

2. Streaming Live Audio/Video: A user listens to broad audio and video through the internet.

e.g. : Internet radio

Some radio stations broadcast their programs only on the internet.

3. Interactive Audio/Video: People use the internet to interactively communicate with one another.

e.g. : Internet teleconferencing and internet telephony.

9.19.1. Streaming Stored Audio and Video *(RGTU, Bhopal, Sem. Exam; June, 2014)*

This is the class of applications which allows the clients to request for compressed audio or video files which are stored on the servers. Stored audio files may contain educational lectures, songs and other recorded audio stuff. Stored video files may contain movies. TV shows, documentaries, cartoons or music video clips. This class of applications has three distinguishing factors as follows :

1. Stored media
2. Streaming
3. Continuous playout.

1. Stored Media: The multimedia content has been prerecorded and stored at the server. So, it is possible for the user to make following operations on the stored contents.

(i) Pause
(ii) Rewind
(iii) Fast forward
(iv) Index through

2. Streaming: In a streaming stored audio/video applications, a client starts playout of the audio/video a few seconds after receiving the file from the server. Hence, the client plays out audio/video from one location in the file when it is receiving later parts of the same file from the server. This technique is called as **streaming.** This avoids need to download the entire file before beginning playout.

3. Continuous Playout: Once the playout of the multimedia content begins, it is necessary to proceed as per the original timing of recording. The requirement puts critical delay constraints on data delivery.

9.19.2. Streaming Live Audio and Video

This is similar to traditional broadcast radio and TV, except for one change that transmission takes place over the internet.

Internet Radio: After it became possible to stream audio over the Internet, commercial radio stations started broadcasting their contents over the Internet. There are two general approaches followed. In the first one, the programs are prerecorded and stored on disk. Listeners can download any of these programs for listening. The other approach is to broadcast live over the Internet. With this technique, buffering on the users side is necessary to smooth out jitter. The Internet radio should use multicasting with RTP/RTSP protocols. The streaming live audio/video is not stored. Hence, a client cannot fast forward through the media. However, local storage is always possible, which makes the other interactive applications possible. Distribution of live audio/video to many receiver scan be accomplished using the IP multicasting technique or using separate unicast streams.

9.19.3. Real Time Interactive Audio and Video

In this class of applications, people are allowed to use audio/video to communicate with each other in real time. Real time interactive audio over the internet is called as **Internet phone**. This can provide local, PBX and long distance telephone service at a very low cost. This service also facilitates some special services such as :

(i) Web phone integration
(ii) Group real time communication
(iii) Directory services
(iv) Caller filtering

Real time interactive video is called as video conferencing. Here, the users communicate visually as well as orally. The delays should not be longer than 400 msec.

9.20. PROBLEMS FOR MULTIMEDIA IN TODAY'S INTERNET

The protocol used by the Internet is IP. Recall that IP provides the **best effort service** to all the datagrams it carries. This means that the Internet makes best effort to move each datagram from sender to receiver as quickly as possible. But, the Internet does not make any promise about end to end delay for a individual packet. Same is true for TCP and UDP as they operate on IP. As there is no guarantee of delivering the packets in time, it is an extremely challenging problem to develop successful multimedia networking applications for the Internet. So, multimedia on Internet has not achieved the success that it is expected to get.

9.21. EXPECTATIONS OF MULTIMEDIA FROM THE INTERNET

Different multimedia types have different expectations from the Internet. Some of the important expectations are as under :

1. Real time characteristics
2. High bandwidth
3. Multicasting support
4. Error characteristics of the medium
5. Session individuality
6. Security
7. Mobility

1. Real Time Characteristics: Most multimedia applications have real time traffic requirements. For example, audio and video data must be played back continuously at the rate at which they are sampled. If the data does not arrive in time, the playing back process will stop and human ears and eyes can easily pick up the artifact. In Internet telephony, human beings can tolerate a latency of ~200 msec. If the latency exceeds this limit, the voice will sound like a call routed over a long satellite circuit and users will complain about the quality of the call. Thus, a real time traffic requires strict bounds on end-to-end delay and jitter. The time taken by the packet to travel from the source to the destination is the end-to-end delay, while the variability in the inter-packet delay at the receiver is the jitter. Performance improves with decrease in both these quantities.

2. High bandwidth requirements : Most multimedia applications demand significantly higher bandwidth. For e.g. a PCM encoded voice grade telephone conversation requires a bandwidth of 64 kbps, which is greater than any of the currently available modem speeds. Similarly unencoded video conferencing with a picture size of 1024 × 1024 pixels at 30 frames/sec will require a bandwidth of ~1Gbps. Of course, there are compression techniques that take advantage of spatial and temporal locality present in voice and video, but they still require significantly higher bandwidth than what we have been using in the past. The best effort Internet model neither provides any mechanism to demand such high bandwidths nor prevents any one from sending at such high rates. If everyone starts sending at a high rate there will be heavy congestion in the network leading to a congestion collapse that can completely halt the Internet. There is no mechanism in the best-effort Internet to prevent this from happening (except using a brute force technique of disconnecting the source of such congestion). It is left to the discretion of the application to dynamically adapt to network congestions. Elastic applications do utilize a closed-loop feedback mechanism (built into TCP) to prevent congestion, but RTI and RTT applications cannot use the same feedback techniques, as the resulting delay will significantly degrade performance. Thus such applications have the capability to create a congestion collapse. The enhanced Internet service models prevent this by employing admission control mechanisms. The application must first get permission from some authority to send traffic at a given rate and with some given characteristics.

If the authority allows the admission, then the application can send data limited at that rate only. The newer service models employ traffic policing mechanisms to ensure that applications do not violate this contract.

Compression techniques: There are two kinds of compression techniques :

Those that loose some information (or lossy compression) and those that do not loose any information (Lossless Compression). In lossless compression, the data received by the user are identical to the original data. With lossy compression, some distortion is introduced. Lossy compression usually yields significantly higher compression ratios than lossless compression. However, lossy compression might not be acceptable to all sections of the society (viz. medical specialist).

Also compression of such huge data sets will increase the packet processing delay.

3. Multicasting Support : Many multimedia applications, like video/audio conferencing will require the communication network to provide a strong multicasting support. The best-effort Internet is sufficiently equipped to handle multicasting , however, we wish to make a point that any enhancements to the Internet must also support multicasting. In the best-effort Internet, multicasting is achieved by sending packets to a multicast IP address (Class D IP addresses are multicast IP addresses). All those receivers that desire to receive the multicast have to register with appropriate routers, which forward any packets that are sent to that address. For the enhanced Internet service models, we expect that the problem of multicasting to be solved along similar principles.

4. Error Characteristics of the Medium: From a receiver's point of view, error occurs when a packet is lost or when it arrives with some bits turned the other way. Both these events are treated as lost packets in most multimedia applications and suitable actions are taken to overcome them. Many algorithms are now available that allow such applications to interpolate lost information. In case of wired networks, the probability of packet loss is less, while unfortunately the same cannot be about wireless networks. So the entire equation of handling lost packets in wireless networks changes. Interesting and complex scenario arise when a multimedia streaming session occurs between wired host and mobile host. Often, the schemes applied to one may not be the optimum choice for the other. For e.g. FEC is most widely used to detect and correct errors in packets. However, FEC requires overhead bits in each packets, which wastes bandwidth. The choice of FEC might seem to be optimum for wireless networks, as this will prevent unnecessary retransmission in case of lost packets (which happen relatively more frequently). However FEC might not be an optimum choice for wired networks, as the lesser probability of lost just not justify the added overhead. The best-effort Internet does not provide any information regarding the path over which the packets might travel and their error characteristics. Hence without the feedback, the sender application has no knowledge of the error characteristics of the medium and ends up using error correction/detection mechanism that may not be optimum. For the newer Internet Service models, as discussed earlier, sender application has to go through admission control. At that time, the sender can specify the maximum error that it can tolerate. If the network is unable to find a path that can satisfy this requirement (assuming some QoS based routing is in use), it will just reject the connection or make a counter offer to the sender specifying the error rate at which it is willing to accept the connection. In other words, the sender is made aware of the error characteristics of the medium.

5. Session individuality: A multimedia session might consist of multiple streams that need to be clearly identified. For e.g. the sender might choose to send the audio and video as two separate streams, which the receiver needs to decode synchronously. Another option might be that the sender might bunch the audio and video streams together, but divide quality into a base layer and some enhancement layers, so that low-bandwidth receivers might be able to receive only the base layer, which high-bandwidth receivers might also receive the enhancement layers. In the best-effort Internet, every session can be identified using the tuple <Src Ip, Src Port, Dst IP, Dst Port, Protocol>. Thus, individual transport layer sockets have to be established for every session. However, if there ever arises a need to bunch sessions together (for cutting costs, say) there is no available mechanism. Hence, there is clearly a need to multiplex different streams into the same transport layer socket. This can be accomplished by having a visible "Session layer" between the application and the transport layer (as in the 7-layer OSI model), which has been notably absent in the best effort Internet. Apart from identifying the individual streams, the session layer might even insert timing and sequencing information into the stream that facilitates playout at the receiver. RTP and RTCP are two protocols that are most widely used for this purpose.

While RTP is concerned with identifying individual streams and inserting time stamps, RTCP is used for exchange of commands and responses between the client and the server. It may be noted that this is not a major issue, as introducing such a session layer in the best-effort Internet is also a trivial task.

6. Security: The security issue has been neglected in almost all discussions of multimedia communication. However, it is now apparent that security needs to be incorporated. Security provides the following three aspects to multimedia data: Integrity (data cannot be changed in mid-flight), Authenticity (Data comes from the right source) and Encryption (Data cannot be deciphered by any third party).

For example, public broadcasts require data integrity and data authenticity, while private communication requires data encryption. All the above aspects can be provided using cryptographic techniques.

7. Mobility: Mobility aspect has added another dimension of complexity to multimedia networks. It opens up questions on a host of complex issues like: routing to mobile terminals, maintaining the Quality of Service (QoS) when the host is in motion, inter-networking between wireless and wired networks etc.

The 3rd Generation cellular systems, now offers data communication in addition to voice at rate approaching 384 kbps and it is quite possible to view a low-quality streaming video or listen to a low quality streaming audio file on your cell phone. Cellular systems have a large coverage area and hence permit higher mobility. Another emerging option is IEEE 802.11b wireless LAN, which can operate at speeds exceeding 54 Mbps. Wireless LANs typically cover a smaller area and hence limit mobility. However, the main advantage of wireless LANs is that they work in the ISM band, which does not require licensing. This eliminates the need to invest significant amounts of money for obtaining a license. Hence, wireless LANs can be easily operated by any start-up company. Also because of its simplicity and availability of cheap products in the market the end cost to the user is lower. It remains to be seen, which system, 3G or wireless LANs will win the race for providing wireless data service. Interesting scenarios arise when we think about the Internetworking between wireless LANs and cellular systems. Due to the vastly different environments on the wired and wireless Internet, application performance degrades and some proxy agents need to be introduced.

Many projects are researching into extending the newer Internet service models to be extended to cover wireless networks. It remains to be seen how effective this models will be.

9.22. AUDIO AND VIDEO COMPRESSION

9.22.1. Digitizing Audio

When a sound wave strikes a microscope, it gets converted in an analog electrical signal which represents the sound amplitude as a function of time. The frequency range of human ears is from 20 Hz to 20 kHz. The ear hears logarithmically. Therefore, the sound intensity is expressed in dB (decibels). Audio waves can be converted into digital form by an ADC (analog to digital converter). An A/D converter receives electrical voltage as input and generates a binary number as output. The A to D conversion process includes two steps namely the sampling and quantization. They are illustrated in figure 9.33. Quantization is a process of approximation and in the process of approximation, there is bound to be some error. Such an error is called as **quantization error** or **quantization noise.** If this noise is too large then it can be detected by ears.

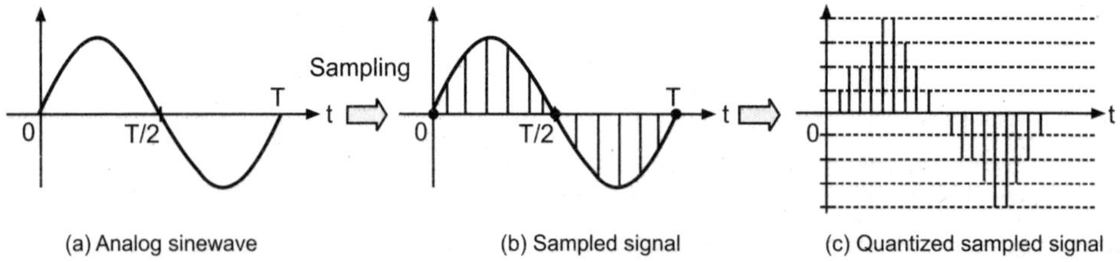

Fig. 9.33.

Two well known examples where sampled sound is used are as under:

(i) Telephone

(ii) Audio compact disc.

Telephone system use 8 bits pulse code modulation to make 8000 samples per second. Such a system gives data rate of 56 kbps or 64 kbps. Audio CDs are digital with a sampling rate of 44,100 samples per second. Digitized sound can be easily processed by computers in software. Many programs exist on PCs to allow users to record display, edit, mix and store sound waves from multiple sources. All professional sound recording and editing are digital now a days.

9.22.2. Digitizing Video

In the normal video reproduction on the TV or computer screen, the picture frames are displayed at a very rapid rate (typically 50 frames per sec.). The interlaced scanning is used in which 25 frames are displayed per second in an alternate manner. Each frame is divided into small grids called pixels. For the black and white TV, each 8 bit pixel represents one of 256 possible different gray levels. For colour TV, each pixel is represented by 24 bits out of which 8 bits each are allotted for one primary colour (Red, Green and Blue). The lowest resolution for a coloured picture is 1024 × 768 pixels per frame. Therefore, the number of bits corresponding to one picture is calculated as under:

Number of bits/pictures = 2 × Number of frames/sec. × Resolution

$$= 2 \times 25 \times 1024 \times 768$$

$$= 944 \text{ Mbps}$$

To send video with minimum resolution, we need to transmit at 944 Mbps which requires a huge bandwidth. This much bandwidth is not available practically. Hence, we have to use the compression technique so as to reduce the number of bits and hence to reduce the bandwidth. Compression is essential to send video over the internet.

9.22.3. Audio Compression

The audio compression is used to send speech or music. For speech, we have to compress the 64 kHz digitized signal. This concept is illustrated as shown in figure 9.34.

For the speech, we have to compress the 1.411 MHz digitized music signal. For audio compression, we need to use either predictive encoding or perceptual encoding.

Predictive Encoding : In this technique, first the audio signal is sampled. Then the differences between the samples are encoded instead of encoding all the samples. This type of encoding is generally preferred for the speech signals. A quality audio on CD requires a transmission bandwidth of 1.411 Mbps. So, a substantial compression is necessary to make transmission over the Internet possible. Hence, various audio compression algorithm are developed. The most popular algorithm is MPEG audio. It has three layers (variants). Out of these layers, MP3 (MPEG layer 3) is the most powerful and best known. MPEG is actually a video compression standard and MP3 belongs to its audio portion. One way of audio compression is to use the waveform coding technique. The other type is perceptual coding. MP3 is based on **perceptual coding.**

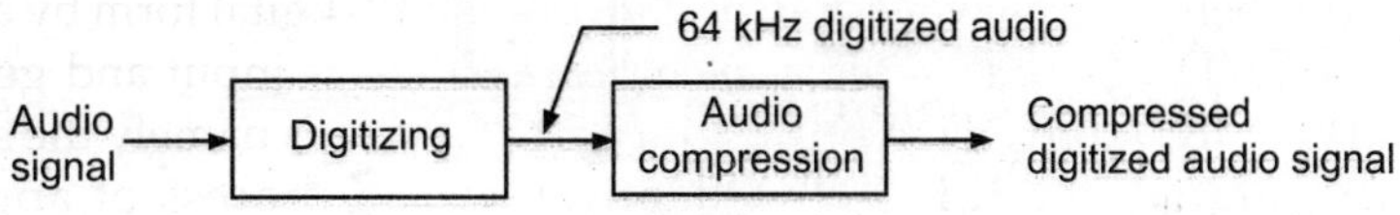

Fig. 9.34.

The perceptual coding is based on **frequency masking principle**. The audio compression is done by sampling the waveform of 32 kHz, 44.1 kHz or 48 kHz. Sampling can be done on one or more channels, in any one of following four configurations :

(i) Monophonic (a single input stream)

(ii) Dual monophonic (e.g. an English and French sound tracks)

(iii) Disjoint stereo (each channel compressed separately)

(iv) Joint stereo.

First the output bit rate is chosen. Standard bit rates are 96 kbps, 128 kbps etc. Then the samples are passed in groups of 1152. Each group of samples is first passed through 32 digital filters to get 32 frequency bands. At the same time, the input is fed into psychoacoustic mode in order to determine the masked frequencies. Then each of the 32 frequency bands is further transformed to provide a finer spectral resolution. The bits are encoded using Huffman encoding various techniques are used for noise reduction, antialiasing and for exploiting the interchannel redundancy.

9.22.4. Video Compression in the Internet

A video is a sequence of images. They are displayed at a constant rate of 24 or 30 images per second.

Video Compression: We can compress video by compressing images. The two standard used for image compression are as under:

(i) Joint Photographic Experts Group (JPEG)

(ii) Moving Picture Experts Group (MPEG)

Out of these, the JPEG is used to compress still images whereas the MPEG is used for compressing moving pictures.

9.22.5. Data Compression

One of the way of using the computer is to download the compressed form of text, audio and video data. Data compression is essential for efficient storage and transmission of this type of data. A data compression system consists of an encoder and a decoder. The encoder performs compression of the incoming data and decoder is used for decompression and recons-truction as shown in figure 9.35.

We need less number of bits to transmit the same information if data compression is exercised. This also reduces the bandwidth requirement.

Types of Compression: Basically, there are two forms of data compression :

1. Lossless compression
2. Lossy compression

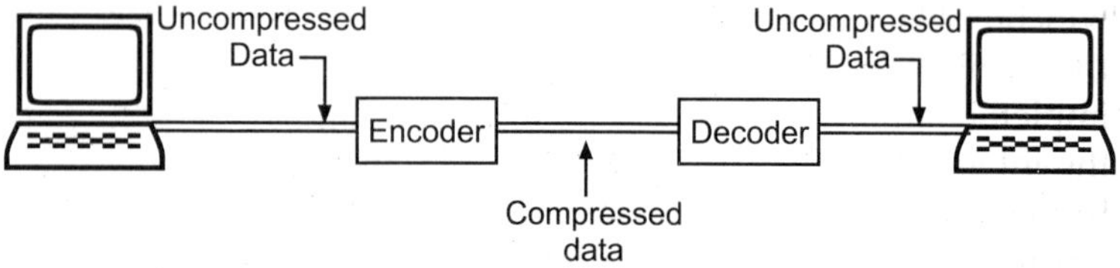

Fig. 9.35. Illustration of data compression

1. Lossless Compression: In lossless compression, the redundant information contained in the data is removed. Due to removal of such information, there is no loss of the data of interest. Hence, it is called as lossless compression. Lossless compression is also known as data compaction.

2. Lossy Compression: In this type of compression there is a loss of information in a controlled manner. The lossy compression is therefore not completely reversible. But the advantage of this type is higher compression ratios than the lossless compression. The lossless compression is used for the digital data. The lampele - ziv algorithm is used for the same. For many applications, the lossy compression is preferred due to its higher compression without a significant loss of important information. For digital audio and video applications, we need a standard compression algorithms.

DO YOU KNOW?

Lossless compression eliminates redundant data bits, thereby, reducing the bit rate with no effect on signal quality.

9.22.6. JPEG Standards*

1. Basic Concepts: Video is simply a sequence of images (plus sound). Hence, video compression is called as image compression. One of the standards used for compression is JPEG i.e. Joint Photographic Experts Group standards. This is a standard for compressing still pictures such as photographs.

2. JPEG Objectives: JPEG undertook to develop a single standard applicable to the still-imaging needs to a wide range of applications in all the different industries that might use digital continuous-tone imaging. Listing the objectives in detail best sees the scope of this as follows :

(i) To be at or near the state of the art for degree of compression versus image quality.

(ii) To be parameterizable so that the user can select the desired compression versus quality trade-off.

(iii) To be applicable to practically any kind of source image, without regard to dimensions, image content, aspect ratio, etc.

(iv) To have computational requirements that are reasonable for both hardware or software implementations, and

(v) To support four different modes of operation :

(a) Sequential encoding, where each image component is encoded in the same order that it was scanned.

(b) Progressive encoding, where the image is encoded in multiple passes so that a coarse

DO YOU KNOW?

Lossy compression compromises signal quality in order to reduce the bit rate. For voice transmissions, vocoders are often used to achieve great reductions in bit rate.

* Discuss different steps of JPEG compression standard.

image is presented rapidly, followed by repeated images showing greater and greater detail;

(c) Lossless encoding, where the encoding guarantees exact reproduction of all the data in the source image ;

(d) Hierarchical encoding, where the image is encoded at multiple resolutions.

These objectives were extremely ambitious, yet they are largely met by the completed standard, which is testimony to the excellent work of the JPEG committee. The four modes of operation and the options within them cause the complete standard to be quite complex. However, most implementations will not use them all, but rather they will focus on the choices that are relevant to the intended application(s). In this way, the standard is more like a shopping list of algorithms, from which you can select the approach, which best suits, your purpose. The JPEG standard defines a Baseline sequential coder, which will be sufficient for a wide range of applications. Many early implementations are only supporting the Baseline.

3. Image Compression Using JPEG: If the picture is not coloured (gray scale), then each pixel can be represented by an 8-bit integer (i.e., $2^8 = 256$ levels). If a coloured picture is to be represented, then we have to use 24 bits (3×8 bits) with each 8 bits representing one of the three basic colours (Red, Green and Blue). As shown in figure 9.36 in JPEG, a gray scale picture is divided into blocks of 8×8 pixels.

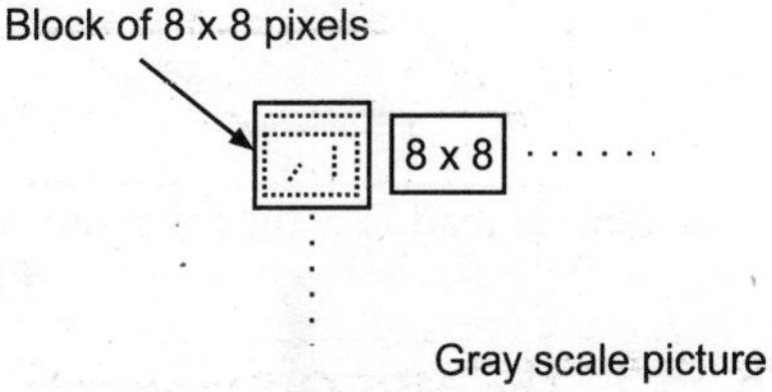

Fig. 9.36.

The JPEG changes the picture into a linear (vector) set of numbers. This process is explained below.

4. Three Steps in JPEG: The three steps in JPEG process are as under :

(i) Discrete Cosine Transform (DCT)

(ii) Quantization

(iii) Data compression

These steps have been illustrated in figure 9.37.

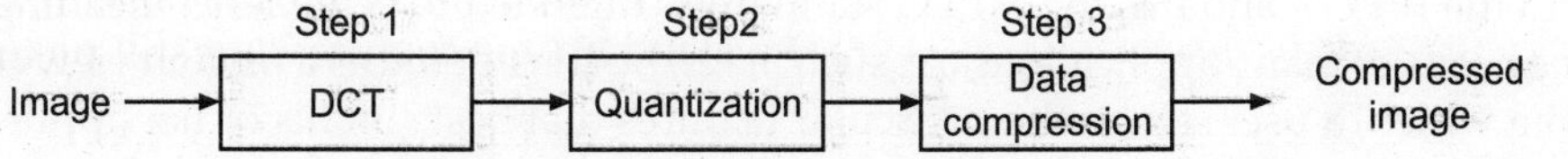

Fig. 9.37. Illustration of JPEG process

(i) Discrete Cosine Transform (DCT): The image in the form of blocks of $8 \times 8 = 64$ pixels is applied to the DCT box. Each 8×8 block will undergo a transformation called discrete cosine transform (DCT).

(ii) Quantization: The transformed 8×8 blocks are then quantized. The process of quantization reduces the number of bits require to be used for encoding. Quantization is the process of approximation. Hence, we lose some information and it is not possible to recover this information at all. This is why JPEG is called as lossy compression.

(iii) Compression: In the first step, (DCT) a transformation table T is created. After quantization, the values in this table are read to remove redundant zero bits. The table is read in a zig-zag manner, as shown in figure 9.38 to produce the compressed digital image. In may be noted that the numbers in the table indicate the gray scale and zeros indicate redundancy i.e. no change in shade. So at the output of the compressor, we get the compressed digitized image.

T Table

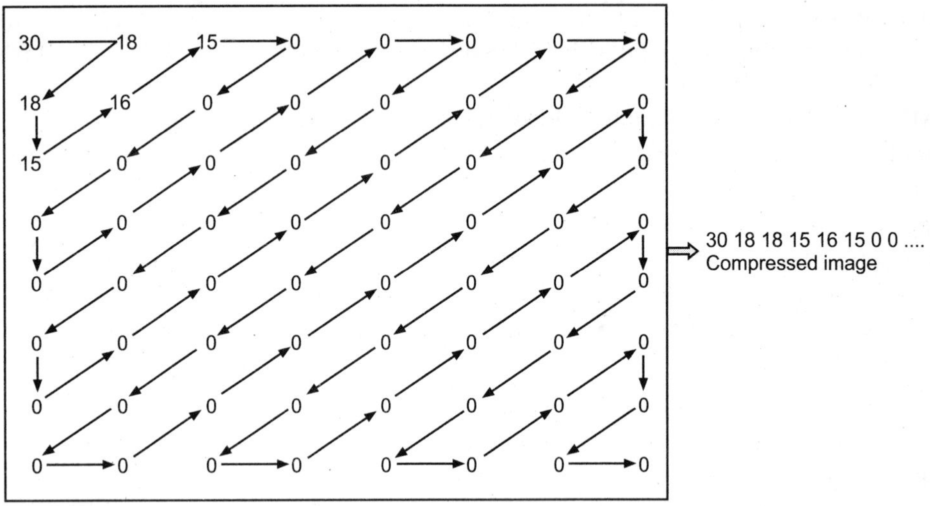

The table is read in a zig-zag manner

Fig. 9.38. Illustration of compression

9.22.7. The MPEG Standards

MPEG is another standard. Long form of MPEG is Motion Picture Experts Group standards. These algorithms are standardized to compress moving pictures. Digital motion video can be accomplished with the JPEG still image standard if you have fast enough hardware to process 30 images per second. However, the maximum compression potential cannot be achieved because the redundancy between frames is not beginning exploited. Furthermore, there are many other things to be considered in compressing and decompressing motion video, as indicated in the objectives.

1. Objectives

As with the JPEG standard, the MPEG standard is intended to be generic, meaning that it will support the needs of many applications. As such, it can be considered as a motion view compression toolkit, from which, a user selects the particular features that best suit his or her application. More specific objectives are as under:

(1) The standard will deliver acceptable video quality at compressed data rates between 1.0 and 1.5 Mbps.

(2) It will support either symmetric or asymmetric compress/decompress applications.

(3) When compression takes it into account, random-access playback is possible to any specified degree.

(4) Similarly, when compression takes it into account, fast-forward, fast-reverse, or normal-reverse playback modes can be made available in addition to normal (forward) playback.

(5) Audio/video synchronization will be maintained.

(6) Catastrophic behaviour in the presence of data errors should be avoidable.

(7) When it is required, compression-decompression delay can be controlled.

(8) Editability should be available when required by the application.

(9) There should be sufficient format (flexibility to support playing of video in windows).

(10) The processing requirements should not preclude the development of low-cost chipsets, which are capable of encoding in real-time.

As you can see, some of these objectives are conflicting, and they all conflict with the objectives of cost and quality. In spite of that, the proposed standard provides for all of the objectives, but of course not all at once. A proposed application has to make its own choices about which features of the standard it requires and accept any trade-off that this may cause. The MPEG compression standards are among the mot popular compression techniques. Various MPEG standards are as follows :

(i) MPEG 1 : For CD-ROM quality video (1.5 Mbps)

(ii) MPEG 2 : For high quality DVD video (3.6 Mbps)

(iii) MPEG 4 : For object oriented video compression.

The H.261 video compression standards are also very popular in the Internet.

2. Principle of Image Compression in MPEG-2: MPEG-2 takes the advantage of high correlation that exists between successive pictures of a moving picture to compress a sequence of moving images. MPEG constructs three types of pictures namely,

(i) Intra pictures (I-pictures)

(ii) Predicted pictures (P-pictures)

(iii) bidirectional predicted pictures (B-pictures)

In MPEG, every Mth picture in a sequence can be fully compressed by using a standard MPEG algorithm, these are I pictures. Then the successive I-pictures are compared and the portion of the image that have moved are identified. The image sections which do not move are carried forward in time domain to intermediate picture by the decoder memory. Then a subset of intermediate pictures is selected and the prediction and correction of locations of the image section which have moved, is carried out. These predicted and corrected images are the P-pictures. The pictures between I and P-pictures are the B-pictures. They incorporate the stationary image sections uncovered by the moving sections. Figure 9.39 shows the relative position of these pictures.

The P and B pictures are allowed but they are not required and their number keeps changing. It is possible to form a sequence without P or B pictures but a sequence without I pictures is not possible.

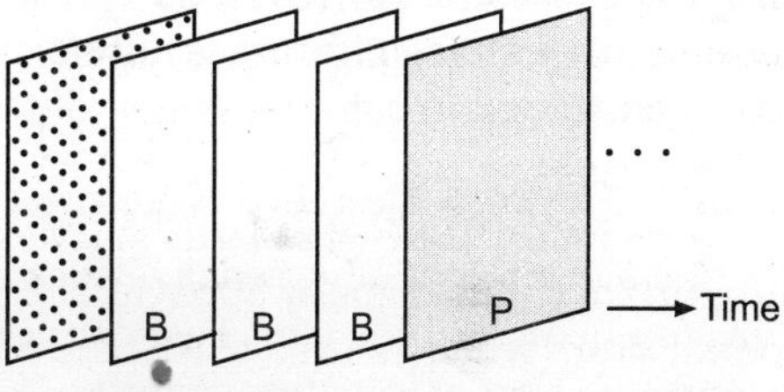

Fig. 9.39. Sequence of pictures in MPEG compression

3. Compression Process: The I-pictures are compressed as if they are JPEG images. The compressed images are then applied to four 8 × 8 blocks called macro blocks. The macro block are down sampled for subsequent compression of chrominance component as shown in figure 9.40. The first step in MPEG is to identify the macroblock moved between the I-pictures. Next step is to form the P-frame between the I-pictures. Each macro block is placed at its predicted location on the P-frame and is cross-correlation in its neighbourhood to determine the true location of the macro block in P-frame.

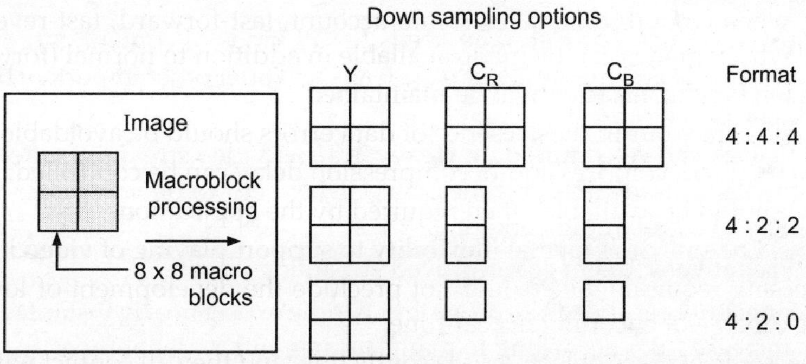

Fig. 9.40. Processing of macroblocks for chrominance down sampling

The difference between the predicted and true position of the macro block represents the error in prediction. This error is compressed using DCT and used for correcting the P-frame. Figure 9.41 shows the macro block shift between I-picture and an intermediate P-picture.

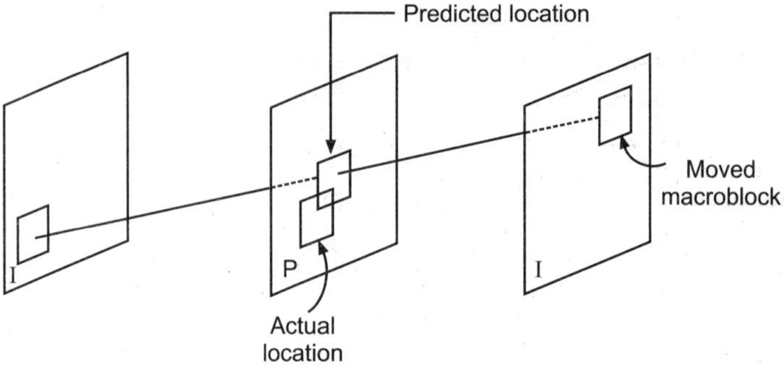

Fig. 9.41. Movement of macroblocks between I-picture and P-pictures

9.23. CONTENT DELIVERY NETWORK?

Content delivery networks use multiple servers in many geographic locations that improve deliveries of static and streaming content. Global content requests automatically get routed to the closest servers, speeding up page loads, maximizing bandwidth and providing identical content regardless of Internet- or site-traffic spikes. Depending on traffic and number of nodes, the network's algorithms select the best routing options to deliver optimum performance and avoid bottlenecks.

Users with high-speed connections often experience choppiness, loading lags and poor quality, especially when viewing live events or if they are located far from the hosting servers. CDNs minimize latency issues that cause image jitters, optimize delivery speeds and maximize available bandwidth for each viewer

9.23.1. Content Delivery Networks Improve Search Engine Rankings

Search engines now penalize sites that load too slowly, and broadband penetration has created unprecedented traffic jams Fast connection times do little good for viewers if the content delivers slowly. Customer attention spans become increasingly short because people spend good money on their connections and equipment, expecting the fastest page loads and highest quality videos and live-streaming events

CDN technology provides redundancy for fail-safe protection during partial Internet malfunctions. Duplication of content also protects against loss of data and image degradation.

9.23.2. Benefits of CDNs

Business owners face increasing storage needs for all kinds of content, but they must deliver these materials quickly. CDNs facilitate faster page loads and offer other important benefits including the following advantages :

1. Eliminate Pauses and Accommodate Heavy Traffic: Video streaming often results in jitters and pauses due to lags in transmission times, but CDNs help to deliver better user experiences when downloading video and audio content.

2. Minimize Packet Loss: Users get improved streaming quality.

3. Faster Loading: Internet speeds improve, but people expect nearly instantaneous page loads. Faster load times increase sales according to many studies and a vast body of anecdotal evidence.

4. File Mirroring: File mirroring protects data if natural disasters affect certain areas of the

Internet. Hurricanes and earthquakes could cause significant disruptions to services in key geographical areas.

5. Optimize Live Delivery: The success of YouTube and social media sharing has made video extremely popular, and most businesses should consider adding video elements to their content.

(i) Live events help to generate increased traffic.

(ii) Small organizations and schools can broadcast live events to strengthen SEO efforts.

(iii) Anyone can broadcast live networks of material as part of a marketing strategy.

6. Enable Linear Networks: Companies can broadcast in the same way that major networks do, creating their own program schedules for 24-hour CDN deliveries around the world.

7. Support Video on Demand: Organizations can enhance their online presence with video libraries, how-to videos, training programs and other marketing tools.

8. Scalability: New technology and advanced mobile applications place increasingly greater demands on servers, but large CDNs can handle new material as companies expand their online presences.

9.23.3. Disadvantages of CDNs

Disadvantages include additional costs, but large companies often save money by delivering their content along shorter routes. Some drawbacks include the following trouble spots:

1. New Points of Failure: CDNs create new points of potential failures along the delivery chains.

2. Sharing Resources: Delivery networks have many clients, and response times could vary due to the volume of website traffic of other CDN customers.

3. Geographical Choice Considerations: Website owners need to research their clients and choose the CDNs that offer the most convenient server locations near where they get the most business. If a website gets most of its customers from North America, then server locations in Europe and Asia would offer fewer benefits.

4. Content Management Problems: Companies might need to consider using content management systems to manage and update content throughout the networks. Some customers might need special widgets, content players or specific applications. Customer interactions could have different requirements in certain regions.

5. Lack of Direct Control: Changes to content must be made through CDN providers instead of directly, which could pose problems for editors and developers.

9.23.4. Choosing a Content Delivery Network

Websites owners face questions about which type of network to choose. Customers can choose from Internet-based or private networks, and some networks integrate with Online Video Platform, enabling features that enhance video distribution. Other factors customers should study include the following:

- High definition video has become increasingly popular, but not all networks support it.
- Transcoding changes code to make videos compatible on certain players like Silverlight, Flash, Quicktime and iDevice. Clients need to choose compatible players for their content.
- Mobile delivery has become very important to consumers, so website owners need to know if their network supports mobile delivery.
- Licensed or protected material, such as Pay-Per-View, requires Digital Rights Management.
- Analytics capabilities of content delivery networks become important in real-time for many applications, so, owners need to assess potential networks for their analytic and reporting abilities.

GLOSSARY

1. **Uniform Resource Locator (URL)** A specific type of URI which is used to identify a resource's location throughout the World Wide Web by specifying the access method needed to acquire the resource.
2. **Very High-Speed Digital Subscriber Line (VDSL)** A DSL variant which provides asymmetric service over fiber. Downstream rates range from 13 to 52 Mbps; upstream rates range from 1.5 to 2.3 Mbps. It is suitable for Internet/intranet access, video-on-domain, database access, remote LAN access, and high-definition TV.
3. **Client** A networked device which requests resources from a *server.*
4. **Client/Server** A model or paradigm which describes network services and the programs used by end users to access these services. A client side (or front end) provides a user with an interface for requesting services from the network, and server side (or back end) is responsible for accepting user requests for services and providing these services transparent to the user.
5. **Domain Name** A logical name assigned to an *IP address* and used as another type of addressing construct for identifying Internet nodes. The translation between logical name and IP address is known as name resolution, which is provided by a *domain name service* (DNS).
6. **Domain Name Service (DNS)** An Internet translation service which resolves *domain names* to *IP addresses*, and vice versa. Domain name service is usually provided by DNS servers.
7. **File Transfer Protocol (FTP)** An TCP/IP application protocol that is used for transferring files between two systems.
8. **Firewall** A device or product which allows systems or network managers to restrict access to components on a network. Five generally accepted types of firewalls are used on Internet connections. They are *frame-filtering, packet-filtering circuit, circuit gateways, stateful and application gateways,* and *proxy servers.*
9. **Multipurpose Internet Mail Extensions (MIME)** It is an extension to the *Simple Mail Transport Protocol (SMTP)* which extends the concept of e-mail by providing support for different data types and for complex message bodies. MIME enables users to exchange nontext files such as graphics, video, or audio in addition to plaintext files.

SUMMARY

- The application layer receives services from the transport layer. For the real applications in the application layer to function, there is a requirement of support protocols.
- The three areas or protocols required for such support may be listed as under :
  (i) Network security. (ii) Domain Name Service (DNS)
  (iii) Network management.
- Basically, security is not a single protocol but it consists of a large number of concepts and protocols used for providing privacy. DNS is used to handle naming or addressing within the Internet. The third support protocol is network management.
- The application layer programs are based upon the concept of clients and servers.
- For communication between client and server, addressing is needed. When a client requests a service from the server, it has to include the servers address as destination address and its own address as source address. When the server responds, it reverses the addresses.
- Some of the important applications may be listed as under :
  (i) Electronic mail. (ii) USENET (net news).
  (iii) World Wide Web. (iv) Multimedia.
  (v) Remote file transfer and access.
- To map a name onto an IP address, an application program calls a library procedure called the resolver. The name is passed onto the resolver as a parameter.
- The process of mapping a name to an address or an address to a name is called as name address resolution.

- One of the most popular network services is electronic mail (e-mail). Simple Mail Transfer Protocol (SMTP) is the standard mechanism for electronic mail in the Internet.
- MIME uses the same RFC 822 format but it adds structure to the message body (In RFC 822 there is no structure to the message body).
- The message transfer system is used to relay the messages from originator (sender) to the recipient. This can be achieved by establishing a transport connection between the source machine and destination machine.
- In Internet the source machine establishes a connection to port 25 of the destination machine so as to deliver an e-mail. An e-mail daemon which speaks SMTP is listening to this port.
- The web is basically a client-server system. The web pages are written in the languages HTML and Java. The growth of the World-Wide Web (WWW or simply Web) today is simply phenomenal. Each day, thousands or more people gain access to the Internet (above 100 million users at recent estimates).
- The Web and the Internet are not the same thing. The Web is a collection of standard protocols or instructions, sent back and forth over the Internet to gain access to information. The Internet, on the other hand, is a **network of networks** — a more physical entity.
- A web page starts with a title and contains the following
  (i) Some information
  (ii) Strings of text, linked to other pages
  (iii) E-mail address of the page's maintainer.
- Strings of text that are links to other pages are called hyperlinks. They are highlightened by underlining, using special colour or both. In order to follow a link, the user has to place the cursor on the highlightened area using the mouse or arrow keys and select it by clicking or hitting ENTER. The browsers can be of two types, namely the graphical browsers and non-graphical browsers. But, the graphical browsers are more popular. Voice based browsers are also being developed.
- All pages may not be viewable in the conventional way, some pages may contain audio tracks, video clips or both. If the hypertext pages are mixed with other media, the result of such a mixing is called as **hypermedia**.
- All the servers cannot speak HTTP, some of them use the FTP, Gopher or some other protocols. A large information is available on FTP and Gopher servers, hence, it must be made available to web users. To do so, one solution can be to have a browser which can use the HTTP as well as FTP, Gopher and other protocols. But, this makes the browser unnecessarily large.
- HTTP is used mainly to access data on WWW. This protocol transfers data in the form of plaintext, hypertext, audio, video etc. The function of HTTP is like a combination of FTP and SMTP. It uses services of TCP. It uses only one TCP connection (port 80). There is no separate control connection. Only the data transfer takes place between the client and server. The data transfer in HTTP is similar to SMTP. The format of the messages is controlled by MIME like headers.
- The principle of HTTP is simple. A client sends a request. The server sends a response. The request and response messages carry data in the form of a letter with a MIME like format.
- Basically, HTTP socket can be implemented in following two programs :
  1. A client's program   2. A server's program
- As a matter of fact, there are following two types of connections in HTTP :
  (i) Non-persistent connections   (ii) Persistent connections
  HTTP use persistent connections in its default mode. But, it is possible to configure HTTP clients and servers to use non-persistent connections.
- Further, the persistent connections can be of following two versions :
  (i) Without pipelining   (ii) With pipelining
  The default mode of HTTP uses persistent connections with pipelining.
- Basically, the HTTP messages are of following two types :
  (i) Request message   (ii) Response message
  The format of both these messages is almost the same.
- This is one of the fields in the request line format. It defines different kinds of messages referred to a request types or methods. The request method is a command or request issued by the client to the server. Following are some of the important methods (request types).

- Following are the four components of the cookie technology :
  - (i) A cookie header line in HTTP response message.
  - (ii) A cookie header line in the HTTP request message.
  - (iii) A cookie file kept on the user's end system and managed by user's browser.
  - (iv) A back end database at web site.
- Even though a number of browsers are available around, the browser architecture is nearly the same. Each browser consists of the following parts :
  - (i) A controller
  - (ii) Client Program
  - (iii) Interpreters
- The multimedia applications can be broadly categorized into three classes as under :
  1. Streaming stored audio/video
  2. Streaming live audio/video
  3. Interactive audio/video.
- Different multimedia types have different expectations from the Internet. Some of the important expectations are as under :
  1. Real time characteristics
  2. High bandwidth
  3. Multicasting support
  4. Error characteristics of the medium
  5. Session individuality
  6. Security
  7. Mobility
- URL has following three parts :
  - (i) Name of the protocol (e.g. http)
  - (ii) Name of the machine where the page is located.
  - (iii) The name of the file containing the page.

SHORT QUESTIONS WITH ANSWERS

Q.1. Write the applications of Application layer.

Ans. Some of the important applications may be listed as under :

- (i) Electronic mail
- (ii) USENET (net news)
- (iii) World Wide Web
- (iv) Multimedia
- (v) Remote file transfer and access.

Q.2. What do you mean by Resolution?

Ans. The process of mapping a name to an address or an address to a name is called as name address resolution.

Q.3. What do you mean by Principle of MIME?

Ans. MIME uses the same RFC 822 format but it adds structure to the message body (In RFC 822 there is no structure to the message body). In addition to this, MIME defines encoding rules for non ASCII message. MIME messages can be sent using the existing mail programs and protocols. The sending and receiving programs need to be changed which users can do themselves.

Q.4. What is the Difference between Web and Internet ?

Ans. The Web and the Internet are not the same thing. The Web is a collection of standard protocols or instructions, sent back and forth over the Internet to gain access to information. The Internet, on the other hand, is a **network of networks** — a more physical entity.

Q.6. What is meant by Uniform Resource Locator (URL) ?

Ans. URL has following three parts :

- (i) Name of the protocol (e.g. http)
- (ii) Name of the machine where the page is located.
- (iii) The name of the file containing the page.

REVIEW QUESTIONS

1. When web pages are sent out, they are prefixed by MIME headers ? Why ? Explain with reasons.
2. Why do HTTP, FTP, SMTP, POP3 and IMAP run on top of TCP rather than UPD? Explain.
3. What do you mean by application layer ? Explain.
4. How does DNS work ? Explain.
5. Which are the generic domains ? Explain
6. Write the types of domain names.
7. Are the domain names case sensitive ? Explain
8. Write few DNS examples.
9. What are the features of e-mail ? Enlist them.
10. Which are the modes to transfer a file ? Explain
11. Briefly discuss the following terms, emphasis more an implementation details.
    (i) DNS (ii) Mail server.
12. Write short technical notes on the following:
    (i) DNS (ii) SMDS.
13. What do you mean by domain name system ? How does it work ? Explain resolution process.
14. What do you mean by mailing list ? Explain with suitable block diagram.
15. Explain about a typical resolution process in DNS.
16. Write a short technical note on Domain name resolution.
17. Explain about addressing used in DNS.
18. Explain about the hierarchy of name servers.
19. Explain about name address resolution.
20. Explain about e-mail architecture and services.
21. Write message format of e-mail.
22. Write a short technical note on MIME.
23. Explain in brief SMTP.
24. Write a short technical note on POP3 server.
25. Explain DNS addressing scheme.
26. Explain the two mail access protocols in brief :
    (i) POP3; (ii) IMAP
27. Describe the principles of application layer protocols.
28. Discuss SMTP.
29. Explain about the following terms, emphasis more on implementation details.
    (i) Proxy server; (ii) Web server
30. Write a short technical note on Internet and web architecture.
31. Write short technical note on Proxy server.
32. What is the difference between Web and Internet ? Explain
33. Explain :
    (i) Hypertext (ii) Links of pointers; (iii) Browser.
34. What is meant by Hyperlinks and Hypermedia ? Explain
35. Explain about URL.
36. What is meant by proxy server ? Why is it used ? Explain
37. Explain about the principle of HTTP operation.
38. State and explain about various types of HTTP messages.
39. State and explain about various request types in HTTP.
40. What is meant by the Response Message in HTTP ? Explain.
41. How can you compare HTTP and SMTP ?
42. State and explain different types of Web documents.
43. What is meant by HTML ? Explain.

PREVIOUS YEAR QUESTION PAPER

B.Tech.

COMPUTER NETWORKS (EC-803)

Eight Semester Examination, June 2014

Time : 3 Hours *Total Marks : 70*

Note : (i) Attempt any one question from each unit. Each unit have internal choice.
(ii) Assume data/value if required.

UNIT-I

1. (a) What are the various protocols at the different layer in the TCP/IP protocol suite?
   (b) What are the different transmission media used in computer communication? Compare with data rates.

OR

2. (a) Compare the different types of switching techniques used in communication networks. Discuss their advantages and disadvantages with the help of timing diagram.
   (b) What is network topology? Explain mesh, ring and star topologies bringing out their merits and demerits.

UNIT-II

3. (a) How does selective-repeat-ARQ protocol differ from G-back-N ARQ ? Discuss with suitable diagrams.
   (b) Name the three types of HDLC frames and give a brief explanation of each.

OR

4. (a) What is CSMA Scheme? Discuss non-persistent, 1-persistent and p-persistent CSMA with a diagram.
   (b) Derive an expression for the throughput of a slotted ALOHA system and compare this with pure ALOHA system.

UNIT-III

5. (a) Draw the frame format of IEEE802.11 standard and briefly discuss the function of each field.
   (b) Explain uses of different network devices.

OR

6. (a) Describe the structure of IPv4 packet with a diagram.
   (b) List the advantages of IPv6.

UNIT-IV

7. (a) Draw the UDP header format and brief the function of each field.
   (b) Discuss the various types of services provided by the transport layer.

OR

8. (a) Draw the TCP header format and explain the function of each field.
   (b) How are transport layer protocols different from data link layer protocols?

UNIT-V

9. (a) What are the different ways of sending e-mail? Dicuss with suitable diagrams.
   (b) What is the function of a browser in WWW? Explain with a diagram.

OR

10. (a) List and explain the various characteristics of real time audio/video communication.
    (b) What is the need for a DNS system? When IP address can be directly used?

INDEX

C...

D...

E...

F...

N...

O...

P...

Q...

R...

S...

T...

U...

V...

W...

Win Prizes

ATTENTION READERS

Win Prizes

We request for your frank assessment regarding various aspects of the book as given below :

ESSENTIALS OF
COMPUTER NETWORKS
(For RGTU)

By: Dr. Sanjay Sharma

Please Fill up the Sheet in Neat and Clean Writing

(i) What topics of your syllabus, important for your examination point of view, are not covered in this book ?

....

....

....

....

(ii) What are the chapters wherein the treatment of the subject matter is not systematic or organised or updated ?

....

....

....

....

(iii) Have you come across misprints/mistakes/factual inaccuracies in the book? Please specify the chapters and the page numbers.

....

....

....

....

(iv) What other books on the same subject have you found/heard better than the present book? Please specify in terms of price and quality

....

....

....

....

(v) Further suggestions and comments for the improvement of this book.

....

....

....

....

(PS: If need be please attach a separate sheet to write your views)

OTHER DETAILS

(i) Who recommended you this book ? (Please tick (✓) in the box near the option relevant to you.

☐ Teacher ☐ Friends ☐ Bookseller

....

....

....

....

(ii) Name of the recommending teacher, his designation and address :

....

....

....

....

(iii) Name and address of the bookseller you purchased the book from :

....

....

....

....

(iv) Name and address of your institution (Please mention the University or Board as the case may be)

....

....

....

....

(v) Your name and complete postal address :

....

....

....

....

(vi) Course you are enrolled in :

....

....

....

....

The best assessment will be awarded each month: The award will be in the form of our publications as decided by the Editorial Board.

Please mail the filled up coupon at your earliest to :

S.K. KATARIA & SONS®
4885/109, Prakash Mahal, Dr. Subhash Bhargav Lane,
Opposite Delhi Medical Association, Daryaganj, New Delhi–110002
☎: +91-11-23243489, 23269384
Email : katariabooks@yahoo.com
katariabook@gmail.com